B. P. Kneubuehl (Editor)
Wound Ballistics: Basics and Applications

Beat P. Kneubuehl (Ed.)
Robin M. Coupland
Markus A. Rothschild
Michael J. Thali

Wound Ballistics

Basics and Applications

Translation of the revised third German edition (2008)
With 234 figures and 156 tables

Beat P. Kneubuehl, Dr in forens. sci., MD h. c. (Editor)
M. in Mathematics
Head, Centre of Forensic Physics and Ballistics
Institute of Forensic Medicine, University of Berne
CH-3012 Berne, Switzerland

Dr Robin M. Coupland MB, BChir, FRCS
Assistance Division, International Committee of the Red Cross
CH-1202 Geneva, Switzerland

Prof. Dr Markus A. Rothschild
Director of the Institute of Forensic Medicine
Medical Centre of the University of Cologne
D-50823 Cologne, Germany

Prof. Dr Michael J. Thali
Director of the Institute of Forensic Medicine
University of Zurich
CH-8057 Zurich, Switzerland

Translation
Steve Rawcliffe, BA, PgDip. Int. Trans.
F-01210 Versonnex, France

ISBN-13   978-3-642-20355-8   Springer-Verlag Berlin Heidelberg New York

Bibliographic information Deutsche Bibliothek
The Deutsche Bibliothek lists this publication in Deutsche Nationalbibliographie;
detailed bibliographic data is available in the internet at <http://dnb.ddb.de>.

This work is subject to copyright. All rights are reserved, whether the whole or part of the material is concerned, specifically the rights of translation, reprinting, reuse of illustrations, recitation, broadcasting, reproduction on microfilms or in any other way, and storage in data banks. Duplication of this publication or parts thereof is permitted only under the provisions of the German Copyright Law of September 9, 1965, in its current version, and permission for use must always be obtained from Springer-Verlag. Violations are liable to prosecution under the German Copyright Law.

The use of proprietary names in this work does not indicate that these are free of intellectual property rights.

The greatest of care has been taken in producing the text, figures and tables in this book.
However, it is not possible to completely exclude the risk of error. Neither the publisher nor the authors can be held responsible for errors in this work, or for the consequences thereof.

Springer Medizin
Springer-Verlag GmbH
ein Unternehmen von Springer Science+Business
springer.de

© Springer-Verlag Berlin Heidelberg 2011

The use of general descriptive names, registered names, trademarks, etc. in this publications does not imply, even in the absence of a specific statement, that such names are exempt from the relevant protective laws and regulations and therefore free for general use.

Product liability: The publishers cannot guarantee the accuracy of any information about dosage and application contained in this book. In every individual case the user must check such information by consulting the relevant literature.

Planning: Hinrich Küster
Project management: Kerstin Barton
Typesetting: the author 's own, print-ready draft

Cover design: deblik Berlin
creditline cover images: left: ©photos.com / ©right: fotolia.com

22/2122 – 5 4 3 2 1 0   SPIN 12745258

Preface

The German original of the present work on wound ballistics is already in its third edition. The first two German editions focused on the behaviour of a bullet in the body and the physical modelling of that behaviour. In parallel with this, a system of simulants was established that allowed gunshot wounds to be simulated empirically.

In the period between the second German edition (2001) and the third (2008), the emphasis in wound ballistics research was on consolidation and application. It was possible to use the results to answer many forensic and surgical questions. Both the physical models and the empirical simulation of gunshot wounds using simulants have been validated by experience. It is therefore logical that the emphasis in this, the third edition, should have shifted towards the practical application of wound ballistics.

The introduction to the basics (physics, ammunition and the principles of ballistics) was previously divided over two chapters; these topics are now covered in one chapter. In a similar manner, the specific wound ballistics of handgun bullets, rifle bullets and fragments are dealt with in one chapter instead of three. These changes created room for three chapters on the use of wound ballistics in forensic medicine, surgery and international instruments. Three leading and highly competent co-authors were recruited to cover the topics of surgery and forensic medicine: Dr Robin Coupland (who for many years was chief surgeon at the International Committee of the Red Cross), Prof. Dr Markus Rothschild (Director of the Institute of Forensic Medicine, Cologne) and Prof. Dr Michael Thali (Director of the Institute of Forensic Medicine, Zurich).

So-called "non-lethal" projectiles have become increasingly important in recent years – not only for police use but also in military contexts. We have therefore devoted a separate section to the ballistics and effectiveness of this type of projectile.

The tables at the end of the book now include a number of projectiles used in sport, as these may be relevant to sports accidents.

Such an interdisciplinary field as wound ballistics requires an intensive exchange of experiences and thoughts. I should like to thank Prof. Dr Michael Thali, Zurich, and the Institute of Forensic Medicine, University of Berne, for their unstinting support. With their extensive surgical experience, Dr Robin Coupland and Dr Wolfgang Titius each made a major contribution to the success of the experimental simulation of gunshot wounds. My thanks go to Dr Ulrich Stoller for his numerous suggestions and pointers regarding medical questions.

Christoph Simon, PhD, of gelatine manufacturer Gelita, was kind enough to check and correct the section on the characteristics of gelatine and the process by which it is produced.

My special thanks go to Steve Rawcliffe, with whom it was a great pleasure to work during the translation of this demanding text. I am extremely grateful to Dr Leslie Payne and Dr Virginia Fitzgerald-Swallow for their thorough and expert review of the English manuscript. I should like to thank two members of my team – Matthieu Glardon, M. in forens. sc. and Lea Siegenthaler, M. in Physics – for their careful checking of the layout.

My thanks go also to the Competence Centre for Science and Technology of armasuisse for their kind permission to use some of their illustrations in this work. Warm thanks go to the publisher, Springer, with whom it was a pleasure to work and who are responsible for the excellent appearance of the book.

Finally, I wish to take this opportunity to remember physicist **Prof. Dr Karl Sellier**, who died in 1997. He laid the foundations of this work with the first, German edition, on which it was a great honour for me to work as co-author. He would have been happy to see how successful this book has been so far.

Thun, April 2011   Beat P. Kneubuehl,

                                                                                                                    Editor

Outline of contents

1 Introduction 1
2 **Basics** 3
   Beat P. Kneubuehl
   2.1 The physics of wound ballistics 3
   2.2 Ammunition and weapons 33
   2.3 Ballistics 65
3 **General wound ballistics** 87
   Beat P. Kneubuehl
   3.1 Introduction 87
   3.2 Processes in the wound channel; the temporary cavity 95
   3.3 Simulants 136
   3.4 Other approaches to simulation 154
4 **Wound ballistics of bullets and fragments** 163
   Beat P. Kneubuehl
   4.1 The effectiveness of bullets 163
   4.2 Wound ballistics of handgun bullets 186
   4.3 Wound ballistics of rifle bullets 212
   4.4 Wound ballistics of fragments 232
   4.5 "Non-lethal" projectiles 240
5 **Wound ballistics and forensic medicine** 253
   5.1 Conventional forensic medicine 253
       Markus A. Rothschild
   5.2 Modern graphical methods 286
       Michael J. Thali
   5.3 Experimental reconstruction 291
       Beat P. Kneubuehl, Michael J. Thali
6 **Wound ballistics and surgery** 305
   Robin M. Coupland
   6.1 The historical connection between wound ballistics and surgery 305
   6.2 Wound ballistics and ballistic trauma – what's the difference? 306
   6.3 Comparing simulated wounds and real wounds 307
   6.4 Clinical features of real wounds 312

6.5   The contribution of wound ballistics to the care of wounded people....313
6.6   Documenting ballistic trauma..317

7   Wound ballistics and international agreements...321
Beat P. Kneubuehl
7.1   Introduction 321
7.2   History of firearms and ammunition 321
7.3   International treaties 334

Appendices

A   Tables 345
   A.1   List of tables in the main text 345
   A.2   Characteristics of materials 347
   A.3   Calibre designations (metric system) 348
   A.4   Ballistic data for cartridges (metric system)..350
   A.5   Calibre designations (British/U.S. system)..355
   A.6   Ballistic data for cartridges (British/U.S. system)...............................357
   A.7   Bullet designations 362
   A.8   Geometric data for selected bullets 363
   A.9   Twist length, angle of twist and rotation 364
   A.10 Ballistics tables (metric system)...366
   A.10 Ballistics tables (British/U.S. system) 384
   A.12 Shotguns and shot...402

B   Glossary..405
   B.1   English ⇒ German ⇒ French 405
   B.2   German ⇒ English ⇒ French 425
   B.3   French ⇒ German ⇒ English 443

C   Bibliography 463
   Photo credits...485

Index..487

Detailed table of contents

Table of symbols XIX
Prefix symbols for decimal multiples and submultiples of SI units XXIII
Conversions XXIII

1 **Introduction** 1
2 **Basics** 3
  2.1 The physics of wound ballistics 3
      2.1.1 Preliminary remarks 3
      2.1.2 Coordinates, systems of units and notation 3
      2.1.3 Mechanics 4
          2.1.3.1 Kinematics 4
          2.1.3.2 Mass, momentum and force 7
          2.1.3.3 Work and energy 8
          2.1.3.4 Rotation 10
          2.1.3.5 Laws of conservation of mass, energy and momentum ... 11
          2.1.3.6 Equations of motion 12
      2.1.4 Fluid dynamics 15
          2.1.4.1 General 15
          2.1.4.2 Basic concepts in thermodynamics 15
          2.1.4.3 Material characteristics 18
          2.1.4.4 Frictionless flow 21
          2.1.4.5 Flow of a viscous fluid 23
      2.1.5 Fluid jets 26
          2.1.5.1 General 26
          2.1.5.2 Exhaust flow from a muzzle 26
          2.1.5.3 De Laval nozzles (converging-diverging nozzles) ... 27
          2.1.5.4 Jet velocity and energy 28
      2.1.6 Measuring techniques for wound ballistics 29
          2.1.6.1 General 29
          2.1.6.2 Dynamic phenomena 30
          2.1.6.3 Physical values 32
  2.2 Ammunition and weapons 33
      2.2.1 Introduction 33
      2.2.2 Ammunition 34

		2.2.2.1	The structure of a cartridge	34
		2.2.2.2	Types of ammunition	41
		2.2.2.3	Blank and irritant rounds	52
		2.2.2.4	Fragmenting ammunition	53
	2.2.3	Weapons		55
		2.2.3.1	Firearm design and typology	55
		2.2.3.2	Handguns	59
		2.2.3.3	Long weapons	61
		2.2.3.4	Alarm pistols and revolvers	65
2.3	Ballistics			65
	2.3.1	Definitions		65
	2.3.2	Interior ballistics		66
		2.3.2.1	General	66
		2.3.2.2	Powder combustion	66
		2.3.2.3	The firing sequence	67
		2.3.2.4	Interior ballistics calculations	68
		2.3.2.5	Energy balance	69
	2.3.3	Muzzle phenomena		69
		2.3.3.1	Muzzle gas flow	69
		2.3.3.2	Flash	70
	2.3.4	Exterior ballistics		71
		2.3.4.1	General; terms used	71
		2.3.4.2	Exterior ballistics calculations	72
		2.3.4.3	Ballistics tables	73
		2.3.4.4	Proper motion of a bullet	73
		2.3.4.5	Disturbances to the trajectory	74
	2.3.5	Stability and tractability		75
		2.3.5.1	Definition of stability	75
		2.3.5.2	Spin-stabilized projectiles	76
		2.3.5.3	Projectiles stabilized by air forces	78
		2.3.5.4	Shoulder stabilization	78
		2.3.5.5	Tractability	79
		2.3.5.6	Stability and ricochets	80
	2.3.6	Fragment ballistics		81
		2.3.6.1	Acceleration of fragments	81
		2.3.6.2	Exterior ballistics of fragments	82
	2.3.7	Terminal ballistics models		83
		2.3.7.1	General	83
		2.3.7.2	The plugging model	83
		2.3.7.3	The displacement model (ductile failure)	84
		2.3.7.4	Bullet passing through a thin layer of material	84

3 General wound ballistics 87
- 3.1 Introduction 87
  - 3.1.1 General...87
  - 3.1.2 The history of wound ballistics 88
  - 3.1.3 Basic relationships 93
- 3.2 Processes in the wound channel; the temporary cavity 95
  - 3.2.1 Preliminary remarks 95
    - 3.2.1.1 The concept of the "temporary cavity"...........................95
    - 3.2.1.2 Different ways of looking at wounding 95
    - 3.2.1.3 Modelling wound ballistics processes............................96
  - 3.2.2 Motion and behaviour of a bullet 97
    - 3.2.2.1 Rifle bullets 97
    - 3.2.2.2 Handgun bullets 103
    - 3.2.2.3 Fragments and fragment-like projectiles..................... 104
    - 3.2.2.4 Possible types of wound channel 107
    - 3.2.2.5 Physical models.. 107
  - 3.2.3 The temporary cavity 110
    - 3.2.3.1 Phenomenology of the temporary cavity 110
    - 3.2.3.2 Quantitative description of the temporary cavity........... 117
    - 3.2.3.3 Influence of impact conditions and bullet characteristics... 118
    - 3.2.3.4 The effect of the sectional density of a bullet on the shape of the temporary cavity 122
  - 3.2.4 The effect of bullet design on behaviour..................................... 126
    - 3.2.4.1 Categories of bullet 126
    - 3.2.4.2 Deformation and fragmentation; general points 126
    - 3.2.4.3 Experimental results.. 128
  - 3.2.5 Patterns in bullet wounds to bones... 131
  - 3.2.6 Bullet temperature and sterility.. 133
    - 3.2.6.1 Historical background... 133
    - 3.2.6.2 Bullet temperature... 134
    - 3.2.6.3 Bullets contaminated with bacteria 135
    - 3.2.6.4 Burns due to bullets... 136
- 3.3 Simulants 136
  - 3.3.1 General... 136
  - 3.3.2 Gelatine 138
    - 3.3.2.1 Characteristics and fabrication..................................... 138
    - 3.3.2.2 Fabrication of gelatine blocks; preparation for experiments 138
    - 3.3.2.3 Evaluating gelatine experiments 140
  - 3.3.3 Glycerine soap (ballistic soap)... 143
    - 3.3.3.1 Characteristics and fabrication..................................... 143

		3.3.3.2	Ageing	144
		3.3.3.3	Evaluating soap experiments	145
		3.3.3.4	Using soap to conduct measurements	146
	3.3.4	Comparison between soap and gelatine		147
		3.3.4.1	General	147
		3.3.4.2	Availability, handling and measuring techniques	147
		3.3.4.3	Reaction to bullets	148
		3.3.4.4	Which simulant for which purpose?	150
		3.3.4.5	Connection between the analysis methods	151
	3.3.5	Bone		151
		3.3.5.1	General	151
		3.3.5.2	Hollow bones	152
		3.3.5.3	Modelling the head	153
	3.3.6	Other simulants		153
3.4	Other approaches to simulation			154
	3.4.1	Experiments on animals and cadavers		154
		3.4.1.1	Animals	154
		3.4.1.2	Cadavers	156
		3.4.1.3	Cell cultures	157
	3.4.2	Physical/mathematical models		157
		3.4.2.1	General	157
		3.4.2.2	SELLIER's velocity profiles	158
		3.4.2.3	Computer Man	159
		3.4.2.4	The "Verwundungsmodell Schütze" (VeMo-S)	160

4 Wound ballistics of bullets and fragments ... 163

4.1	The effectiveness of bullets			163
	4.1.1	Effectiveness versus effect		163
		4.1.1.1	Definitions	163
		4.1.1.2	Factors that contribute to the effect of a bullet	163
	4.1.2	Measures of effectiveness		165
		4.1.2.1	Historical background	165
		4.1.2.2	The "stopping power" fallacy	166
		4.1.2.3	Traditional measures of effectiveness	167
		4.1.2.4	Summary and conclusions	176
	4.1.3	Determining the effectiveness of a bullet		178
		4.1.3.1	Definition of effectiveness	178
		4.1.3.2	Measuring effectiveness	179
	4.1.4	Military effectiveness criteria		179
		4.1.4.1	Definitions of effectiveness	179
		4.1.4.2	Probability of incapacitation	181

4.2 Wound ballistics of handgun bullets 186
    4.2.1 Penetration depth of handgun bullets and ability to pass through gelatine, soap, muscle and bone 186
        4.2.1.1 General 186
        4.2.1.2 Penetration depth in gelatine, soap and muscle 187
        4.2.1.3 Penetration capacity in bone 195
        4.2.1.4 Threshold velocities for eyes 200
    4.2.2 Characteristics of handgun bullets 201
        4.2.2.1 General 201
        4.2.2.2 Bullets with good penetration properties 202
        4.2.2.3 Bullets designed for maximum effectiveness 202
        4.2.2.4 Unconventional bullet design 207
    4.2.3 Gas and fluid jets as projectiles 208
        4.2.3.1 General 208
        4.2.3.2 Liquid jets 209
        4.2.3.3 Gas jets 209
        4.2.3.4 The effects of gas jets in the case of gas and alarm pistols 210

4.3 Wound ballistics of rifle bullets 212
    4.3.1 Introduction 212
    4.3.2 Remote effects 213
        4.3.2.1 General 213
        4.3.2.2 Shock waves 214
        4.3.2.3 Biological/pathological consequences of shock waves . 218
        4.3.2.4 Pressure changes in blood vessels 223
        4.3.2.5 The effects of pressure pulses on blood vessels 224
        4.3.2.6 Bone fractures at locations remote from the wound channel 225
    4.3.3 Wound ballistic characteristics of rifle bullets 226
        4.3.3.1 Bullets designed for military use 226
        4.3.3.2 Hunting bullets 228
        4.3.3.3 Shot and slugs 229

4.4 Wound ballistics of fragments 232
    4.4.1 General 232
        4.4.1.1 Frequency of fragment wounds 232
        4.4.1.2 Wounds caused by fragments and similar projectiles 232
    4.4.2 Equations of motion and energy for fragments 234
        4.4.2.1 Hypotheses 234
        4.4.2.2 The geometrical form of the wound channel 234
        4.4.2.3 Equation of energy and motion 235
        4.4.2.4 Entry wound diameter and penetration depth 235
    4.4.3 Experimental verification of the models 236

		4.4.3.1 Method	236
		4.4.3.2 Entry wound diameter	236
		4.4.3.3 Penetration depth	238
		4.4.3.4 Comparison with other studies	239
		4.4.3.5 Applications	240
4.5	"Non-lethal" projectiles		240
	4.5.1	General	240
	4.5.2	Projectile design	241
		4.5.2.1 Projectiles with low sectional density	241
		4.5.2.2 Expanding bullets	241
		4.5.2.3 Rubber shot	244
		4.5.2.4 Special projectiles for handguns	246
	4.5.3	Wound ballistics of "non-lethal" projectiles	246
		4.5.3.1 Penetrating projectiles	246
		4.5.3.2 Non-penetrating projectiles	248
	4.5.4	Dangerosity of projectiles	250
		4.5.4.1 Criteria of dangerosity	250
		4.5.4.2 Determining hazard areas	251
		4.5.4.3 Danger area for persons wearing protective equipment	251

5 Wound ballistics and forensic medicine 253

5.1 Conventional forensic medicine 253
    5.1.1 General 253
    5.1.2 Crime-scene investigation 253
        5.1.2.1 Bullet damage at the crime scene 253
        5.1.2.2 Examination of the body at the scene 254
        5.1.2.3 Bloodstain pattern analysis 255
    5.1.3 Morphology of entry and exit wounds 257
        5.1.3.1 Entry wounds 257
        5.1.3.2 Exit wounds 260
        5.1.3.3 Grazing shots 261
        5.1.3.4 Indicators of muzzle-target distance 262
    5.1.4 The wound channel 265
        5.1.4.1 Wound morphology 265
        5.1.4.2 The relationship between the wound channel and the direction of shot 266
    5.1.5 Bullet wounds to the head 267
        5.1.5.1 Brain injuries 267
        5.1.5.2 Skull injuries 268
    5.1.6 Bullet wounds to the trunk 270
        5.1.6.1 The ribcage 270
        5.1.6.2 Abdomen 271
    5.1.7 Bullet wounds to bones 272

		5.1.7.1	General 272
		5.1.7.2	Flat bones 273
		5.1.7.3	Long hollow bones 274
		5.1.7.4	Vertebrae 275
	5.1.8	Peculiarities of shotgun wounds 275
		5.1.8.1	General 275
		5.1.8.2	Morphology of entry wounds 276
		5.1.8.3	Internal morphology of shotgun wounds 276
	5.1.9	Causes of death and incapacitation 277
		5.1.9.1	Causes of death 277
		5.1.9.2	Incapacitation 279
	5.1.10	Particular projectiles 281
		5.1.10.1	Gas-powered weapons 281
		5.1.10.2	Alarm pistols and weapons firing irritants 282
		5.1.10.3	Arrow wounds 283
		5.1.10.4	Captive bolt pistols and bolt-firing tools 284

5.2 Modern graphical methods 286
    5.2.1 Surface documentation 286
    5.2.2 Radiological documentation 286
    5.2.3 Combining surface and radiological documentation 289
    5.2.4 Documenting crime scenes using modern graphics techniques ... 289

5.3 Experimental reconstruction 291
    5.3.1 Introduction 291
    5.3.2 Reconstructing shooting incidents 292
        5.3.2.1 Preliminary remarks 292
        5.3.2.2 Points to bear in mind 292
        5.3.2.3 Examples 293
    5.3.3 Blunt force 296
        5.3.3.1 Equipment used and experimental options 296
        5.3.3.2 Examples 297
    5.3.4 Using virtopsy in practice 298
        5.3.4.1 Documentation and visualization 298
        5.3.4.2 Example 301

6 Wound ballistics and surgery 305
6.1 The historical connection between wound ballistics and surgery 305
6.2 Wound ballistics and ballistic trauma – what's the difference? 306
6.3 Comparing simulated wounds and real wounds 307
    6.3.1 Preliminary remarks 307
    6.3.2 Case studies 307
    6.3.3 Conclusions 311
6.4 Clinical features of real wounds 312
6.5 The contribution of wound ballistics to the care of wounded people 313

        6.5.1   The "wound profile" 313
        6.5.2   What causes tissue damage? 313
        6.5.3   Gas in tissues on a clinical x-ray... 314
        6.5.4   The "hot bullet" theory.. 314
        6.5.5   Long bone fractures.. 315
        6.5.6   Cranio-cerebral wounds 316
        6.5.7   Unresolved issues... 316
   6.6 Documenting ballistic trauma.. 317
        6.6.1   Overview 317
        6.6.2   Scoring wounds in the field... 318
        6.6.3   The role of surgeons and the application of international
                humanitarian law 319
        6.6.4   Documenting ballistic trauma – a wider responsibility for health
                professionals?.. 319

7 **Wound ballistics and international agreements**... 321
   7.1 Introduction 321
   7.2 History of firearms and ammunition 321
        7.2.1   General 321
        7.2.2   The development of ammunition 322
                7.2.2.1   The situation in 1800.. 322
                7.2.2.2   The elongated bullet... 322
                7.2.2.3   The primer... 323
                7.2.2.4   The metal cartridge 323
                7.2.2.5   Smokeless powder.. 324
                7.2.2.6   Bullets 325
                7.2.2.7   "Dum-dum" bullets... 326
        7.2.3   The development of firearms in the 19th century 329
                7.2.3.1   Muzzle loaders and their problems 329
                7.2.3.2   Breech-loaders 329
                7.2.3.3   Repeaters.. 330
                7.2.3.4   Handguns 330
        7.2.4   The 20th century.. 331
                7.2.4.1   Ammunition 331
                7.2.4.2   Weapons... 333
   7.3 International treaties 334
        7.3.1   Basic principles 334
        7.3.2   The instruments.. 334
                7.3.2.1   The original Geneva Convention (1864) 334
                7.3.2.2   The St Petersburg Declaration (1868)...................... 335
                7.3.2.3   The Brussels Conference (1874).............................. 335
                7.3.2.4   The Hague Convention (1899)................................. 336

       7.3.2.5 The Regulations concerning the Laws and Customs of
War on Land (The Hague, 1907) 337
       7.3.2.6 The Geneva Conventions of 1949 337
       7.3.2.7 The 1977 protocols additional to the Geneva
Conventions 338
       7.3.2.8 The United Nations Conference (Geneva, 1980) 339
       7.3.2.9 The relevance of international instruments to wound
ballistics 339
  7.3.3 A basis for formulating future instruments of international
humanitarian law 340
       7.3.3.1 The disadvantages of the wording of existing
conventions 340
       7.3.3.2 Projectile-independent assessment processes 341
       7.3.3.3 Formulation of standards 342

Appendices

A Tables 345
  A.1 List of tables in the main text 345
  A.2 Characteristics of materials 347
      A.2.1 Fluids and materials that behave like fluids 347
      A.2.2 Solid materials 347
  A.3 Calibre designations (metric system) 348
      A.3.1 Handguns 348
      A.3.2 Military rifles 349
      A.3.3 Hunting and sporting rifles 349
  A.4 Ballistic data for cartridges (metric system) 350
      A.4.1 Handgun cartridges 350
      A.4.2 Military ammunition 351
      A.4.3 Hunting and sporting ammunition 352
      A.4.4 Pre-1900 weapons and ammunition 353
      A.4.5 Ballistic performance of certain bows and crossbows 354
            A.4.5.1 Technical data 354
            A.4.5.2 Ballistic data 354
      A.4.6 Ballistic data for various types of projectiles used in sport 354
  A.5 Calibre designations (British/U.S. system) 355
      A.5.1 Handguns 355
      A.5.2 Military rifles 356
      A.5.3 Hunting and sporting rifles 356
  A.6 Ballistic data for cartridges (British/U.S. system) 357
      A.6.1 Handgun cartridges 357

```
        A.6.2   Military ammunition ................................................................. 358
        A.6.3   Hunting and sporting ammunition ............................................. 359
        A.6.4   Pre-1900 weapons and ammunition............................................ 360
        A.6.5   Ballistic performance of certain bows and crossbows............... 361
                A.4.6.1  Technical data ................................................................ 361
                A.4.6.2  Ballistic data ................................................................. 361
        A.6.6   Ballistic data for various types of projectiles used in sport ....... 361
  A.7   Bullet designations ............................................................................... 362
        A.7.1   Bullet form ................................................................................. 362
        A.7.2   Bullet material............................................................................ 362
        A.7.3   Bullet structure........................................................................... 362
  A.8   Geometric data for selected bullets ....................................................... 363
        A.8.1   Military bullets........................................................................... 363
        A.8.2   Other bullets............................................................................... 363
  A.9   Twist length, angle of twist and rotation ............................................... 364
        A.9.1   Handguns .................................................................................. 364
        A.9.2   Rifles.......................................................................................... 364
                A.9.2.1  Military rifles ................................................................ 364
                A.9.2.2  Hunting and sporting rifles ........................................... 365
  A.10  Ballistics tables (metric system)............................................................ 366
        A.10.1 Notes ......................................................................................... 366
        A.10.2 Handguns .................................................................................. 366
        A.10.3 Rifles.......................................................................................... 372
        A.10.4 Old rifles ................................................................................... 379
        A.10.5 Various...................................................................................... 381
  A.11  Ballistics tables (British/U.S. system) .................................................. 384
        A.11.1 Notes ......................................................................................... 384
        A.11.2 Handguns .................................................................................. 384
        A.11.3 Rifles.......................................................................................... 390
        A.11.4 Old rifles ................................................................................... 397
        A.11.5 Various...................................................................................... 399
  A.12  Shotguns and shot ................................................................................. 402
        A.12.1 Calibres of shotgun barrels ....................................................... 402
        A.12.2 Ballistic data for shot pellets..................................................... 402
        A.12.3 Designations for buckshot pellets ............................................. 402
        A.12.4 Designations for normal shotgun pellets: British/U.S. system .. 403
        A.12.5 Designations for normal shotgun pellets: metric system .......... 403

B  **Glossary**............................................................................................................**405**
  B.1   English $\Rightarrow$ German $\Rightarrow$ French ............................................................. 405
  B.2   German $\Rightarrow$ English $\Rightarrow$ French ............................................................. 425
  B.3   French $\Rightarrow$ German $\Rightarrow$ English ............................................................. 443
```

C Bibliography 463
   General 463
   Articles and papers 465
   Photo credits 485

Index 487

Table of symbols

This book uses SI units and units derived from them (some tables are also printed in British/U.S. units). Dimensionless quantities are indicated by [-]. Where no dimension is possible for a quantity, the corresponding space is left blank.

A	Area	[m²]
C	General proportionality factor (e.g. specific heat capacity)	
C/L	Measure of effectiveness (CARANTA and LEGRAIN)	
C_D	Drag coefficient	[-]
C_{dr}	Pressure coefficient	[-]
C_F	Coefficient of friction	[-]
C_L	Lift force coefficient	[-]
C_M	Overturning moment coefficient	[-]
C_p	Pressure coefficient	[-]
D	Plate thickness (terminal ballistics)	[m]
E	Energy	[J]
E'	Energy density	[J/mm²]
E'_{ab}	Wounding potential (energy deposited per cm travelled)	[J/cm]
E'_{gr}	Threshold energy density	[J/mm²]
E_a	Impact energy	[J]
E_{ab}	Energy transferred	[J]
E_{ad}	Entry energy (the energy of the projectile as it enters a layer, having passed through another)	[J]
E_{dr}	Pressure energy	[J]
E_{ds}	Energy expended in passing through a layer	[J]
E_e	Exit energy	[J]
E_{gr}	Threshold energy	[J]
EKE	Expected kinetic energy	[J]
E_{kin}	Kinetic energy	[J]
E_{mech}	Mechanical energy (= $E_{kin} + E_{pot} + E_{rot}$)	[J]
E_{pot}	Potential energy	[J]
E_{rot}	Energy of rotation	[J]
E_{rst}	Residual energy of the projectile after it has exited the target (e.g. the body)	[J]

E_{stk}	Residual energy of the projectile at the moment when it comes to rest in the target	[J]
F	Force	[N]
F_D	Resisting force (flow resistance)	[N]
F_Q	Lateral force	[N]
F_R	Resultant force	[N]
F_W	Air resistance	[N]
G	Weight	[N]
G_K	Gurney constant	[m/s]
I	Momentum	[N·s]
J	Moment of inertia	[kg·m^2]
J_a	Axial moment of inertia	[kg·m^2]
J_q	Lateral moment of inertia	[kg·m^2]
KO	Taylor's Knockout Value	
L	Angular momentum	[N·m·s], [kg·m^2/s]
L	Centre of pressure (in graphics)	
M	Torque	[N·m]
Ma	Mach	[-]
NC	Narrow channel	[cm]
P(I|H)	Conditional probability of incapacitation (assuming a hit occurs)	[-]
PIR	Power Index Rating (MATUNAS)	
Q	Quantity of heat (heat energy)	[J]
Q_{ex}	Specific explosion heat of an explosive	[J/g]
ℜ	Retardation coefficient	[1/m]
R	Special gas constant	[J/(kg·K)]
Re	Reynolds number	[-]
RII	Relative Incapacitation Index	
RSP	Relative Stopping Power	
S	Centre of gravity (in graphics)	
StP	Stopping Power	
T	Temperature	[K]
T_C	Temperature in degrees Celsius	[°C]
U	Internal energy	[J]
V	Volume	[m^3]
VI	Vulnerability Index (used in conjunction with RII)	
V_{TH}	Volume of the temporary cavity	[cm^3]
W	Work	[N·m], [J]
W_H	Weigel's measure of effectiveness	
W_{TH}	Sellier's measure of effectiveness	
Y	Young's modulus	[N/mm^2]
Z	Characteristic wave impedance of a medium	[kg/(m^2·s)]
a	Acceleration	[m/s^2]
c	Speed of sound	[m/s]
c_p	Specific heat (at constant pressure)	[J/(kg·K)]

c_V	Specific heat (at constant volume)	[J/(kg·K)]
d	Diameter (generally)	[m]
e	Euler's number (basis of natural logarithms: 2.71828 …)	[-]
f	Form factor or tractability number (external ballistics)	[-]
fw	Angle of descent	[mil]
g	Acceleration due to gravity	[m/s^2]
k	Calibre	[mm]
ℓ	Length (generally)	[mm, cm]
ℓ_g	Length of bullet	[mm]
ℓ_{NC}	Length of narrow channel	[cm]
ℓ_s	Length of wound channel	[cm]
ln	Natural logarithm (based on e)	[-]
m	Mass	[g, kg]
m_c	Load mass (interior ballistics)	[g]
p	Pressure	[N/m^2], [bar]
q	Sectional density	[g/mm^2], [kg/m^2]
r	Correlation coefficient (statistics)	
r	Radius, distance from axis of rotation	[m]
r	Reflection coefficient (shock wave)	[-]
s	Gyroscopic stability number (exterior ballistics)	[-]
s	Standard deviation (statistics)	
s	Distance	[m]
s_{ad}	Distance penetrated at an entry velocity v_{ad}	[cm]
sw	Angle of shot	[mil]
t	Time	[s]
t_e	Time of flight	[s]
v	Velocity	[m/s]
v_0	Muzzle velocity	[m/s]
v_a	Impact velocity	[m/s]
v_{ad}	Entry velocity (the velocity of the projectile as it enters a layer, having passed through another)	[m/s]
v_{ds}	Decrease in velocity of the projectile on passing through a layer	[m/s]
v_{gr}	Threshold velocity	[m/s]
v_{rst}	Exit velocity	[m/s]
v_{stk}	Velocity of a projectile immediately before coming to rest in a target	[m/s]
x,y,z	Cartesian coordinates	[m]
x_e	Range	[m]
x_s	Vertex distance	[m]
y_s	Vertex height of a trajectory	[m]
z	Combustion rate of a propellant (interior ballistics)	[-]
α	Angular acceleration	[rad/s^2]
β	Reflection coefficient (shock wave)	[-]
γ	c_p/c_V	[-]
δ	Angle of attack	[°]

η	Dynamic viscosity	[Pa·s]
θ	Trajectory angle	[rad]
θ_0	Angle of departure	[rad]
κ	Compressibility	[1/Pa]
λ_1	First proportionality number (wound ballistics of fragments)	[s/m]
λ_2	Second proportionality number (wound ballistics of fragments)	[kg/m^3]
μ	Proportionality factor linking volume and energy transferred	[cm^3/J]
ν	Speed of rotation (kinematics)	[s^{-1}]
ν	Kinematic viscosity (fluid dynamics)	[m^2/s]
ρ	Density	[kg/m^3]
σ	Tension	[N/m^2]
τ	Half-life of amplitude (shock wave)	[s]
τ_M	Time required for a signal to reach maximum amplitude (rise time)	[ms]
τ_{TH}	Duration of a pulse (temporary cavity)	[ms]
φ	Angular displacement	[rad]
ψ	Angle of incidence	[rad]
ψ_e	Angle of incidence at point of impact	[rad]
ω	Angular velocity	[rad/s]
Γ	Angle of twist	[°]
Λ	Twist length	[mm]
Σ	Sum	
Ψ	Measure of effectiveness	[J/cm]

Relationships

∝	Proportional to		≈	Approximately equal to
<	Less than		<<	Much less than
>	Greater than		>>	Much greater than

Prefix symbols for decimal multiples and submultiples of SI units

Factor	Name	Symbol	Factor	Name	Symbol
10^{12}	tera	T	10^{-1}	deci	d
10^{9}	giga	G	10^{-2}	centi	c
10^{6}	mega	M	10^{-3}	milli	m
10^{3}	kilo	k	10^{-6}	micro	µ
10^{2}	hecto	h	10^{-9}	nano	n
10^{1}	deca	da	10^{-12}	pico	p

Conversions

U.S. → metric

	U.S.	Symbol	Definition	Metric	
Length	inch	in		25.4	mm
	foot	ft	12 in	0.3048	m
	yard	yd	3 ft	0.9144	m
Area	square inch	in²		645.16	mm²
Velocity	foot/second	ft/s		0.3048	m/s
Mass	grain	gr	$1/7000$ lb	0.0647989	g
	pound	lb		0.4535924	kg
Energy	foot pound force	ft lbf		1.35582	J
Force	pound force	lbf		4.448221	N
Pressure	pound/square inch	psi		0.0689476	bar

Metric → U.S.

	Metric	Symbol	Definition	U.S.	
Length	millimetre	mm	$1/1000$ m	0.03937	in
	metre	m	Base unit	3.28084	ft
	metre	m		1.0936133	yd
Area	square millimetre	mm²		0.001550	in²
Velocity	metre/second	m/s		3.28084	ft/s
Mass	gram	g	$1/1\,000$ kg	15.43236	gr
	kilogram	kg	Base unit	2.204622	lbs
Energy	joule	J	1 N·m	0.737561	ft·lbf
Force	newton	N	1 kg·m/s²	0.224809	lbf
	bar	bar	10^5 Pa	14.503774	psi

1   Introduction

Ballistics is the science of bodies in flight, encompassing the physical phenomena involved and the movement of the projectile. It is divided into a number of areas, based on where the projectile is.

Interior ballistics is the study of the acceleration of the bullet in the weapon and the related processes. The domain of interior ballistics ends where the bullet leaves the barrel. However, the weapon can continue to influence the flight of the bullet even after this point, e.g. through oscillations or via the gases that follow and overtake the bullet. This phase of the bullet's motion is known as *intermediate ballistics*.

Between the moment at which the bullet escapes the influence of the weapon and the moment at which it strikes its target, the bullet obeys the laws of *exterior ballistics*. This part of ballistics involves determining the changes over time and space of the trajectory of the bullet, its velocity and the movements it describes about its centre of gravity, taking into account all the forces acting upon it.

The study of the phenomena occurring when a bullet strikes and penetrates an object is termed *terminal ballistics*. If the object is a person or an animal, we speak of *wound ballistics*.

Interior, intermediate and exterior ballistics can all affect wound ballistics, depending on the distance between muzzle and target. The structure of the bullet and certain aspects of the weapon may also play a role. As a result, one can only understand what happens to a bullet in a living being if one has a basic understanding of the physics involved (mechanics, thermodynamics and fluid dynamics), of ballistics and of arms and ammunition. We shall cover these aspects in Chapter 2.

Chapter 3 – General wound ballistics – examines the phenomenon of the wound channel, describes simple physical models of velocity and energy over time and distance and provides an overview of the simulants generally used in wound ballistics.

Building on Chapter 3, Chapter 4 will introduce the concept of a projectile's "effectiveness" and its "effect," and will look in some detail at the wound ballistic aspects of handgun and rifle bullets. This chapter devotes a number of sections to the wound ballistics of fragments, as fragments are the most frequent cause of injury in armed conflict and in bomb attacks. A separate section is devoted to so-called "non-lethal" projectiles, as these are becoming increasingly important.

The remaining three chapters apply the basic knowledge acquired in Chapters 3 and 4.

Forensic medicine (Chapter 5) uses the laws of wound ballistics to derive ballistic data (type of weapon and ammunition, direction of shot, range, etc.) required to elucidate killings involving firearms. Simulants can readily be used for dynamic reconstructions of events, and are particularly valuable as part of the Virtopsy concept (see 5.2 and 5.3).

In war surgery (Chapter 6) wound ballistics is of interest primarily as a diagnostic tool. If a surgeon knows how a bullet behaves in the human body, he or she is in a better position to manage the resulting wound.

Even though this book focuses on scientific facts and phenomena, on pathology and on the changes that bullet wounds cause, we must not lose sight of the human aspect. Doctors have repeatedly spoken out against undesirable developments in the field of arms and ammunition – generally with little success, unfortunately. The International Committee of the Red Cross (ICRC), with its headquarters in Geneva, has played a major role in this area. It therefore seemed appropriate to outline the historical development of ammunition and the efforts made in parallel to render war more humane, efforts that are reflected in a number of international agreements. However, we also discuss ways of formulating such agreements much more precisely, using the knowledge acquired from the study of wound ballistics (Chapter 7).

Annex A contains comprehensive tables of ballistic data for a wide variety of ammunition and bullets. This data is essential to the study of ballistics in general and wound ballistics in particular. This annex also contains a number of ballistics tables for contemporary and historical bullets and for other projectiles (such as fragments and arrows). These tables are followed by similar tables for shotgun cartridges and pellets.

Annex B consists of a glossary of ballistics and wound ballistics terms. This is followed by a bibliography and index.

2 Basics

B. P. KNEUBUEHL

2.1 The physics of wound ballistics

2.1.1 Preliminary remarks

Wound ballistics is an inter-disciplinary science, involving a wide range of specialists – doctors, physicists, lawyers, weapons experts, etc. There is therefore a need for a concise introduction to the basic physics involved. Readers with a good knowledge of physics may wish to skip Section 2.1.

2.1.2 Coordinates, systems of units and notation

To describe physical phenomena simply, one needs a suitable system of coordinates. For ballistics, we generally use the ballistic coordinate system: the x- and y-axes between them define a vertical plane, with the y-axis pointing in the opposite direction to the earth's gravitational pull. The z-axis completes the system, creating a right-handed three-dimensional system (see Fig. 2-1). The movements of a body (a bullet in this instance) are described by reference to a Cartesian system fixed in the body. The origin of that system is located at the centre of gravity of the body concerned and its principal axis is aligned with the direction of movement of the centre of gravity at any given point in time.

The units are those of the SI system (Système International d'Unités), which is the official system in many countries. Length is measured in metres, mass in

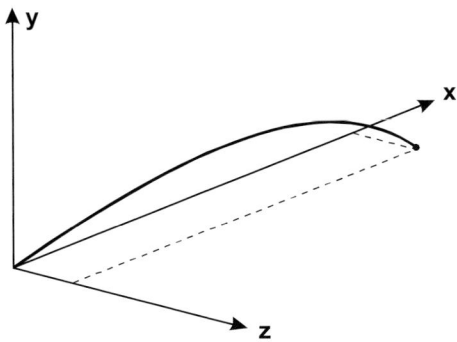

Fig. 2-1. Ballistic coordinate system x-axis in the direction of the shot, y-axis upwards, z-axis to the right.

kilograms and time in seconds, often modified by prefixes such as milli-, kilo- or mega-. A summary of the decimal prefixes appears at the end of the table of symbols (p. XXIII). Other, derived units will be explained the first time the corresponding parameter occurs.

Imperial/US units are still widely used in the fields of weapons, ammunition and ballistics. Again, the formulas for converting between these units and SI units are at the end of the table of symbols (p. XXIII).

Most phenomena of interest to us are three-dimensional, and must therefore be described in terms of three components. For the sake of clarity, however, we shall in many cases use only one component. This makes little difference in practice, as it is possible to study many processes one-dimensionally by selecting a system of coordinates appropriately.

Some definitions and equations include differential quotients. In such cases, we shall follow the usual convention: derivatives over time are written with a dot over the symbol, while derivates over distance are indicated using a prime, thus:

$$(2.1:1) \qquad \frac{dx}{dt} \Leftrightarrow \dot{x}, \quad \frac{dv_x}{dt} \Leftrightarrow \dot{v}_x, \quad \frac{dv_x}{dx} \Leftrightarrow v'_x .$$

2.1.3 Mechanics

2.1.3.1 Kinematics

Kinematics is the study of the motion of a body in space, disregarding the specific characteristics of the body concerned. The primary role of kinematics is to describe the path, or trajectory of the body. For these purposes we ignore the size of the body, treating it as a point. The most important kinematic parameter is velocity. Velocity is a vector value, of which the components are defined as distance covered per unit time along three axes:

$$(2.1:2) \qquad v_x = \frac{dx}{dt}, \quad v_y = \frac{dy}{dt}, \quad v_z = \frac{dz}{dt} . \qquad [m/s]$$

Speed (as opposed to velocity) is given by:

$$(2.1:3) \qquad v = \sqrt{v_x^2 + v_y^2 + v_z^2} . \qquad [m/s]$$

Table 2-1 gives a rough indication of typical speeds encountered in ballistics.

Change in velocity per unit time is termed *acceleration* if velocity increases and *negative acceleration*, or *deceleration*, if velocity decreases:

$$(2.1:4) \qquad a_x = \frac{dv_x}{dt}, \quad a_y = \frac{dv_y}{dt}, \quad a_z = \frac{dv_z}{dt} . \qquad [m/s^2]$$

Table 2-1. Typical velocities

Object/medium		Velocity	
		[m/s]	[ft/s]
Projectile	Air rifle/air pistol	100 ... 250	330 ... 820
	Handguns	250 ... 400	820 ... 1,310
	Rifles	600 ... 1,000	1,970 ... 3,280
	Flechette	1,500 ... 1,800	4,920 ... 5,900
	Fragment	< 2,000	< 6560
Sound waves	In air (15 °C)	340	1,115
	In water (20 °C)	1,483	4,865
	In steel	5,180	16,995
	In glass	5,225	17,142

We can express the magnitude of acceleration in a manner similar to that already discussed for speed:

$$(2.1:5) \qquad a = \sqrt{a_x^2 + a_y^2 + a_z^2} \, . \qquad [m/s^2]$$

Table 2-2 lists some typical acceleration values encountered in ballistics.

Because velocity is a vector, a change in direction is also an acceleration, even in cases where speed does not change.

If acceleration is constant, we can readily calculate speed and distance using the following formula:

$$(2.1:6) \qquad v = v_0 + a \cdot t \, , \qquad [m/s]$$

$$(2.1:7) \qquad x = x_0 + v_0 \cdot t + \tfrac{1}{2} \cdot a \cdot t^2 \, , \qquad [m]$$

where v_0 is the speed at time $t = 0$ and x_0 is the distance that the object has already moved at that moment in time.

Movement along a curved path can always be expressed as movement along a segment of a circle. The rotational movement is described in kinematic terms by

Table 2-2. Typical acceleration values

Object	Acceleration	
	[m/s²]	[ft/s²]
Acceleration due to gravity (standard value)	– 9.80665	– 32.1740
Pistol bullet in air	– 200	– 6,560
Rifle bullet in air	– 400	– 1,310
Fragment in air	– 60,000	– 197,000
Rifle bullet in barrel on firing, max.	100,000	– 328,100
Deformation projectile penetrating a dense medium (mean value) up to	– 800,000	– 2,625,000

the angular velocity, with angular displacement φ measured in radians (one complete revolution corresponds to 2π radians, which equates to 360°).

(2.1:8) $\qquad \omega = \dfrac{d\varphi}{dt} .$ [rad/s]

We obtain the rotational speed ν (the number of revolutions per second) from the angular velocity, thus:

(2.1:9) $\qquad \nu = \dfrac{\omega}{2 \cdot \pi} .$ [1/s]

Rotation at a constant angular velocity is known as *uniform circular motion*. The tangential velocity is given by:

(2.1:10) $\qquad v = r \cdot \omega ,$ [m/s]

where r is the radius of rotation, i.e. the distance between the point under consideration and the centre of rotation.

This type of motion involves acceleration, because while the speed of the object does not change, its direction is continuously changing. The acceleration is perpendicular to the direction of travel at any given moment, and is directed away from the centre of the circle. Acceleration is given by the following equation:

(2.1:11) $\qquad a = v \cdot \omega = \dfrac{v^2}{r} = r \cdot \omega^2 .$ [m/s²]

Conversion between the various representations is performed using Eqn 2.3:9. Changes in angular velocity are expressed as angular acceleration α:

(2.1:12) $\qquad \alpha = \dfrac{d\omega}{dt} .$ [rad/s²]

One case of rotation in ballistics is that of a spin-stabilized bullet, which rotates about its longitudinal axis. The angular velocities and rotational speed involved are quite high, and although the radius of a bullet is small, the circumferential velocities attained are considerable (see Table 2-3).

Table 2-3. Typical values for angular velocity ω, rotational speed ν and circumferential velocity v

Type of bullet	ω [rad/s]	ν [1/s]	v [m/s]	v [ft/s]
Pistol (9 mm Luger)	8,800	1,400	40	130
Revolver (.44 Rem. Mag.)	5,440	866	30	100
Rifle (7.62 mm NATO)	17,100	2,721	67	220

2.1.3.2 Mass, momentum and force

Mass is one of the basic characteristics of a body. It expresses itself in two ways:

- a body resists any change in state of movement (inertia);
- a body alters the state of movement of other bodies (gravitation).

Physical phenomena are often independent of the dimensions of a body. In such cases, it is easier to describe the phenomenon if we treat the body as a point, with the entire mass of the body concentrated in that point. This point – the centre of gravity – is chosen such that even when forces act upon the body there is no difference between the path of the actual body (of non-zero dimensions) and that of the point mass.

In SI units, mass is expressed in kilograms or grams. Grains and pounds are still used in some English-speaking countries (see conversion factors on p. XXIII).

The velocity of a body, v, multiplied by its mass, m, is its *momentum*, M. Momentum is a vector, of which the direction is identical to that of the body's velocity:

(2.1:13) $\quad I = m \cdot v$. $\quad$ [kg·m/s]

The physical parameter responsible for a change in the state of movement of a body, or deformation of that body, is known as *force*. Force is defined by *Newton's laws*.

1. The velocity of a body (or, more accurately, its momentum) remains constant unless a force acts upon it.
2. Force equals change in momentum over time, so:

(2.1:14) $\quad F = \dfrac{dI}{dt} = \dfrac{d}{dt}(m \cdot v)$. $\quad$ [N]

The unit of measurement for force is the Newton, and 1 Newton [N] = 1 kg·m/s^2.

Though not an SI unit, the kilopond [kp] is also in widespread use. In some English-speaking countries, force is also measured in pounds (or pounds-force). Conversion formulas for units of force are to be found on p. XXIII.

If the mass of a body remains constant, force equals mass times acceleration:

(2.1:15) $\quad F = \dfrac{d}{dt}(m \cdot v) = m \cdot \dfrac{dv}{dt} = m \cdot a$. $\quad$ [N]

3. For every force, there is an opposing force of the same magnitude (action/reaction). Please see Table 2-4 for a list of typical forces encountered in ballistics.

As acceleration is a vector, it follows from Eqn 2.1:15 that force also behaves as a vector. In order to describe a force, we must therefore know three things: its *mag-*

Table 2-4. Typical forces acting on a bullet

Type of force	Force	
	[N]	[lbf]
Force of air acting on a rifle bullet	4	0.9
Mean drag acting on a bullet fired into water	5,000	1,125
Force on a bullet in a rifle barrel	12,000	2,700

nitude, its *direction* and its *point of application*. The best-known force is weight, G. Weight is the force that occurs when a body is subjected to the acceleration due to the earth's gravity, g. From Eqn 2.1:15:

(2.1:16) $\qquad G = m \cdot g$. $\hfill$ [N]

In physics, it is often useful to express force as a function of the area on which it is acting. When a force acts perpendicular to a surface, it is termed either a *tensile* or a *compressive* force. If a force acts parallel to a surface, it is known as a *shear* force. Force per unit area is known as *stress*.

Stress (generally represented by the symbol σ) is often more important than the corresponding force. As a result, it has its own unit of measurement, the Pascal (Pa). 1 Pa ⇔ 1 N/m². Compressive stress may also be expressed in bar. This unit is widely used in ballistics:

$\qquad$ 1 bar ⇔ 10⁵ Pa .

In some English-speaking countries, pressure is also measured in pounds per square inch (lbf/in²). Conversion formulas for units of pressure are to be found on p. XXIII.

2.1.3.3 Work and energy

Work and energy, and their relationship to velocity, force, momentum and power, play a crucial role in ballistics. We shall therefore examine these terms in detail, the more so as they are not always used correctly in everyday speech.

$\quad$ Work W is defined as the force F acting in a given direction, multiplied by the distance covered under the influence of that force, s:

(2.1:17) $\qquad W = F \cdot s$. $\hfill$ [J]

Any irregular movement, and any change in the structure of a material (deformation, destruction, etc.) requires work. The amount of work required depends not only on the force, but also on the distance covered. Even if the force is large, the work done will be small if the force acts over a short distance (which often equates to a short exposure time).

Given that work is defined as force times distance, the unit is the Newton-metre [N·m]. However, as work is such an important term it has its own unit, the Joule [J]:

$\qquad$ 1 J ⇔ 1 N·m .

Certain English-speaking countries still use the foot-pound-force [ft·lbf]. See p. XXIII for the conversion formula.

In order to move a body against the acceleration due to the earth's gravity, a force is required that corresponds to the weight of the body. From Eqn 2.1:17, we can calculate the work done W using the following formula:

(2.1:18) $\qquad W = m \cdot g \cdot y$, $\qquad$ [J]

where y is the distance that the body is moved against the force of gravity.

If, however, the body undergoes acceleration as a result of the force acting upon it, the work done results in movement. Substituting Equations 2.1:15, 2.1:7 and 2.1:6 in Eqn 2.1:17 (the latter two equations with $x_0 = 0$ and $v_0 = 0$) we obtain:

(2.1:19) $\qquad W = \frac{1}{2} \cdot m \cdot v^2$. $\qquad$ [J]

If we ignore non-mechanical phenomena, the work done on a body therefore results either in an increase in distance within the gravitational field, or in movement. In the first instance, this can be expressed in terms of the weight of the body and the increase in distance (Eqn 2.1:18) and in the second case in terms of the mass of the body and the speed it attains (Eqn 2.1:19).

Both a body that has been raised from its initial position and a body in motion are capable of performing work. The ability to perform work is generally referred to as *energy*. In the case of a raised body we speak of *potential energy*, while in the case of a moving body the term is *kinetic energy* (see Table 2-5 for examples). In physics, therefore, work and energy are equivalent. Neither work or energy can be created or destroyed. All that can happen is that the one is converted into the other. From Eqn 2.1:19, we can therefore derive an analogous formula for the kinetic energy of a body in motion, with the same unit [J]:

(2.1:20) $\qquad E_{kin} = \frac{1}{2} \cdot m \cdot v^2$. $\qquad$ [J]

Using Eqn 2.1:18, we can express potential energy thus:

(2.1:21) $\qquad E_{pot} = m \cdot g \cdot y$. $\qquad$ [J]

Wounding only occurs when energy is converted into work. However, this process is only partially mechanical, as the energy taken from the projectile is used primarily to deform and destroy tissue. In other words, this energy performs work

Table 2-5. Typical kinetic energy levels encountered in ballistics

Type of weapon		Muzzle energy	
		[J]	[ft·lbf]
Air rifle	over	10	7.4
Pistol		500	370
Modern military rifle (5.56 mm)		1,600	1,180
Older military rifle (7.62 mm)		3,000	2,215
Hunting rifle	up to	10,000	7,375

on the molecular structure of the material. Clearly, however, the severity of a wound (the quantity of tissue destroyed) can depend only on the energy taken from the bullet, and not on the total energy that the bullet possessed.

2.1.3.4 Rotation

If a force acts on a system capable of rotation, at a point away from its axis of rotation, the system will begin to rotate. The angular velocity will depend on the force and on the distance between the point of application and the axis of rotation (the lever arm). This phenomenon can be represented as:

(2.1:22) $\qquad M = F \cdot r$, $\qquad$ [N·m]

where M is the torque and r is the distance between the point of application and the centre of rotation (i.e. the length of the lever arm).

If we consider torque as a vector, it always acts perpendicular to both the direction of the force and the lever arm, and hence is always parallel to the axis of rotation.

From Equations 2.1:15, 2.1:10 and 2.1:12 we obtain:

(2.1:23) $\qquad M = m \cdot \dfrac{dv}{dt} \cdot r = m \cdot \dfrac{r \cdot d\omega}{dt} \cdot r = m \cdot r^2 \cdot \alpha$. $\qquad$ [kg·m²/s²]

From this equation, we can see that there is a linear relationship between the torque and the angular acceleration it produces, as is the case for force and acceleration. The factor that links the two is the moment of inertia J, a characteristic of the body related to the axis of rotation:

(2.1:24) $\qquad M = J \cdot \alpha$. $\qquad$ [kg·m²/s²]

The moment of inertia J therefore has the same relationship to rotation as does mass to linear motion.

The moment of inertia of a body can be calculated by analytical or numerical integration over the volume:

(2.1:25) $\qquad J = \int r^2 \cdot dm$. $\qquad$ [kg·m²]

The moment of inertia can also be measured using a "moment of inertia" measuring instrument. The formulas for basic, regular bodies are in some cases quite simple.

Using the definition of force in Eqn 2.1:14, we can derive the following relationship from Eqn 2.1:24:

$$M = F \cdot r = \dfrac{dI}{dt} \cdot r = J \cdot \dfrac{d\omega}{dt} = J \cdot \alpha \ . \qquad [\text{kg·m}^2/\text{s}^2]$$

Integrating the two corresponding terms, we obtain an equation with, on one side, the product of the torque and its distance from the axis of rotation. This product is the *angular momentum* or *spin*:

(2.1:26) $\qquad L = I \cdot r = m \cdot v \cdot r = J \cdot \omega$. $\qquad$ [kg·m²/s]

Table 2-6. Comparison between linear and rotational motion

	Linear motion (translation)		Rotational motion (rotation)
Time	t	t	Time
Velocity	v	ω	Angular velocity
Acceleration	a	α	Angular acceleration
Mass	m	J	Moment of inertia
Force	$F = m \cdot a$	$M = J \cdot \alpha$	Torque
Momentum	$M = m \cdot v$	$L = J \cdot \omega$	Angular momentum
Kinetic energy	$E_{kin} = \frac{1}{2} \cdot m \cdot v^2$	$E_{rot} = \frac{1}{2} \cdot J \cdot \omega^2$	Rotational kinetic energy

In determining the kinetic energy of a rotating mass (its energy of rotation, or spin energy), we must remember that each mass particle has a velocity that depends on its distance from the axis of rotation. We must therefore integrate the differential energy over the entire volume:

$$E_{rot} = \frac{1}{2} \cdot \int_V v^2 \cdot dm = \frac{1}{2} \cdot \int_V \omega^2 \cdot r^2 \cdot dm = \frac{1}{2} \cdot \omega^2 \cdot \int_V r^2 \cdot dm ,$$

(2.1:27) $\qquad E_{rot} = \frac{1}{2} \cdot J \cdot \omega^2 .$ [J]

The analogy between the formula for linear motion and that for rotation is so striking that it is worth placing them alongside one another (see Table 2-6).

It is often forgotten that yaw and partial rotation are also forms of rotation. In addition to rotating rapidly about its longitudinal axis, a moving bullet makes other rotational movements, such as those about a horizontal transverse axis. See Fig. 2-2.

2.1.3.5 Laws of conservation of mass, energy and momentum

All physical phenomena are controlled by a number of "laws of conservation." This is particularly true of mechanics. The analysis of any physical process starts with the application of these fundamental laws, which generally make it possible to understand a phenomenon quite easily. However, it is important first to delimit the physical system in such a way that one can ignore any interaction with the

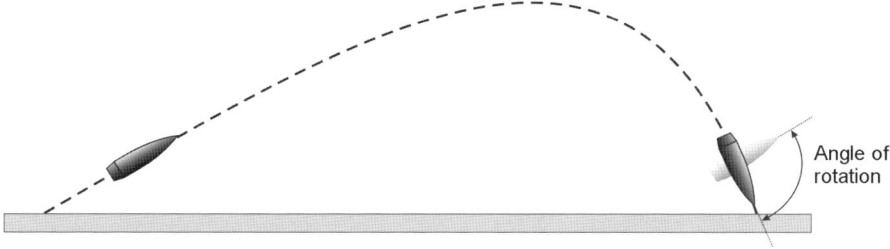

Fig. 2-2. Rotation of a bullet about a transverse axis along its trajectory.

outside world. Such a system is described as "closed." Laws of conservation apply only within closed systems.

1. *The law of conservation of mass.* The total mass within a closed system remains constant. Mass is neither created nor destroyed:

$$(2.1\!:\!28) \qquad m_{tot} = \sum m_i = \text{const}. \qquad [\text{kg}]$$

In a system consisting of several masses, the centre of gravity of the system also remains constant, regardless of the movements of the individual masses (such as in the case of a shot sheaf).

2. *The law of conservation of momentum.* The total momentum within a closed system (i.e. one on which no external forces are acting) remains constant:

$$(2.1\!:\!29) \qquad I_{tot} = \sum I_i = \text{const}. \qquad [\text{N·s}]$$

3. *The law of conservation of angular momentum.* The total momentum within a closed system (i.e. one on which no external moment is acting) remains constant:

$$(2.1\!:\!30) \qquad L_{tot} = \sum L_i = \text{const}. \qquad [\text{kg·m}^2/\text{s}]$$

4. *The law of conservation of energy for mechanics.* The sum of the potential, kinetic and rotation energy in a closed, frictionless system (with no energy entering or leaving the system) remains constant:

$$(2.1\!:\!31) \qquad E_{mech} = \sum E_{pot} + \sum E_{kin} + \sum E_{rot} = \text{const}. \qquad [\text{J}]$$

However, the law of conservation of energy goes beyond mechanics. It also applies to other, non-mechanical forms of energy.

5. *The general law of conservation of energy.* The total energy (i.e. the sum of all forms of energy) in a closed system remains constant.

The conversion of energy from one form to another within a system is not necessarily reversible.

2.1.3.6 Equations of motion

In physics, the motion of a body is described fully if its position, velocity and position in space are known at every point in time. The motion of a body is determined exclusively by the forces acting upon it and the accelerations those forces produce.

Linking the acceleration of a body to its position and velocity yields *equations of motion*. These form a system of differential equations that describe the time-dependent velocity and position functions of a body. In the case of a point mass (e.g. the centre of gravity of a body), the system consists of six equations – three for the spatial coordinates and three for velocity. A rigid body of non-zero volume requires six further equations, which describe the body's position in space and the rate at which its position changes. In many cases, however, the system of equations of movement can be considerably reduced. This is the case, for instance, when no force is acting in a particular direction.

D'Alembert's principle. Equations of movement can be derived from the law of conservation of energy or of momentum by differentiation. D'Alembert's principle allows us to take a more direct approach. This principle involves replacing the acceleration of a body by a fictitious force F_{fict} which, in accordance with Newton's Third Law of Motion, has to equal the sum of the forces applied:

$$(2.1:32) \qquad m \cdot a = F_{fict} = \sum F_i \, . \qquad [N]$$

The equations of motion utilize another basic principle of mechanics: the 'superposition principle'. This principle states that different movements of a body occurring simultaneously do not influence each other. Any movement can therefore be divided into separate components, one for each of the axes of the coordinate system. We can study each component independently of the others and hence can describe each component using simple equations.

Trajectory in a vacuum. One typical example of how equations of movement are developed and solved – and one of particular relevance to ballistics – is that of the trajectory of a body in a vacuum. No forces act along the x-axis, and the only force acting along the y-axis is the weight of the body. In accordance with d'Alembert's principle, we can write the following two equations:

$$(2.1:33a) \qquad m \cdot \dot{v}_x = 0 \qquad \text{(x-axis)}, \qquad [N]$$

$$(2.1:33b) \qquad m \cdot \dot{v}_y = -m \cdot g \qquad \text{(y-axis)} \, . \qquad [N]$$

These two differential equations form the system of equations of movement describing the *trajectory of a body in a vacuum*. We can readily integrate the two equations, using the initial values below, derived from the initial velocity v_0 and angle of departure θ_0:

$$v_{0x} = v_0 \cdot \cos \theta_0 \, , \qquad [m/s]$$

$$v_{0y} = v_0 \cdot \sin \theta_0 \, . \qquad [m/s]$$

Integration yields the following two equations:

$$(2.1:34a) \qquad x(t) = v_0 \cdot \cos \theta_0 \cdot t \, , \qquad [m]$$

$$(2.1:34b) \qquad y(t) = v_0 \cdot \sin \theta_0 \cdot t - \tfrac{1}{2} \cdot g \cdot t^2 \, . \qquad [m]$$

If we eliminate t from both equations, we obtain the equation for the (parabolic) trajectory of a body in a vacuum:

$$(2.1:35) \qquad y(x) = \tan \theta_0 \cdot x - \frac{g \cdot x^2}{2 \cdot v_0^2 \cdot \cos^2 \theta_0} \, . \qquad [m]$$

Taking y = 0 (base of the trajectory), the maximum range x_e for a given angle of departure θ_0 is:

(2.1:36) $$x_e = \frac{v_0^2}{g} \cdot \sin(2 \cdot \theta_0) \ .$$ [m]

Similarly, we can obtain the corresponding time of flight from Eqn 2.1:34b:

(2.1:37) $$t_e = \frac{2 \cdot v_0}{g} \cdot \sin \theta_0 \ .$$ [s]

We can readily determine the highest point along the trajectory (the peak) if we bear in mind that the vertical component of the velocity must be zero at that point. This gives us the time of flight to the peak of the trajectory. Substituting this into Eqn 2.1:34b we obtain:

(2.1:38) $$y_s = \frac{v_0^2}{2 \cdot g} \cdot \sin^2 \theta_0 \ .$$ [m]

Combining Equations 2.1:37 and 2.1:38 gives us the principal equation for the vertex height. This equation is very useful in practice, and to a close degree of approximation is also valid for the trajectory of an object subject to air resistance:

(2.1:39) $$y_s = \tfrac{1}{8} \cdot g \cdot t_e^2 \ .$$ [m]

Velocity as a function of distance. It is often clearer to show velocity as a function of distance instead of time. For these purposes, "distance" generally means the distance the projectile has travelled along its trajectory, although the term may refer to the distance travelled along the x-axis in some cases.

For example: we can say that a light, high-speed hunting bullet with a muzzle velocity of 1050 m/s undergoes a deceleration of 1700 m/s^2, but it is difficult to translate this information into anything practical. If, however, we say that the bullet loses 1.6 m/s per metre travelled, we can understand what is happening much more easily.

We can write the change in distance along the trajectory thus:

(2.1:40) $$ds = \sqrt{dx^2 + dy^2 + dz^2} \ ,$$ [m]

from which we can calculate velocity as follows:

$$v = \frac{ds}{dt} = \sqrt{\left(\frac{dx}{dt}\right)^2 + \left(\frac{dy}{dt}\right)^2 + \left(\frac{dz}{dt}\right)^2} \ ,$$ [m/s]

$$v = \sqrt{v_x^2 + v_y^2 + v_z^2} \ .$$ [m/s]

This yields the following formula for converting between velocity as a function of time and velocity as a function of distance (using the typographical convention specified in 2.1.2):

(2.1:41) $\quad \dot{v} = \dfrac{dv}{dt} = \dfrac{dv}{ds} \cdot \dfrac{ds}{dt} = v' \cdot v \;.$ $\hfill [m/s^2]$

2.1.4 Fluid dynamics

2.1.4.1 General

The materials involved in wound ballistics (certain types of tissue, together with the soap and gelatine used as simulants) act upon a bullet in a manner similar to that of a viscous liquid. It is considerably easier to understand such phenomena if we model them as such. On the other hand, the behaviour of a bullet in tissue or in a simulant depends to a large degree on the impact conditions. In turn, these conditions are directly linked to the movement of the bullet in air. The mechanics of fluids and gases – known as fluid dynamics – therefore play a major role in wound ballistics.

When we are studying the movement of a body in a medium it makes no difference whether we consider the medium to be stationary and the body in motion, or the body stationary and the medium in motion (in the opposition direction and at the same speed). In terms of physics, the two approaches are equivalent. We will therefore use whichever is clearer in a given context.

Motion in a liquid or gaseous medium always generates heat, whether through friction or as a result of subjecting a gas to increased pressure. It is therefore necessary to include thermodynamic processes in any consideration of fluid dynamics.

2.1.4.2 Basic concepts in thermodynamics

Temperature. Temperature is one of the basic parameters required to describe processes that involve heat. Temperature describes the thermal state of a body and serves as a measure of the total kinetic energy of the molecules that make up that body. It is independent of the mass and the material composition of the body.

If the thermal state of a material changes, so do a number of its physical characteristics, such as its dimensions (length, volume, etc.), its colour or its electrical conductivity. We can use this behaviour to measure temperature. However, it is only possible to conduct relative measurements. As a result, we are free to select the zero point and the units of a scale at will.

The SI unit for temperature is the Kelvin [K]. The only difference between the Kelvin and the older degree Celsius [°C] is the zero point used for the two systems. On the Celsius scale, 0 °C corresponds to the freezing point of water, whereas 0 K corresponds to "absolute zero," the lowest temperature possible. Degrees Fahrenheit are in widespread use in some English-speaking countries. All scales are calibrated using internationally agreed reference points, corresponding

Table 2-7. Temperature reference points

		[K]	[°C]	[°F]
Water	Freezing point	273.15	0.00	32.00
	Boiling point	373.15	100.00	212.00
Oxygen	Boiling point	90.18	−182.97	−297.35
Gold	Melting point	1336.15	1063.00	1945.40

to the boiling and freezing points of various materials at normal pressure (1013.25 mb). See Table 2-7.

Conversion formulas for the various units appear after the table of symbols (p. XXIII).

Temperature and heat. If the temperature of a given quantity of a material rises during a physical process, then heat has been introduced. However, an increase in temperature means that the total kinetic energy of the molecules in the quantity of material under consideration has also increased. Heat is therefore a form of energy and is hence measured in Joule. The quantity of heat added is proportional to the mass and to the increase in temperature. The proportionality factor C is known as the *specific heat capacity* of the material:

(2.1:42) $$\Delta Q = C \cdot m \cdot \Delta T \ . \qquad [J]$$

States of matter. The state of a material can be *solid, liquid* or *gaseous*. The more precise terms for these three states are *crystalline, amorphous* and *gaseous*. In the solid state, the molecules are arranged in a crystal lattice, and are held in place not only by intermolecular forces of attraction but also by lattice linkage forces. Solid matter has a specific geometrical form and a fixed volume.

If so much heat is applied to a body that its molecules acquire enough kinetic energy to overcome the lattice forces, the material enters the amorphous (liquid) state. All that is keeping the material together at this point is the intermolecular force. Liquid matter has a fixed volume, but generally has no specific geometrical form.

If the intermolecular forces are sufficiently strong, amorphous materials may be shape stable. Examples of this include glass and wax.

If we add enough heat to raise the kinetic energy of the molecules to the point at which the intermolecular forces are overcome, the molecules become free to move. The material enters the gaseous state, in which its volume is limited not by the quantity of matter present, but by the total space available to it. From the above, we see that it is only possible to change the state of a material to the next higher level by adding energy in the form of heat. Similarly, the corresponding quantity of energy is released as the material returns to a lower level.

2.1 The physics of wound ballistics   17

Equation of state for gases. Because of the kinetic energy of its free molecules, a gas exerts a pressure p on the surfaces that delimit its volume. The impacts of the particles against these surfaces result in a mean force per unit area, which can be explained by the law of conservation of momentum. Any change in temperature (i.e. in kinetic energy) or in volume (i.e. in surface area) will affect the pressure. Pressure, temperature and volume are the *thermodynamic state variables* of the gas in question. The equation that links these variables is known as an *equation of state*. The best-known equation of state is Boyle's law:

(2.1:43) $p \cdot V = m \cdot R \cdot T$.   [J]

Where m is the mass of the gas and R is the special (material-dependent) gas constant. If a gas obeys Boyle's law (Eqn 2.1:43), it is known as an *ideal gas*. If the density of the gas is sufficiently low (i.e. if the pressure is low), many gases behave almost as ideal gases.

The special gas constant of air is 287.05 J/(kg·K).

In gas dynamics, we generally use density (mass per unit volume) rather than mass. Eqn 2.1:43 can then be written:

(2.1:44) $\dfrac{p}{\rho} = R \cdot T$.   [J/kg]

If temperature remains constant, the right-hand side of the equation remains constant, and hence so does the left-hand side.

Heat, work and internal energy. As the volume of a material can only change considerably when it is in a gaseous state, it follows that materials can only convert heat energy into mechanical work (energy) when they are in that state. Devices that perform this conversion are known as *heat engines*.

Heat engines – which include firearms – always use materials in gaseous form. Those materials are usually created by heating or by combustion.

If the volume of a gas is increased by heating it, then the gas performs work. According to Eqn 2.1:17, the work done is proportional to the force exerted (pressure · area) and the distance moved (see Fig. 2-3):

(2.1:45) $\Delta W = p \cdot A \cdot \Delta s = p \cdot \Delta V$.   [J]

As we know, the ability to do work is *potential energy*.
   The potential energy of a gas of volume V at a pressure p is therefore:

(2.1:46) $E_{dr} = p \cdot V$.   [J]

In this context, the potential energy of the gas is termed *pressure energy*.
   When heat energy is added to a system, two things happen. The first is that the total kinetic energy of the molecules in that system (its internal energy U) increases, which means that its temperature also increases. The second is that the

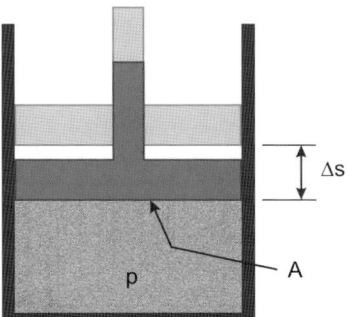

Fig. 2-3. Calculating the work done as a gas expands over a distance Δs.

system performs mechanical work. The following energy balance therefore applies:

(2.1:47) $dQ = dU + dW = dU + p \cdot dV$.   [J]

If no heat is added to the system (dQ = 0) then, from Eqn 2.1:47:

$$dU + p \cdot dV = 0 ,$$

and hence:

(2.1:48) $U + p \cdot V = \text{const}$.   [J]

The sum of internal energy and pressure energy therefore remains constant.

This statement, taken with the corresponding law of mechanics (Eqn 2.1:31) constitutes the mechanical thermodynamic law of conservation of energy:

(2.1:49) $E_{tot} = E_{mech} + E_{dr} + U = \text{const}$.   [J]

In order for the temperature of a gas to rise by a specific amount, the gas must absorb more heat at a constant pressure than at a constant volume. This is a direct consequence of Eqn 2.1:48.

If volume remains constant (dV = 0), all of the additional heat is used to raise the internal energy of the gas (i.e. its temperature), as no mechanical work can be done. At constant pressure, however, heat is "lost" to work, and the temperature increase is smaller.

This means that the specific heat capacity in Eqn 2.1:42 is greater at constant pressure than at constant volume:

(2.1:50) $\gamma = \dfrac{c_p}{c_v} > 1$.   [-]

The quotient γ is known as the *adiabatic coefficient*.

2.1.4.3 Material characteristics

Density and compressibility. When liquids and gases (fluids) are subjected to a force they change their form and, in many cases, their volume. In order to describe motion, we therefore need physical parameters that describe the main characteristics of the materials involved. In fluid dynamics, density ρ is more important than mass:

(2.1:51) $$\rho = \frac{m}{V} \, . \qquad [kg/m^3]$$

The compressibility of a material is a measure of the relative change in the volume of that material in response to a change in pressure:

(2.1:52) $$\kappa = -\frac{1}{V} \cdot \frac{dV}{dp} \, . \qquad [1/Pa]$$

The minus sign indicates that an increase in pressure brings about a decrease in volume.

Liquids are generally considered incompressible. At low speeds, we can make the same assumption for gases.

Viscosity. The absence of lattice force means that a fluid changes its form when subjected to an external force. However, the fluid reacts to that force with a certain inertia. This resistance, caused by the intermolecular forces of attraction, is termed *viscosity*. If a shear stress is applied to the fluid, with a shear force F_R, (see Fig. 2-4), a velocity gradient is created in the area under stress. In Newtonian fluids, this gradient is proportional to the stress:

(2.1:53) $$\sigma_s = \frac{F_R}{A} = \eta \cdot \frac{dv}{dy} \, . \qquad [N/m^2]$$

The proportionality factor is the *dynamic viscosity*, η:

(2.1:54) $$\eta = \sigma_s \cdot \frac{dy}{dv} \, . \qquad [Pa \cdot s]$$

Dynamic viscosity as a function of the density of the fluid is known as *kinematic viscosity*:

(2.1:55) $$\nu = \frac{\eta}{\rho} \, . \qquad [m^2/s]$$

Viscosity depends very much on temperature, with the viscosity of a gas changing with temperature in exactly the opposite manner to that of a liquid.

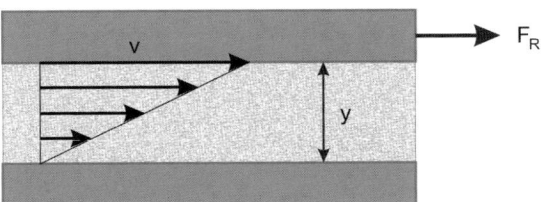

Fig. 2-4. Dynamic viscosity.

As the temperature of a liquid (and hence its internal energy) increases, the effect of the intermolecular forces decreases, and it becomes less viscous. In the case of a gas, internal friction is caused by the exchange of momentum between freely moving molecules. As temperature increases, the speed of those molecules increases, causing the gas to become more viscous.

If the mass remains constant, a fluid also resists any change in volume, as a change in volume can only occur if the form of the fluid also changes. This effect is known as *bulk viscosity*, which is to be distinguished from dynamic viscosity.

Speed of sound. Changes in pressure and density (disturbances) propagate through a fluid in the form of longitudinal waves.

In the case of longitudinal waves, the disturbances oscillate in the same direction as that in which the wave is moving.

If the disturbances remain small, they propagate at the speed of sound. In liquids, this speed depends on density and compressibility:

$$(2.1:56) \qquad c = \sqrt{\frac{1}{\kappa \cdot \rho}} \; . \qquad [m/s]$$

In gases, it depends on density and pressure:

$$(2.1:57) \qquad c = \sqrt{\frac{\gamma \cdot p}{\rho}} = \sqrt{\gamma \cdot R \cdot T} \; . \qquad [m/s]$$

The symbol γ represents the ratio of the specific heat capacities, c_p/c_v (Eqn 2.1:50). For air, its value is 1.4.

The general equation for the propagation speed of small disturbances in a gas is:

$$(2.1:57a) \qquad c^2 = \left(\frac{\partial p}{\partial \rho}\right)_{\Delta S = 0} . \qquad [m^2/s^2]$$

The significance of $\Delta S = 0$ is that changes in temperature have no effect on the heat (energy). This is known as *constant entropy*. In the case of an isentropic gas, this gives us the first part of Eqn 2.1:57.

In accordance with Eqn 2.1:43, the second part of Eqn 2.1:57 only applies to ideal (or nearly ideal) gases. In such cases, the speed of sound depends solely on temperature. As a result, this speed is a suitable reference speed for compressible media. This leads us to the definition of the Mach number, Ma:

$$(2.1:58) \qquad Ma = \frac{v}{c} \; . \qquad [-]$$

The behaviour of a flow is closely related to the Mach number, which gives us a simple classification criterion.

If $Ma < 1$, the flow is described as *subsonic*. If $Ma = 1$, the flow is described as *transonic*. If $Ma > 1$, the flow is said to be *supersonic*.

2.1.4.4 Frictionless flow

Definition. Laws of conservation. The study of flow is much simplified if one ignores internal friction. This implies that no shear stresses occur, and hence that the fluid does not deform. This assumption also ignores the fact that every fluid adheres to a fixed surface. As a result, external friction is also ignored. Despite this limitation, a surprising number of phenomena can be described and presented in the form of laws.

This is because when a fluid flows over a body, the effect of friction is often limited to a thin layer at the surface of the body (known as the *boundary layer*). If the thickness of this boundary layer is very small compared to the other dimensions of the flow, we can safely ignore the effects of friction.

A fluid in which no internal friction forces occur is termed *frictionless*. If, in addition, the fluid is incompressible, it is said to be *ideal*. Flow is also described in terms of its relation to time. If the speed and thermodynamic characteristics of the flow depend only on position, and do not change over time, the flow is described as *stationary*.

The laws of conservation of mass, energy and momentum set out in 2.1.3.5 also apply to the movements of fluids. Here again, the conservation of mass plays a central role: the change in mass in any fixed volume element corresponds to the difference between the mass entering the system (subscript 1 in the equation below) and the mass leaving it (subscript 2):

(2.1:59) $$\frac{\Delta m}{\Delta t} = \rho_1 \cdot v_1 \cdot A_1 - \rho_2 \cdot v_2 \cdot A_2 \ . \qquad [kg/s]$$

This relationship is known as a *continuity equation*. For an incompressible fluid (i.e. $\Delta m / \Delta t = 0$, $\rho = $ const), Eqn 2.1:59 becomes:

(2.1:59a) $$v \cdot A = \text{const} \ . \qquad [m^3/s]$$

Speed and flow cross-section are inversely proportional to one another. The law of conservation of energy applies. If no heat is added, the sum of pressure energy, kinetic energy, potential energy and internal energy remains constant (see Eqn 2.1:49):

$$E_{dr} + E_{kin} + E_{pot} + U = \text{const} \ , \qquad [J]$$

(2.1:60) $$p \cdot V + \tfrac{1}{2} \cdot m \cdot v^2 + m \cdot g \cdot y + U = \text{const} \ . \qquad [J]$$

Momentum and angular momentum are also conserved in a flow. As in the case of mechanics, we obtain the equations of motion for flow from the law of conservation of momentum, using d'Alembert's principle. In the case of flow, the formulas are known as Euler's formulas. The question of angular momentum arises only in turbulent flow, which we shall ignore in this context.

Bernoulli's equation. In theory, Bernoulli's equation is an equation of motion, and has to be derived from the equilibrium of forces acting on a liquid element. However, in the case of ideal flow (i.e. where ρ and U are both constant) it is easier to derive this equation from the law of conservation of energy (Eqn 2.1:60) by dividing by volume:

(2.1:61) $\quad\quad p + \tfrac{1}{2} \cdot \rho \cdot v^2 + \rho \cdot g \cdot y = \text{const}$, $\quad\quad$ [N/m²]

or, if we also assume a constant flow height:

(2.1:61a) $\quad\quad p_1 + \tfrac{1}{2} \cdot \rho_1 \cdot v_1^2 = p_0 + \tfrac{1}{2} \cdot \rho_0 \cdot v_0^2 = \text{const}$. $\quad\quad$ [N/m²]

This means that pressure and flow velocity are directly related to one another. If pressure is high, fluid velocity is low. If pressure is low, fluid velocity is high. Acceleration occurs when there is a pressure gradient.

Flow forces. If a flow strikes a flat, stationary plate at right angles, the flow velocity drops to zero at the surface of the plate. If we use $_0$ to indicate the values for undisturbed flow and $_1$ for those at the plate surface, then from Eqn 2.1:61a, for $v_1 = 0$ we have:

$$p_1 = p_0 + \tfrac{1}{2} \cdot \rho_0 \cdot v_0^2 ,$$

(2.1:62) $\quad\quad p_1 - p_0 = \tfrac{1}{2} \cdot \rho_0 \cdot v_0^2 .$ $\quad\quad$ [N/m²]

The left-hand side of Eqn 2.1:62 corresponds to the increase in pressure at the plate surface by comparison with the pressure in undisturbed flow. This is termed the *stagnation pressure* of the plane plate.

In the case of a real body, the streaming pressure depends not only on velocity but also on the form and dimensions of the body. The actual values will differ from those calculated using the equation above.

The streaming pressure on a specific body is described using the ratio of the actual increase in pressure to the plane stagnation pressure. This (dimensionless) quantity is known as the *pressure coefficient* of the body:

(2.1:63) $\quad\quad C_p = \dfrac{p_{\text{eff}} - p_0}{\tfrac{1}{2} \cdot \rho \cdot v^2} .$ $\quad\quad$ [-]

The force that a flow exerts on a body (the flow resistance) is obtained by multiplying the force by the effective area. However, it is often difficult to determine this area. The effective area is therefore replaced by a fixed reference area A_0 which, together with the plane stagnation pressure from Eqn 2.1:62, yields a reference force that is independent of the shape of the body. Dividing the effective resistance F_D by this reference force gives another dimensionless quantity that takes account of all form-specific influences. This quantity is known as the *drag coefficient*:

$$\text{(2.1:64)} \qquad C_D = -\frac{F_D}{\frac{1}{2} \cdot \rho \cdot v^2 \cdot A_0}, \qquad [-]$$

$$\text{(2.1:64a)} \qquad F_D = -C_D \cdot \tfrac{1}{2} \cdot \rho \cdot v^2 \cdot A_0. \qquad [N]$$

The minus sign indicates that the force acts in the opposite direction to the movement.

A body subject to a flow experiences a lift force (F_D) – a force perpendicular to the direction of flow – when the pressure on its two sides is unequal as a result of the flow velocity on the two sides being unequal (Bernoulli's equation). The lift force is calculated in relation to the same reference force as the resistance/drag. The corresponding ratio is known as the *lift force coefficient*:

$$\text{(2.1:65)} \qquad C_L = \frac{F_L}{\frac{1}{2} \cdot \rho \cdot v^2 \cdot A_0}. \qquad [-]$$

As a lift force generally acts at a point away from the centre of gravity, it will also induce a torque M, with its own reference distance d. The corresponding parameter is known as the *overturning moment coefficient*:

$$\text{(2.1:66)} \qquad C_M = \frac{M}{\frac{1}{2} \cdot \rho \cdot v^2 \cdot A_0 \cdot d_0}. \qquad [-]$$

The reference area and distance chosen differ according to the area of study. In the aerodynamics of missiles, one often uses either the largest cross-section in the plane of the flow or the wing surface/wingspan. In ballistics, however, one generally uses the calibre cross-sectional area (which is generally smaller than the cross-sectional area of the bullet) and the calibre. It is therefore important to specify the reference value when stating a coefficient.

Furthermore, coefficients are always expressed as a function of the Mach number.

2.1.4.5 Flow of a viscous fluid

Forces and equations of motion. In accordance with d'Alembert's principle, equations of motion are derived by equating inertial force and the forces acting on a body. Because density and volume are subject to change, both are expressed as a function of mass.

The following discussion applies to a one-dimensional flow (a streamline). This makes the relationships considerably simpler and clearer, and facilitates comprehension of the subject. For complete versions of the general equations of motion for fluids, please consult the literature on the subject.

A particle in a flow can undergo a change in velocity under two circumstances:

- when the velocity field changes over time (unsteady flow);
- when, in a (stationary) velocity field, it is displaced to a position at which the flow velocity is higher or lower.

Total acceleration is derived from the combination of the two sources. This acceleration is known as the *substantial acceleration*:

(2.1:67) $$\frac{dv}{dt} = \frac{\partial v}{\partial t} + v \cdot \frac{\partial v}{\partial x} .$$ [m/s²]

The two terms on the right can be interpreted as time-dependent or position-dependent inertial forces, expressed as a function of mass.

In calculating the forces, we assume that the streamline under consideration is oriented along the x-axis and has a cross-section of dy × dz (see Fig. 2-5). The compressive force per unit mass on a flow element of length dx is then:

(2.1:68) $$\frac{(p + \partial p) \cdot dy \cdot dz - p \cdot dy \cdot dz}{\rho \cdot dx \cdot dy \cdot dz} = -\frac{1}{\rho} \cdot \frac{\partial p}{\partial x} ,$$ [N/kg]

and the frictional force is:

(2.1:69) $$\frac{(\sigma + \partial \sigma) \cdot dy \cdot dz - \sigma \cdot dx \cdot dz}{\rho \cdot dx \cdot dy \cdot dz} = \frac{1}{\rho} \cdot \frac{\partial \sigma}{\partial y} .$$ [N/kg]

In the case of a Newtonian fluid, we can substitute the shear stress from Eqn 2.1:53:

(2.1:70) $$\frac{d\sigma}{dy} = \frac{\partial}{\partial y} \cdot \left(\eta \cdot \frac{\partial v}{\partial y} \right) = \eta \cdot \frac{\partial^2 v}{\partial y^2} = v \cdot \rho \cdot \frac{\partial^2 v}{\partial y^2} .$$ [N/m³]

Weight is omitted, as it does not exert a force along the x-axis, nor will we include any external forces, as these would only introduce an additional summand. From Equations 2.1:67, 2.1:68 and 2.1:69, and taking into account Eqn 2.1:70, we obtain:

(2.1:71) $$\frac{\partial v}{\partial t} + v \cdot \frac{\partial v}{\partial x} = -\frac{1}{\rho} \cdot \frac{\partial p}{\partial x} + v \cdot \frac{\partial^2 v}{\partial y^2} .$$ [m/s²]

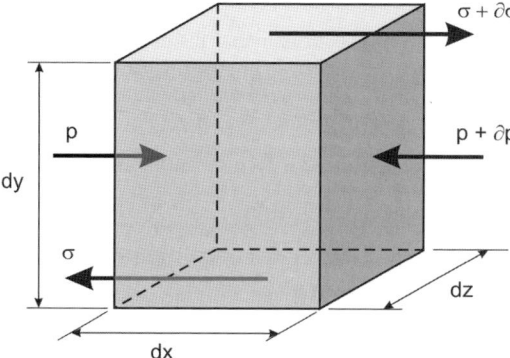

Fig. 2-5. Streamline element, with compressive stresses (*p*) and shear stresses (*σ*).

The three equations obtained by extending Eqn 2.1:71 to three dimensions are known as the *Navier-Stokes equations* for (incompressible) Newtonian fluids. They only produce unique solutions under very special conditions. To solve real problems, we must use numerical methods (e.g. finite elements).

For stationary ($\partial v/\partial t = 0$) and frictionless ($\nu = 0$) flow, given that

$$v \cdot \frac{\partial v}{\partial x} = \frac{\partial}{\partial x} \cdot (\tfrac{1}{2} \cdot v^2),$$

we obtain the following equation from Eqn 2.1:71:

$$\frac{\partial}{\partial x} \cdot (\tfrac{1}{2} \cdot \rho \cdot v^2 + p) = 0.$$

This corresponds to Bernoulli's equation (Eqn 2.1:61a) which, as mentioned above, is derived from the equations of motion.

Taking account of viscosity. Reynolds number. From the parameters of a flow – length s, velocity v, time t, pressure p and density ρ – it is possible to derive a term for each of the mass-related forces in Eqn 2.1:71. By forming quotients, it is possible to derive a number of dimensionless values from these four terms, which can then be used to describe the forces in the flow.

In order to decide whether it is possible to ignore friction in a flowing fluid, we must know the ratio of inertia forces to friction forces. This (dimensionless) quantity is called the *Reynolds number*. After the Mach number, the Reynolds number is the most important parameter for describing a flow:

(2.1:72) $\qquad \mathrm{Re} = \dfrac{v^2/s}{\nu \cdot v/s^2} = \dfrac{v \cdot s}{\nu}.$ $\qquad$ [-]

If Re $\gg$ 1, e.g. if speed is high or viscosity is low, it is possible to ignore friction.

Since every flowing fluid adheres to a solid body, there is a layer at the surface of the body within which the velocity of the flow rises from zero to flow velocity, and within which friction occurs. For large Reynolds numbers this layer (the boundary layer) is very thin, and friction can be ignored.

If Re $\approx$ 1 or Re < 1, the friction forces must be taken into account in all parts of the flow. This is the case for low speeds, small dimensions or high levels of viscosity.

From the values in Table 2-8, we can see that the flight of a bullet in air can be treated as frictionless flow, whereas friction cannot be ignored when a bullet passes through a dense medium, such as glycerine soap or gelatine.

Table 2-8. Examples of Reynolds numbers for bullets

Bullet		Re
Rifle bullet, v = 800 m/s, s = 30 mm;	in air	$1.8 \cdot 10^6$
	in glycerine soap	0.05
Pistol bullet, v = 300 m/s, s = 15 mm;	in air	$3.4 \cdot 10^5$
	in gelatine	1.1

26   2 Basics

2.1.5   Fluid jets

2.1.5.1 General

A jet occurs when a fluid escapes from an opening in a container and penetrates another, stationary fluid, at a certain velocity. The density, state and composition of the two fluids can be the same or different. Such a jet always forms a mass flow. At high speeds or high densities, it can possess considerable kinetic energy. Under such circumstances, and up to a certain distance, a jet can be seen as a projectile, and can have a corresponding effect.

Exit flow phenomena occur when there is a pressure differential between the container and its surroundings. It is this differential that is responsible for the acceleration of the mass flow. The cross-sectional form of the opening plays a decisive role. If the cross-section of a tube varies, it is termed a nozzle. If the tube is at its narrowest at its exit, it is known as a muzzle.

2.1.5.2 Exhaust flow from a muzzle

In order to calculate the energy flux and the energy-flux density (which ultimately determines the wounding potential of the jet), we must first determine the exhaust velocity and the mass flow rate through the nozzle.

The process by which a gas leaves a nozzle is adiabatic, or nearly so. This means that heat exchange with the surroundings can be ignored. The equation of state for adiabatic processes is:

(2.1:73) $$\frac{p}{p_0} = \left(\frac{\rho}{\rho_0}\right)^\gamma,$$ [-]

where γ is the adiabatic coefficient.

The exhaust velocity is calculated using Bernoulli's equation, with density as per Eqn 2.1:73. After some calculation, we obtain:

(2.1:74) $$v_m = \sqrt{\frac{2\cdot\gamma}{\gamma-1}\cdot\frac{p_i}{\rho_i}\cdot\left[1 - \left(\frac{p_m}{p_i}\right)^{\frac{\gamma-1}{\gamma}}\right]},$$ [m/s]

where v_m is the exhaust velocity, p_m the exit pressure and p_i and ρ_i the pressure and density of the fluid in the container.

Using Eqn 2.1:59, and taking Eqn 2.1:73 into account, the mass flow rate for an exhaust cross-sectional area A is:

$$\text{(2.1:75)} \quad \dot{m} = A \cdot \sqrt{\frac{2 \cdot \gamma}{\gamma - 1} \cdot p_i \cdot \rho_i \cdot \left[\left(\frac{p_m}{p_i} \right)^{\frac{2}{\gamma}} - \left(\frac{p_m}{p_i} \right)^{\frac{\gamma+1}{\gamma}} \right]} \quad \text{[kg/s]}$$

The mass flow rate therefore depends upon the pressure and density of the fluid in the container, the cross-sectional area of the opening and the ratio of the pressure inside the container to the pressure at the muzzle. From Eqn 2.1:75, it is apparent that there exists a pressure ratio p_m/p_i at which the mass flow rate reaches a maximum.

Eqn 2.1:75 is a continuous function, with a value of zero both at $p_m/p_i = 1$ and at $p_m/p_i = 0$. A maximum must therefore occur somewhere between these two values.

That pressure ratio is known as the *critical pressure ratio* and the corresponding muzzle pressure is the *de Laval pressure*, p_L. Substituting extreme values into Eqn 2.1:75 we obtain:

$$\text{(2.1:76)} \quad \frac{p_L}{p_i} = \left(\frac{2}{\gamma + 1} \right)^{\frac{\gamma}{\gamma - 1}} \quad \text{[-]}$$

The corresponding velocity is called the *de Laval velocity*. This is the maximum possible exhaust velocity for a gas leaving a muzzle:

$$\text{(2.1:77)} \quad v_{max} = \sqrt{\frac{2 \cdot \gamma}{\gamma - 1} \cdot \frac{p_i}{\rho_i} \cdot \left[1 - \left(\frac{2}{\gamma + 1} \right) \right]} \quad \text{[m/s]}$$

The corresponding mass flow rate is given by the following equation:

$$\text{(2.1:78)} \quad \dot{m} = A \cdot \sqrt{\gamma \cdot p_i \cdot \rho_i \cdot \left(\frac{2}{\gamma + 1} \right)^{\frac{\gamma+1}{\gamma-1}}} \quad \text{[kg/s]}$$

If the ambient pressure is greater than or equal to the de Laval pressure, the escaping gas can expand fully in the muzzle. The jet will be focused and will be capable of transferring kinetic energy. If the ambient pressure is lower than the de Laval pressure, the gas cannot expand fully in the muzzle. Expansion from muzzle pressure to ambient pressure will occur explosively, causing completely turbulent flow that is incapable of doing work.

2.1.5.3 De Laval nozzles (converging-diverging nozzles)

From the continuity equation and Bernoulli's equation, we can now show that there is a close relationship between the Mach number of the flow Ma and the form of a nozzle.

From the equation of motion, Eqn 2.1:71, we obtain Bernoulli's equation in differential form for a stationary, frictionless flow (the precondition for a jet):

(2.1:79) $$\frac{1}{\rho} \cdot \frac{\partial p}{\partial x} + v \cdot \frac{\partial v}{\partial x} = 0 \, . \qquad [m/s^2]$$

Multiplying the equation by ∂x and inserting $\partial \rho$ in the first term and v in the second, we obtain:

(2.1:80) $$\left(\frac{\partial p}{\partial \rho}\right) \cdot \frac{\partial \rho}{\rho} + v^2 \cdot \frac{\partial v}{v} = 0 \, . \qquad [m^2/s^2]$$

However, from Eqn 2.1:57a, the first factor in the first summand is the speed of sound squared, which means that we can insert the Mach number Ma in Eqn 2.1:80:

(2.1:81) $$\frac{\partial \rho}{\rho} + Ma^2 \cdot \frac{\partial v}{v} = 0 \, . \qquad [-]$$

As we can assume that the flow is stationary, the mass flow rate over time is constant. We can therefore express Eqn 2.1:59 in differential form, thus:

(2.1:82) $$\partial (\rho \cdot v \cdot A) = \frac{\partial \rho}{\rho} + \frac{\partial v}{v} + \frac{\partial A}{A} = 0 \, . \qquad [-]$$

Replacing the first summand in Eqn 2.1:81 with the aid of Eqn 2.1:82, we obtain the following after calculation:

(2.1:83) $$\frac{\partial v}{v} = -\frac{1}{1 - Ma^2} \cdot \frac{\partial A}{A} \, . \qquad [-]$$

From this equation, it is apparent that acceleration only occurs at Ma < 1 (i.e. when $\partial v/v > 0$) with a convergent nozzle ($\partial A/A < 0$) and only occurs at Ma > 1 with a divergent nozzle ($\partial A/A > 0$).

In other words, subsonic flows will be accelerated by a convergent nozzle, supersonic flows by a divergent nozzle. With a combination of these two types of nozzle, it is therefore possible to achieve gas velocities well over the speed of sound. This type of nozzle is called a *convergent-divergent nozzle*, or *de Laval nozzle*. By choosing the length of the divergent section appropriately, it is possible to ensure that the pressure in the exit section is equal to ambient pressure. The nozzle is then said to be adjusted. It is under these conditions that the jet attains the maximum possible speed.

The exhaust gas velocity is calculated using Eqn 2.1:74, taking the pressure at the exhaust cross section for p. Eqn 2.1:78 gives the mass flow rate for the smallest cross section.

The exhaust gas velocity and the mass flow rate form the basis for calculating the jet. These two variables also determine the maximum energy-flux density (i.e. the energy hitting a given area per unit time) that can occur in the jet.

2.1.5.4 Jet velocity and energy

Except at very low velocities (which are unlikely to cause jet injuries), a free jet becomes turbulent very shortly after emerging from the nozzle.

In turbulent flow, irregular fluctuating motion is superposed upon the basic longitudinal motion. As a result, the fluid particles are able to move transverse to the direction of flow. In laminar flow, this does not occur.

The velocity of a turbulent flow must therefore be seen as a mean value over time. Because of the different directions in which the particles are moving, there is an apparent increase in internal friction by comparison with the mean movement (see 2.1.4.3). This *eddy kinematic viscosity* is an important descriptor of turbulent flow.

As a result of the turbulence, the jet partly mixes with the surrounding fluid, dragging particles of that fluid along with it. The mass of fluid picked up by the jet increases with distance, and the jet becomes broader. At the same time, its velocity decreases, as total momentum remains constant.

For the purposes of wound ballistics, we can limit ourselves to studying round jets, as both the gases that exit the muzzle of a weapon and the propulsion jet of a rocket belong to this category. One advantage of round jets is that the eddy kinematic viscosity is constant throughout the jet. This being so, the equation for the mean velocity distribution in the jet is as follows (only the component acting in the direction of the jet being relevant):

(2.1:84) $$v_x(x,r) = \frac{3}{8 \cdot \pi} \cdot \frac{I}{\rho \cdot \varepsilon \cdot x} \cdot \left(\frac{1}{1 + \frac{1}{4} \cdot \beta^2}\right)^2,$$ [m/s]

where β is calculated as follows:

(2.1:84a) $$\beta = \sqrt{\frac{3}{16 \cdot \pi}} \cdot \sqrt{\frac{I}{\rho}} \cdot \frac{r}{\varepsilon \cdot x}.$$ [-]

Symbols for the formulas above:

x	distance from opening	r	distance from axis of jet
I	total momentum (per unit time)	ρ	density
ε	apparent kinematic viscosity		

Along the axis of the jet (r = 0), β is zero, which simplifies Eqn 2.1:84 considerably. If we know the velocity, we can use the mass flow rate from Eqn 2.1:59 to calculate the energy-flux density (energy density E′ per unit time) as follows:

(2.1:85) $$\frac{E'}{\Delta t} = \tfrac{1}{2} \cdot \rho \cdot v_x^3.$$

The main difficulty in performing such calculations lies in determining the eddy kinematic viscosity. This can be accomplished by taking pressure measurements along the axis of the jet. From the pressure, we can determine the velocity, which in turn allows us to calculate ε from Eqn 2.1:84.

2.1.6 Measuring techniques for wound ballistics

2.1.6.1 General

Wound ballistics is very much an experimental science. The starting point is always the observation of phenomena, which are generally preceded by events of which the preconditions are unknown. In order to understand these phenomena, we perform experiments. That simply means repeating the events under known, defined conditions. This is the only way of understanding the sequence "*preconditions* ⇒ *event* ⇒ *phenomenon*" and, if possible, creating a physical model.

Scientific experiments must be reproducible. This means that the entire sequence of events must meet certain conditions. An experiment therefore involves measuring not only the events (the "results") but all the preceding processes and preconditions that could affect the results. Measurements are hence of vital importance to empirical science.

From a mathematical standpoint, measurements are a type of mapping. Numerical values are assigned in a physical "model space" to a time- and space-related situation observed in "real space." This implies that the observer conducting the measurements always has a model of the process in his head, even if he is unaware of this. As a result, there is the danger that he will attempt to use the results of measurements to confirm his preconceptions. It is important to adopt a critical attitude and to be prepared to revise one's assumptions, both in designing the experiment and when interpreting the results. *The aim of an experiment is not to confirm one's assumptions, but to discover "reality."*

2.1.6.2 Dynamic phenomena

Recording the motion of a bullet – especially in a dense medium – poses a number of problems. The most significant problem is that all the processes occur within a very short time. For instance, at an impact speed of 900 m/s it takes a bullet approximately 1 ms to penetrate to a depth of 30 cm. In the process, very high decelerations occur, the axis of the bullet rotates about a transverse axis and the bullet often deforms or breaks up into several pieces. Regardless of the type of measuring system, all measurements must be recorded at a frequency of a few μs.

Velocity and deceleration. In ballistics, the most relevant parameters are the velocity and deceleration of a bullet. If these two values are known for all points, the forces can be calculated from Eqn 2.1:14 and the energy present from Eqn 2.1:19. These are the physical parameters of greatest importance in describing the processes.

The most common approach to determining velocity is to measure distance travelled over time using photoelectric barriers, radar or high-speed cameras. A first differentiation gives speed over time, a second gives deceleration. In order to determine speed over time to a reasonable degree of accuracy, the measuring points need to be as close together as possible. A high sampling rate is therefore required.

High-speed video cameras operate at speeds of tens of thousands of frames per second, as long as there is sufficient light. Special cameras designed for very fast processes can record over 100 million frames per second, with up to 16 images being available for each process. It is possible to choose the interval between these images as required. In the case of opaque materials, x-ray flash photography is used. This generally produces a maximum of six images.

2.1 The physics of wound ballistics   31

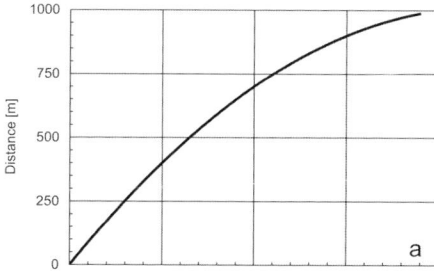
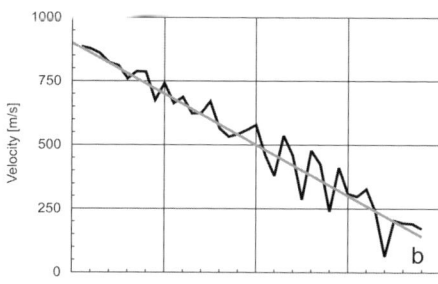
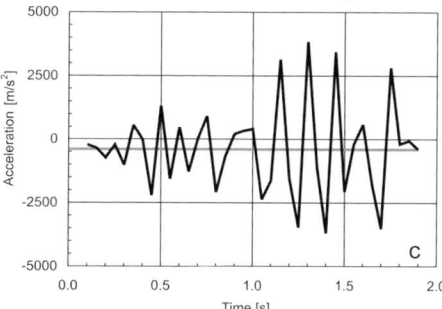

Fig. 2-6. Effect of measurement error on velocity and acceleration. The "correct" lines are shown in grey. See text for details.

Regardless of the method used, the double numerical differentiation required to calculate deceleration from time/distance measurements means that measurements must be extremely accurate. Even small errors in measuring distance and time will produce errors in the drag coefficient approximately two orders of magnitude greater.

Example: An object undergoes a constant deceleration of -600 m/s^2. A statistical error of just a few percent in the time/distance measurements will already cause visible discrepancies following differentiation over time (see Fig. 2-6b). The deceleration graph produced from a second numerical differentiation (see Fig. 2-6c) almost completely hides the fact that the object was undergoing constant deceleration.

Using pairs of photoelectric barriers and time measuring devices, it is possible to determine velocity directly. Two measuring points of this type make it possible to measure the change in velocity over a known distance, from which it is possible to derive deceleration by single differentiation (KNEUBUEHL 1982). This makes it possible to obtain a much higher degree of accuracy. Radar devices that utilize the Doppler principle can also be used to measure speed over time.

Doppler principle: A radar beam of a given frequency striking a moving body will be reflected at a different frequency. The change in frequency is proportional to the speed at which the body is moving.

However, neither method can be used for wound ballistics studies. Pairs of photoelectric barriers only allow us to measure the entry and exit speeds, and hence to calculate the drop in speed between the two points. They are not suitable for measuring a fast process that occurs over a short distance. Modern radar equipment can take measurements through simulants, which makes it possible to calculate the impact speed and residual speed of a bullet passing through a block of simulant. However, the sampling rate of such equipment is currently still too low

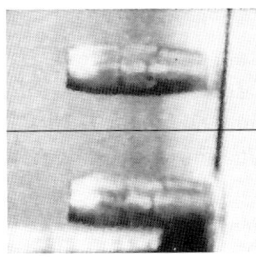

Fig. 2-7. Image of a bullet penetrating a block of soap, made using a mirror angled at 45°. *Upper picture:* plan. *Lower picture:* elevation. The angle of incidence can be determined from the two pictures (see also KNEUBUEHL 2004).

to record the change in speed over distance as the bullet passes through the material.

Proper motion of a bullet The proper motion of a bullet (the movement of the axis of the bullet about its centre of gravity) involves more than one plane. This motion can only be recorded using a measuring system that operates in two planes.

That involves taking a photograph of the bullet using two photographic plates at the same time, with the two plates arranged at right angles to each other. The photographic technology used (visible light or x-ray) is of secondary importance. From the pair of images produced, it is then possible to determine the orientation of the bullet's axis. This procedure is appropriate when one wishes to measure the angle of incidence of the bullet at a particular point (e.g. the point of impact) or to measure the change in angle of incidence along a section of the bullet's trajectory.

If the bullet is fired parallel to a mirror angled at 45° to the camera, it is possible to record both the side elevation and the plan view of the bullet on a single image. This method has the advantage that both images are created at exactly the same instant. See Fig. 2-7.

Another technique, which can only be used in the case of ferrous objects (e.g. bullets with a steel core or a thick steel jacket), utilizes the fact that a magnet rotating in a coil generates a potential difference. The change in potential difference (voltage) is proportional to the speed of rotation. If the bullet is magnetized and the target medium (soap or gelatine) placed inside a large coil, the rotating bullet will induce a voltage in the coil as it passes through. By recording this voltage, it is possible to determine the change in speed of rotation over distance (KNEUBUEHL 1990c; measuring method based on that of R. CATTIN).

2.1.6.3 Physical values

Centre of gravity. Moments of inertia. Determining the mass and the external dimensions of a bullet is both simple and commonplace. Determining the moments of inertia and the centre of gravity is less common, but is necessary in the field of ballistics, especially when one is studying stability.

The moment of inertia can be determined empirically using the principle of the physical (rotating) pendulum. The period of such a pendulum is directly related to the moment of inertia of the oscillating body relative to the centre of rotation. If a torsion wire is used to impart a retarding moment, measurements can be conducted through a bullet's centre of gravity, along its longitudinal or transverse axis. The moment of inertia of the measuring apparatus is measured beforehand using an object with known moments of inertia (see KNEUBUEHL 2004 for further details).

One way of finding the centre of gravity is to use a center of gravity measurement instrument. This involves sliding the bullet across a sharp edge until it is more or less in equilibrium. It is then held steady while the point in contact with the blade is marked.

If the structure of the bullet is not too complex, it is also possible to determine the centre of gravity and moments of inertia by calculation. This is the case for many handgun and rifle bullets. The process consists of dividing the bullet into thin slices, calculating the moment of inertia for each slice and calculating the sum of the moments.

Characteristics of materials. It is worth mentioning just a few specific measurement issues of relevance to wound ballistics. These include measuring viscosity and the speed of sound in dense media of the types used as simulants. Both parameters have a significant influence on shock waves.

Using Eqn 2.1:2, the speed of sound in a medium is determined by measuring the time a wave takes to pass through a layer of known thickness. Ultrasonic waves are generally used, although it is difficult to filter out the main signal.

Measuring the viscosity (and other mechanical parameters) of a simulant is important if fluid dynamics modelling is to be carried out. Measuring devices (cone viscometers) are available that allow standard measurements to be conducted even on very viscous materials.

2.2 Ammunition and weapons

2.2.1 Introduction

To understand wound ballistics, one needs a sound knowledge of ammunition in general and bullets in particular. We shall therefore be looking at the design of contemporary ammunition in some detail (2.2.2). Our focus will be on bullet design, as this has a major influence on wounding.

Ammunition and weapon are two halves of the same system. Sub-section 2.2.3 therefore gives an overview of key weapon characteristics and common types of weapon. This book is not the place for a detailed description of individual weapons, but exterior ballistics plays a major role in wound ballistics, as it tells us a great deal about the impact conditions. We shall be looking at exterior ballistics in sub-section 2.3.4.

2.2.2 Ammunition

2.2.2.1 *The structure of a cartridge*

Cartridge parts and nomenclature. A cartridge generally consists of four parts. We shall be discussing the structure and characteristics of these parts in the sections below. The four parts are:

- bullet;
- propellant;
- case;
- primer.

See Fig. 2-8.

There are two systems for identifying cartridges. One is of European origin and uses metric units. The other originated in the English-speaking world and uses inches. Both systems indicate not only the dimensions of the cartridge itself, but also the relevant dimensions of the weapon, such as those of the chamber or the bore – the inside of the barrel. In many cases, we can refer to the same cartridge using both a metric designation and a designation in inches.

The metric designation consists of:

- calibre;
- (case length);
- additional indicator.

The calibre and case length are in millimetres. The calibre is often a nominal value, corresponding only approximately to the diameter of the bore (measured from land to land, see 2.2.3.1) or that of the bullet.

Examples: 9 mm Luger
7 mm Rem. Mag. (Remington Magnum)
7.62 × 54 R

The Anglo-American cartridge naming convention also includes the calibre (as a decimal fraction of an inch), but the calibre in the designation may not correspond exactly to a physical dimension. The format is:

-   calibre;
-   additional indicator.

The calibre and the additional indicator between them define all the dimensions of the cartridge. Older calibres often include a second

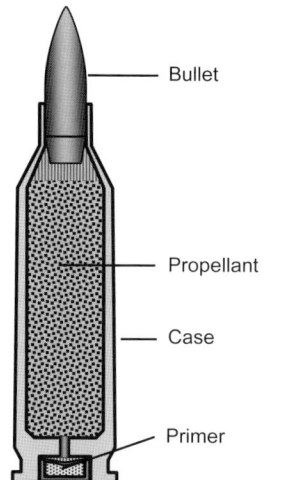

Fig. 2-8. The parts of a cartridge.

number, separated from the first by a dash (-). This is the powder load in grains of black powder, indicating the approximate power of the cartridge. In some cases, the number after the dash had a different meaning, as explained in the examples below. A slash (/) followed by a number indicates a later reduction in calibre, achieved by changing to a drawn case.

Examples: 357 Mag. (same bore diameter as 38 Spl)
30-30 Win. (with NC propellant)
30-06 Springfield (in this instance, the second number indicates the year of introduction, 1906)
450/400 Magnum

Annex A.4 contains tables of the most common ammunition types, including the calibre of the bore (the inside diameter, measured land to land) the diameter of the bullet and the length of the case.

There are often several cartridge calibres for a given bore calibre, e.g. there are at least 10 different 9 mm cartridges. It is therefore important to use the exact calibre designation.

The bullet. The role of the bullet is to transport to the target the energy needed to achieve a given effect and to convert that energy into work on reaching the target. Almost all small arms use kinetic energy, and that largely determines the form and structure of the bullet. The bullet must survive the forces it undergoes in the barrel as it acquires its energy and must lose as little energy as possible between muzzle and target. The physical processes related to these requirements are fairly well understood, which limits the scope for major differences in bullet design.

Bullets can be divided into the following basic categories:

– *solid bullets* (see Fig. 2-9 a), which generally consist of a single material (lead, brass, plastic, etc.)
– *jacketed bullets*, which contain a core, generally of lead, and are hence known as soft-core bullets. The core is covered by a thin layer of some other material, the jacket. Typical jacket materials include copper alloys and plated steel. In

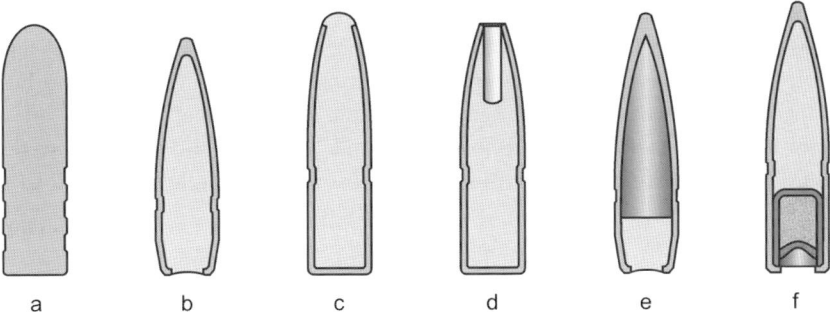

Fig. 2-9.   Cross-sections of various types of bullet: **a.** Solid. **b.** Full metal-jacketed. **c.** Semi-jacketed. **d.** Semi-jacketed hollow-point. **e.** Hard-core. **f.** Tracer.

most cases, the jacket does not completely enclose the core, as this would complicate the manufacturing process. However, if the entire bullet is enclosed, including the tip (but often except the base), and the jacket is crimped shut at the base of the bullet, one speaks of a *full metal-jacketed bullet* (see Fig. 2-9 b). If the jacket leaves part of the tip uncovered, the bullet is designated a *semi-jacketed* bullet (SJ); see Fig. 2-9 c. If, in addition, the tip is hollowed out, the bullet is described as a *semi-jacketed hollow-point* bullet (SJHP); see Fig. 2-9 d.

- *Hard-core bullets* are special jacketed bullets designed to penetrate hard materials. They contain a core of very hard material (such as hardened steel, tungsten or a tungsten alloy) (see Fig. 2-9 e) and are mainly used by the armed forces.
- *Tracer bullets* are solid or full metal jacketed bullets with a pyrotechnic charge in the tail. The charge burns during flight, rendering the trajectory of the bullet visible (see Fig. 2-9 f).

There has been (and still is) considerable disagreement regarding the way energy is transferred from bullet to target and the effect of bullets on the body. The result is a wide variety of bullet types for hunting and handgun ammunition. In terms of their penetration behaviour, we can divide such bullets into three categories:

- *Shape stable* bullets, which maintain their shape in the target. This type of bullet only undergoes minor changes in form (such as bulges) and loses no mass.
- *Deforming* bullets, which deform in the target. This type of bullet loses only a small percentage of its mass.
- *Fragmenting bullets*, which are designed to fragment as they pass through the target medium.

Bullets do not always fall neatly into one or other of the above categories, as their behaviour depends not only on their design but also on impact velocity and target medium. A bullet may fragment at high velocity, deform at medium velocity and be shape-stable at low velocities, for instance. The most important specialized bullet designs are described in 2.2.2.2, "Types of ammunition."

Bullets are also classified according to shape, materials and structure. A number of abbreviations are used to provide additional information about the bullet and these form part of the cartridge designation. For instance, "FMJ RN" designates a full metal-jacketed round-nose bullet and "LHP" a solid lead hollow-point bullet. Please see Annex A.5 for a list of the most common bullet designations.

Propellant. The propellant is the firearm's energy source. Its task is to accelerate the bullet to a particular velocity. There are various ways of generating the necessary force; it is possible to use a spring (as in a spring pistol), a rubber cord (as in a catapult) or electromagnetic force (as in a rail gun). In the various majority of cases, however, the required force is obtained by applying the pressure of a compressed gas to the base of the bullet. This pressure can come from a number of sources:

- mechanical compression (in the case of an airgun);
- air reservoir or gas cartridges (e.g. in the case of a CO_2 weapon);
- combustion in a confined space.

The first two sources produce only modest levels of muzzle energy. Any more powerful weapon has to use the third source of gas – combustion in a confined space. The substances used must be highly flammable and must release a large quantity of gas during combustion. In order to generate sufficient pressure, combustion must take place in a small space, but such a small space will contain too little oxygen for combustion. Combustible propellants must therefore include a component capable of supplying the oxygen they need for their own combustion.

The standard composition of the oldest known propellant, black powder, is as follows:

75% saltpetre (potassium nitrate, KNO_3);

15% charcoal (carbon, C);

10% sulphur (S).

The saltpetre supplies the oxygen, while the carbon and the sulphur constitute the combustible elements.

To produce black powder, the raw materials are ground down, mixed and formed into cakes using a hydraulic press. The cakes are then crushed and the resulting grains are sieved and sorted by size (see Fig. 2-10). As we will see, the entire surface of the propellant plays a decisive role. The finer the powder, the greater its surface area for a given mass and hence the faster it will burn. As fine grains were used in the days of the musket, such propellant has acquired the name "powder."

When black powder burns, it produces the gases required – carbon dioxide and nitrogen (CO_2 and N_2) – but also potassium sulphide. This is a solid compound that produces visible smoke and leaves unwanted deposits in the barrel. Its mass is equal to 40% of the mass of the powder that produced it, which explains the relatively low efficiency of black powder.

Black powder is rarely used as a propellant in firearms today, except in historical or reproduction muzzle-loaders. However, it is still used as an explosive, for pyrotechnics (detonating cord, caps, fireworks, etc.) and as a booster in the primer screws and priming cartridges of heavier weapons.

So-called "smokeless" powders, which have almost completely replaced black powder in firearms, are based on a mid-19th-Century invention; nitrocellulose (guncotton, see 7.2.2.5). Nitrocellulose is produced by adding a nitrating acid (a mixture of nitric and sulphuric acid) to cotton linters (short-fibered cotton). The nitrocellulose is gelatinized using solvents and becomes a homogenous, kneadable mass. This mass can then be extruded and cut into "powder grains" of the form required (see Fig. 2-11). The solvent is then removed by drying. Nitrocellulose

Fig. 2-10. Black powder, Grain Size 2, as used for long weapons.

Fig. 2-11. Nitrocellulose powder. *Left:* Fine flake powder, as used in handguns and shotguns. *Right:* Coarse tubular powder, as used in large-calibre long weapons.

(NC) powders are known as single-base powders, as they contain one substance that supplies energy.

However, nitrocellulose can also be dissolved in nitro-glycerine, a highly impact-sensitive explosive. When warm, the resulting mass is kneadable and can readily be formed as required. Nitroglycerine (NG) powders are known as double-base powders, as they contain two substances that supply energy.

On combustion, smokeless powders are transformed almost entirely into gas: carbon dioxide (CO_2), carbon monoxide (CO), water vapour (H_2O), hydrogen (H_2) and nitrogen (N_2). These powders produce approximately three times as much gas as the same mass of black powder, making it possible to obtain higher pressures – and hence greater accelerating forces – with less powder.

The parameters that can be used to describe a powder include the following:

- explosion heat Q_{ex}: the quantity of heat released by the rapid (explosive) combustion of 1 kg of a powder;
- specific gas volume: the quantity of gas produced by the combustion of 1 kg of a powder;
- explosion temperature: the temperature at which the powder burns in a space of fixed volume;
- auto-ignition temperature: the temperature at which the powder will ignite spontaneously.

Table 2-9 lists the data for the three most important types of powder.

Primer. In order to ignite the propellant as evenly as possible, and to ensure that all of the propellant ignites at the same time, a cartridge requires a primer. A primer contains a small quantity of an impact- and friction-sensitive high explosive (not propellant). This explosive must generate both sufficient heat and a sufficiently powerful flame. Originally, a mixture of mercury fulminate and potassium nitrate was used, but this led to severe corrosion of the barrel. Around 1920, the SINOXID primer was invented, which used lead styphnate and barium nitrate

Table 2-9. Powder data

Type of powder		Black powder	NC	NG
Explosion heat	[J/g]	2650	3500	4800
Specific gas volume	[ml/g]	280	900–970	800–860
Explosion temperature	[°C]	approx. 2400	2500–3000	3000–3800
Auto-ignition temperature	[°C]	300	170	160
Solid residues	[%]	40	approx. 1	approx. 1

instead of mercury nitrate and potassium nitrate. However, SINOXID primers had a high lead content, and have now been replaced by lead-free primers such as the SINTOX primers produced by RUAG (formerly Dynamit Nobel AG). These primers use diazole instead of lead styphnate.

Leaving aside proprietary designs, there are three primer systems. In a rimfire cartridge, the primer is incorporated into a bead that forms a rim around the circumference of the case head (see Fig. 2-12 a). The primer is fired by striking the rim, hence the name. Today, virtually the only rimfire cartridges are the Flobert, the 22 calibres and the 5 mm Rem. Mag. In order to achieve the required degree of impact sensitivity, the case must be fairly thin at the rim, so rimfire cartridges are only suitable for low gas pressures.

The other two primer designs involve pressing the primer into a depression (the primer cup) in the centre of the case head. Both Boxer and Berdan cartridges are hence termed "centrefire" cartridges. When the firing pin strikes the base of the primer, it forces it up against the anvil. The resulting impact and friction energy detonate the primer. The principal difference between the two types of centrefire primer lies in the position of the anvil. In the case of a Boxer primer, the anvil is built into the detonator cap (see Fig. 2-12 b). The primer thus forms a single unit, which simply has to be inserted into the primer cup in the head of the case. The powder is ignited via a central flash hole. In the case of the Berdan primer (see Fig. 2-12 c), the anvil is part of the case, rendering a central flash hole impossible. The Berdan primer therefore uses two or more flash holes arranged radially.

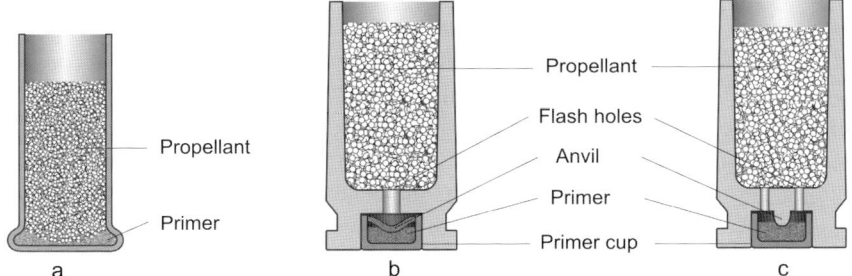

Fig. 2-12. Types of primer: **a.** Rimfire, **b.** Boxer, **c.** Berdan.

Case. The introduction of breech-loading weapons posed a number of problems, and these were only resolved by the invention of the cartridge case. The case fulfils a number of important functions during the firing process.

As the propellant burns, high pressures are created in the weapon – it is this pressure that accelerates the bullet. It is therefore necessary to seal the space in which combustion takes place. This function is performed by the case, as the pressure causes the cartridge to expand and become a tight fit against the inside of the chamber. The tight seal also protects the firing mechanism against the corrosive effect of the hot gases.

Deflagration of the powder also generates large quantities of heat, raising the temperature of the weapon and especially that of the barrel. When the case is ejected after firing, a large quantity of heat goes with it, slowing down the rate at which the temperature of the weapon rises. In addition to sealing the chamber and dissipating heat, the case fulfils the following functions:

- it acts as a means of transporting the propellant and primer;
- it holds the bullet and primer, and positions them in the weapon;
- it provides the bullet with initial resistance against the pressure of the expanding gases, which plays an important role in ensuring that the propellant burns evenly;
- it centres the bullet and guides it as it starts to accelerate.

In order to maximize the volume of propellant available, the diameter of the case body is often greater than that of the bullet. The point at which the case diameter is brought down to the diameter of the bullet is called the shoulder. The case must be seated securely in the weapon to ensure that the bullet and primer are always in the correct position. The various ways of seating the case in the chamber and the different pressures used have resulted in a variety of case designs.

The rimmed case is an immediate descendant of the rimfire cartridge. Its head extends beyond the maximum diameter of the body and sits against the rear opening of the chamber. In the case of European cartridges, this type of case is often identified by the letter R (rim), e.g. 9.3×74 R.

There are rimmed cases with a shoulder and there are those without. Most revolver cartridges, for instance, have a rim but no shoulder (see Fig. 2-13 a). The French 8 mm Lebel is a typical rimmed cartridge with shoulder (see Fig. 2-13 b) and is probably one of the oldest cartridges using a metallic case and NC powder. The Russian 7.62×54 R is another example.

The junction between the head and the body is one of the weak points of a cartridge case. Cartridges for particularly high pressures are therefore reinforced by a bead at this point, which also constitutes the contact surface between the cartridge and the rear of the chamber. Cases of this design are said to be "belted" and are designed for high muzzle energies. Belted cases are most often used for high-powered hunting ammunition (see Fig. 2-13 c).

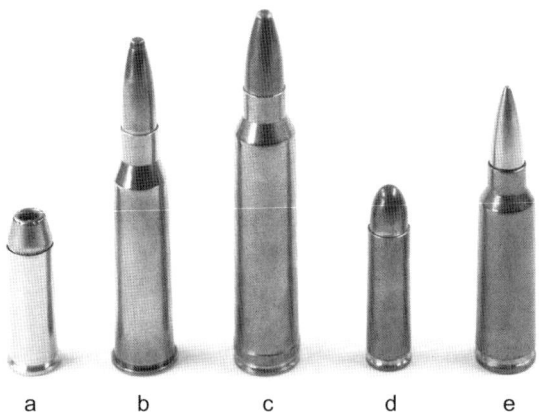

Fig. 2-13. Cartridges with …
a. Rimmed, no shoulder,
b. Rimmed and shoulder,
c. Belted rimless and shoulder,
d. Rimless, no shoulder,
e. Shoulder, no rim.

Rimless cartridges are the most common type. The case head does not project beyond the body. They are easier to insert into a magazine, which accounts for their widespread use. When chambered, this type of cartridge is seated against its shoulder (in the case of a shouldered case) or otherwise against its mouth. Rimless cases are used mainly for military and pistol ammunition (see Fig. 2-13 d, e).

2.2.2.2 Types of ammunition

General. We can make a basic distinction between two types of ammunition: ammunition for handguns (pistols and revolvers) and ammunition for long weapons (rifles and shotguns). The calibre standards of the C.I.P. (*Commission Internationale Permanente pour l'Epreuve des Armes à Feu portatives*, Permanent International Commission for the Proof of Small-arms) in Brussels list some 90 handgun calibres and over 300 for long weapons. This total does not include the additional calibres listed by the US standardization body SAAMI *(Sporting Arms and Ammunition Manufacturer's Institute)*. As each cartridge case can be combined with a number of different bullets, the number of ammunition types is very large. To provide an overview, we shall look at each ammunition type separately for each category of weapon, as it is the weapon that has the greatest influence on the characteristics of the ammunition.

Pistol ammunition. With a few exceptions, pistol cartridges are of the rimless design, with no shoulder, and are slightly conical in shape. These features simplify the design of the feed mechanism and magazine of semi-automatic pistols and submachine guns. Muzzle energy and moment must be kept low, both to allow a pistol is to be fired one-handed and because pistols and submachine guns are comparatively light. As a result, a small quantity of propellant is sufficient. This in turn means that the volume of the case can be kept small and the cartridge short. Fig. 2-14 shows a number of typical pistol cartridges from across the calibre spectrum (5 mm to 13 mm).

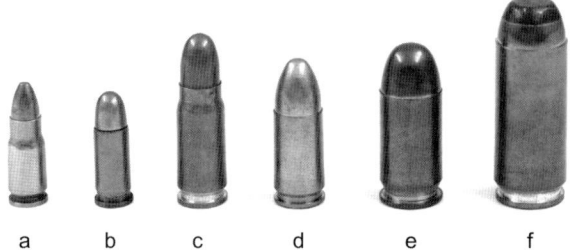

Fig. 2-14. Pistol ammunition:
a. 5.45 × 19,
b. 6.35 Browning,
c. 7.63 Mauser,
d. 9 mm Luger (previously known as Parabellum),
e. 45 Auto,
f. 50 AE (Action Express) with semi-jacketed bullet.

Pistol ammunition generally uses full metal-jacketed round-nose bullets, as these chamber smoothly (see Fig. 2-14 b–e). There is little justification for using pointed bullets, given the low muzzle velocities and short ranges involved.

The relatively low energy levels limit the effectiveness of handguns and many attempts have been made to compensate for this through bullet design. As a result, many types of bullet exist; there are over 60 types of bullet for 9 mm Luger ammunition, for example.

There are two principal approaches to the issue. One is to use deforming solid and full metal-jacketed hollow-point bullets, which deform well in a soft medium. Examples include the MEN QD P.E.P. bullet and the RUAG Action 4 (see Fig. 2-15 a), the Federal Hydra-Shok (which has a lead spike, known as a "centre post," inside the hollow tip, see Fig 2-15 b) and Remington's Golden Saber (see Fig. 2-16 a).

The other approach aims to maximize effectiveness against body armour and hard materials (such as steel and aluminium panels) using extremely light brass, copper or steel bullets and high muzzle velocities (up to 600 m/s). These include the French THV (Très Haute Vitesse, very high speed) (see Fig. 2-15 c) and Arcane; the American KTW (see Fig. 2-16 b) and the British sabot-guided Alpha.

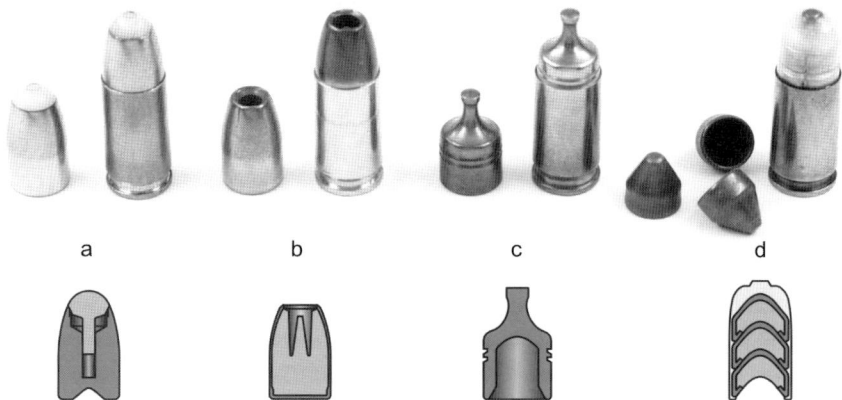

Fig. 2-15. 9 mm Luger pistol cartridges and bullets: a. RUAG Action 4, b. Federal Hydra-Shok, c. SFM THV, d. Triplex (three separate bullets) Bottom row: cross-sections.

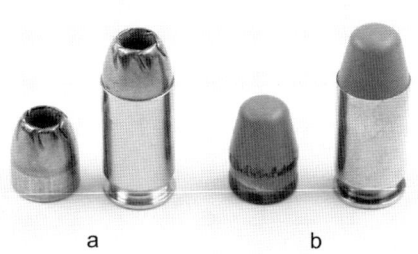

Fig. 2-16. 45 Auto bullets:
a. Golden Saber (SJHP),
b. KTW (brass, teflon-coated).
Both bullets are available in other calibres (e.g. 9 mm Luger).

a   b

The many proprietary designs include tracer, plastic bullets that release a shot payload on impact (the Glaser Safety Slug, see 4.2.2.3), fragmenting projectiles, three-part bullets (Triplex, see Fig. 2-15 d) and exploding bullets, which incorporate a small explosive charge.

In addition to ammunition designed for maximum effectiveness, manufacturers also produce training ammunition. Examples include the DNAG PT (plastic training) cartridge. The bullet has a muzzle velocity of approx. 950 m/s but slows to approx. 250 m/s after just 20 m. Short-range ammunition also exists, with ballistic characteristics similar to those of full metal-jacket bullets at expected engagement ranges (out to 50 m) but a significantly shorter maximum range.

Because of the materials used (compressed metallic powder and plastic), these bullets fragment on hitting hard surfaces, even at an angle, thus preventing ricochets. These bullets are therefore known as *frangible* bullets and are used for training.

However, the most widely-used pistol cartridge is probably the 22 L.R. (Long Rifle), known in the German-speaking world as 5.6 lfB (*lang für Büchsen*, "long for rifles"). This is a rimfire cartridge, originally developed on the basis of the 22 Long for use in small bore rifles. The cartridge was later adopted by pistol shooters, and is in widespread use for target shooting. It is also used in many other pistols, revolvers and rifles. Two types of bullet are available. Competition shooters use lead round-nose bullets (LRN, see Fig. 2-17 a), while hollow soft-point bullets are available for varminting (LHP, see Fig. 2-17 b). There are other cartridge standards for this calibre, such as 22 extra long (see Fig. 2-17 c) 22 extra

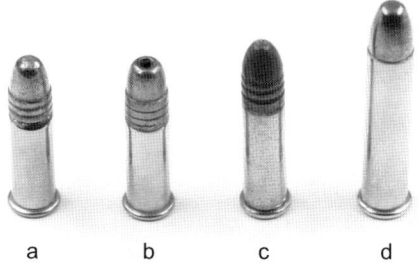

Fig. 2-17. Small-calibre cartridges:
a. 22 L.R. with lead round-nose bullet (LRN),
b. 22 L.R. with lead hollow-point bullet (LHP),
c. 22 extra long, with longer case,
d. 22 Winchester Magnum with copper-plated lead hollow-point bullet.

a   b   c   d

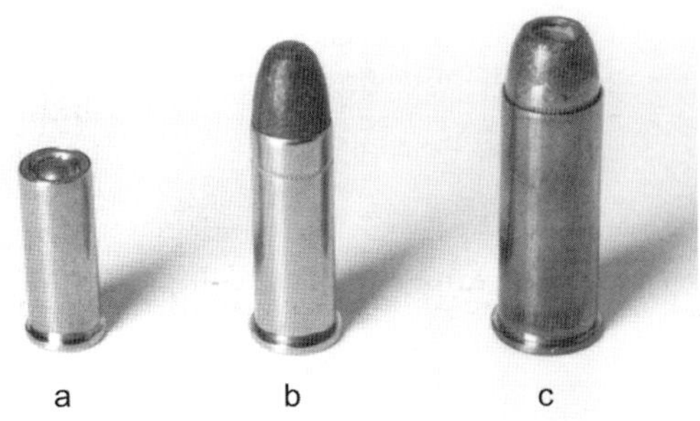

Fig. 2-18. Revolver cartridges:
a. 32 S & W Long, Wadcutter (WC),
b. 38 Special LRN,
c. 44 Rem. Mag. SJHP.

L.R. and 22 Winchester Magnum (see Fig. 2-17 d). The last of these delivers almost as much energy as a 9 mm Luger bullet.

Revolver ammunition. Revolver cartridges differ from pistol cartridges in a number of basic respects. Almost all revolver cartridges are rimmed, and cases are significantly longer than those of pistol cartridges (see Fig. 2-18). Rimmed cases sit much better in the cylinder and are easier to extract after firing.

The longer case has a historical background. Unlike semi-automatic pistols, revolvers were introduced when black powder was still the usual propellant, and black powder cartridges require approximately three times as much propellant as those loaded with NC powder. In addition, gas escapes through the gap between the cylinder and the barrel, and this loss must be compensated by additional powder. A fairly large load volume was therefore required, which meant that the case had to be relatively long.

It was not possible to modify existing weapons when smokeless powder was introduced, so the long cases remained, and today they are only partly filled with propellant. This makes it possible to produce cartridges with the bullet positioned right inside the case (such as the 38 Spl Wadcutter (WC) and the 32 S & W Long WC, see Fig. 2-18 a). The large case can tempt inexperienced reloaders to use excessive loads, and this has often led to weapons being destroyed and to accidents.

Revolver bullets differ from those used in pistol ammunition, in that while pistols generally use fully-jacketed or (more recently) solid bullets, most revolvers fire semi-jacketed or solid lead bullets. Here again, there is a historical reason. The gas pressure and acceleration in a revolver were not so high as to make jacketed bullets essential, and it was easier for the bullet to exit the cylinder and enter the barrel if it was either solid lead or soft-tipped. The problem of lead deposition in the bore was overcome by greasing the bullet – the cylindrical part of a solid lead bullet incorporates one or more grease grooves. In addition to solid lead bullets, almost exactly the same bullet types are available for revolvers as for pistols (Fig. 2-19 a, c–e).

Ammunition for military rifles. Almost all contemporary military small arms are designed for automatic or semi-automatic fire. The ammunition used must

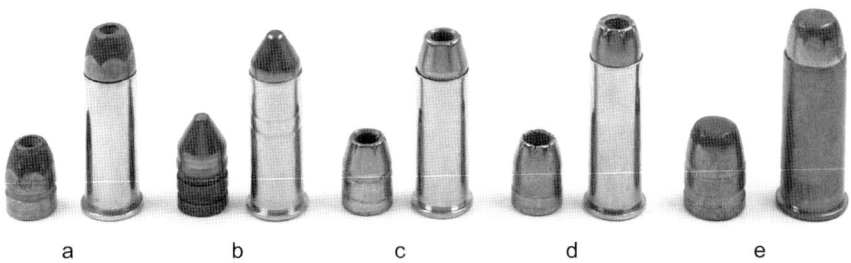

Fig. 2-19. Revolver cartridges and bullets: **a**. 38 Spl SJHP **b**. 38 Spl Metal Piercing **c**. 38 Spl Winchester Hydra-Shok **d**. 357 Magnum, Remington Golden Saber **e**. 44 Rem. Mag. SJFP

therefore meet a number of requirements. Cartridges fed automatically are subjected to significantly higher forces than those fed by manually operating a bolt or similar device. The inertia of the bullet must not cause it to slide into the case during loading, and nor must it fall out of the case. The bullet must therefore be fixed to the case, which is generally achieved by crimping the mouth of the case into a groove in the bullet, known as a cannelure. To comply with international treaties, military bullets are generally of the fully-jacketed type and because of the comparatively long ranges at which they are used they are generally aerodynamic in form, i.e. pointed. See Figs 2-20 and 2-21.

When firing on fully automatic, the barrel of a rifle rapidly attains temperatures of well over 100°C. Despite this, the weapon is expected to fire many thousands of rounds. As a result, military ammunition is not designed to produce the maximum power possible at a given calibre. This minimizes the risk of a cartridge "cooking off" in the barrel (firing spontaneously) when the weapon is running hot.

Since World War Two, four calibres have dominated the field of military ammunition. Eastern armed forces adopted the short 7.62 × 39 (Kalashnikov) (see

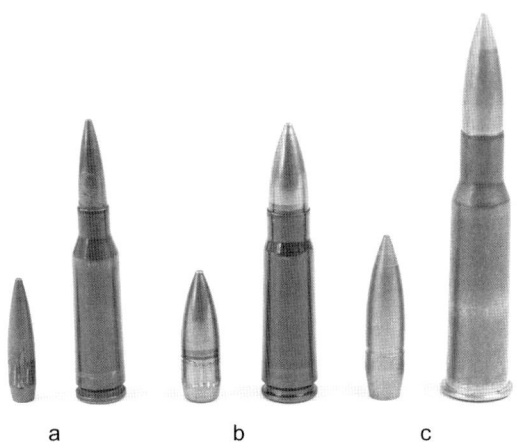

Fig. 2-20. Military cartridges and bullets (eastern):
a. 5.45 × 39 (Kalashnikov),
b. 7.62 × 39 (Kalashnikov),
c. 7.62 × 54 R (Dragunov) with steel-core bullet (yellow tip).

Fig. 2-21. Military cartridges and bullets (western):
a. 5.56 mm NATO (5.56 × 45, also designated 223 Remington), SS 109 bullet,
b. 7.62 mm NATO (7.62 × 51, also designated 308 Winchester).

Fig. 2-20 b), which was joined in the mid-1970s by the smaller 5.45 × 39 (Kalashnikov) (see Fig. 2-20 a). The 7.62 × 54 R (see Fig. 2-20 c) is widely used in machine guns and sniper rifles. After the Second World War, Western forces initially used the commercial 308 Win. calibre, designated 7.62 mm NATO (see Fig. 2-21 b). In the mid-1960s, the United States introduced a new, smaller calibre: the 223 Rem. (5.56 × 45) and some 15 years later this calibre became the NATO standard, albeit with a special bullet and shorter twist length. See Fig. 2-21 a. Ballistic data for the various military calibres are in Annex A.4.

In addition to the normal full metal-jacketed lead-core bullet, armed forces use a number of special bullets. *Tracer bullets* include a pyrotechnic charge in the tail (see Fig. 2-9 b). The charge burns for a certain time during flight (generally up to 1.5 s, depending on calibre), showing both the trajectory of the bullet and the point of impact. To enhance penetration of hard materials (steel, ceramics, etc.), the lead core can be partly or completely replaced by a core of hardened steel (see Fig. 2-9 e). The result is known as a steel-core or hard-core bullet, the nomenclature varying from one country to another. Newer types of bullet use tungsten or tungsten-alloy cores. These include the Nammo AP-8 Armour Piercing bullet (Sweden) and the RUAG Ammotec Styx (Switzerland), both of which are available in a number of calibres.

Opportunities for live firing during training are steadily dwindling and short-range bullets have been developed in response. These bullets follow approximately the same trajectory over short ranges (50–100 m) as normal bullets, but have a much shorter maximum range, making it possible to reduce danger zones considerably. This type of ammunition generally uses very light plastic bullets, which are fired at very high muzzle velocities and therefore undergo rapid deceleration.

There is a trend towards using larger calibres in sniper rifles in order to achieve longer ranges and greater effect, with a number of armies introducing 338 Lapua

Fig. 2-22. Sizes of military cartridges:
a. 5.56 mm NATO (5.56 × 45, 223 Rem.),
b. 7.62 mm NATO (7.62 × 51, 308 Win.),
c. 338 Lapua Magnum (8.4 × 69),
d. 50 Browning (12.7 × 99) with solid brass A-Max bullet.
The second number of the metric designation is the case length in mm.

Magnum (8.4 × 69) and 50 Browning (12.7 × 99) sniper rifles. Fig. 2-22 shows the corresponding cartridges in comparison with standard military ammunition.

Hunting ammunition. Unlike military ammunition, hunting ammunition is designed for single-shot shooting. There is no risk of barrel temperatures rising significantly during hunting, so there is no danger of a cartridge cooking off. Furthermore, hunting rifles are not expected to fire thousands of rounds. As a result, hunting cartridges can produce higher muzzle energies. This is apparent when one sees the relatively large cases (compare the hunting cartridges in Fig. 2-23 a, in military 308 Win. calibre, with the 7 × 64 und 7 mm Rem. Mag. hunting cartridges in Fig. 2-23 b-d). Depending on range and purpose, the aim is either to achieve very high muzzle velocities (up to 1150 m/s) using light bullets or else to achieve good penetration at long range (in the case of big game). Ballistic data for the common hunting calibres are given in Annex A.4.

These bullets are required to be highly effective when they hit the animal, so a wide range of bullets is produced. In addition to the conventional semi-jacketed and hollow-tip bullets, examples include the RUAG Ammotec partition bullet (Germany, formerly DNAG) (see Figs. 2-23 d and 2-24 e) and the Brenneke Torpedo (see Fig. 2-24 g, Torpedo-Universal, TUG). The newer designs are not in fact jacketed bullets; they consist of an H-shaped body made of copper or a copper alloy, which divides the bullet into two parts. Depending on the design, either both parts are filled with lead (Winchester Partition Gold, Fig. 2-24 f), or just the front part (Hirtenberger ABC and Nammo Forex, Figs 2-24 b and c) or just the rear part (Winchester Fail-Safe, Fig. 2-24 d). The principle is the same as that of a

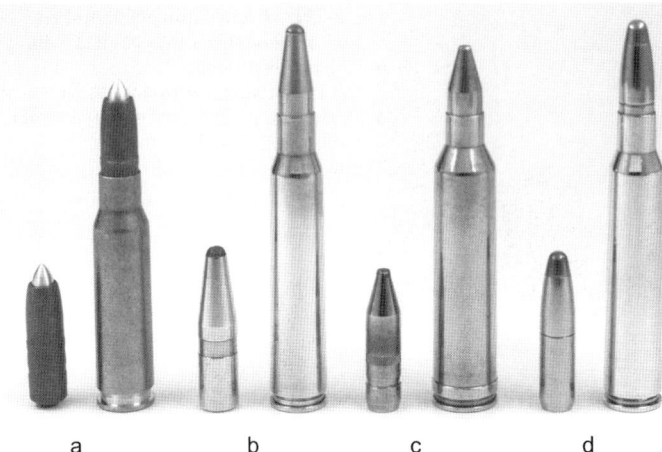

Fig. 2-23. Hunting cartridges and bullets:
a. 308 Win. with Brenneke TAG,
b. 7 × 64 with Hirtenberger ABC,
c. 7 mm Rem. Mag. with RWS conical spitzer bullet,
d. 7 × 64 with RWS partition bullet. Same scale as Figs. 2-20 and 2-21.

conventional hunting bullet; the front part provides the deformation, while the rear section ensures that the bullet retains sufficient residual mass and passes through the animal.

Many new bullets use no lead at all, for environmental reasons. These bullets generally consist of pure copper or copper alloy (MEN bevel bullet, Brenneke TAG, Fig. 2-24 a, Barnes Triple-Shock or the non-deforming Impala).

Shotgun ammunition. A shotgun generally has a smooth bore. It is therefore unable to impart spin to a projectile. As a result, shotgun projectiles have to be capable of stable flight without spin. The most common form of shotgun cartridge contains balls made of hardened lead (lead with antimony or arsenic added) or soft iron. The number of balls ranges from single figures to several hundred, depending on their diameter (see Fig. 2-25 a). On leaving the muzzle, the shot sheaf expands and lengthens. This can be controlled by incorporating a choke into the muzzle design. The purpose is to increase the chances of a hit, with the characteristics of the choke reflecting the nature of the target.

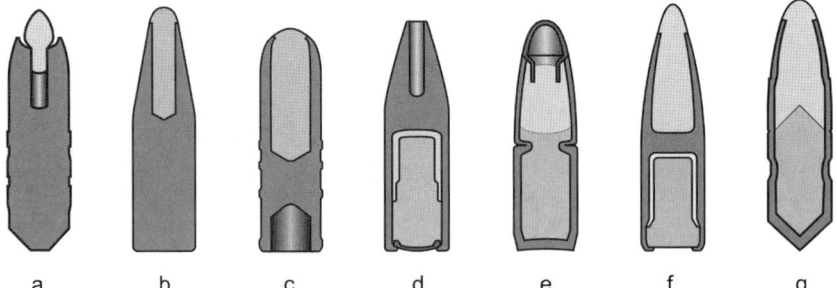

Fig. 2-24. Types of hunting bullet: **a.** Brenneke TAG (Turbo-Alternativ-Geschoss, see also Fig. 2-23 a), **b.** Hirtenberger ABC (see also Fig. 2-23 b) **c.** Nammo Forex, **d.** Winchester Fail Safe, **e.** RUAG (RWS) partition bullet (see also Fig. 2-23 d), **f.** Winchester Partition Gold, **g.** Brenneke TUG (Turbo-Universal-Geschoss); Geschoss = bullet.

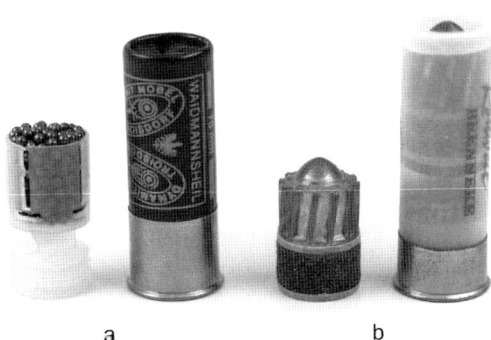

Fig. 2-25. Shotgun cartridges:
a. Shotgun cartridges and shot in plastic cup,
b. Shotgun cartridge with Brenneke slug (older type with felt wad, see also Fig. 2-26 b).

Older shotgun cartridges used a felt wad between the shot and the powder, both to prevent gas passing between the pellets and to ensure that the entire load accelerated homogenously. Newer designs include a plastic cup (see Fig. 2-25 a), which also prevents the outer pellets from rubbing against the barrel. Please see Annex A.9 for typical pellet dimensions and the primary roles of each.

The *slug* was developed in order to obtain the advantages of a single, heavy projectile when using a shotgun. Slugs are usually made of copper alloys or lead alloys and weigh between 20 g and 32 g, depending on calibre. A number of designs have emerged in recent years (see Fig. 2-26). In Europe, the best-known slug is produced by Brenneke (see Fig. 2-25 b).

The diagonal ribs on the Brenneke slug are not intended to impart spin to the projectile but act as a deformation zone when using a choked barrel. The diagonal arrangement of the ribs allows them to be crushed more easily.

No manufacturer has found a fully satisfactory way of stabilizing these projectiles in the absence of spin, and their effective range is similar to that of a shot sheaf. The newest slugs allow a good degree of accuracy out to 100 m, but these slugs (such as the Brenneke Super Sabot, Fig. 2-26), do require a slight degree of spin. As a result, shotguns with rifled barrels are now available.

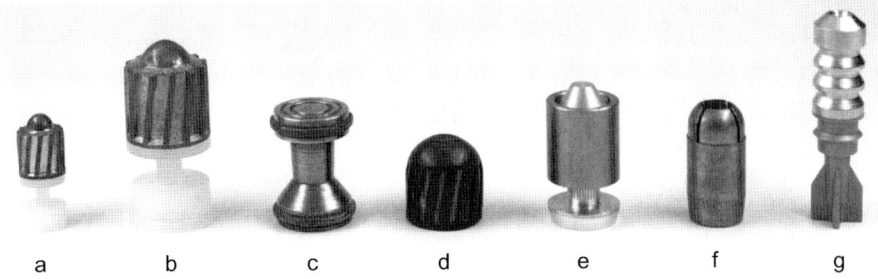

Fig. 2-26. Shotgun slugs: a. Brenneke FLG with plastic sabot, 20 bore, b. Same as a., 12 bore, c. Balle Blondeau slug, d. Winchester Foster, e. Brenneke Super Sabot, f. Remington Copper Solid, g. Sauvestre Balle Fleche (sub-calibre) Slugs e, f and g are fired using a sabot that separates from the slug on leaving the barrel.

In principle, a shotgun with a rifled barrel is a contradiction in terms, as the term "shotgun" implies a smooth bore. Strictly speaking, a shotgun with a rifled barrel is a rifle, of which the calibre, maximum gas pressure and chamber correspond to that of a shotgun.

The use of shotguns in military roles has led to the development of special types of ammunition, although these are not yet in widespread use. Here again, fin-stabilized projectiles (flechettes) are used, and the smooth barrel of a shotgun is ideally suitable for this type of projectile. Projectiles are also under development that comprise seven flechettes held together by a sabot, and designs incorporating a single flechette would also be possible.

Another, typically military projectile proposed for shotguns is based on the principle of the hollow charge.

A hollow charge consists of explosive with a conical hollow in the tip. This is covered with a layer of tough metal (generally copper). On detonation, the copper becomes a very long, thin projectile, moving at high speed.

These projectiles are designed to penetrate hard materials, and are perfectly capable of penetrating 60 mm of high-tensile steel.

Because they use cardboard or plastic cartridge cases, shotguns are only designed for relatively low gas pressures (650 bar to 1050 bar). As a result, one cannot expect to achieve high muzzle velocities. Furthermore, shot and slugs decelerate rapidly on leaving the barrel, causing shot to spread and slugs to deviate from the point of aim. The maximum practical range of a shotgun is therefore limited – approx. 30 m to 50 m.

The practical range of military shotgun ammunition is greater than that of hunting ammunition. There are two reasons for this. First, the weapons are proofed for higher gas pressures (the cartridge cases are made entirely out of brass), and can therefore achieve higher muzzle energies. Second, the design of the projectiles enables them to achieve stable flight, and hence to retain a higher percentage of their initial energy. An effective range of 150 m is possible, the limit being set primarily by spread.

Sub-calibre ammunition and flechettes. Flatter trajectories and shorter flight times demand higher muzzle velocities, and lighter projectiles offer a simple way of achieving these aims. The disadvantages – in exterior ballistics terms – of using a lighter bullet can be mitigated by reducing the cross-sectional area, resulting in a bullet with a diameter smaller than that of the bore. The bullet is mounted in a sabot, which is also responsible for providing spin (see Fig. 2-27), and the process of separating the bullet from the sabot after they leave the barrel has a decisive effect on spread.

It is possible to further increase the ratio of mass to cross-sectional area by using fin-stabilized flechettes. These too must be guided along the barrel using a sabot, although in this instance it must not impart spin. Here again, it is important that projectile and sabot separate cleanly, without disturbing the flight of the projectile. Muzzle velocities for this type of projectile are very high – in excess of

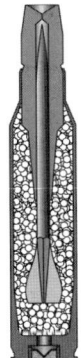

Fig. 2-27. 7.62 × 33 cartridge with sub-calibre spin-stabilized projectile.

Fig. 2-28. 5.56 × 45 cartridge with flechette, *left:* cross-section, *above:* flechette, with and without sabot.

1400 m/s. See Fig. 2-28 for an example. This type of projectile has not entered widespread use as yet.

Trends. Cartridge ammunition has been in use for over 100 years. In that time there have been a number of trends, with the resulting products either establishing themselves on the market or remaining at the experimental stage. At the end of the 20th Century, for instance, various types of caseless ammunition were developed, together with the weapons to fire them. These included the military 4.73 mm G 11, the 9 mm AUPO for handguns and the 5.7 mm and 6 mm UCC for hunting rifles. However, none of these became truly successful.

One type of ammunition that is currently highly popular, at least in Europe, is lead-free ammunition, generally using copper or copper-alloy bullets. Examples include the RUAG Action, the MEN QD P.E.P. for handguns, the Brenneke TAG, the Impala, the MEN bevel bullet and steel shot for hunting.

For hunting ammunition, there is a trend towards short cartridges (e.g. the 300 WSM and the 6 mm Norma BR, see Fig. 2-29 a), which have certain advantages in terms of ballistics. In the military sphere, there is a trend towards larger-calibre ammunition for assault rifles (e.g. the 6.8 mm Rem. SPC) and smaller cartridges for PDW (personal defence weapons) such as the FN 5.7 × 28 (see

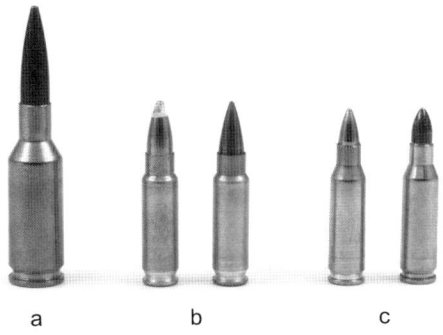

Fig. 2-29. Recent developments:
a. 6 mm Norma BR for target shooting,
b. 5.7 × 28 cartridge for the FN P90,
c. 4.6 × 30 cartridge for the HK MP-7.

Fig. 2-29 b) and the 4.6 × 30 produced by Heckler & Koch and by RUAG (see Fig. 2-29 c). Pistols have also been developed for these two calibres.

2.2.2.3 Blank and irritant rounds.

General. Cases do arise in wound ballistics in which blank cartridges have been used. Blank cartridges, or "blanks", consist of a cartridge case, propellant and primer. Unlike normal ammunition, blank cartridges do not include a projectile, though in certain instances a projectile may be loaded separately.

Artillery guns often use ammunition consisting of a propellant charge and a separate projectile. This makes it possible to choose the charge when loading, and hence to adjust the range.

Muzzle-loaders also used a separate bullet and propellant charge, although the two were often carried together in a paper cartridge (see for instance the cartridges from the period 1856 to 1863 in Fig. 7-1).

As far as the weapons of interest to us are concerned, blank cartridges are most often to be found in alarm pistols. Such weapons are in fairly widespread use, especially in Germany. For legal reasons, they must be built in such a fashion that it is impossible to load a projectile via the muzzle.

For a comprehensive treatment of blank cartridges and irritant rounds, please see ROTHSCHILD (1999).

Calibre and design. Alarm pistols are so designed as to imitate real pistols or revolvers, and blank ammunition is produced for both. Blank cartridges for revolvers are rimmed, whereas blank pistol cartridges are generally rimless, as in the case of ammunition for the corresponding real firearms (see Fig. 2-30). There are four calibre groups, with designations based on cartridge case diameter:

– 22 long and 6 mm Flobert, both rimmed, case diameter approx. 5.7 mm;
– 315, 8 mm und 320 short, case diameter 8 mm;
– 35, 35 R, 9 mm PA and 380 R (9 mm R), case diameter approx. 9.6 mm (corresponds to 9 mm short);
– 45 R, case diameter 12 mm.

The official designation used by the C.I.P. (*Commission Internationale Permanente pour l'Epreuve des Armes à Feu portatives*, Permanent International Commission for the Proof of Small-arms)

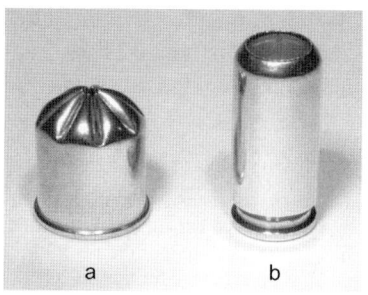

Fig. 2-30. Blank cartridges:
a. 45 R, rimmed, crimped,
b. 9 mm PA, rimless, plastic seal.

adds the word "blanc," i.e. blank (e.g. 380 short blank). In the German-speaking world, however, the cartridge designation generally makes no explicit mention of "blank."

There are two systems for sealing blank cartridges. In the case of the 22 long, 6 mm Flobert, 320 short, 380 R and 45 R, the mouth of the brass or copper case is crimped shut (Fig. 2-30 a). Other cases are closed off by a plastic disc, scored in such a way as to fragment on firing. The disc must be so designed that no solid fragments emerge from the muzzle. See Fig. 2-30 b.

The effects of blank cartridges in terms of blast and gas jet depend not only on the calibre but also on the design of the weapon.

Types of ammunition. In addition to blank cartridges, which contain only propellant, there are also irritant rounds, which contain both a propellant and an irritant, in powder form. The irritants used are either CN (Chloracetophenone) or CS (2-chlorobenzalmalononitrile in crystalline form, or Capsaicin.

2.2.2.4 Fragmenting ammunition

General. The role of ammunition is to achieve a specific destructive effect on a target at a distance. Two packages of energy are required: one to transport the projectile to its target and one to achieve the effect. The small arms ammunition described so far combines the two, with all the energy required being produced in the weapon and the projectile carrying the destructive energy to the target in the form of kinetic energy.

Work must be performed along the path to the target to overcome air resistance (drag), and kinetic energy is taken from the projectile to achieve this. The projectile must therefore be given enough energy on firing to both reach the target *and* achieve the required effect.

In the case of fragmentation munitions, however, the two packages of energy are created separately. As a result, the transport energy can be adapted to the range required without altering effectiveness.

For short ranges, e.g. in the case of a hand grenade, human energy is sufficient. The range of mortars and artillery guns can be adjusted over a wide spectrum by adapting the charge – and hence the initial velocity – to the range required.

The energy needed to achieve the effect on the target is transported by the projectile as chemical energy, generally in the form of a high explosive. This energy is converted into pressure (via combustion or detonation) in the vicinity of the target, and it is this pressure that creates and accelerates the fragments. These fragments may be compared to bullets, in that the kinetic energy imparted is used to get them to the target *and* to achieve the effect.

Structure and nomenclature. The overall structure of fragmentation munitions is similar to that of small arms ammunition (see Fig. 2-31). Like small arms ammunition, fragmentation shells have a projectile (the fragment), a propellant (the high explosive) and a firing device (the fuse and the booster).

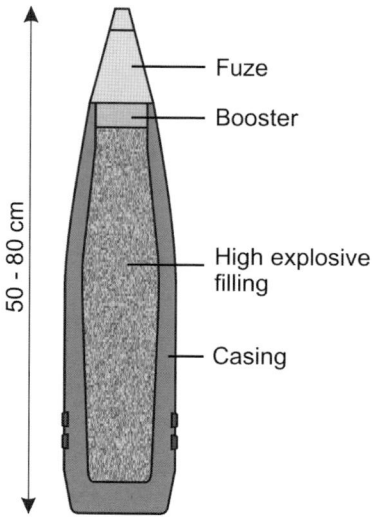

Fig. 2-31. Schematic of an artillery shell.

The casing of a fragmentation shell both provides the fragments and contains the high explosive. A fuse is built into either the tip or the base of the projectile, depending on the role of the shell. The purpose of the fuse is to detonate the explosive at the appropriate moment.

The body of the shell or grenade is generally made of steel. In order to produce fragments of regular shape, it is often prefragmented to a geometric pattern on the inside or outside. In some cases, the fragments are prefabricated and embedded in a substrate such as plastic or aluminium (see Fig. 2-32). These prefabricated fragments generally consist of balls or cubes made of steel or tungsten. Such munitions are also classed as prefragmented shells.

Tungsten is often used for the fragments, as its high density yields heavy fragments with lower air resistance.

As the fuse has the task of detonating the high explosive when the shell arrives in the vicinity of the target, there has to be some way of determining when that is. There are many ways of achieving this, using mechanical or electronic means. The simplest device is the impact fuse, which detonates on impact with the ground. However, if the shell detonates on impact, a large percentage of the fragments ends up in the ground, so the shell is often detonated a few metres above the ground.

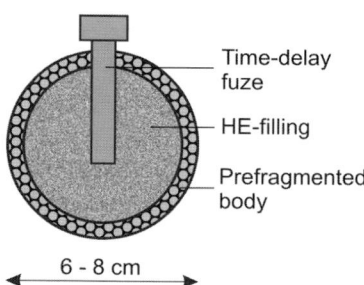

Fig. 2-32. Structure of a hand grenade. Delayed detonation is achieved by means of a pyrotechnic charge with a predefined burning rate.

In the past, attempts were made to achieve this using time fuses, which involved setting the time of flight before firing. However, the wide disparity in times of flight led to considerable scatter in the height at which the shell detonated. The modern approach is to use electronic proximity fuses, which measure the distance from the ground and detonate at a predetermined height.

If it is not possible to predict the orientation of a device on impact (as is the case with hand grenades, for instance) they are fitted with pyrotechnic time fuses.

Pipe bombs are often used in civilian contexts (i.e. in terror attacks). A pipe bomb consists of a length of gas or water

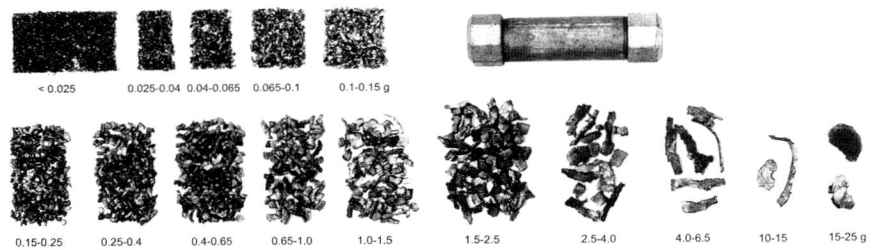

Fig. 2-33. Fragments from a single pipe bomb, 20 cm in length. This is an example of an improvised explosive device, or IED. Top right: the device before detonation.

pipe filled with a high explosive and equipped with a detonator. These simple devices can be very effective, as the fragments from a 20 cm bomb illustrate (see Fig. 2-33).

2.2.3 Weapons

2.2.3.1 Firearm design and typology

Components. Every firearm consists of certain components, the actual design of which may vary considerably depending on the type of weapon and the individual model. The components are:

- barrel and chamber;
- bolt, locking mechanism and cartridge feed;
- firing mechanism and safety device.

In dividing firearms into categories, we consider both their role and the characteristics of these components.

Barrel. The barrel is the central element of a firearm, providing the path over which the bullet accelerates. At the rear of the barrel we find the chamber, a widening of the barrel in the form of the cartridge to be used. Because the chamber matches the shape of the cartridge, a weapon is usually designed for one specific cartridge type.

There are two basic types of barrel. Spherical projectiles (and, more recently, flechettes) are generally fired from a smooth-bore barrel. Long projectiles without fins must be made to spin around their longitudinal axis in order to achieve stability in flight (see 2.3.5). This is achieved by incorporating spiral grooves into the barrel, which guide the bullet and impart the required spin (see Fig. 2-34). This type of barrel is described as *rifled*. The raised areas of metal between the grooves are referred to as *lands*. The internal diameter of the barrel – measured between the lands in the case of a rifled barrel – is the *calibre*. The measured calibre of the bore is often not identical to the nominal calibre used to identify the weapon.

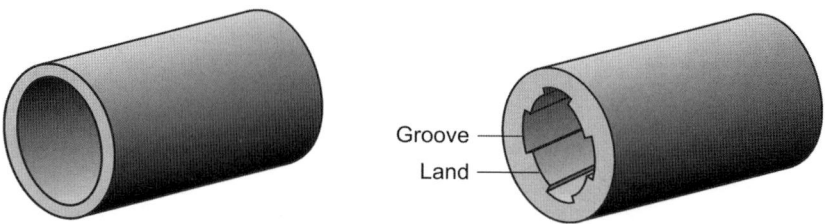

Fig. 2-34. Barrels (in cross-section). *Left:* Smooth-bore (shotgun) barrel. *Right:* Rifled (rifle) barrel. The lands and grooves are shown in exaggerated form.

The distance that a bullet moves along the barrel as it makes one full rotation is the *twist length* and is a key characteristic of a rifled barrel. Instead of the twist length, one can specify the helix angle of the grooves, known as the *angle of twist*. The relationship between twist length Λ, calibre k und angle of twist Γ is given by:

(2.2:1) $$\Gamma = \arctan\frac{\pi \cdot k}{\Lambda} .$$ [°]

The speed of rotation of the bullet ν is determined by the twist length and the muzzle velocity v:

(2.2:2) $$\nu = \frac{v}{\Lambda} .$$ [1/s]

Often – especially on military weapons – the muzzle is fitted with a muzzle brake, a device that disperses the combustion gases that follow the bullet out of the muzzle (see Fig. 2-35). Depending on the design of the muzzle brake, it may serve either or both of two purposes: to reduce recoil and the upwards deflection of the barrel and/or to prevent secondary flash (see 2.3.3.2).

Bolt or breech block. The combustion gases generated when the weapon is fired act in all directions. To prevent the cartridge case from moving in the opposite direction to the bullet (owing to the law of conservation of momentum), the cartridge must be held firmly in place. This is the job of the bolt or breech block. In the closed position, this device seals off the chamber from behind. In the open position, it must ensure that the cartridge loads reliably and that the spent case is removed after firing. It also houses those parts of the firing mechanism (firing pin, etc.) that must be able to reach the head of the cartridge. It is generally closed

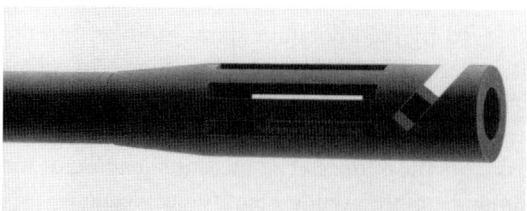

Fig. 2-35. Muzzle brake on a sniper rifle (Sauer 2000). The angled slot reduces the upwards deflection of the muzzle on firing.

when the weapon is ready to fire, except in the case of weapons that operate at relatively low gas pressures. There are many different bolts, breech blocks and locking mechanisms. The most important are as follows:

- *Break action*. Most often found on single- and double-barrel shotguns, but also used in single-shot pistols.
- *Bolt action* or *cylinder lock*. The breech block moves in a straight line along the axis of the barrel, in a lock-box. The action is generally locked by a rotating movement, for instance by lugs on the breech block engaging with corresponding slots. Weapons designed for low-energy cartridges (e.g. 22 L.R.) often use the *blowback system*, and have no locking mechanism at all.
- *Falling block*. The breech block moves vertically in the receiver, behind the chamber. This system is found mainly on single-barrelled hunting rifles with fixed barrels.

The action can be operated by hand or automatically. In the case of automatic operation, the energy required is obtained either from the recoil (in the case of a *blowback operated weapon*) or from the combustion gases (in the case of a *gas operated weapon*). On a gas operated weapon, some of the combustion gas is diverted onto the head of a piston via an aperture, the gas port. The piston operates a mechanism that unlocks the breech block and forces it back. The breech block therefore remains locked in the closed position until the gas has passed the gas port. A blowback operated weapon uses momentum. To ensure that the gas pressure acts on the bullet for as long as possible, the weapon is so designed that there is a certain delay before the breech opens.

The action of a weapon falls into one of two categories with respect to the sequence of movements. On a weapon that "fires from an open bolt," the breech block is to the rear when the weapon is ready to fire. When the trigger is pulled, the breech block flies forward, picking up a cartridge from the magazine as it does so, and a fixed firing pin fires the cartridge as soon as it is seated in the chamber. The recoil from the cartridge the forces the breech block back to its initial position. On a weapon that "fires from a closed bolt," however, the breech block is closed when the weapon is ready to fire. Squeezing the trigger causes a floating firing pin to strike primer, firing the cartridge. The breechblock is forced to the rear, ejects the spent case and flies forward again, sliding a new cartridge into the chamber. This means that when a weapon that fires from a closed bolt is ready to fire there is a cartridge in the chamber, whereas in the case of a weapon that fires from an open bolt the chamber is empty when the weapon is ready to fire, thus avoiding all risk of a cartridge cooking off.

Firing mechanism and safety devices. Every weapon needs a mean whereby the user can fire a round at the moment chosen. This is the firing mechanism, and is generally linked closely with the bolt or breech block. Its main components are the firing pin, striker (or hammer), main spring, trigger bar, trigger and trigger spring. The main spring supplies the energy required to detonate the cartridge and is generally compressed by the cocking action.

To prevent the weapon going off unintentionally, one or more safety devices are incorporated. First, every trigger has a certain resistance (10 N to 40 N), which prevents it being operated too easily. To allow the weapon to be fired without operation of the trigger affecting accuracy, the *double set trigger* was introduced, whereby one action sets the trigger and a second one fires the weapon, the second action requiring a pull of only a few decinewton (dN).

Most safety devices stop the trigger from operating the firing mechanism at some point within the mechanism. The closer to the cartridge that happens, the more reliable the safety device; so modern weapons often use hammer/striker or firing pin safety devices.

Weapon names and categories. Firearms are generally categorized according to their design and the way they are operated:

- One-handed versus two-handed operation.

    Weapons designed to be fired with one hand are generally known as *handguns*. Weapons designed to be fired with two hands are known as *long weapons*.

- Single-shot, repeater, semi-automatic or automatic.

    Weapons that must be loaded before each shot are called single-shot weapons. If a weapon has a magazine, from which a cartridge can be loaded by a simple hand movement (cocking action), it is known as a repeater. If the weapon cocks automatically after each shot, but only fires one shot each time the trigger is pulled, it is described as semi-automatic, self-loading or autoloading. If it is possible to fire more than one shot by pulling the trigger just once, the weapon is said to be automatic or fully automatic.

- Long weapons are further categorized according to whether the bore is smooth or rifled.

    A weapon with a smooth bore is known as a shotgun, while a weapon with a rifled bore is called a rifle.

- Handguns are categorized according to their design.

    If the barrel and chamber are fixed together, the weapon is a pistol. A handgun with a number of chambers in a revolving cylinder mounted behind the barrel is designated a revolver.

In many cases, more than one of these terms may apply to the same weapon. For instance, both shotguns and rifles can be single-shot or self-loading. Most pistols are semi-automatic, but there is no such thing as a semi-automatic revolver. A fully automatic pistol is known as a *submachine gun*.

"Shotguns" with rifled barrels have appeared in recent years, designed to fire slugs with a higher degree of accuracy. However, according to the generally ac-

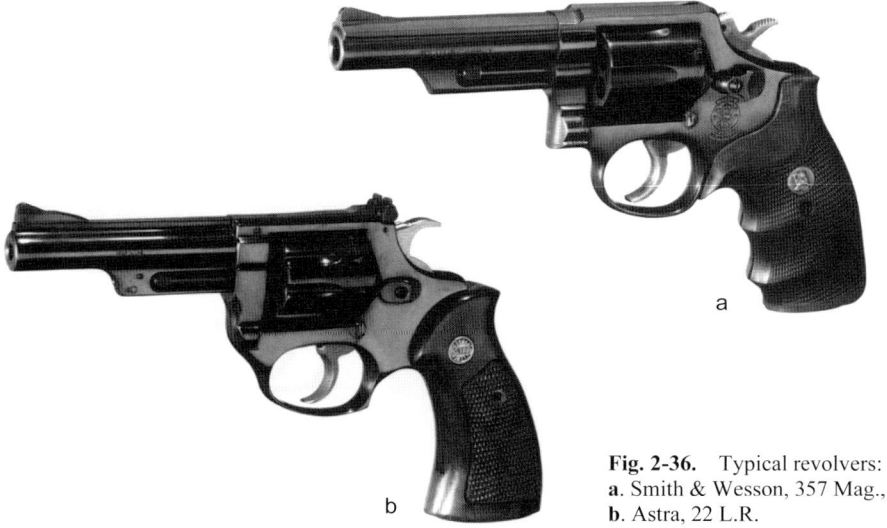

Fig. 2-36. Typical revolvers:
a. Smith & Wesson, 357 Mag.,
b. Astra, 22 L.R.

cepted criteria above, such a weapon is a rifle, of which the calibre, maximum gas pressure and chamber correspond to that of a shotgun.

2.2.3.2 Handguns

Revolvers. Revolvers (see Fig. 2-36) are inherently single-shot weapons. The hammer, which strikes the firing pin every time the trigger is pulled, has to be cocked after every shot. A revolver of which the hammer can only be cocked by pulling it back manually is known as a single action (SA) revolver. If the hammer can also be cocked by the action of the trigger, the weapon is known as a double-action (DA) revolver. The action of cocking the hammer rotates the cylinder by one chamber, placing the cartridge that was previously to the left or right of the

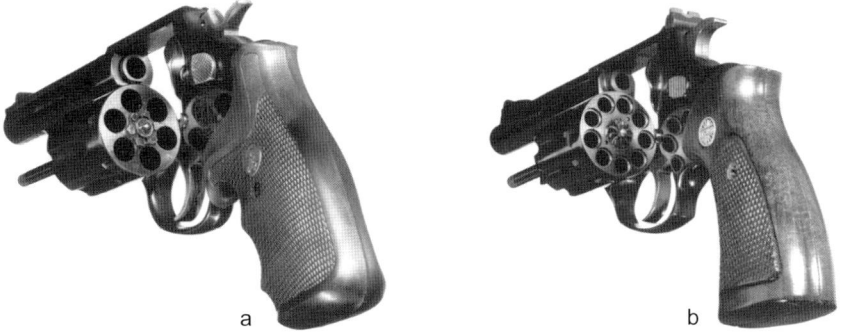

Fig. 2-37. a. Same (6-shot) weapon as in Fig. 2-36 a, with cylinder swung out, b. Same (9-shot) weapon as in Fig. 2-36 b, with cylinder swung out.

Fig. 2-38. Typical large-calibre pistol (9 mm Luger): SIG-Sauer P 228, muzzle energy approx. 500 J.

Fig. 2-39. Typical small-calibre pistol (7.65 Browning): Walther PPK, Manurhin licence. Muzzle energy approx. 220 J.

barrel in position to be fired. The direction of rotation varies according to the model.

The cylinder generally holds five to six cartridges, depending on the size of the weapon and of the cartridge. The cylinders of weapons designed for 22 calibre ammunition (22 L.R. or 22 Mag.) often hold nine cartridges or more. On most modern revolvers, the cylinder swings out to the side for loading (see Fig. 2-37). However, certain revolvers have a special loading position or use a break action.

Pistols. With the exception of a small number of single-shot and two-shot designs, pistols are generally semi-automatic and capable of firing several shots. The cartridges are stored in a magazine, which usually slides into the grip. The current trend is towards more compact pistols, but with greater magazine capacity even in the case of larger calibres. There are already surprisingly small pistols available with 12- or 16-round magazines (see Fig. 2-38). To ensure that the user is ready to fire the first shot as rapidly as possible, double-action (DA) or, more rarely, double-action only (DAO) has become most popular. The hammer of a DA pistol only has to be cocked by the trigger before the first shot, while the hammer of a double-action-only weapon is cocked by the trigger before every shot.

Pistols designed for low-power cartridges (with a muzzle momentum of less than approx. 2 N·s; see Fig. 2-39) generally have an unlocked blow-back system, with the breech block simply held against the chamber by the spring. On a pistol with a higher muzzle momentum the breech block has to be locked, as for a rifle.

Most single-shot pistols are made for target shooting, where accuracy is particularly important; an automatic loading mechanism, however well made, will reduce the accuracy of the weapon. Where possible, such mechanisms are avoided in competition pistols.

Fig. 2-40. Typical submachine gun (Uzi, 9 mm Luger).

Submachine guns. Submachine guns are automatic weapons that fire pistol ammunition. Most are 9 mm Luger, although a few fire 45 Auto. They are generally very simple in design, using a blowback mechanism firing from the open bolt and no locking system. In many cases, they are incapable of firing single-shot. Famous examples of submachine guns include the Thompson (45 Auto), the British Sten Gun and the Israeli Uzi (see Fig. 2-40). The Sten and the Uzi both fire 9 mm Luger.

The capacity of a submachine gun depends on the type of magazine, which is normally 30 to 50 rounds, though drum magazines may hold up to 100. The rate of fire lies between 600 and 800 rounds per minute. Submachine guns are not usually used for aimed shots, and the sights are correspondingly basic.

2.2.3.3 Long weapons

Military rifles. The majority of rifles used by armies today are automatic. They are known as *assault rifles* and form part of the soldier's personal weaponry. While older models may well be blowback operated, virtually all modern assault rifles are gas operated. They have magazines holding 20 to 30 cartridges and can fire single-shot, fully automatic and, in many cases, in bursts of three rounds.

Some types of assault rifle have become popular the world over. Probably the most frequently encountered military weapon in the world is the Russian 7.62 × 39 AK-47, designed by M. T. KALASHNIKOV and introduced in 1947 (see Fig. 2-41 a). Almost 100 million of these weapons have been produced, and armies that use this weapon as standard include those of the former Warsaw Pact and a number of eastern countries. A 5.45 × 39 successor was introduced in 1974, the AK-74 (see Fig. 2-41 b).

After the Second World War, Western forces initially standardized on the commercial 308 Win. calibre, designated 7.62 mm NATO, with several assault rifles being designed for this ammunition. The American M14, German G 3 and Belgian FAL were all in fairly widespread use. Fig. 2-42 a shows a modern assault rifle produced by SIG. In the 1960s, the US Army introduced a new,

Fig. 2-41. Typical military weapons of eastern origin:
a. 7.62 × 39 AK-47, (Kalashnikov)
b. 5.45 × 39 AK-74, (Kalashnikov).

smaller-calibre rifle, which was produced in large numbers. This was the M16, chambered for 5.56 × 45 (223 Rem.). The M16 was designed by Armalite but manufactured by a number of different companies and initially fired the M193 cartridge. This rifle formed the basis for several new rifle and cartridge designs in a number of western countries. In time, this led to the emergence of a new NATO standard 5.56 × 45 round, designated 5.56 mm NATO, based on the SS 109 cartridge produced by Belgian manufacturer FN. The SS 109 uses a different bullet from the 223 Rem., with a much shorter twist length (178 mm instead of 305 mm). This new version of the M16 (see Fig. 2-42 b) was designated M16A4 and the US designation for the corresponding cartridge is M855. The Israeli Galil also became very well known, and is available in all the western calibres mentioned above.

To enhance accuracy, sniper rifles (see Fig. 2-43) are generally repeater, although semi-automatic sniper rifles do exist. They are equipped with optical sights for day and night use. Because of the range required in a military role,

Fig. 2-42. Typical military weapons of western origin:
a. 7.62 mm NATO SG 542 (SIG-Manurhin),
b. 5.56 × 45 M16A4 Bushmaster.

Fig. 2-43. SIG SG 550 semi-automatic sniper rifle, 5.56 × 45 (223 Rem.).

western countries' sniper rifles are designed for 7.62 mm NATO, 338 Lapua Magnum or 12.7 × 99 (50 Browning). Eastern countries use 7.62 × 54 R for the same reason, although there is a trend towards larger calibres (up to 14.5 mm). At shorter ranges, as required by police sharpshooters in urban contexts, 5.56 × 45 is sometimes used.

Most recently, armed forces have been adopting autoloading and automatic shotguns. These are generally chambered in 18.5 × 59 (more usually known as 12/70, see "Hunting weapons" below) and are usually designed to deliver higher performance than hunting shotguns of the same calibre. Nevertheless, their use remains limited to the shorter ranges encountered in rough country or in heavily built-up areas.

Machine guns. Machine guns are automatic weapons designed to fire large quantities of ammunition. As a result, their design differs from that of assault rifles (see Fig. 2-44). It is difficult to feed large numbers of rounds from a magazine, so most machine guns use belted ammunition, with each belt holding up to 250 rounds. A heavy, stable tripod is required to maintain stability during sustained fire. As overheating of the barrel is a problem when the weapon is used for long periods, machine guns often have interchangeable barrels or cooling systems (air or water).

These weapons are generally used at longer ranges and therefore tend to be chambered in larger calibres (7.62 mm to 12.7 mm).

Hunting rifles and shotguns. Most hunting rifles and shotguns are single-shot. As hunters often wish to be able to use more than one type of ammunition, designs with two to four barrels of differing calibres have become popular. A weapon with two smooth-bore barrels is known as a *double-barrelled shotgun*, while a weapon with two rifled bores is called a *double rifle*. A weapon with one rifled barrel and one smooth barrel is known as a *rifle-shotgun*. If the barrels are arranged one above the other (see Fig. 2-45 a) the weapon is referred to as a *su-*

Fig. 2-44. Machine gun on tripod (SIG 710-3, 7.62 mm NATO).

Fig. 2-45. Typical hunting weapons:
a. Superposed rifle-shotgun,
b. Drilling, with two shotgun barrels and one rifled barrel.

perposed rifle-shotgun. A weapon with three barrels is called a drilling (also the German word for "triplet"). See Fig. 2-45 b.

The naming convention for shotgun calibres has a historical basis. The "bore" number is determined by the weight, in fractions of a pound, of a solid sphere of lead with a diameter equal to the inside diameter of the barrel. The greater the bore diameter, the smaller the number. The nominal inside diameter of a shotgun barrel can be calculated using the following formula:

(2.2:3) $$d = 42.431 \cdot \sqrt[3]{\frac{1}{k_s}} \qquad [\text{mm}]$$

where k_s is the pellet diameter and d the nominal inside diameter of the barrel. Please see Annex A.9 for tables of common shotgun ammunition dimensions. A second number is used to indicate the length of the chamber and hence the maximum length of cartridge that can be used. This figure is given in mm in mainland Europe and in inches in the English-speaking world, the 12/70 (12/2³/₄") being one of the most common.

Competition weapons. Effect on the target and a flat trajectory are of secondary importance in weapons designed exclusively for competition shooting. What is required is that spread be minimized. This is achieved by manufacturing the relevant parts of the weapon to a high degree of precision. Competition weapons are therefore single-shot or repeating weapons, with the exception of those used for rapid-fire competitions. The stock, butt and grips can be modified to suit the user and ensure a comfortable shooting position.

Typical competition weapons include match rifles, while in southern Germany, Austria and Switzerland, a type of short rifle known as a *stutzen* or *stutzer* is also popular. At short ranges (i.e. up to 50 m) competitors use 22 L.R. At longer ranges (300 m), rifle calibres such as 6 mm Norma BR or 308 Win. are popular, as these are less sensitive to crosswinds than 223 Rem., for instance. Special shotguns are available for clay pigeon shooting and similar competitions.

2.2.3.4 Alarm pistols and revolvers

Alarm pistols and revolvers are generally replicas of real weapons. Their mode of operation is based on that of the weapons they imitate. Alarm pistols are therefore semi-automatic weapons, with a manual cocking action required for the first round only. Both single-action and double-action revolvers are available. To prevent the weapon being made capable of firing real ammunition, the barrel must be fixed to the receiver or grip and must incorporate hardened metal baffles, both to prevent its being drilled out to a normal calibre and to prevent a muzzle-loaded projectile accelerating to a dangerous velocity (see ROTHSCHILD 1999).

When a blank cartridge is fired, gas can leave the muzzle at velocities of over 3000 m/s (see 4.2.3.4). Barrel length has an influence on the velocity of the gas jet and hence on the danger that it poses.

2.3 Ballistics

2.3.1 Definitions

The word ballistics comes from Greek, and means the study of objects that are thrown and of their trajectories. As firearms became more common, the term came to mean all processes related to the motion of a bullet. Four sub-areas have developed:

- *Interior ballistics* is the study of the acceleration of the bullet in the weapon and the related processes.
- *Intermediate ballistics* is the study of the process by which the bullet leaves the muzzle, and how this is influenced by the weapon and by combustion gases.
- *Exterior ballistics* is the study of ballistics in the original sense of the term – the path of the bullet through the air.
- *Terminal ballistics* is the study of the penetration of the target by the bullet, the target generally being a material substantially denser than air.

Wound ballistics is a sub-domain of terminal ballistics, concerned primarily with the penetration of a bullet into persons and animals, and the effects that result.

Sectional density is one of the most important physical quantities in all areas of ballistics, and is defined as the ratio of mass to cross-sectional area A. The cross-sectional area equates to that of the projection of the bullet onto a plane perpendicular to the direction of movement. In the normal case, that of a bullet in stable flight, we generally use the cross-sectional area of the calibre, which is often slightly smaller than the cross-sectional area of the bullet itself:

(2.3:1) $$q = \frac{m}{A} = \frac{4 \cdot m}{k^2 \cdot \pi} \qquad [kg/m^2]$$

A heavy, slender projectile (such as an arrow) therefore has a high sectional density, whereas a light projectile with a large cross-section has a low sectional density. For projectiles of similar geometry and design, q increases linearly with calibre. For smaller calibres, q is generally measured in $[g/mm^2]$.

In most ballistic processes, it is the sectional density that plays the decisive role, rather than the calibre or mass of the bullet.

Example: A 22 L.R. bullet weighing 2.6 g has a slightly higher sectional density than a 9 mm Browning bullet weighing 6.1 g. As a result, it is deflected slightly less by crosswinds. The effect of the crosswind depends not on the mass of the bullet but on its sectional density.

2.3.2 Interior ballistics

2.3.2.1 General

Interior ballistics starts when the firing pin hits the primer and ends when the bullet leaves the barrel. This field of study is of only peripheral relevance to wound ballistics. However, when a shot is fired at short range or with the muzzle in contact with the body, interior ballistics does play a decisive role in wounding and in the visible characteristics of the wound. A knowledge of interior ballistics is therefore of value in the field of forensic ballistics when it is necessary to assess gunshot residues, determine muzzle-target distances, etc.

2.3.2.2 Powder combustion

The main difference between combustion (or deflagration) and detonation lies in the speed at which oxidation propagates. If this *burning rate* is of the order of millimetres or centimetres per second, we speak of combustion or deflagration. A burning rate of kilometres per second is classed as detonation. Normally, the powder burns inside the weapon, but the burning rate is determined to a large degree by pressure (see Table 2-10). This is known as the law of powder combustion.

At any given moment, powder will burn perpendicular to its surface, at its current burning rate. The quantity of gas produced, and hence the pressure developed, therefore depends on both the pressure itself and on the geometric form of the powder particle. If the surface area of the powder reduces as it burns, the powder is described as degressive. If the surface area increases, it is known as progressive. If it remains constant, it is known as neutral. Ball powder and flake powder (see Fig. 2-11, left) are therefore degressive powders. Tubular powder (see Fig. 2-11, right) is neutral (if we ignore the ends of the tubes). Cylindrical

Table 2-10. Burning rates

Propellant	Pressure [bar]	Burning rate [mm/s]
Black powder	1	1.8
Nitrocellulose powder	1	0.06–0.09
Nitroglycerine powder	1	0.07–0.25
do	10	1.5
do	500	75
do	4000	600
Explosives		[m/s]
Trinitrotoluene (TNT)	1	6900
Hexogen	1	8700

powder with seven holes (known as seven-hole powder) is progressive, as the sum of the expanding surfaces burning from inside to outside is greater than the sum of the reducing surfaces burning from outside to inside. The burning rate is often modified by treating the surface with a retardant or accelerator.

The combustion properties of a powder depend not only on the pressure and on the form of the particles but also on the *vivacity of the powder*. This is a constant that governs the relationship between burning rate and pressure.

2.3.2.3 The firing sequence

The firing sequence is highly complex and takes place under extreme physical conditions. High pressures and high temperatures are generated together, in a very brief timeframe.

The process begins when the primer is ignited, the hot gases and flames partially burning the surface of the powder. The bullet has a certain initial resistance, causing the pressure in the cartridge case to rise rapidly, which ensures that the entire surface of the powder ignites as evenly as possible. As a result, the bullet only begins to move when the pressure exceeds a certain level, the shot start pressure. As it does so, the volume available to the gas increases. From this point on, increase in volume and the quantity of gas produced determine the change in gas pressure, which reaches a maximum when the two are equal. Thereafter, pressure drops until the bullet leaves the muzzle. The pressure remaining at this point (the *muzzle pressure*) determines what happens at the muzzle. Fig. 2-46 a is a typical curve of gas pressure over time. Fig. 2-46 b shows pressure plotted against the distance that the bullet has moved and the velocity of the bullet.

Powder combustion must be regulated in such a fashion as to ensure that the maximum gas pressure is not too high and the muzzle pressure as low as possible. Combustion should have finished before the bullet leaves the muzzle. In addition, it is generally desirable for muzzle velocity to be as high as possible. This means

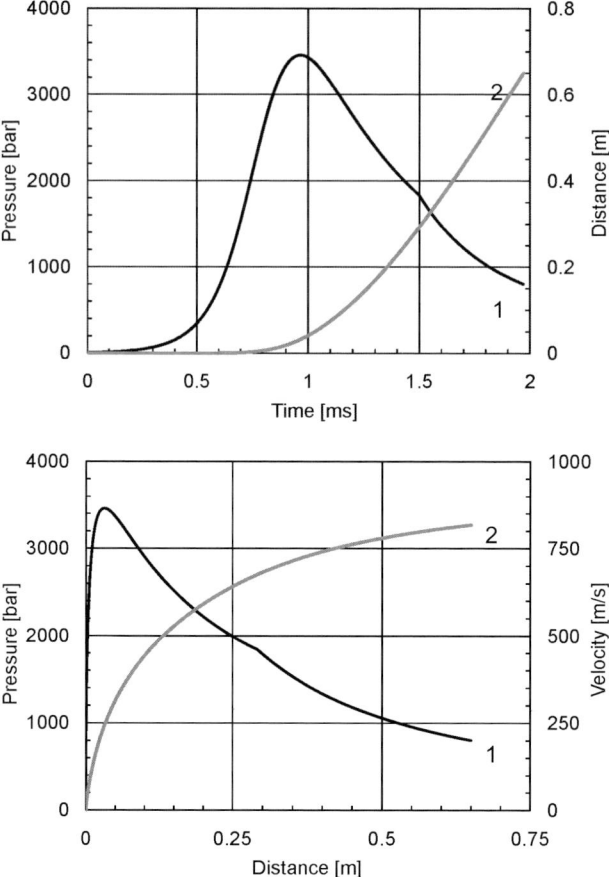

Fig. 2-46. Typical curves of gas pressure (1):
a. Over time. Curve 2 shows the distance travelled by the bullet.
b. Over distance. Curve 2 shows the bullet velocity.
The kink in the curves corresponds to the point at which combustion ends.

that a short barrel and a light bullet require degressive powder, whereas a long barrel and a heavy bullet require progressive powder.

Heavy bullets accelerate more slowly, which means that the gas volume initially rises slowly. If large amounts of gas are produced at the beginning of the process, maximum pressures will be high. In the case of lighter bullets, the volume available to the gas will increase rapidly, which means that large quantities of gas are required at an early stage. If the barrel is longer, the bullet will remain in it for longer. This allows the production of gas to be delayed somewhat by comparison with short-barrelled weapons.

2.3.2.4 Interior ballistics calculations

Despite the exceptional complexity of interior ballistics processes, mathematical models can be used to calculate the relevant physical data (maximum pressure, barrel pressure as the bullet passes through the muzzle, muzzle velocity, time from primer firing to muzzle exit, etc.) to a sufficient degree of precision. These calculations are based on the thermodynamic model of interior ballistics, which in turn is based on the basic laws of physics (conservation of energy and equations of motion) and on the law of powder combustion (see 2.3.2.2).

Factors of relevance to forensics and wound ballistics include the drop in muzzle velocity and increase in muzzle pressure that occur when a barrel is shortened, as sometimes happens (see for instance GROSSE PERDEKAMP et al. 2006).

2.3.2.5 Energy balance

From an energy point of view, firearms are heat engines with a surprisingly high degree of efficiency: 30–40 %. The following types of energy are involved:

Kinetic energy

- The bullet's flight energy and energy of rotation
- The energy of motion of the gases
- The recoil energy of the weapon

Thermal energy

- The heat transferred to the cartridge case and the weapon on firing
- Friction between the bullet and the bore
- The internal energy of the gases

Please see Table 2-11 for an example of the approximate percentages of each type of energy in the case of a 308 Win. (7.62 mm NATO) cartridge.

2.3.3 Muzzle phenomena

2.3.3.1 Muzzle gas flow

Gases already escape from the muzzle before the bullet leaves the barrel. These consist in part of the air column pushed out of the barrel by the bullet and in part

Table 2-11. Energy balance. 7.62 mm NATO (308 Win.)
$v_0 = 830$ m/s, m = 9.5 g, $m_c = 3$ g, $Q_{ex} = 3200$ J/g

Form of energy	[J]	[%]
Translational energy of the bullet	3270	34
Rotational energy of the bullet	9	–
Kinetic energy of the gases	1640	17
Recoil energy of the weapon	13	–
Thermal energy of the weapon	approx. 1150	12
Residual energy	approx. 3520	37
Total	9600	100

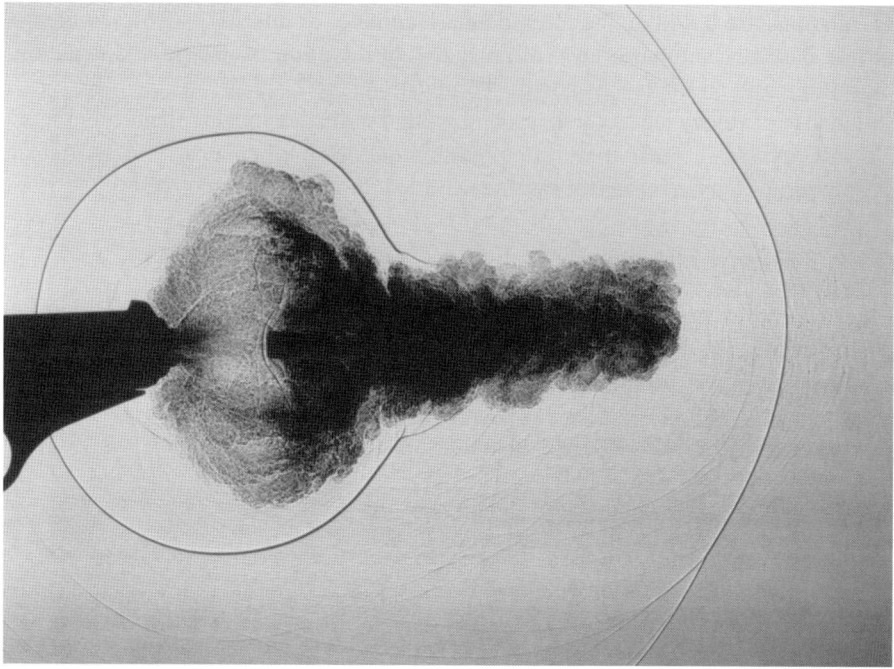

Fig. 2-47. Shadowgraph image of a bullet leaving a barrel (exposure time approx. 1 μs). The bullet has just left the muzzle and the combustion gases are flowing over it from behind. The dark lines are shock waves, which are perceived as muzzle blast.

of combustion gases that escape past the bullet and overtake it, as the bullet does not make a complete seal with the bore.

After the bullet has left the barrel, the gases are still under considerable pressure (several hundred bar; see Table 2-12). They expand, accelerating rapidly. Their velocity then exceeds that of the bullet and they flow past and around it. The flow surrounding the bullet is not symmetrical, so in addition to imparting some slight additional acceleration to the bullet, the flow subjects it to irregular lateral forces, causing it to oscillate about its centre of gravity. This effect is particularly marked in the case of long bullets and those with a long boattail. After a few tens of centimetres the gases have slowed to the point at which the bullet overtakes them again. Fig. 2-47 shows a shadowgraph of a pistol muzzle just after a bullet has left it.

The information in 2.1.5 makes it possible to make a rough calculation of the exhaust velocity and energy of the combustion gases, if required.

2.3.3.2 Flash

In addition to gases, flash is visible at the muzzle. This comes from two sources. In the case of a short-barrelled weapon, the bullet often leaves the barrel before

Table 2-12. Typical gas pressures at the muzzle

Calibre	Barrel length [mm]	v [m/s]	Muzzle pressure [bar]
9 mm Luger	120	350	155
38 Special	51	230	365
38 Special	102	265	160
44 Rem. Mag.	102	410	615
44 Rem. Mag.	152	440	350
5.56 × 45 mm (223 Rem.)	450	960	660
7.62 × 51 mm (308 Win.)	500	830	470

the gas has finished burning, and the burning powder carried by the gases causes *primary flash*. Additionally, the gases following the bullet can ignite on contact with the oxygen contained in the air. This is known as *secondary flash* and always occurs a short distance in front of the muzzle.

2.3.4 Exterior ballistics

2.3.4.1 General; terms used

Terminal ballistics – and especially wound ballistics – is determined to a large extent by the state of movement and orientation of the bullet at the time of impact. The bullet's state of movement is defined by its impact velocity and the speed at which it is rotating about its longitudinal and lateral axes. Its orientation is defined by its angle of attack and its angle of incidence.

The angle of attack is the angle between the direction in which the bullet's centre of gravity is moving and the target surface. The angle of incidence is the angle between the direction of movement and the axis of the bullet (see Fig. 2-48).

However, these values are not immediately accessible. They must be calculated from the basic ballistic data for the bullet in question (mass, moments of inertia and form factors) and the weapon (muzzle velocity and twist length) for the distance between the weapon and the target. Simple mathematical models allow us to calculate both the impact velocity and the angle of attack. Speed of rotation and

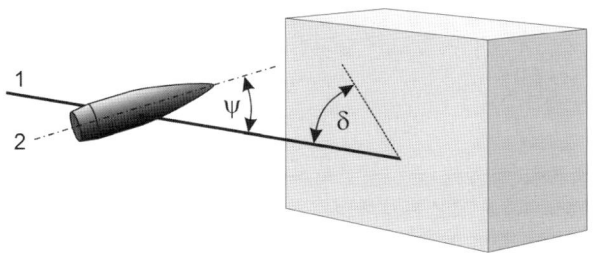

Fig. 2-48. Angle of incidence and angle of attack. *1*: Direction of flight. *2*: Axis of bullet. ψ: Angle of incidence. δ: Angle of attack.

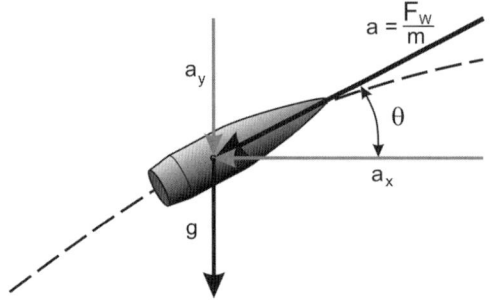

Fig. 2-49. Acceleration acting on a bullet:
a: Due to air resistance,
g: Due to gravity.

angle of incidence cannot readily be determined, which means that we must often manage without these values.

2.3.4.2 Exterior ballistics calculations

In order to calculate the change in velocity and the trajectory from the initial values, we treat the bullet as a point mass. The motion of the bullet is determined entirely by three forces: air resistance (drag) and weight (see Fig. 2-49), plus a lateral force component. Drag acts in the opposite direction to the motion of the bullet and hence also in the opposition direction to the bullet's velocity. The angle between the horizontal and the direction of flight is the trajectory angle, θ. From Fig. 2-49 and 2.1.3.6, we obtain the following formulas for acceleration (one formula for each of the three axes):

$$(2.3{:}2\mathrm{a}) \qquad \dot{v}_x = -a_x = -\frac{F_w}{m} \cdot \frac{v_x}{v}, \qquad [\mathrm{m/s^2}]$$

$$(2.3{:}2\mathrm{b}) \qquad \dot{v}_y = -a_y - g = -\frac{F_w}{m} \cdot \frac{v_y}{v} - g, \qquad [\mathrm{m/s^2}]$$

$$(2.3{:}2\mathrm{c}) \qquad \dot{v}_z = -a_z + a_d = -\frac{F_w}{m} \cdot \frac{v_z}{v} + a_d, \qquad [\mathrm{m/s^2}]$$

where F_w is the air resistance (or drag), calculated according to Eqn 2.1:64a and a_d is the possible lateral acceleration.

What is not immediately apparent from Equations 2.3:2a-c is the effect of height y. The resistance of the air F_w is directly related to air density and indirectly to temperature, both of which change with height ($\rho = \rho(y)$). The effect of height means that the system of equations to determine change in velocity cannot be solved analytically; they can only be solved numerically, step by step. This requires computer software.[1] If we ignore height (which is legitimate in the case of flat-trajectory weapons) then we can derive approximation equations from

---

[1] e. g. k-ballistics-Software 4, see Annex A.8.

Equations 2.3:2a-c that allow us to calculate trajectory data directly (KNEUBUEHL 1998b).

The question is therefore under which conditions we need to carry out exterior ballistics calculations for forensic investigations. The first question is whether we are trying to calculate point of impact, speed or energy. At short ranges (up to about 10 m), we can treat the trajectory of bullets with velocities in excess of 200 m/s as straight without introducing major errors. For light bullets, however, we will need to calculate the velocity even at this range.

A light, solid bullet, such as the 9 mm Luger THV (see Fig. 2-15 c) loses approximately 70 m/s of its velocity (approx. 90 J of its energy) in the first 10 m. Light, fast-moving fragments can lose up to 80 m/s per metre at the start of their trajectory.

Such calculations become essential if we are attempting to identify the origins of stray rounds, which may involve reconstructing trajectories of hundreds of metres or even a number of kilometres. Ballistics tables are useful in making first approximations, but they take no account of meteorological conditions.

2.3.4.3 Ballistics tables

The exterior ballistics data of bullets are presented in the form of ballistics tables. These tables show the relevant ballistic characteristics (see Fig. 2-50) as a function of distance, the values being calculated using software of the type mentioned in 2.3.4.2. Annex A.8 contains ballistics tables for the most common types of bullet, compiled under International Standard Atmosphere (ICAO-Atmosphere) sea level conditions (air density $\rho = 1.225$ kg/m^3) with a 10 m/s crosswind. As there is a linear relationship between lateral deflection and wind velocity, it is a simple matter to calculate deflections for other wind velocities.

2.3.4.4 Proper motion of a bullet

Wound ballistics is influenced not only by impact speed and energy but also by the proper motion of the bullet. Proper motion includes the speed at which the bullet rotates about its longitudinal axis (which determines stability), the angle of

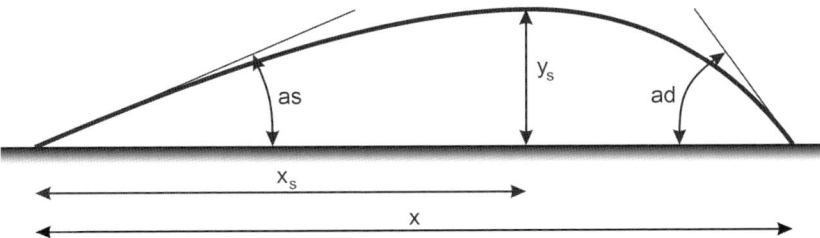

Fig. 2-50. Trajectory data used in ballistics tables: x = range, as = angle of shot, ad = angle of descent, x_s = vertex distance, y_s = vertex height.

74   2 Basics

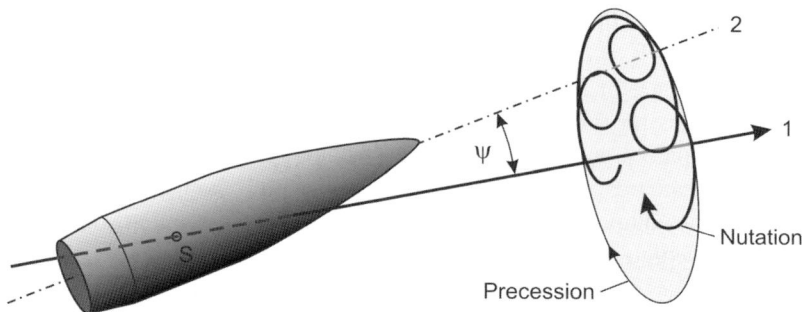

Fig. 2-51. Proper movement of a spin-stabilized bullet: 1: Direction of flight. 2: Axis of bullet. ψ: Angle of incidence. The axis of the bullet rotates about the direction of flight (precession) and the angle of incidence varies between a minimum and a maximum (nutation). If the bullet is spinning clockwise the movement is to the right. If it is spinning anti-clockwise, the movement is to the left.

incidence (see Fig. 2-48) and the pitch speed (the speed at which the bullet oscillates about a lateral axis).

A bullet fired from a rifled barrel rotates about its longitudinal axis at high speed. In terms of physics, the bullet behaves as a gyroscope, in that under the influence of a constant external torque (due to drag acting upon it at a point other than its centre of gravity) it undergoes *precession*. This means that the axis of the bullet describes a cone, of which the point lies at the tip of the bullet and of which the axis coincides with the bullet's direction of flight. If the bullet is disturbed briefly (e.g. by the combustion gases as it passes through the muzzle, or by contact with a fixed object) a further movement occurs, known as *nutation*, which is superimposed upon precession and generally decays slowly (see Fig. 2-51). The bullet therefore moves about its centre of gravity throughout its flight.

2.3.4.5 Disturbances to the trajectory

Wind. A sidewind causes lateral deflection, of which the magnitude is proportional to the wind velocity v_w and the time of flight. Precise calculations are possible if we substitute the relative velocity $(v - v_w)$ for the bullet velocity in the differential equations for ballistics.

Rain. Rain has two effects on the trajectory. First, passing through a raindrop slows the bullet and hence increases the time of flight. Second, the moment acquired from the raindrop deflects the bullet downwards, albeit slightly. At long ranges, these two effects will cause the point of impact to be perceptibly lower. Nutation will not increase significantly, as the mass of a raindrop is much smaller than that of a bullet.

Impacts. If a spin-stabilized bullet encounters an obstruction on its trajectory, it will be subjected to a moment and a torque. The moment will modify the bullet's speed and direction. The torque will modify the magnitude of the bullet's torque (thereby altering its angular velocity) and the direction in which the torque acts, such that it no longer coincides with the axis of rotation. The bullet will react by undergoing nutation (see Fig. 2-51), with the axis of rotation describing a cone around the direction of torque. The accompanying increase in aerodynamic forces and moments acting on the bullet causes its precession to become more marked, increasing the angle of incidence and, in many cases, causing it to tumble ("keyhole") (KNEUBUEHL 1999c).

2.3.5 Stability and tractability

2.3.5.1 Definition of stability

Non-ballisticians often misunderstand stability as it applies to wound ballistics, there being a tendency to describe as "unstable" a bullet that is merely exhibiting the usual phenomena of precession and nutation (oscillation). At the same time, stability is one of the factors with the greatest influence on the wounding potential of a bullet, so it is worth looking at the topic in more detail.

Generally speaking, every physical system has a normal rest position. This state of repose corresponds to a particular equilibrium of forces, and the system will remain in this condition unless that equilibrium is disturbed. If an outside force acts on the system, causing movement, the system can react in three different ways:

- If the induced motion becomes less pronounced with time, the system is said to be *stable*. In such a case, there will be an opposing force (generally induced by the outside force).
- If the movement becomes more pronounced with time, the system is said to be *unstable*. In such cases forces are often present that amplify the disturbance.
- If the movement continues, becoming neither more nor less pronounced with time, the system is said to be *neutral*. When the source of the disturbance is removed, the system enters a new state of repose.

These three cases can be illustrated by a ball. In the first case, it lies on the con-

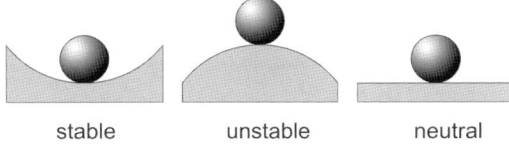

stable   unstable   neutral

Fig. 2-52. The behaviour of a ball when subjected to a small deflection on different surfaces illustrates the difference between stability and instability.

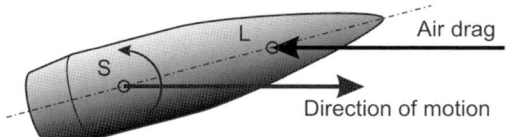

Fig. 2-53. The position of the centre of pressure (L) and centre of gravity (S) of a spin-stabilized bullet. Air drag causes the bullet to rotate about its centre of gravity.

cave side of a hemisphere, in the second case on the convex side of a hemisphere and in the third case on a plane surface (see Fig. 2-52).

The state of repose for a bullet in flight is that in which its longitudinal axis coincides with the direction of flight (the tangent to the trajectory). The centre of pressure (the point on the bullet on which air pressure is acting) always lies somewhere between the centre of gravity and the tip of the bullet. As a result, a torque (or overturning moment) acts on the bullet, turning its axis away from the rest position (see Fig. 2-53). The angle between the direction of flight and the axis of the bullet is the *angle of incidence*.

It is only possible to achieve stable flight – i.e. a situation in which the angle of incidence becomes smaller as the bullet progresses – if the overturning moment is counteracted by a stabilizing moment. There are two quite different ways of generating this stabilizing moment:

- Cause the bullet to rotate about its longitudinal axis at high speed, creating a gyroscope. The gyrostatic moment will then stabilize the bullet.
- Design the bullet in such a way that air pressure has a stabilizing effect on it.

To look at stability in more detail, we shall study spin-stabilized and non-spin-stabilized projectiles separately.

2.3.5.2 Spin-stabilized projectiles

As soon as it enters free flight, a spinning projectile behaves like a heavy spinning top. Its proper motion is centred on its centre of gravity, which can be compared to the centre of rotation of a top (see Fig. 2-54). Air pressure induces a torque (overturning moment) about the centre of gravity, which is analogous to the weight of a spinning top inducing a moment about its centre of rotation.

If the angular velocity is high (as is the case for a spin-stabilized bullet), the direction of the torque more or less coincides with the axis of the projectile. The direction of change in torque is the same as the direction of the moment, which is to say perpendicular both to air pressure and to the axis of the projectile (see 2.1.3.4). The torque (and hence the axis of the projectile) therefore describes a circle about the direction of flight – the phenomenon of precession mentioned earlier. This stabilizing gyrostatic moment prevents the axis of the projectile from tipping.

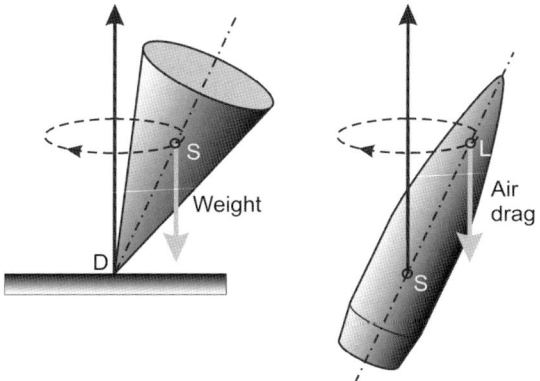

Fig. 2-54. Analogy between a spinning top and a bullet in free flight.
S: Centre of gravity,
D: Centre of rotation of the top,
L: Centre of pressure.

A spin-stabilized projectile is said to be stable if the ratio of the stabilizing gyrostatic moment M_S to the moment of the air pressure (the overturning moment) M_L is greater than 1:

(2.3:3) $$s = \frac{M_S}{M_L} > 1,$$ [-]

where s is the *gyroscopic stability number* of the bullet. Equation 2.3.3 is an essential precondition for stable flight; it ensures that the angle of incidence does not increase during flight. In practice, values of s of between 1.3 and 1.6 are considered desirable, depending on whether the bullet is intended for flat or steep trajectories.

The angle of incidence will only decrease during flight if the Molitz stability equation is also fulfilled (see KNEUBUEHL 1982).

Spin-stabilized bullets execute a decaying precession motion, on which nutation is superimposed in most cases as a result of disturbances (see Fig. 2-51). At close range, this can have a major influence on the behaviour of the bullet as it enters the body.

Under normal circumstances, the angle of incidence due to precession and nutation may amount to several degrees just after the bullet leaves the muzzle. The nutation component falls almost to zero during the first 100 m. From that point on, the angle of incidence is almost equal to the precession angle, which is determined by air pressure. This angle becomes smaller during flight, as air pressure reduces with speed. As the bullet's speed of rotation (and hence the stabilizing gyroscopic force) only decreases slightly, the gyroscopic stability of the bullet increases with distance from the muzzle. The bullet therefore becomes increasingly stable along its trajectory. Experiments with soap blocks confirm this in qualitative terms, as the length of the narrow channel increases with distance from the muzzle (see Fig. 3-23).

There are exceptions to this rule, however. As the speed of the bullet decreases, the airflow around it alters. The centre of pressure moves towards the tip, increasing the overturning moment. In most cases, this effect is negligible by comparison with the increase in gyroscopic stability. However, the overturning moment of certain bullets increases markedly around Mach 0.9 (approx. 300 m/s) due to the way they are affected by flow. It is perfectly possible for such bullets to become unstable after several hundred metres of stable flight.

Bullets that become unstable as a result of striking an object (e.g. ricochets) only remain unstable for a few hundred metres. The angular momentum is altered by the impact and then remains virtually constant at its new value, allowing stable gyroscopic motion to resume after a certain time. It is possible that the bullet, while stable, will continue its flight tail-first.

2.3.5.3 Projectiles stabilized by air forces

If a projectile is fired with little or no spin, the stabilizing moment must be provided by air force. In such cases, the relative positions of the centre of pressure (L) and centre of gravity (S) play a decisive role. If L lies in front of S, the moment produced will destabilize the projectile, as we have already seen in the section on spin-stabilized bullets. However, if L lies behind S, the air pressure has a stabilizing effect (this is the principle by which a weathervane operates: see Fig. 2-55). Stabilizing a projectile without using spin therefore involves designing it in such a way as to place the centre of pressure behind the centre of gravity. This can be achieved as follows:

- By so distributing the mass as to move the centre of gravity forward (*arrow stabilization*, see Fig. 2-56 a).
- By applying drag to the tail, thereby moving the centre of pressure towards the rear. This can be achieved in either of two ways:
    - by increasing the lift force at the tail (*fin stabilization*, see Fig. 2-56 b);
    - by increasing the drag at the tail (*drag stabilization*, see Fig. 2-56 c).

Fin stabilization is the most widely-used method, and the easiest to control from a technical point of view. This technique has been introduced for small arms ammunition in recent years, in the form of flechettes.

2.3.5.4 Shoulder stabilization

Another type of stabilization comes into play in the case of projectiles with a larger frontal surface area, regardless of whether the projectile is spinning or not. At

Fig. 2-56. Stabilizing a projectile without the use of spin: **a.** Arrow stabilization, **b.** Fin stabilization, **c.** Drag stabilization.

Fig. 2-57. Shoulder stabilization. **a.** The resultant of the asymmetric distribution of pressure on the face of the projectile generates a stabilizing moment. **b.** At larger angles of incidence, this moment has a destabilizing effect (S = centre of gravity).

small angles of incidence, the asymmetric distribution of pressure on the face of the projectile due to stagnation pressure generates a stabilizing moment (see Fig. 2-57 a). Normally, the flow separates on the edge of the face. As a result, there are no compressive forces acting on the rest of the bullet. At larger angles of incidence this moment has a destabilizing effect (see Fig. 2-57 b), with the angle of transition depending on the diameter of the face and its distance from the centre of gravity. This type of stabilization plays an important role in the behaviour of a bullet in a dense medium, as stagnation pressure, and hence the stabilizing moment, can be fairly large. This provides a plausible explanation for many phenomena.

2.3.5.5 Tractability

The question of overstabilization of rifle bullets is frequently raised in wound ballistics. It is often claimed that the axis of a bullet with a high stability number does not follow the tangent to the trajectory and therefore presents a higher angle of incidence on the descending branch of the trajectory.

In stability theory, the ability of a bullet to keep its axis aligned with the tangent to the trajectory is known as *tractability*.

Drag moment causes the axis of the bullet to describe a cone around the tangent to the trajectory (precession, see 2.3.5.2). If the direction of the tangent to the trajectory only changes slowly, and if the angular momentum is not too great, the axis of the bullet will follow the tangent to the trajectory.

The more tractable a bullet, the smaller the angle between the tangent to the trajectory and the axis of the bullet at all points along the trajectory. This is the case if the change in the angle of precession is at all times greater than the change in the trajectory angle, i.e. if:

(2.3:4) $$f = \left| \frac{d\psi/dt}{d\theta/dt} \right| > 1 ,$$ [-]

where f is the tractability number.

It is therefore clear (and can of course be proved mathematically) that tractability is inversely proportional to stability. Excessive stability can therefore lead to

Table 2-13. Extract from ballistics table, showing stability and tractability numbers. Full metal-jacketed bullet, 7.62 mm NATO, mass: 9.5 g, v_0: 830 m/s

x [m]	v [m/s]	ω [rad/s]	s [–]	f [–]
0	830	2729	1.37	34021
50	791	2711	1.49	29649
100	754	2692	1.62	25837
150	718	2673	1.76	22485
200	683	2654	1.92	19519
250	649	2634	2.09	16883
300	616	2614	2.29	14538

tractability problems. Having said that, critical values only arise in the vicinity of the apex of a steep trajectory. The velocity of the projectile is low at that point (often less than 100 m/s) and it is therefore extremely stable. At the same time, the trajectory angle is changing rapidly, and the tractability number becomes small. Trajectory calculations that include stability and tractability reveal that critical values only arise if the projectile is fired at an elevation of more than 70°.

These calculations are confirmed by a forensic case of which the author has direct experience. A man fired a rifle into the air at an outdoor event and a girl was hit by the falling bullet 1.4 km away (angle of attack approx. 85°). The bullet entered tip first, causing a wound of which the severity was such that it could only have been caused by a bullet in stable flight. As the man came forward it was possible to fully reconstruct the trajectory, knowing his position at the time, that of the victim and the meteorological conditions obtaining. The bullet was fired at an angle of 69.2° and reached a vertex height of approx. 1.8 km.

In the case of a weapon with a flat trajectory, the bullet is always tractable. There is no possibility of overstabilization leading to tractability problems (i.e. of f approaching 1). See data in Table 2-13.

2.3.5.6 Stability and ricochets

Only by carrying out complex calculations is it possible to say how large an impact a bullet can undergo and still be able to recover stability in the sense described in 2.3.5.1.

Many years of experience indicate, however, that normal bullets become unstable if the angle of incidence exceeds approximately 15°.

The exact angle naturally depends on the design of the bullet, and the speed of rotation will be an important factor. The figure of 15° assumes a normal design of bullet and a gyroscopic stability number of between 1.5 and 2.

For a first approximation, however, we can assume that the angle of incidence immediately after the impact is equal to the nutation angle and ignore the precession component. Under these assumptions, we can establish a relationship be-

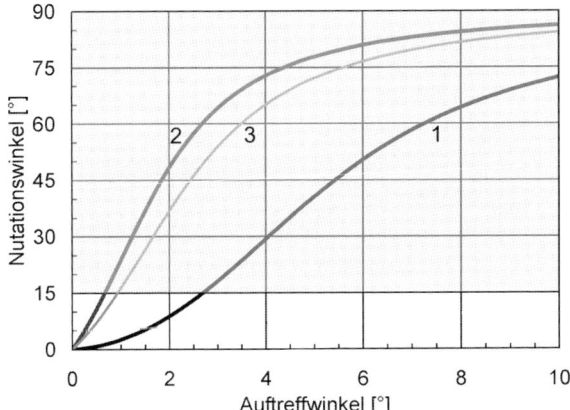

Fig. 2-58. Relationship between angle of attack and nutation angle (it being assumed that the angle of incidence after impact is very similar to the nutation angle). 1: 9 mm Luger FMJ RN, 2: 7.62 mm NATO FMJ spitzer, with twist length of 305 mm, 3: 7.62 mm NATO FMJ spitzer, with twist length of 203 mm.

tween the angle of impact and the angle of incidence immediately after the impact, taking into account bullet geometry and angular velocity (KNEUBUEHL 1999c). This approach has been used to calculate the relationship between angle of attack and nutation angle for a handgun bullet (9 mm Luger FMJ RN) and a rifle bullet (7.62 mm NATO). The results correspond closely to those observed in practice. See Fig. 2-58.

From the graph, we can see that the 9 mm Luger bullet can resume stable flight after impact if the angle of attack does not exceed approx. 2.5°. A 7.62 mm bullet will fail to recover stability even at an angle of attack of less than 1°. It is interesting to note the influence of rotation speed in the case of the rifle bullet; the higher rotation speed (weapon with a twist length of 203 mm as opposed to 305 mm) delays the onset of destabilization considerably.

2.3.6 Fragment ballistics

2.3.6.1 Acceleration of fragments

Fragments are generally accelerated using explosives. Their initial velocity (and hence their initial energy) therefore depends on the efficiency with which the energy in the explosive is utilized.

Fragments may also be created when a bullet chips pieces out of a hard material or when the bullet itself breaks up. In such cases the fragments acquire most of their energy from a transfer of momentum.

The initial velocity of a fragment accelerated by an explosive can be calculated fairly accurately if the geometry of the casing and the explosive are known. This velocity is calculated using the GURNEY formula, which uses an explosive constant and the ratio of the mass of explosive to the mass of fragmentation material (see KNEUBUEHL 1999b for a comprehensive description of the derivation and use of this formula):

(2.3:5) $$v_{Sp} = \frac{G_K}{\sqrt{\left(\dfrac{m_f}{m_c} + C_G\right)}}, \quad [m/s]$$

where G_K is the GURNEY constant (see Table 2-14), m_f is the mass of the fragments accelerated m_c is the mass of the explosive and C_G is a form factor (see Table 2-15).

Fragments with a significant effect on the human body have a mass of between a few hundredths of a gram and a few tens of grams. They may have an initial velocity of over 2000 m/s, and their energy may therefore range from less than 1 J to 60 J.

In terms of shape, fragments can be divided into two basic categories:

- *Natural fragments* are created on detonation by the destruction of the body of the projectile. They generally consist of steel. Copper, aluminium or plastic fragments may also be present, depending on the device.
- *Prefabricated fragments* are embedded in a base material. Such fragments are generally spherical or cuboid, and are usually made of steel or tungsten.

2.3.6.2 Exterior ballistics of fragments

In forensics, we are usually dealing with fragments that cover short distances, e.g. fragments from a hand-grenade in a pub or a bomb in a department store. As the initial velocity of the fragment is generally high, it has a short flight time and loses little height. It is therefore reasonable to assume the trajectory of a fragment to be a straight line in such cases.

For instance, a small cuboid fragment with a mass of 0.1 g and an initial velocity of 1000 m/s loses only 10 mm in height under the influence of gravity over a distance of 25 m.

For greater distances, such as those involved when calculating danger zones on shooting ranges, assessing hazards and reconstructing incidents (e.g. car bombs), it will be necessary to calculate the trajectory using the equations of motion (see 2.3.4.2), as the trajectory of a fragment will be markedly asymmetrical over longer distances, with the apex situated well after the half-way point.

Table 2-14. Gurney constant G_K for a number of common explosives

Explosive	Density [kg/m³]	G_K [m/s]
Hexogen (RDX)	1.77	2830
Octogen (HMX)	1.89	2970
Trinitrotoluene (TNT)	1.63	2370
Black powder[a]	1.30	950

[a] Used as an explosive

Table 2-15. Constant C_G for the Gurney formula

Shape of fragment	C_G
Sphere	$3/5$
Cylinder	$1/2$
Symmetrical sandwich	$1/3$

A fragment will hardly ever maintain its initial orientation all the way to the target. It is reasonable to suppose that the lines of the accelerating forces will not pass through its centre of gravity. The fragment therefore undergoes torque, causing it to rotate about an axis, and this axis will generally not be fixed. In calculating the sectional density, we generally take the mean face area of the fragment which, in the case of a convex shape, corresponds to ¼ of the total surface.

We use the drag coefficient for either a sphere, a cube or a cylinder, depending on the shape of the fragment. Selecting the appropriate form does require a certain amount of experience.

There is a tendency to grossly underestimate the deceleration that a fragment undergoes along its trajectory. As a rule of thumb, one can say that a 0.1 g cuboid fragment loses about $^1/_{20}$ of its velocity per metre and a 1 g fragment approximately $^1/_{40}$.

2.3.7 Terminal ballistics models

2.3.7.1 General

From a physical point of view, an impact between a bullet and a solid (or liquid) object is a highly complex process. It takes place over a very short timespan and involves large forces and pressures. As deformation of the bullet and of the target will inevitably occur, the dynamic characteristics of the materials will play an important role.

The behaviour of a material under large loads of short duration differs from that of the same material under static or quasistatic loading. Materials are often stronger under dynamic load.

A number of penetration models have been developed on the basis of observation and experience, of which two are quite useful in understanding the processes associated with handgun and rifle bullets.

2.3.7.2 The plugging model

The plugging model is useful in the case of thin layers of material where the bullet removes material as it passes through but causes little deformation. Here, it is assumed that the bullet removes a disc or plug from a sheet of the material.

The shear work can be derived from the shear stress and the area involved, and equated to the energy used by the bullet, E_s. We obtain the following equation:

(2.3:6) $$E_S = C_S \cdot k \cdot D^2 ,$$ [J]

where k is the calibre, D the thickness of the plate and C_S a material-specific shear constant.

Rearranging and extending Eqn 2.3:6, we obtain an equation for the thickness penetrated as a function of energy density E´:

(2.3:7) $$D = \frac{\pi}{4 \cdot C_S} \cdot \sqrt{E' \cdot k} .$$ [m]

The thickness penetrated is therefore proportional to the square root of energy density times calibre. For a given energy density and bullet design, a larger-calibre bullet has better penetration characteristics.

2.3.7.3 The displacement model (ductile failure)

If the target is capable of undergoing deformation, it is reasonable to assume that the bullet will displace material as it passes through. According to MARTEL's theory, the volume displaced will be proportional to the energy used, E_V.

The validity of this rule can be demonstrated easily and accurately by empirical means for materials capable of undergoing plastic deformation (plasticine, soap, etc.).

If the target is in the form of a sheet, one can assume as a first approximation that the volume displaced will correspond to the volume of the hole created by the bullet:

$$(2.3:8) \qquad E_V = C_V \cdot \frac{\pi}{4} \cdot k^2 \cdot D = C_V \cdot A \cdot D, \qquad [J]$$

where k is the calibre, D the thickness of the plate, C_V a proportionality factor associated with the material and A the cross-sectional area of the bullet.

If we divide Eqn 2.3:8 by the calibre cross-sectional area, we obtain the relationship between thickness penetrated and energy density:

$$(2.3:9) \qquad D = \frac{1}{C_V} \cdot \frac{E_V}{A} = \frac{1}{C_V} \cdot E' = \frac{1}{C_V} \cdot \frac{q \cdot v^2}{2}. \qquad [m]$$

It is interesting to note that the thickness penetrated depends not on the calibre of the bullet but on its sectional density, q.

The displacement model is valid for most materials that are capable of deformation. In practice, we see that the direct relationship between energy density and penetration capacity holds good for bullets that behave in a similar manner in the target material.

2.3.7.4 Bullet passing through a thin layer of material

When a bullet passes through a hard material, especially in the form of a thin sheet, something surprising often occurs. Let us say that a sheet of a material just stops a bullet with impact energy E_{gr}. At a slightly higher impact energy E_a the bullet passes through the sheet. The residual energy behind the sheet is E_{rst}. The energy E_{ds} required to perforate the sheet is therefore:

$$(2.3:10) \qquad E_{ds} = E_a - E_{rst}, \qquad [J]$$

and, surprisingly,

(2.3:11) $$E_{ds} < E_{gr}. \qquad [J]$$

In other words, the sheet absorbs less energy when the bullet passes through than when it remains in the sheet. This behaviour can be explained as follows: when the bullet passes through the sheet, a rupture line forms around the channel created by the bullet, preventing further energy being transferred from the bullet to the sheet. As a result, the bullet retains more residual energy. The faster the rupture line forms, the less energy the sheet is able to absorb. This is indeed what happens. Increasing the impact energy reduces the energy required for the bullet to pass through the material.

3 General wound ballistics
B. P. KNEUBUEHL

3.1 Introduction

3.1.1 General

Wound ballistics is the sub-domain of terminal ballistics that addresses the behaviour and effects of a bullet in a person or an animal.

In this book, we shall divide wound ballistics into three areas: handgun bullets, bullets from long weapons and fragments. The logic behind this is that there are fundamental differences between the three types of projectile in terms of form, structure and energy. These differences affect their behaviour in the target and the formation of the wound channel.

The determining factor is not the weapon, but the ammunition. For instance, a 22 L.R. is classed as a handgun cartridge, even when fired from a rifle, because of its form and the energy it carries. Conversely, a 223 Rem. remains a rifle cartridge, even when fired from a Contender pistol.

The most significant factors with regard to rifle bullets are the temporary cavity and the effect it has on the body. Rifle bullets carry some 1500 to 4500 J of energy – more than enough to cause serious, life-threatening injury.

Handgun bullets are generally more compact (no more than two calibres long) and carry significantly less energy (250 to 750 J). However, the most powerful of handgun bullets deliver more energy than the least powerful rifle bullets – a 454 Casull handgun round delivers 2500 J, while a 222 Rem. rifle bullet carries only 1200 J, for instance. Handgun ammunition with high energy levels is extremely rare, however.

Fragments behave very differently to bullets in the body. Bullets in unstable flight, and those with a large angle of incidence, behave in a manner similar to fragments. We shall therefore discuss such projectiles in a separate section.

Wound ballistics is first and foremost an empirical science, based on experience and experiment. Knowledge is acquired by conducting test firings, observing the results and formulating explanations. Naturally, one also attempts to model the experimental results using mathematics and physics, in order to derive quantitative rules and make predictions.

Section 3.3 discusses the materials used for simulation shots. These materials must reproduce as closely as possible the behaviour of biological tissue subjected to bullet wounds. Section 3.4 examines other ways of simulating bullet wounds.

3.1.2 The history of wound ballistics

In the present context, we can only consider the history of wound ballistics in the narrower sense. The development of techniques for treating bullet wounds (i.e. war surgery) lies outside the scope of the present book, if only for reasons of space. Not surprisingly, however, military surgeons were the first to examine the effects of bullets. They were effectively studying wound ballistics, even when the term did not exist.

There were (and still are) two main reasons for studying wound ballistics. On the one hand there were the doctors, who studied the effects of bullets in order to develop improved methods of treatment and – at another level – to warn the armed forces against producing bullets of which the effects would contravene the principle of proportionality. On the other hand, there were the armed forces, who observed the effects of bullets and wished to achieve maximum effect with minimum (technical) effort.

As early as 1830, DUPUYTREN and his students (cited in BIRCHER 1896) conducted experiments using cloth, wood and cadavers, comparing the results with observations from the field.

Later, the main point of interest was the effect on men and horses of the lead balls in use at the time, horses being the soldier's constant companion on the battlefield, either as a mount or as a means of moving artillery. Lead balls of various calibres were fired from rifles or projected in the form of shrapnel shells, which carried a large number of balls and ejected them on detonation. The shrapnel shell was invented by Lieutenant (later Major-General) SHRAPNEL of the British Army from 1784 onwards. The world's armies wanted to know the energy such balls required in order to achieve a given effect.

In the second half of the 19th Century, all armies used solid lead round-nosed bullets weighing between 20 and 25 g, with initial velocities of 400 to 450 m/s. Examples include the Vetterli rifle in Switzerland and Italy, the Springfield 45-70 in the USA, the Chassepot in France and the Dreyse needle gun in Germany.

At the time, there were three main theories explaining the effects of solid lead bullets:

- partial melting of the bullet on impact;
- centrifugal force on tissue due to the rotation of the bullet;
- hydraulic pressure.

BUSCH in Bonn (1873, cited in KOCHER) was the first to show that a bullet fired at short range by a Chassepot rifle did not leave a clean hole but rather had a "substantial crushing effect" on, for instance, the epiphyses (ends) of the large hollow bones. He ascribed this effect to molten lead particles "spraying" on impact due to the sudden transfer of so much energy. In a later work (BUSCH 1874) the author concluded on the basis of his experiments that, particularly in the case of a shot to the skull or a diaphysis (the middle section of a hollow bone) the hydraulic pressure generated by the bullet must also play a significant role. This conclusion was prompted by the difference between the effect on an empty skull and that on a skull filled with brain. The ejection of particles from the entry wound was also interpreted as being the effect of hydraulic pressure.

The funnel-shaped defects pointing towards the exit wound, observed in the liver, muscle, etc., were explained as being due to the rotation of the bullet about its longitudinal axis dragging the surrounding tissue with it in a circular movement, the resulting centrifugal force widening the wound channel.

It was at this stage of research into the effects of bullets that KOCHER (1841-1917) arrived on the scene. He laid the foundations for studying the effects of bullets on the body in a scientific manner. His publications in this field appeared between 1874 and 1895.

In 1909, he became the first surgeon to win the Nobel Prize for Medicine, for his work in the fields of thyroid physiology and surgery.

After conducting numerous experiments, and having calculations carried out by physicists and mathematicians, he concluded that neither partial melting of the bullet nor centrifugal force could have a significant effect on tissue, and that hydraulic pressure was the decisive factor.

For instance, KOCHER had physicist FORSTER calculate the velocity (and hence the energy) that a lead ball would need in order for the sudden impact to generate sufficient heat to just bring the entire mass of lead to melting point (i.e. to raise the temperature from 15 °C to 326 °C). The answer was 353 m/s. KOCHER acknowledged that some of the lead could melt on impact with a bone, but stated that this was impossible when a bullet passed through soft tissue. He was also the first to use soap and gelatine for modelling purposes. Because of the mushrooming observed in the lead bullets then in use (a phenomenon that others had already noted) he called for harder, non-deformable bullets to be used, as he had realized that mushrooming produced larger wounds. As a result, the Swiss federal munitions factory in Thun, under the leadership of RUBIN, began to produce a full metal-jacketed bullet in 1880. The new bullet formed part of a cartridge, the other half being a newly patented bottleneck cartridge case.

KOCHER must take most of the credit for the introduction of this jacketed bullet, as it was he who persuaded RUBIN to start producing it (see Fig. 3-1).

By the beginning of the 20th Century, the armies of all the major States had introduced this type of bullet.

KOCHER also recommended a reduction in calibre, to minimize hydraulic pressure, and an increase in rotation speed, to increase stability.

His three-part work "On the explosive effect of modern small-calibre bullets" closes with the following, somewhat gloomy comment:
  "*The only way to keep hydraulic pressure as low as possible is to keep the size of the bullet to*

Fig. 3-1. Prof. KOCHER (seated, wearing a black hat) during firing tests at the Swiss federal munitions factory, Thun.

a minimum, as it is patently impossible to alter the fluid content of the human body (*) and the armed forces will scarcely be willing to forego the propulsive force of the newer weapons. Even with the smallest of bullets, there will be enough of those extensive destructions of bone that are the chief aim wherever differences are settled by force of arms.

(*) Although an empty stomach is still to be recommended."

The footnote refers to the practice of soldiers at the front, who avoided eating and drinking before an attack so as to facilitate treatment in the event of a stomach wound.

In 1896, Swiss Army doctor H. BIRCHER published a paper entitled "New studies on the effects of handguns," together with an atlas (which is still worth consulting) containing over 40 illustrations of bullet deformation and of bullet wounds to various organs. Please see Fig. 3-2 for a small selection of these illustrations. Some ten years earlier, there had been a significant change in the field of ammunition; jacketed bullets had replaced solid lead, calibres had decreased and velocities had increased. The view at the time was that preventing mushrooming by using full metal-jacketed bullets, together with a decrease in calibre, would reduce the destructive effect of bullets on the human body and render them more "humane," insofar as one can ever use this term in connection with bullets. The participants at a medical conference in Rome were therefore astonished to hear Prussian army doctor VON COLER put forward a totally contradictory view. On the basis of numerous experiments, he maintained that the "reputation of this new, humane bullet has been forever destroyed." BIRCHER and his team undertook numerous experiments in an attempt to resolve the differences between the viewpoints.

The results of these experiments – including shots fired into tin cans with a range of different contents – are shown in very revealing drawings, which are included in the atlas mentioned above.

Naturally, discussion of the effects of the "old" and "new" types of ammunition form a large part of his book. In particular, the author points out the explosive effect of the new bullets – the result of hydraulic pressure. Two years later, US Army doctor WOODRUFF published a paper on the explosive effect of modern small-calibre ammunition. This author likewise worked with physicists and mathematicians to conduct a theoretical analysis of the phenomena. In his paper, WOOD-

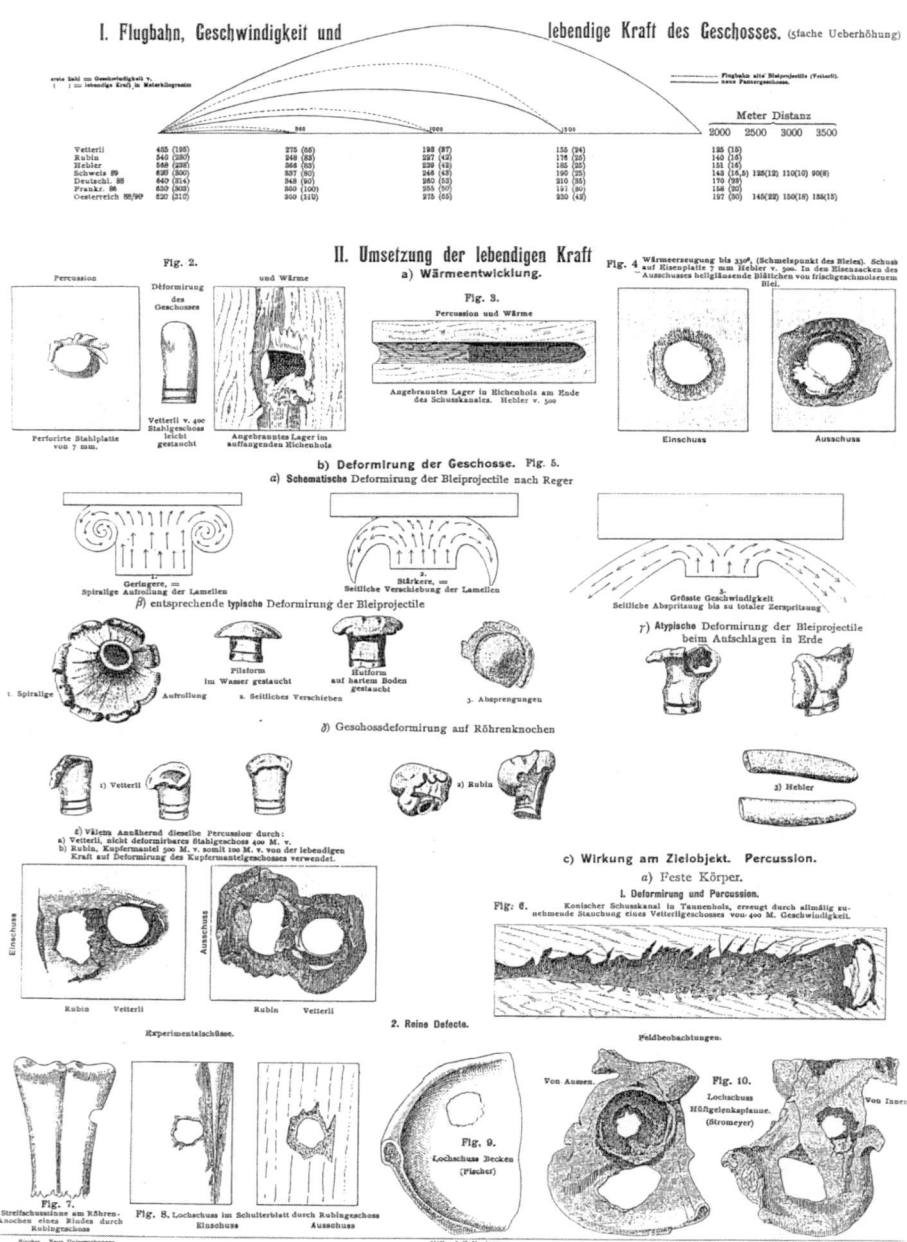

Fig. 3-2. Extract from BIRCHER's illustrations.

RUFF used five images to present the formation of the temporary cavity, albeit without using that term. His pictures would not be out of place in a modern book on the subject. The only mistake is that the bullet is represented as continuing in stable flight after passing through the target. The author fired bullets through empty tomato tins and tins filled with water, dry sand and damp sand. He also used the bladders of recently slaughtered cattle (empty, half-full and full).

The soldiers who fired at the bladders reported seeing them swell up briefly before collapsing.

SIMPSON (1868, cited in JOURNÉE 1907) observed the following: a bullet (of unspecified calibre) that penetrated a piece of wood (fir) to a depth of 0.8 cm caused slight bruising to the skin of a horse. A bullet that penetrated to a depth of 1.5 cm caused a substantial wound, without disabling the horse, and a bullet that penetrated to a depth of 3.0 cm caused a very dangerous wound. JOURNÉE pointed out, quite rightly, that the characteristics of the projectile (ball/bullet, calibre) are not specified. Nevertheless, SIMPSON's experiments appear to be the first published wound ballistics experiments.

QUESNAY (1875) writes in his book that a ball that penetrates fir by 8 mm causes a "simple bruise," whereas a ball that penetrates to twice that depth causes a dangerous wound. Here again, there is no mention of calibre.

Around 1900, experiments in Germany showed that lead balls with a mass of 12.5 g, travelling at 112 m/s (energy approx. 80 J) could cause serious injuries. ROHNE (1906) and PANGHER (1909) also published on this subject.

The work by JOURNÉE (1907) was among the most significant of the early 1900s, covering the effects of lead balls, rifle bullets and shrapnel balls.

The publications cited above focused on the relationship between penetration depth and the damage that a projectile causes to skin, muscle and bone, and the relationship between penetration depth and calibre, mass and velocity. The work of LAGARDE (discussed in detail in HATCHER (1935)) looked at the effect on the person, and more particularly at the question of how handgun bullets should be designed in order to be as effective as possible. In other words, what bullet characteristics yielded maximum "stopping power."

At the turn of the century, the US armed forces were using a 38 revolver that fired 38 Long Colt cartridges (bullet data: m = 10 g, v = 230 m/s, E = 260 J). Because of the cartridge's unsatisfactory performance in a number of theatres (including the Philippines), the US War Department set up a commission in 1904 tasked with choosing the most effective cartridge/handgun combination from among those available at the time.

The task was given to Colonel LAGARDE, assisted by Colonel THOMPSON (inventor of the eponymous sub-machine gun). Firing was conducted at close range (approx. 1 m) using the then common cartridges listed in Table 3-1. The targets consisted of 10 human cadavers, 16 live oxen and 2 horses. Longer ranges (approx. 35 m and 70 m) were simulated by reducing the propellant loads.

Table 3-1. Weapons used by LAGARDE for his experiments (from HATCHER, 1935)

Weapon	Cal.	Bullet	Mass [g]	v_0 [m/s]	Energy [J]	I [N·s]
7.65 mm Luger pistol		Jacketed	6.0	425	552	2.55
9 mm Luger pistol		Jacketed	8.0	315	400	2.52
Colt Army revolver	38	Solid lead	9.6	230	254	2.21
Colt automatic pistol	38	Jacketed	8.4	332	471	2.79
Colt New Service Revolver	45	Solid lead	16.1	216	383	3.48
Colt New Service Revolver	45	Hollow-point	14.2	210	318	2.98
Colt New Service Revolver	455	Man Stopper	14.1	240	383	3.38
Colt New Service Revolver	476	Solid lead	18.6	220	452	4.09

Cal.: nominal calibre, v_0: Muzzle velocity, I: Momentum.

The cadavers were suspended by the neck, and the experiment consisted of measuring the displacement (which LAGARDE termed the "shock") when the cadaver was struck by a bullet.

In physics terms, what he measured was the momentum of the bullet. The cadaver effectively formed a ballistic pendulum, of which the displacement was proportional to the momentum of the bullet. If the bullet passed through the cadaver, the displacement was proportional to the momentum transferred to the pendulum.

The live animals were shot in an abattoir and the effects of the various bullets were observed. LAGARDE concluded that the 476 bullet was the most effective. Given the way the experiment was set up, this was the obvious conclusion, as that bullet had the greatest momentum.

3.1.3 Basic relationships

The kinetic energy of the bullet on impact with a body, E, is of fundamental importance in wound ballistics. This kinetic energy is the sole source of energy for the destruction of tissue. However, it is not the total energy of the bullet that is decisive, but rather the energy transferred as the bullet passes through a medium. We shall label the energy transferred E_{ab} (also measured in Joule). If we relate the transferred energy to the distance penetrated, s (the gradient of the distance/energy function), we obtain the key parameter for determining wounding potential, E'_{ab}. This parameter is the local energy transfer, and is generally measured in J/cm.

If one assumes (as is generally the case today) that the deceleration of the bullet in a dense medium (simulant, muscle, most organs) is proportional to the square of its velocity (the same law of deceleration as for air), then we can derive the following equation for wounding potential E'_{ab}:

(3.1:1) $\qquad E'_{ab} = -2 \cdot \mathfrak{R} \cdot E$. $\qquad$ [J/m]

E'_{ab} is therefore proportional to the instantaneous energy of the bullet and to its retardation coefficient $\Re$. For a given value of instantaneous energy, it is therefore $\Re$ that determines the amount of energy transferred. This means that bullets with the same energy, but different values of $\Re$, transfer different amounts of energy. Or, to put it differently, the energy transferred by a bullet per unit distance is determined solely by the retardation coefficient times energy ($\Re \cdot E$), not by the energy itself. $\Re$ is given by:

(3.1:2) $$\Re = \tfrac{1}{2} \cdot C_D \cdot \rho \cdot \frac{A}{m} = \tfrac{1}{2} \cdot C_D \cdot \rho \cdot \frac{1}{q} ,\qquad [1/m]$$

where A is the surface area of the bullet in contact with the medium, ρ is the density of the medium, C_D is a form-dependent drag factor and q is the corresponding sectional density. It therefore follows that this parameter influences the amount of energy transferred. If we assume that ρ and the drag coefficient C_D remain constant, then from Equations 3.1:1 and 3.1:2 we obtain the following:

(3.1:3) $$E'_{ab} \propto \frac{E}{q} . \qquad [J/m]$$

Energy transfer per unit distance is hence directly proportional to the instantaneous energy of the bullet and inversely proportional to its sectional density.

The value of A is at its smallest when the bullet is in stable flight, tip first. As the bullet begins to yaw, i.e. as the direction of flight and the longitudinal axis of the bullet begin to form an angle ψ, the sectional density q decreases (because the surface exposed to flow increases) and E'_{ab} increases, in accordance with Eqn. 3.1:3. The increase in energy transferred resulting from yaw (up to and including the point at which the longitudinal axis of the bullet becomes perpendicular to its direction of flight) can be considerable. The range of values for E'_{ab} is particularly wide in the case of rifle bullets, as they have a large longitudinal cross-sectional area (in one plane of their axis) in comparison with their frontal area. Handgun bullets are no more than twice as long as their own calibre, and the effect of yaw is considerably less significant.

The sectional density also has a major influence on penetration depth ℓ, i.e. the total length of the wound channel. As for Eqn 2.3:9, we can derive the following equation ($\ell \Leftrightarrow D$):

(3.1:4) $$\ell \propto q \cdot v^2 . \qquad [m]$$

Penetration depth is hence directly proportional to sectional density and inversely proportional to energy transfer.

3.2 Processes in the wound channel; the temporary cavity

3.2.1 Preliminary remarks

3.2.1.1 The concept of the "temporary cavity"

The *temporary cavity* is the most important concept in the wound ballistics of higher-velocity bullets. Almost all biological phenomena can be traced back to the temporary cavity. In the case of living beings, it would be more accurate to speak of a "temporary wound cavity," as the creation of the temporary cavity creates a wound. The word "temporary" emphasizes that this cavity exists only briefly after the bullet passes through, except when using certain types of simulant.

The process of forming a temporary cavity is sometimes compared – wrongly – with that of cavitation. Cavitation bubbles appear on bodies that move through a fluid at a velocity sufficiently high for the critical pressure to drop below the vapour pressure (Bernoulli's equation, see 2.1.4.4). When these bubbles burst, smooth surfaces (such as those of a ship's screw) may suffer erosion due to the energy released. WOODRUFF (1898) was the first doctor to use the term "cavitation" in connection with wound ballistics.

The temporary cavity in the wound channel is created as the medium flows away from the contact surface. At the same time, the medium is accelerated radially away from the channel (see Fig. 3-3) and undergoes deformation – elastic, plastic or both, depending on the medium. This creates a space behind the bullet and, initially, a vacuum. The vacuum and the elastic energy stored in the medium then cause the temporary cavity to collapse, which is indeed comparable with the bursting of a cavitation bubble.

3.2.1.2 Different ways of looking at wounding

The processes involved in the formation of a bullet wound can be seen from two points of view, depending on whether the problem is surgical or physical/forensic. The two aspects can be distinguished by the questions associated with each:

– What happens when a bullet enters the body (of an animal or a human be-

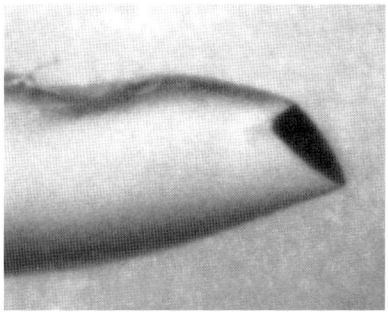

Fig. 3-3. Rifle bullet (7.62 × 39) passing through gelatine. The photo clearly shows the medium flowing away from the bullet at the meridian line in contact with it. A vacuum is forming behind the bullet (creating the temporary cavity). Shutter speed: 1 µs.

ing) and creates a wound channel? ("Actio," the physical ballistic aspect).
- How does the body react to this interaction between bullet and body? ("Reactio", the medical/biological aspect).

The two points of view are closely related, making wound ballistics very much an interdisciplinary science. They will play a significant role in the sections that follow.

We shall start by describing the movement of a bullet in a soft, dense medium, from a physical point of view (3.2.2). We shall then discuss the temporary cavity from a biological point of view, in some detail (3.2.3), and will look at how it can be quantified (3.2.3.2). The next step will be to examine the important relationship between the size of the temporary cavity and the characteristics of the bullet (3.2.3.3 and 3.2.3.4). Wounding potential E'_{ab} plays the primary role here, as this is what determines the volume of the temporary cavity and hence the severity of the damage done to the biological structure. Further sections will deal with the effect of the medium on the bullet, patterns in bullet damage to bones and the question as to whether bullets are sterile.

3.2.1.3 Modelling wound ballistics processes

The interplay between bullet and medium is highly complex, and takes place under exceptional physical conditions. For approximately 1 ms, decelerations of over 1 million m/s^2 and forces of well over 10,000 N are present. The physical characteristics of a material under such dynamic loads – especially the characteristics of biological structures – are virtually unknown. As a result, it would be difficult to produce a mathematical/physical model of the processes involved, and one would lack confidence in the results. Conducting experiments is therefore the only way to gain an understanding of the wounding process.

If such experiments are to produce a usable physical representation of real-life processes, they must fulfill a number of conditions:

- The experiments must be reproducible and must always produce the same results.
- It must be possible to observe the process.
- Bullet dynamics and the behaviour of the medium must correspond very closely to the real-life situation of a bullet in biological tissue.
- The values of the physical parameters along the wound channel (deceleration, force, penetration depth and timing) must be close to those encountered in real life.

Numerous experiments in various countries have resulted in the widespread adoption of two materials for simulating soft tissue in wound ballistics experiments over the last few decades: gelatine and glycerine soap.

Both consist of water and organic matter (fats and alcohol in the case of soap and proteins in the case of gelatine) and are not homogenous. Nevertheless, experience has shown these materials to possess the physical characteristics needed in order to simulate soft biological tissue, including the required flowability under high pressure. Gelatine is used in the majority of cases, at various concentrations and temperatures.

In the past, these two materials have been validated primarily by means of experiments on animals (e.g. JANZON 1982a for glycerine soap). However, it is also possible to verify the validity of models for determining the physical behaviour of bullets by comparing results from such models with real bullet wounds, of which war surgery and forensics provide all too many examples. As a result, there is no need for experiments on animals or cadavers.

There is one fundamental difference between the two simulants; soap mainly deforms plastically, whereas gelatine deforms elastically. Very little of the deformation that a bullet causes in soap disappears, whereas gelatine returns to almost the same state as before. This means that the temporary cavity is "frozen" into soap and can readily be studied and analysed. In gelatine the temporary cavity collapses, as it does in tissue.

As bullets behave in a very similar manner in the two media (in terms of orientation, penetration depth, deformation and velocity), it seems reasonable to assume that the elasto-plastic behaviour of the two materials is of no more than secondary relevance to the *motion of the bullet*. The motion of the bullet is hence determined primarily by density, viscosity and flowability, and these are similar in the two media and in muscle.

Bone and other irregularities can be simulated using plastics, ensuring that the physical behaviour of the bullet is very similar to that observed in real life (see 3.3).

3.2.2 Motion and behaviour of a bullet

3.2.2.1 Rifle bullets

Full metal-jacketed and solid bullets. Full metal-jacketed and solid bullets for rifles are generally 3 to 5 calibres long. If they hit a soft medium while in stable flight, they cause a wound channel that can be divided into three clearly-distinguishable sections. This behaviour – which is particularly clear in soap, as the temporary cavity is frozen – is largely independent of bullet design.

The first section (see Fig. 3-4) consists of a straight entry channel, known as the *narrow channel*, or NC. At its narrowest point, the diameter of this channel generally corresponds to approximately 1.5 to 2.5 times the calibre of the bullet. The blunter the bullet and the higher the energy, the wider the narrow channel at its narrowest point.

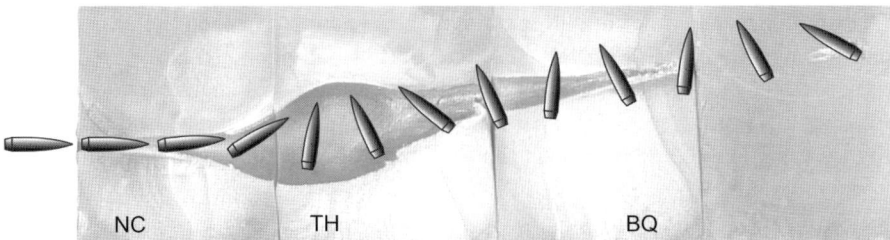

Fig. 3-4. Motion of a non-deforming rifle bullet in gelatine and in glycerine soap (and also in soft tissue). NC: narrow channel. TH: temporary cavity at its largest. BQ: bullet rotating into a position perpendicular to its direction of travel. (Illustration: full metal-jacketed steel-core bullet, 7.62 mm NATO, shown larger than actual size.)

The narrow channel is created as follows: When a bullet enters a soft target medium, extremely high pressure is created at the tip, owing to the high density of the medium which, from the bullet's point of view, is flowing towards it. However, the viscosity and inertia of the medium cause the flow to break away from the surface of the bullet at an early stage, which means that only a small part of the bullet's tip is in contact with the medium and hence exposed to this pressure. A large percentage of the bullet's surface is not in contact with the medium and is hence subjected to virtually no forces. At this point, inertia forces predominate, and friction can be ignored.

If the entire surface of the bullet were in contact with the medium, the forces acting upon it would immediately cause it to become unstable and no narrow channel would be created.

Because the surface in contact with the medium is limited, the force acting upon the bullet is comparatively small, and the contact point (or pressure point) is close to the tip. The overturning moment, which is what causes the bullet to yaw, depends mainly on the angle of incidence at the point of impact. If the bullet is in sufficiently stable flight, the angle of incidence is small and the overturning moment is hence smaller than the stabilizing gyrostatic moment.

The bullet, which acts like a gyroscope, executes a precession movement under the influence of this overturning moment (see 2.3.4.4 and Fig. 2-51). As a result, the angle of incidence does not decrease. As velocity decreases, the surface in contact with the medium increases. This leads to an increase in force and hence to an increase in the moment applied. As a result, the angle of incidence increases and causes the overturning moment to increase still further. This positive feedback rapidly causes the bullet to yaw.

Using various methods of recording the motion of the bullet in the medium, it has been proven that at this point the angle of incidence is indeed increasing very rapidly (AEBI et al. 1977, JANZON et al. 1979, WATKINS et al. 1982 using x-ray flash photography; FACKLER 1987b using high-speed photography; KNEUBUEHL and FACKLER 1990 using electromagnetic induction).

The length of the narrow channel hence depends on the angle of incidence at the point of impact, on gyroscopic stability and on the form of bullet tip.

3.2 Processes in the wound channel; the temporary cavity   99

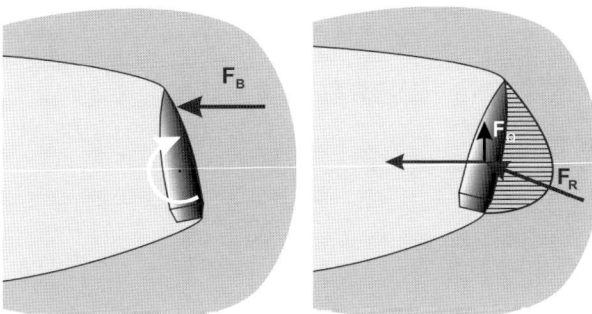

Fig. 3-5. Left:
Force acting against yaw. F_B: Drag. S: centre of gravity.

Fig. 3-6. Right:
Lateral force. The shaded surface represents the pressure distribution. F_R: Resultant force. F_Q: Lateral force.

The second section (see Figs 3-4 and 3-8) begins with the bullet yawing very rapidly. As the angle of incidence increases, the bullet surface in contact with the medium also increases rapidly, until the bullet is in contact with the medium over its entire length. Because the bullet rotates about its centre of gravity, the base of the bullet (or the tip, if the bullet rotates in the opposite direction) is forced into the medium at high speed (see Fig. 3-5). The resultant force is applied away from the centre of gravity, producing a torque that tends to oppose the yaw.

This behaviour is clearly visible in various measurements of the angle of incidence (see for example KNEUBUEHL and FACKLER 1990).

At the same time, the bullet undergoes very rapid deceleration and the medium flowing away from it forms a temporary cavity. The bullet is subjected to bending and compressive stresses (see Fig. 3-6, hatched area), which can compress the bullet or even cause it to fracture (see Fig. 3-7). When this happens, lead particles are often squeezed out at the tail of the bullet and at the point of fracture.

At short ranges (where the energy is high and the angle of incidence is relatively large), this effect is observed in many bullets, especially those of smaller calibre. Bullets with a soft tombac or copper jacket are especially prone to this behaviour.
    At impact velocities of less than 600 m/s, however, full metal-jacketed bullets undergo virtually no deformation.

If the bullet does not fragment, and if it is perpendicular to its direction of travel, the asymmetrical pressure distribution along a very small strip along the jacket produces a resultant force F_R. This force will rarely act exactly along the bullet's line of movement. As a result, there are two components (see Fig. 3-6): one component acts in the opposite direction to that in which the bullet is moving and slows it down, while the other acts perpendicular to the first component and is in equilibrium with the friction between bullet and medium along the small strip

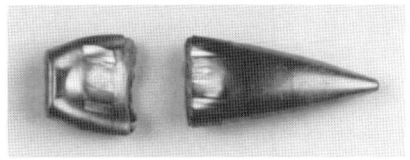

Fig. 3-7. Full metal-jacketed bullet (7.62 × 51) after being fired into a block of soap. Part of the lead core was forced out of the jacket and the bullet has fractured at the cannelure.

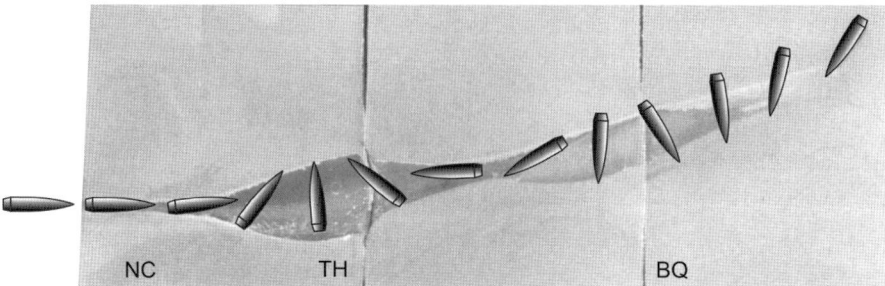

Fig. 3-8. Track of a slender full metal-jacketed bullet. The bullet has deviated substantially from its original path and has rotated through more than 270°. 5.45 × 39 (Kalashnikov), steel core. Bullet shown larger than actual size. See Fig. 3-4 for explanation of symbols.

mentioned previously. If lateral force exceeds friction, lateral acceleration will occur, causing the bullet to depart from a straight path (see Figs 3-4 and 3-8).

The angle between the bullet's direction at the time of impact and the direction in which it moves from the end of the temporary cavity onwards is closely related to the slenderness ratio of the bullet (the ratio of the bullet's length to its diameter). For very slender bullets, this angle can easily exceed 30°. Short bullets deviate little from their initial path.

As a result of inertia (the lateral moment of inertia being the determining factor), the bullet will continue to yaw beyond the point at which it is perpendicular to its own direction of movement, but will do so increasingly slowly. This process can be seen as a kind of strongly damped oscillation, the damping being due to pressure and to friction resistance. The maximum angle through which the bullet rotates depends on its instantaneous velocity and on its geometric characteristics (form, slenderness and moments of inertia). This angle generally lies somewhere between 90° and 180°, but may well exceed 270° for a slender bullet (see Fig. 3-8 and SIEGMUND 2006, p. 60). This marks the end of the second section of the wound channel.

Several authors have described these first two sections, including AEBI et al. 1977, BERLIN et al. 1982 and TIKKA 1982. FACKLER was the first to preserve the entire wound channel in a simulant (FACKLER and MALINOWSKI 1985). It was FACKLER who pointed out that wound channels in gelatine are entirely analogous to those observed in soap (FACKLER and MALINOWSKI 1988a).

At the beginning of the third section (see Figs 3-4 and 3-8, BQ) the bullet yaws

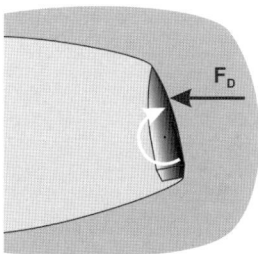
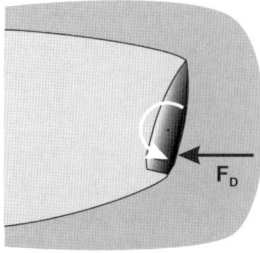

Fig. 3-9. Stabilization in the third section. Every change in direction of rotation produces a damping force F_D.

under the influence of the damping force until it is once again perpendicular to its own direction of travel. If it was in that position when it stopped yawing, it remains in it. The bullet then rocks back and forth in this position, with movement in any given direction immediately producing a damping force that opposes it (see Fig. 3-9). The bullet therefore moves through the medium oriented approximately perpendicular to its direction of travel, rocking back and forth about its centre of gravity.

At the same time, it creates a second temporary cavity. This second cavity is considerably smaller than the first, as less energy is available. Because the velocity of the bullet has also decreased considerably, the medium flows closer to its surface until, shortly before it comes to rest, the bullet is completely embedded in the medium. As a result, the remainder of the wound channel in soap is narrower than the dimensions of the bullet and in gelatine (and in tissue) is often no longer visible. The bullet comes to rest perpendicular to its direction of travel, but the partial vacuum in the temporary cavity usually pulls it back a short distance. This final movement always causes the bullet to come to rest with the tail forwards (relative to the original direction of travel).

Deforming and fragmentation bullets. The behaviour of deforming and fragmentation bullets (generally semi-jacketed or hollow-point bullets) in a soft medium differs substantially from that of full metal-jacketed bullets – even FMJ bullets that break up when perpendicular to their direction of travel. The main difference is that the narrow channel is almost entirely absent; the temporary cavity starts immediately after penetration. This type of bullet deforms or fragments 2 cm to 4 cm after penetrating the medium (see Fig. 3-10).

This has been demonstrated for various deforming bullets by BERLIN et al. (1988a) using X-Ray flash photography and by KNEUBUEHL using high-speed video (up to 4000 f/s).

Deformation is extremely rapid, taking place within approximately 0.1 ms. The forces required to accomplish this are produced by the very high pressure at the tip. In the case of semi-jacketed bullets, this pressure acts on the exposed lead at

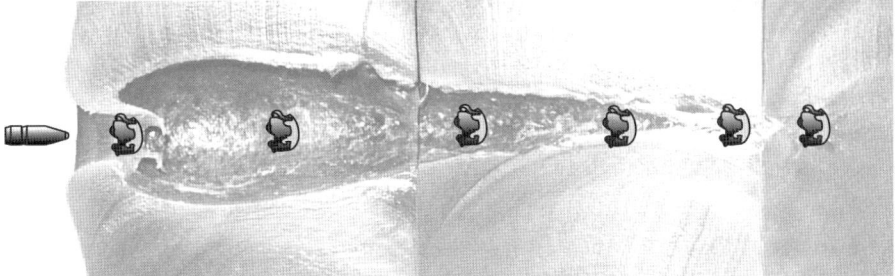

Fig. 3-10. Track of a deforming bullet. The bullet follows a straight line (being shoulder-stabilized). At the entrance we can clearly see the material sucked in by the vacuum of the temporary cavity. When the bullet strikes the surface, this material sprays outwards in a hollow cone pattern in the direction from which the bullet has come.

Fig. 3-11. Semi-jacketed bullet, calibre 7.62 × 51 (308 Win.) and the deformed remains of the bullet after being fired into a soft medium.

the tip of the bullet, which behaves in a manner similar to a fluid under such conditions, forcing it into the interior of the bullet and causing it to burst. In the case of jacketed hollow-tip bullets, the medium has to be able to penetrate the hollowed-out tip in order to burst the bullet. If the opening is too small, a hollow-tip bullet will behave like a full metal-jacketed bullet.

Deformation causes the leading surface of the bullet in contact with the medium to increase suddenly. Sectional density decreases and energy transfer increases (see Fig. 3-11).

The *size* of the temporary cavity is not affected by whether the increase in surface area (and the accompanying decrease in sectional density) is caused by deformation or by the bullet yawing so that it is perpendicular to its direction of movement. The cause of the decrease in sectional density does, however, have a decisive effect on the *position* of the temporary cavity in the body.

The increase in the area of the leading surface of the deformed bullet results in shoulder stabilization, as described in 2.3.5.4. As a result, a deforming bullet always creates a straight wound channel in a homogenous medium. However, the diameter of the channel decreases steadily after attaining its maximum value, as the sectional density remains virtually constant from that point on. If deformation is asymmetrical, or if the tissue is not homogenous, the bullet may deviate substantially from a straight line towards the end of the channel.

Generally, we distinguish between deforming and fragmentation bullets by looking at the mass of the largest fragment remaining after impact. Deforming bullets lose very little material, so a deforming bullet that has been fired weighs practically the same as one that has not. By contrast, the largest fragment of a fragmentation bullet may weigh less than half as much as the original bullet. At the same time, the fragments increase the overall cross-sectional area. These fragments often create their own wound channels, branching off the main channel. Fragmentation bullets are primarily used in hunting rifles. It is interesting to note that both types of bullet deform in a fairly regular manner, both in gelatine and in glycerine soap.

Explosive bullets constitute a particular category of fragmentation bullet. This type of bullet contains a small amount of explosive, which detonates on impact and causes the bullet to fragment. Explosive bullets can be designed in such a way as to explode only when they strike a hard surface, behaving like normal full metal-jacketed bullets in a soft medium.

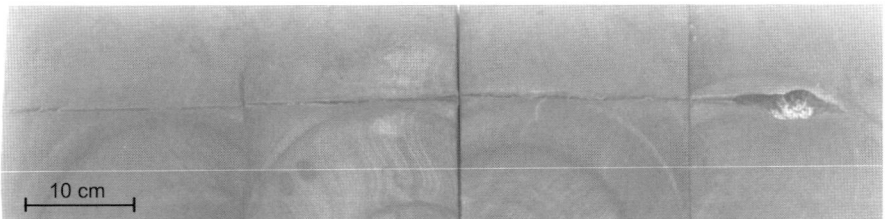

Fig. 3-12. The wound channel of a flechette in soap. Impact velocity approx. 1300 m/s. Projectile fired from left.

Flechettes. Little has so far been published concerning the behaviour of fin-stabilized projectiles. Experiments conducted by the author (KNEUBUEHL 1991) have shown that if the flight of the flechette is sufficiently stable – a precondition for achieving sufficient accuracy – then stable flight will continue in a dense medium. Only after penetrating to a substantial depth (> 60 cm) did the projectiles begin to yaw, at which point the fins were also ripped off. At such penetration depths, temporary cavities were formed comparable with those caused by handgun bullets (see Fig. 3-12). In one case, a flechette passed right through a 1 m block of soap. In the meantime, experiments in other ballistic laboratories have confirmed the behaviour observed over the first 40 cm.

3.2.2.2 Handgun bullets

Full metal-jacketed and solid bullets. With a few exceptions, handgun bullets carry substantially less energy than rifle bullets (approximately a quarter or even a sixth as much). As energy transfer depends on the total instantaneous energy of the bullet (see Eqn 3.1:1), handgun bullets generally cause smaller wound channels than rifle bullets. The movement of such bullets in a soft, dense medium is the same, as is their behaviour. However, some special points are worth to be noted.

For a variety of technical and ballistic reasons, handgun bullets are substantially shorter than rifle bullets, and generally have blunter tips (often round or conical). There are exceptions, such as bullets for silhouette pistols, but these are rare. Because handgun bullets are shorter, the surface in contact with the medium increases only negligibly if the bullet yaws, and the sectional density only decreases by a small amount. Energy transfer hence varies little along the length of the wound channel and the temporary cavity remains small (see Fig. 3-13). As a result, handgun bullets can easily penetrate to depths of 70 cm and more.

Deformation and fragmentation bullets. Deformation and fragmentation bullets for handguns carry substantially less energy than do rifle bullets. This is also true of full metal-jacketed handgun bullets. Energy transfer is hence also considerably less, which means that one cannot compare handgun bullets with deforming rifle bullets (often referred to – erroneously – as "dumdum" bullets, see 7.2.2.7).

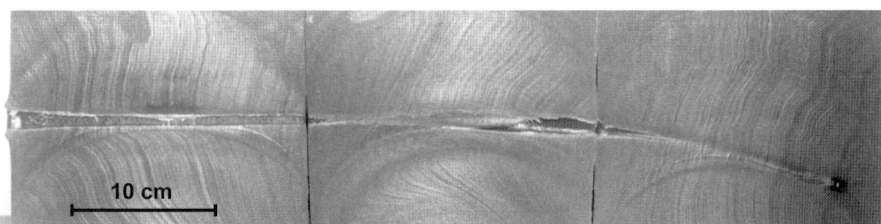

Fig. 3-13. Wound channel of a full metal-jacketed 9 mm Luger bullet. Bullet fired from left.

These bullets generally deform very rapidly, with solid copper and copper-alloy bullets reaching maximum deformation (and hence maximum energy transfer) even earlier than conventional lead-cored semi-jacketed bullets. After 1 cm to 3 cm, most deforming handgun bullets have fully mushroomed (see Fig. 3-14).

SELLIER and KNEUBUEHL noted that a 9 mm Luger Action 1 bullet fired through a 1 cm sheet of ballistic soap had completely deformed on exit.

Once the bullet has deformed, the increased area of the leading surface induces shoulder stabilization. The wound channel is therefore fairly straight, and narrows conically after reaching its maximum diameter. The bullet often comes to a halt in the medium at a slight angle to its direction of travel.

Cylindrical bullets. Cylindrical handgun bullets are also available. These are used primarily for target shooting (in the case of lead wadcutter bullets) or for finishing shots when hunting. Such bullets are very stable in a soft, dense medium, on account of shoulder stabilization. As a result, their sectional density remains unchanged and they leave a straight (and therefore long) wound channel. The diameter of the wound channel is greatest at the point of impact, decreasing exponentially (in accordance with Eqn 3.1:1 with $\Re$ constant).

Rifle bullets that strike their target tail first behave in a similar manner, as the tail of a rifle bullet is approximately cylindrical. This situation can arise with ricochets that regain stability at some point after striking a hard surface, but with the tail leading.

3.2.2.3 Fragments and fragment-like projectiles

Spheres. Spheres are particularly suitable for model experiments. They remain stable in flight and their sectional density remains constant regardless of yaw.

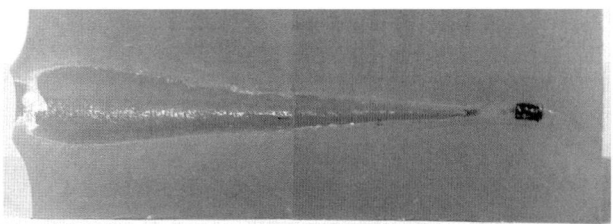

Fig. 3-14. Wound channel of a 9 mm Luger deforming bullet (QD P.E.P.) in soap. Same scale as Fig. 3-13.

This means that it is possible to ignore many parameters, such as bullet length, tip form, angle of incidence at point of impact and speed of rotation about the longitudinal axis. Experiments that involve firing spheres into gelatine and soap targets have been described in many publications.

SCEPANOVIC (1979) used 6 mm steel spheres, fired at different velocities, and established a relationship between impact velocity on the one hand and penetration depth and wound channel volume on the other. The mean deceleration is determined by keeping the drag coefficient constant. It is interesting to note that the diameter of the channel decreases along its length, becoming significantly less than the diameter of the sphere towards the end. JEANQUARTIER and KNEUBUEHL (1983) fired 6, 8 and 10 mm spheres into ballistic soap with a somewhat reduced water and alcohol content (and hence higher viscosity) and observed reduced penetration depths. However, the geometric form of the wound channel and its other characteristics – especially the reduction in diameter to less than the diameter of the sphere – remained analogous.

NENNSTIEL (1990) went a step further in studying the passage of a sphere through a dense medium, determining the relationship between drag and instantaneous velocity using high-speed photography. These experiments revealed that as velocity decreased, the drag coefficient related to the diameter of the sphere increased sharply. It is interesting to note that in many cases the sphere did not come to a standstill at the end of the effective wound channel. WATKINS et al. (1982) conducted comparison experiments using gelatine and soap (manufactured to Swedish specifications). They discovered a close correlation between time/distance measurements in the two materials.

Spheres produce wound channels that correspond very closely to those that one would expect on the basis of Eqn 3.1:11. As sectional density remains constant, one can also take $\mathfrak{R}$ as constant. For such cases, Eqn 3.1:11 predicts an exponential decrease in the volume of the channel, and hence of its diameter. This is largely confirmed by the experiments (see Fig. 3-15).

Natural and pre-formed fragments. To minimize energy loss along their trajectory, pre-formed fragments are generally similar in shape to a sphere, taking the form of cubes or short prisms.

Natural fragments, however, are created by the random disintegration of the body of a shell or bomb, the body usually being made of steel. If large quantities of explosive are used, and the body of the shell fragments sufficiently, the fragments are often cuboid (i.e. all the edges of the fragment are approximately equal in length to the thickness of the walls of the shell). If less explosive is used, the fragments are often longer, and may well have a length-to-width ratio of between 4:1 and 6:1.

As no stabilizing forces act upon fragments in flight, their orientation on impact with the body is random. However, as soon as a fragment touches the target, the same stabilization mechanisms comes into play as with the full metal-jacketed

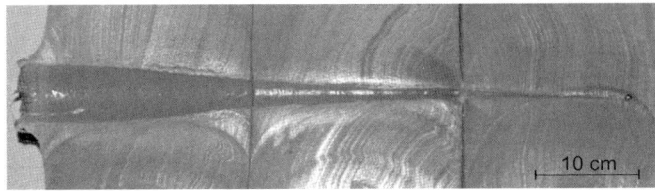

Fig. 3-15. Wound channel created by a 5.54 mm steel sphere. Fired from left. Impact velocity approx. 1 500 m/s.

bullet in the third section of the wound channel (see Fig. 3-8).

Fragments therefore maintain a virtually constant sectional density in the body, generally with the largest surface facing forwards. As a result, the diameter of a fragment wound channel is greatest at the entry wound and becomes progressively smaller.

The largest wound area is therefore visible from the outside in the case of fragment wounds. No major tissue damage resulting from a temporary cavity is to be expected inside the body.

For a given mass and impact energy, spherical fragments will penetrate most deeply and create the smallest entry wounds, as they have the greatest sectional density. Longer fragments create large entry wounds but do not penetrate as far (see also 3.2.3.4 and Fig. 3-27).

Unstable bullets. There are two types of "unstable" bullet:
- previously stable bullets that have ricocheted off an object;
- bullets fired from a shot-out barrel.

In the first case, the bullet executes a periodic movement with a large angle of incidence after the initial impact (nutation, see Fig. 2-51). This angle may well be close to 90°. After a few hundred metres (the exact distance depending on the calibre) gyroscopic stabilization causes these ricochets to resume stable flight, though possibly with the tail forwards. In the second case, the bullet receives too little rotation in the barrel, and is hence unable to attain stable flight. It therefore tumbles throughout its trajectory, the axis of rotation being maintained constant by the gyroscopic effect.

In both cases, the bullet is incapable of creating a narrow channel (see Fig. 3-16). It turns perpendicular to its direction of travel immediately on penetration and moves through the medium as described in 3.2.2.1 (third section, BQ). Temporary cavities are created immediately after impact, and if the impact energy is high these cavities can be extremely large.

The wound channels of unstable bullets are very similar to those caused by fragments of similar energy and sectional density.

Deforming bullets that have ricocheted do not mushroom, but behave like full

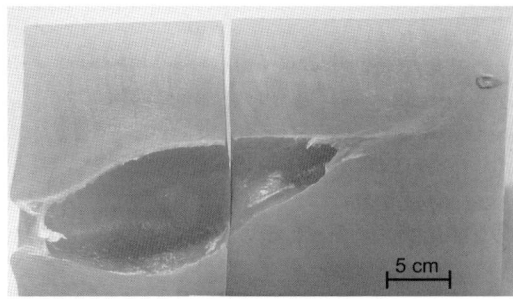

Fig. 3-16. The wound channel of a ricochet in soap. Bullet fired from left. The bullet struck the block with a large angle of incidence and immediately turned so that it was perpendicular to its direction of travel.

metal-jacketed bullets that have ricocheted, as it is no longer possible to create the pressure at the tip that would be required to cause them to deform.

3.2.2.4 Possible types of wound channel.

From 3.2.2.1 to 3.2.2.3, we can distinguish between five types of wound channel. These are shown in Table 3-2. The illustrations are self-explanatory.

Table 3-2. Possible types of wound channel (illustrations approximately to scale).

Type of weapon	Full metal-jacketed and solid bullets	Deforming and fragmentation bullets
Rifle		
Handgun		
Fragment		

3.2.2.5 Physical models

Aims. While qualitative descriptions are useful as a means of understanding a physical process, predicting the behaviour of a particular bullet in a soft medium requires a physical/mathematical model. The first step is to take known bullet parameters, such as mass, moments of inertia, position of centre of gravity, form and structure, together with such ballistics data as velocity, speed of rotation and angle of incidence at point of impact, and use these to derive equations describing the orientation and movement of the bullet within the medium. In a second phase, it could be possible to use material-specific values to calculate stresses in the jacket of the bullet, which in turn would enable us to predict the deformation and fragmentation of the bullet.

We shall now discuss known approaches to modelling from the literature, and indicate opportunities for building on these and improving them.

Simple laws of deceleration. As the energy transferred from the bullet to the medium (the tissue) is clearly of primary importance for the effectiveness of the bullet, most existing models limit themselves to describing the change in velocity of the bullet (or, more accurately, of its centre of gravity).

As we have already indicated in 3.1.3, one obvious approach is to use the principles of drag from fluid dynamics. Drag is assumed to be proportional to velocity squared, to density and to a reference area:

(3.2:1) $$a = -C_D \cdot \tfrac{1}{2} \cdot \rho \cdot v^2 \cdot \frac{A}{m} \; . \qquad [m/s^2]$$

If we can ignore internal friction, the dimensionless drag coefficient C_D becomes a function of the Mach number, i.e. of velocity. If, however, inertial forces are very small compared to friction, then we generally use Stokes' approach, with a linear relationship between drag on the one hand and velocity and viscosity on the other.

(3.2:2) $$F = -a \cdot m = -C^* \cdot \tfrac{1}{2} \cdot \eta \cdot v \cdot r \; . \qquad [N]$$

By introducing the Reynolds number, we can write the equation as in Eqn 3.2:1, with C_D expressed as:

(3.2:3) $$C_D = \frac{4 \cdot C^*}{\pi \cdot Re} \; . \qquad [-]$$

Most authors take the approach represented in Eqn 3.2:1, generally with the cross-sectional area of the calibre as the reference area. If we determine the relationship between time and distance, it is possible to calculate the drag coefficient, at least approximately. It is to be expected that this value will not be constant. Rather, it will depend on both the Mach number (i.e. on velocity) and on the Reynolds number (viscosity), as in aerodynamics. In a simple model, however, all parameters are assumed to be constant, with the exception of velocity. We therefore obtain the following equation:

(3.2:4) $$a = \frac{dv}{dt} = -\mathfrak{R} \cdot v^2 \; , \qquad [m/s^2]$$

with retardation coefficient

(3.2:4a) $$\mathfrak{R} = \frac{C_D \cdot \rho \cdot A}{2 \cdot m} \; . \qquad [m^{-1}]$$

From this we can derive the equation in which we are interested – that of energy transfer per unit distance:

(3.2:5) $$E'_{ab} = \frac{dE}{ds} = -m \cdot a = -2 \cdot \mathfrak{R} \cdot \tfrac{1}{2} \cdot m \cdot v^2 = -2 \cdot \mathfrak{R} \cdot E \; . \qquad [J/m]$$

We shall discuss the solutions of these equations in 3.2.3.2.

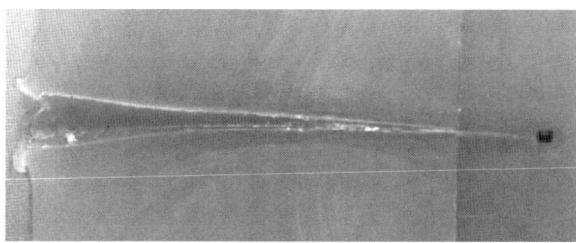

Fig. 3-17. Idealized conical form of a temporary cavity created by a cylinder. Tumbling absent due to shoulder stabilization.

If the sectional density of the projectile remains constant (as with a sphere or a cylinder of which the axis coincides with the trajectory), the diameter of the cavity will decrease along its length (see Fig. 3-17), and its boundary will follow an exponential curve, except in the area immediately following the entry wound. This allows us to conclude that $\Re$ is constant. Using Eqn 3.2:4a, this in turn makes it reasonable to assume that the drag coefficient C_D is also constant.

More advanced models. The above models are highly simplified. As a result, they are unable to predict such important elements as penetration depth or the formation of the temporary cavity. PETERS (1990a,b) therefore proposed extending these models by including the forces that cause irreversible deformation of the material, in addition to the inertial forces. His approach therefore includes not only fluid dynamics equations but also mechanical equations, some of which were derived empirically:

$$(3.2:6) \quad a = -C_D \cdot \tfrac{1}{2} \cdot \rho \cdot \frac{A}{m} \cdot \left[v^2 + \left(\frac{v}{k}\right)^{\frac{1}{2}} \cdot \left(u_6^3 \cdot k_6\right)^{\frac{1}{2}} \right]. \quad [m/s^2]$$

The second addend, which is proportional to the square root of v/k, describes the deceleration component responsible for deformation. The corresponding proportionality factor is derived empirically for a 6 mm projectile, while u_6 is a material-dependent parameter and is known as the characteristic velocity. Using further models for the angle of incidence, for the formation of the primary cavity and for tissue damage, it is possible – for simple projectile shapes – to obtain adequate correlation between calculations and measurements.

A further extension to the model relies entirely on a fluid dynamics approach to the motion of a bullet, this time taking account of viscosity. Unlike the equations discussed above, this model only includes that part of the bullet's surface that is in contact with the medium, as pressure and friction forces can only act on this surface. Separating the two corresponding deceleration components gives:

$$(3.2:7) \quad a = \frac{1}{2 \cdot m} \cdot \rho \cdot v^2 \cdot \left[C_D \cdot A_q(v,\eta) + C_F \cdot A_0(v,\eta) \right]. \quad [m/s^2]$$

Where C_D is a function of the Mach number and C_F is a function of both the Mach number and the Reynolds number.

The part of the bullet tip in contact with the medium determines both the cross-sectional area A_q that determines the compressive forces and the area A_0 that determines the friction forces. Both are dependent on the flow-off angle of the medium which, in turn, is determined by velocity and viscosity. If it were possible to calculate this angle as a function of v and η, it would be possible to calculate the magnitudes of the compressive and friction forces, together with their contact points, and so to determine the yaw angle of the bullet.

3.2.3 The temporary cavity

3.2.3.1 Phenomenology of the temporary cavity

The energy that the bullet transfers to the medium accelerates the medium surrounding the path of the bullet away from it radially. This creates a hollow space behind the bullet and, initially, a vacuum. Because of inertia, the cavity only reaches its maximum diameter at any given point when the bullet has already passed that point.

The cavity is filled with air from the entry wound (and from the exit wound if there is one). It also contains water vapour from the medium (water, gelatine, soap, muscle, etc.).

The front part of the bullet starts to separate the material and push it apart (see rifle bullet in Fig. 3-3). Initially, the rear part of the bullet has no contact with the medium and is therefore not subject to compressive forces. However, as a rifle bullet yaws in the wound channel, the rear part does eventually come into contact with the medium.

The maximum radial extent of the cavity clearly depends on the elasticity of the medium or, more precisely, on its Young's modulus. The higher the Young's modulus, the greater the forces tending to return the particles to their original position.

In the case of water, Young's modulus is replaced by hydrostatic pressure. If all other conditions remain the same, then the higher the water pressure (i.e. the greater the depth), the smaller the cavity. OTTOSON (1964) observed this phenomenon during his experiments.

The cavity collapses after a few milliseconds, as part of the bullet's energy is converted into the elastic energy of the tissue. In most cases, that energy is not fully used up in creating the cavity. A second cavity is created. This second cavity is smaller than the first, as less elastic energy is available. This process of creating cavities is repeated until all the energy has been used up; the cavity pulsates. Finally, the energy is converted into heat via pressure waves and internal friction.

The duration of a pulse τ_{TH} depends on the energy transferred to the medium from the bullet, E_{ab}. Duration is given by:

$$(3.2{:}8) \qquad \tau_{TH} = 2 \cdot \sqrt[3]{E_{ab}} \: . \qquad\qquad [ms]$$

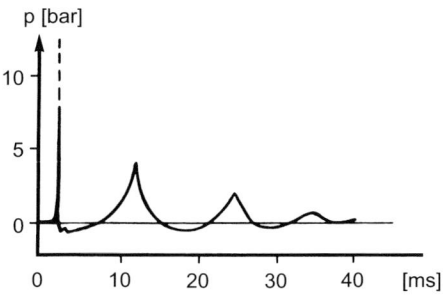

Fig. 3-18. Typical graph of pressure against time for pulses in tissue associated with the temporary cavity (after HARVEY et al. 1946).

If E_{ab} = 450 J, for instance, τ_{TH} = 15 ms, while if E_{ab} = 3600 J, τ_{TH} = 31 ms.

Liu et al. (1988b) carried out experiments with a 7.62 mm NATO bullet (approx. 3200 J) in water, recording an interval of 24 ms between the first and second pressure peaks and 20 ms between the second and third. The corresponding values for a 5.56 mm M193 bullet were 22 ms and 16 ms respectively.

The number of pulses clearly depends on the internal friction of the medium. Seven to eight pulses were observed in water, and a smaller number in tissue. The authors cited above observed three pressure maxima in water, on the basis of the peaks on the oscillogram.

Pulses associated with the passage of a bullet follow a characteristic pattern, as measured by HARVEY et al. (1946).

The pulses can be divided into two phases (see Fig. 3-18):

1. An initial shock wave with a very high pressure peak, which initially propagates at a velocity greater than the speed of sound c in tissue (approx. 1400 m/s to 1500 m/s).

The propagation velocity of a shock wave increases with pressure and hence exceeds the speed of sound c, which is defined as the propagation velocity of small disturbances. For instance, the propagation velocity of a shock wave in water (c = 1465 m/s at 20°C) is 2230 m/s for a pressure peak of 5 kbar (500 MPa) and 2755 m/s for a peak of 10 kbar.

This shock wave is created by the sudden displacement of tissue in the wound channel. Initially, because of the inertia of the tissue, only the area at the point of impact is subjected to a high degree of compression, creating a shock wave. The wave front is very steep; the time τ_a required for the pressure to rise from zero to a maximum is of the order of a few *micro*seconds. The pressure amplitude of the wave can be calculated approximately as described in 2.1.4.4 (Eqns 2.1:62 and 63).

The shock wave does not actually move tissue from one place to another, unlike the temporary cavity, which displaces large masses of tissue. Nevertheless, the shock wave can cause certain effects, which we shall discuss later.

2. The *pressure fluctuations in the second phase* are caused by the pulsations of the temporary cavity. The particles are accelerated radially. A cavity is therefore created, of which the diameter depends on the magnitude of the energy transferred

by the bullet E'_{ab}. A virtual vacuum forms behind the temporary cavity, which is then filled by air entering via the entry wound.

The temporary cavity causes a number of changes in the body of the person or animal. We can distinguish between:

1. The temporary (wound) cavity.
2. The extravasation zone.
3. The permanent wound channel.

The extravasation zone and the permanent wound channel result from the temporary cavity. The permanent wound channel is not the same as the geometrical wound channel – it does not form a "tube" with the same cross-sectional area as the bullet. Rather, the diameter of the permanent wound channel at any given point corresponds to the energy transferred by the bullet at that point and is hence related to the size of the temporary cavity at that point. The permanent wound channel is, however, smaller than the temporary cavity.

The wound channel is marked by an area of crushed and torn tissue containing large amounts of blood. The reason for this damage becomes clear when one looks at what happens to the area when the temporary cavity is created (see Fig. 3-19 c). The tissue close to the wound channel is stretched considerably by the formation of the temporary cavity. Geometrically, these areas form hollow cylinders. These "tubes" of tissue expand considerably, and the tissue is torn (Zone 2 in Figs 3-19 b and c). Not surprisingly, the stretching of the tissue decreases with radial distance from the wound channel; the stretching is inversely proportional to distance from the wound channel. There is therefore a ring of tissue around the wound channel, within which the tissue is torn through having been overstretched. The radius of this zone depends on the radius of the temporary cavity and, of course, on the resistance of the tissue to tearing. There is a fixed numeric ratio between the radius of the temporary cavity and that of the permanent wound channel, as we shall see later.

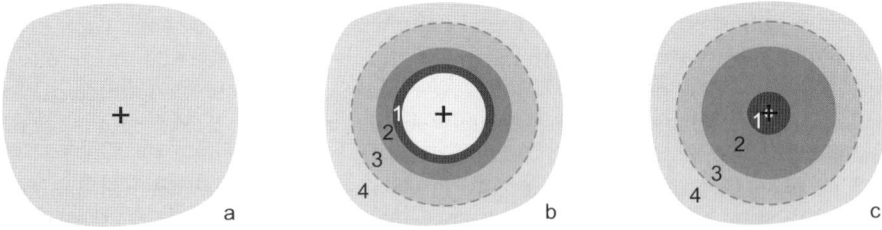

Fig. 3-19. Schematic illustration of tissue displacement due to formation of the temporary cavity. Section through a block of tissue: **a.** before impact (showing the position of the geometric wound channel), **b.** shortly after impact (when the temporary cavity is at its maximum) and **c.** after the bullet has come to rest (permanent wound channel).
+: Geometric wound channel. 1: Destroyed tissue. 2: Extravasation zone. 3: Stretched but undamaged tissue. 4: Unaffected tissue.

The permanent wound channel is surrounded by the extravasation zone (Zone 2 in Fig. 3-19). This zone is characterized by small areas of bleeding (those closer to the wound channel being more intense than those further away) and the absence of macroscopic destruction. The term *extravasation* comes from Latin *extra* ⇒ out of and *vasa* ⇒ vessels and here refers to blood leaking out of blood vessels. Here again, the mechanism is easy to understand. In this zone, the stretching caused by the temporary cavity is not sufficient to tear tissue, but is sufficient to affect those elements most sensitive to stretching, i.e. the capillaries. There is a mathematical relationship between this zone and the size of the temporary cavity.

If one considers the geometry, it is clear that the circular stretching of the skin around the entry wound is inversely proportional to distance from the centre of the wound. This circular area of stretched skin forms the "lid" of the temporary cavity. This conclusion is supported by observation of the skin around the entry wound; the wound is usually surrounded by a clearly visible zone of micro-tears, forming a circular patch of red skin. The colour is more intense towards the centre, where damage to the capillaries is greatest.

The volumes of these zones are proportional to the energy transferred to the tissue from the bullet, E_{ab}. We can therefore write:

(3.2:9) $$V = \mu \cdot E_{ab} \,. \qquad [\text{cm}^3]$$

Values of μ:
  Temporary cavity   0.77 cm³/J
  Extravasation zone   0.35 cm³/J
  Permanent wound channel   0.03 cm³/J

These values are for muscle tissue. Measurements conducted by FACKLER et al. (1986) give a μ of 1.26 cm³/J for the temporary cavity in gelatine (firing 6 mm steel spheres into 10% gelatine at 4°C) and 0.16 to 0.20 cm³/J for soap. See 3.3 for further details. In water, close to the surface, the value of μ in the temporary cavity is 8.7 cm³/J. The reason for the value of μ in water being 11 times that encountered in muscle is that water has a lower cohesive force. The temporary cavity is approximately 26 times larger than the permanent wound channel (0.77/0.03). The ratio between the diameter of the permanent wound channel and that of the temporary cavity is approximately 1:3.

In the case of simulants, the value of μ also depends on how the experiment is set up – the physical properties of the material, the size of the block, etc. Maintaining the same arrangement and keeping conditions constant can produce surprisingly accurate results (see 3.3.2, 3.3.3 and 3.3.4). It is important that the block be large enough that μ remain unaffected by such peripheral effects as swelling around the temporary cavity or cracking of the block.

The temporary cavity is far less able to "breathe" if the medium through which the bullet passes is enclosed. In the human body, this may be the case when a bullet passes through the head, the heart or a full bladder. In the case of these fluid-filled hollow organs (brain can be seen as a fluid from a ballistics point of

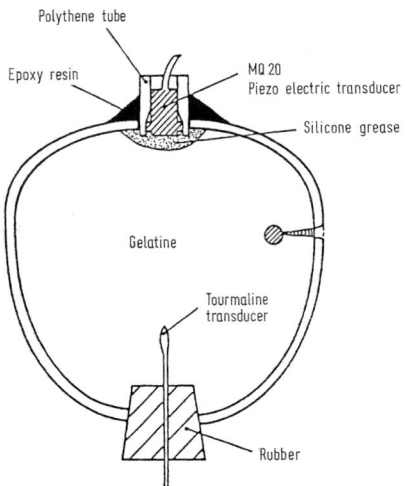

Fig. 3-20. Head model for measuring pressure as a bullet passes through (after WATKINS 1988).

view) the casing allows significant build-up of pressure, which can burst the organ if sufficient energy is present. This is referred to as hydraulic shock. As the pressure builds up in a virtually immobile fluid, the terms "hydrodynamic" or "hydrostatic" shock are inappropriate. The pressure in the fluid in these hollow organs acts in all directions and if the organ bursts, pieces indeed fly in all directions. Fragments are no more likely to be propelled in one direction than another – not even in the direction in which the bullet is travelling.

Of all the hollow organs, the skull is of particular interest. Unfortunately, few authors have conducted pressure measurements in the skull during and after the passage of a bullet. This may be partly due to the practical difficulty of producing a head model capable of representing the human skull with a reasonable degree of accuracy.

WATKINS et al. (1998) looked at the problem of skull pressure measurements. We shall consider the most significant of their results. Their head model is shown in Fig. 3-20.

Dried Asiatic skulls were used that had been kept in a 0.9% saline solution for seven days prior to the experiment. The skulls were then filled with 20% gelatine and covered with two layers of leather soaked in gelatine.

Fig. 3-21 shows a typical pressure curve obtained as a bullet passes through one of these skulls. At the beginning of the process (see Fig. 3-21 for details) we observe a pressure peak of just under 70 bar, which corresponds to the shock wave emanating from the tip of the bullet. This is followed by a series of pressure peaks, which correspond to reflections off the walls of the skull. In order to obtain the pressure due to the pressure wave produced by the temporary cavity, the high-frequency signals of the shock wave were filtered out using a low-pass filter. The result is shown magnified in Fig. 3-21. In addition to the maximum, of just under

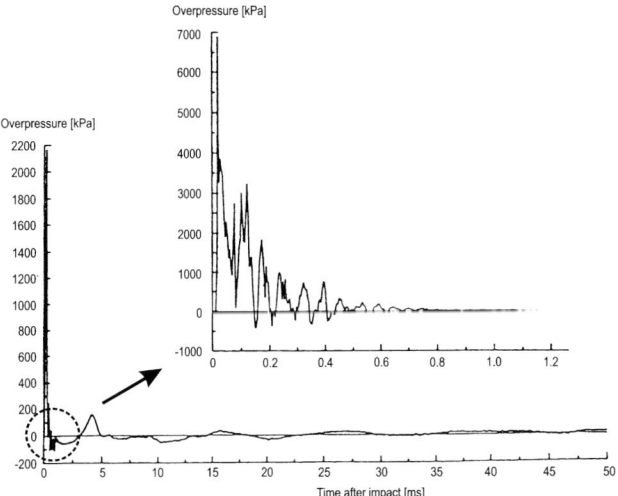

Fig. 3-21. Pressure in a skull. Pressure sensor positioned as shown in Fig. 3-20. The x-axis shows time, in ms.

22 bar, we can identify pressure peaks caused by the pulsation of the temporary cavity. Table 3-3 presents a summary of the values measured.

Fig. 3-22 shows maximum pressure against energy and indicates that there is a linear relationship between maximum pressure and the energy of the bullet. The table also shows the pressures produced when a bullet ricochets off the skull (when the bullet strikes it at an angle). Not surprisingly, the pressure values in such cases are lower than when the bullet passes through the skull, as a bullet that ricochets only transfers a small part of its energy to the skull, with the amount of energy transferred depending on the angle.

It would be useful at this juncture to mention experience from forensic practice regarding the effects of bullets on the human skull.

9 mm Luger bullets always cause a large number of significant bursting injuries to both the calvarium and the base of the skull. A 7.65 Browning bullet causes only fracture lines and bullets from 6.35 Browning and 22 rimfire cartridges leave only entry holes. On close examination, however, one usually finds fracture lines in the anterior cranial fossa. This is the site of the roof of the orbital cavity, which often consists only of a paper-thin bone. These fractures cause bleeding in the soft parts of the orbital cavity, producing the "panda eyes" effect.

Table 3-3. Pressure in a skull as a bullet passes through

Projectile (steel sphere) Size, channel	v [m/s]	E [J]	Maximum pressure	
			Shock [bar]	Quasistatic [bar]
6 mm transverse	205	18.7	7.90	0.75
6 mm transverse	450	90	16	4.30
6 mm transverse	806	290	69	22
3 mm transverse	459	11.8	7	0.65
3 mm transverse	1316	97	45	8.5

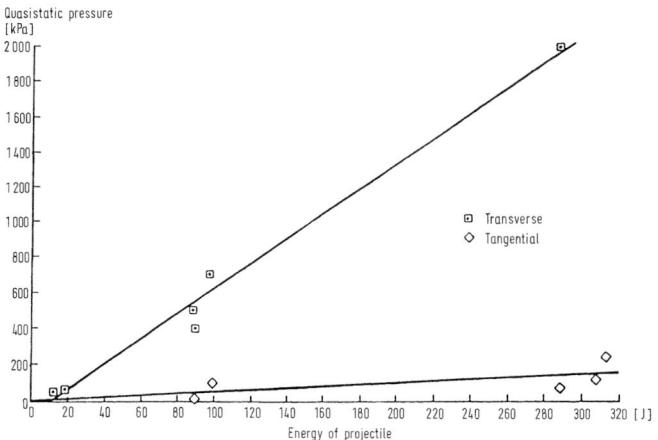

Fig. 3-22. Maximum pressure (y-axis) against bullet energy (x-axis).

As one would expect from theoretical studies and from the experiments described above concerning the pressure inside the skull, rifle bullets, with their high energy levels, have a far greater effect than handgun bullets. A direct hit to the skull at short range – not a ricochet – very often leads to bursting of the skull and destruction of the brain. Tangential hits often destroy the top of the skull but leave the brain lying relatively undamaged next to the corpse (this is known as a "Krönlein shot").

HARVEY et al. (1946) described pressure fluctuations in the *abdomens* of anaesthetized cats as a result of the temporary cavity created by a shot fired into the *upper thigh*. TIKKA et al. (1982a) conducted experiments on pigs.

A Kistler 42213 pressure transducer was implanted in the abdominal cavity via a small incision. The bullets were fired at the middle part of the upper thigh. The cartridges used were as follows: 7.62 × 39 S 309 (Finland), 7.62 × 39 ts D pv (USSR) and 5.56 × 45 M 193 (USA). The ranges were 30 m and 100 m.

On the basis of their measurements, the authors came to the following conclusions:

The pressure fluctuations in the abdominal cavity as a result of shots to (or through) the upper thigh depended on:

– the amount of energy transferred from the bullet to the tissue;
– the distance between the temporary cavity and the abdominal cavity;
– the anatomy of the target.

Maximum pressures of up to just under 1 bar were measured.

There was only a very loose relationship between energy transferred and maximum pressure fluctuation Δp. For instance, $E_{ab} = 107$ J: $\Delta p = 0.98$ kPa. $E_{ab} = 200$ J: Δp only 0.1 kPa (extreme example).

These pressure values are for the abdominal cavity, not the blood vessels. It is reasonable to assume that the pressures in the blood vessels were of the same order of magnitude.

3.2.3.2 Quantitative description of the temporary cavity

OTTOSON (1964) established that if sectional density remains constant, then the diameter d of the temporary cavity at a given penetration depth s along the bullet track is proportional to the velocity v at that point:

(3.2:10) $\quad d(s) = C \cdot v(s)$, $\quad$ where C is constant. $\quad$ [m]

From the above and from the following equation for velocity (derived from Eqn 3.2:5)

(3.2:11) $\quad v(s) = v_a \cdot e^{-\Re \cdot s}$, $\quad$ [m/s]

introducing energy in place of velocity allows us to derive the following equation for the volume of the temporary cavity:

(3.2:12) $\quad V = \dfrac{\pi \cdot C^2}{4 \cdot \Re \cdot m} \cdot (E_a - E_{rst})$. $\quad$ [m^3]

Where E_{rst} is the residual energy after the bullet exits the body and the term in parentheses corresponds to energy transferred E_{ab}. The expression to the right of the equals sign is summarized as the constant μ, thereby proving Eqn 3.2:9.

OTTOSON's empirical approach therefore confirms MARTEL's theory (see 2.3.7.3).

Eqn 3.2:12 only holds good if the retardation coefficient $\Re$ (which depends on the angle of incidence of the bullet, and hence on its instantaneous sectional density) remains constant. This equation is therefore valid only for bullets with a constant sectional density. If $\Re$ changes as the bullet passes through the target, it will not be possible to establish a straightforward relationship between volume and energy transferred.

For the next approach, we shall once again use Eqn 3.2:9:

$$V = \mu \cdot E_{ab} .\quad [\text{cm}^3]$$

E_{ab} is replaced by the sum of the products of E'_{ab} and the corresponding sections of the bullet track Δs:

(3.2:13) $\quad V = \mu \cdot \sum E'_{ab} \cdot \Delta s$. $\quad$ [cm^3]

If we consider only one section Δs at position s_0, then from 3.2:13 we obtain:

(3.2:14) $\quad \left(\dfrac{\Delta V}{\Delta s}\right)_{s_0} = \mu \cdot E'_{ab}(s_0)$. $\quad$ [cm^2]

This equation has serious consequences for the wounding potential (*effectiveness*) of a bullet. It shows that the cross-sectional area of the volumetric element (disc of thickness Δs) at a point s_0 along the track depends on the energy transferred from the bullet at that point. As the value of E'_{ab} at a given point along the track determines the magnitude of the cross-sectional area, the curve of E'_{ab} corresponds to the form of the temporary cavity.

The curve of E'_{ab} against distance along the bullet track depends on the form and design of the bullet. Spheres, and other projectiles of constant sectional density, transfer their energy to the medium in a regular fashion, in accordance with an exponential function. From Eqn 3.2:5, we can derive the following equation:

(3.2:15) $$E(s) = E_a \cdot e^{-2 \cdot \Re \cdot s} ,$$ [J]

using the above, together with Eqn 3.2:9, we obtain the following after rearranging the terms (where d is the diameter of the temporary cavity and C^* is a constant):

(3.2:16) $$d(s) = 2 \cdot C^* \cdot e^{-2 \cdot \Re \cdot s} .$$ [m]

Projectiles with a constant sectional density therefore cause temporary cavities (and hence permanent wound channels) of which the diameter decreases exponentially. For further details on this relationship, please see 4.4.2.2, which deals with the geometric form of the wound channel caused by a fragment of approximately constant sectional density.

3.2.3.3 Influence of impact conditions and bullet characteristics

General. The proper motion of a bullet in flight (the movement of the axis of the bullet about its centre of gravity) was discussed in 2.3.4.4. Under the constant effect of drag, the axis of the bullet describes a hollow cone about the direction of flight (precession). In the event of a sudden disturbance to the flight of the bullet (e.g. if it strikes an object) a second conical movement, known as nutation, is superposed on the first (see Fig. 2-51), causing the axis of the bullet to rotate about the direction of flight, at a constantly changing angle of incidence. As long as the bullet is spinning sufficiently fast, it will not tumble (i.e. yaw).

Motion inside a dense medium is quite different. As mentioned previously, the bullet initially remains stable under the influence of the gyroscopic effect and transfers comparatively little energy to the target. This is the "narrow channel" phase. The angle of incidence then increases extremely rapidly over a relatively short distance due to positive feedback. The bullet yaws and then continues its forward movement while undergoing heavily damped oscillation about a transverse axis. The bullet always stabilizes perpendicular to its direction of movement, as long as it is shape stable.

We shall now see how the length of the narrow channel (NC) is determined by the ballistic characteristics of the bullet at the point of impact and by its dimensions.

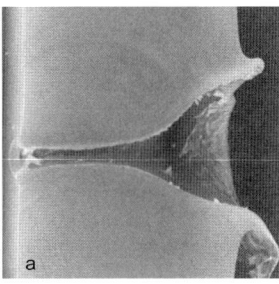 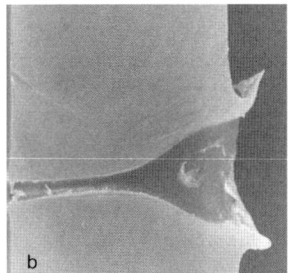 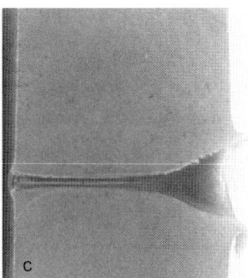

Fig. 3-23. Relationship between the angle of incidence at the moment of impact and the length of the narrow channel. As the distance between muzzle and target (SE) increases, the bullet becomes more stable, the angle of incidence decreases and the length of the narrow channel increases. Experiments conducted by the author (KNEUBUEHL). Bullet fired from left.
a. SE = 30 m, **b.** SE = 100 m, **c.** SE = 300 m.

Relationship between the length of the narrow channel and the angle of incidence at the moment of impact. The angle of incidence ψ_0 of a bullet as it strikes a dense medium has a decisive effect on the way it moves within the medium, and especially upon yaw. A large angle of incidence causes the bullet to yaw rapidly and results in a short NC. If the angle of incidence is small, this sudden yaw occurs later and the NC will be longer. This can be demonstrated empirically by increasing the distance between muzzle and target, as the angle of incidence decreases with increasing distance, as long as the bullet is in stable flight. See 2.3.4.4 and Fig. 3-23 a-c.

The question arises as to what defines the NC, i.e. as to how one defines its end. An angle of incidence of as little as 10° results in a significant lateral force (approx. 17% of the drag) and a correspondingly large overturning moment. Under such conditions, one can fairly confidently assume that yaw has commenced. Geometrically speaking, this angle of incidence corresponds to an in-

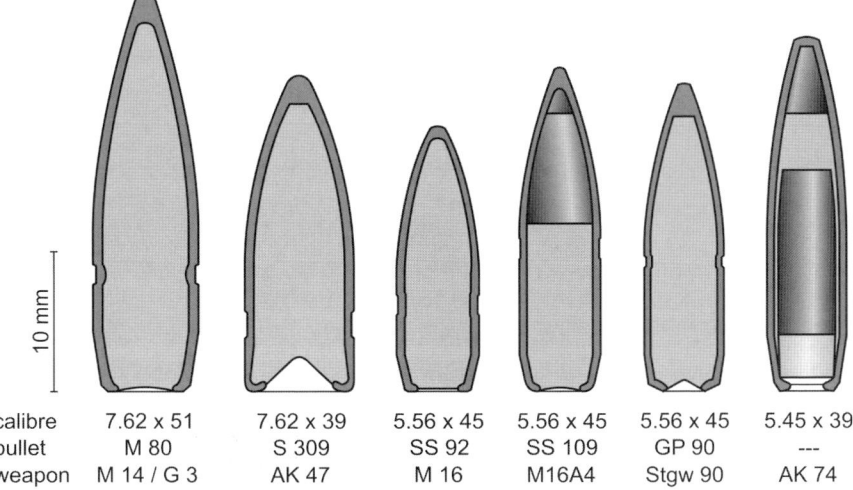

Fig. 3-24. Bullets for military weapons.

crease in the diameter of the wound channel of approximately 20%, which is the criterion defining the end of the NC. If all bullets are assessed in the same manner, there can be no objection to this approach.

The author's own experiments using a variety of bullets (see Fig. 3-24) fired at ballistic soap made it possible to investigate the relationship between the angle of incidence and the length of the NC more closely (KNEUBUEHL and MAISSEN 1978).

In order to determine the influence of the angle of incidence at the point of impact, a simple system of equations of motion was set up that describes the motion of the bullet's centre of gravity and its rotation about a transverse axis through its centre of gravity (yaw). This system of equations involves a drag coefficient C_D and a normal force coefficient C_N. See "Flow forces" in 2.1.4.4. The normal force coefficient is defined in a manner analogous to the lift force coefficient C_L of Eqn 2.1:65 except that lift acts perpendicular to the direction of movement, while the normal force acts perpendicular to the axis of the bullet. With the aid of this system of equations, best-fit calculations were performed using the narrow channel lengths determined empirically. These calculations yielded the initial angle of incidence for each length of NC[1].

Table 3-4. Length of the narrow channel as a function of the angle of incidence ψ_0 at the point of impact.

No.[a]	Calibre	v_a [m/s]	Λ[b] [-]	C_{N0} [-]	ℓ_{NC}[c] [mm]	ψ_0[d] [°]	ℓ_{NC}[e] [mm]	ℓ_g/J_q[f] [1/m·kg]
1	7.62 × 51	760	40	2.30	81	1.45	92	6.9
					102	0.73		
2	7.62 × 39	680	36	1.75	84	0.94	83	9.08
					94	0.65		
3	5.56 × 45	930	55	2.10	64	1.15	67	24.90
					80	0.55		
		930	32	2.10	74	0.73		
					93	0.31		
4	5.56 × 45	905	32	2.65	53	2.20	72	20.57
		905	55	2.65	30	5.30		
					44	3.20		
5	5.56 × 45	875	45	2.50	81	0.77	75	19.13
6	5.45 × 39	870	36	3.00	62	1.55	73	20.74

[a] The numbers correspond to the drawings of bullets in Fig. 3-24.
[b] Twist length, measured in calibres.
[c] Determined empirically (range: 30 m).
[d] Determined from the data obtained empirically.
[e] Calculated taking $\psi_0 = 1°$.
[f] Length of bullet divided by lateral moment of inertia multiplied by 10.

---

[1] A full explanation of this system of equations is to be found in the previous edition of the present work. SELLIER K.†, KNEUBUEHL B., Wundballistik und ihre ballistischen Grundlagen, Springer-Verlag, Berlin, 2nd edition, 2001

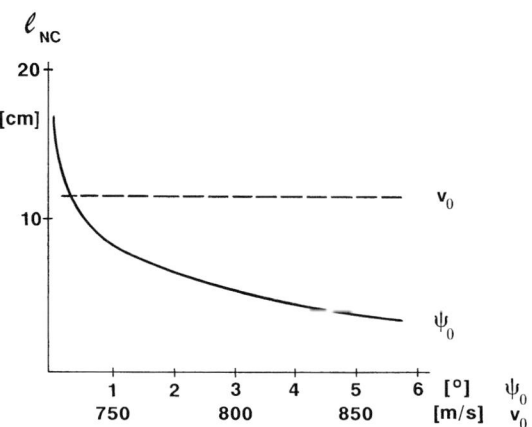

Fig. 3-25. Relationship between the length of the narrow channel and the impact conditions. *y axis:* Length of the narrow channel (ℓ_{NC}). *x axis:* v_0, and angle of incidence (ψ_0) on impact.

Table 3-4 lists the results of the experiments and of the calculations based on them. For bullets 1 to 4, those shots that had yielded the longest and shortest straight narrow channels were selected from six shots. The values for bullets 5 and 6 are typical individual values.

Plausible values were obtained for the angle of incidence, even when (in the case of bullets 3 and 4) the same bullets were fired with varying degrees of stability (achieved by using barrels with different twist lengths).

Using this system of equations, it is also possible to calculate the influence of the impact conditions and the bullet parameters, if the latter are more or less constant. From Fig. 3-25, we can see that – under this condition – the length of the narrow channel is greatly influenced by the angle of incidence but – within a certain range – is affected only minimally by the impact velocity. Likewise, the drag coefficient of the bullet C^*_{D0} can vary across a wide range without significantly affecting the length of the narrow channel, whereas C^*_{N0}, the bullet parameter that determines lateral force, has a very pronounced effect (see Fig. 3-26).

Example: According to WARKEN (1982), the mean angle of incidence at 30 m for a 7.62 mm NATO round is 0.95°. An extensive series of experiments conducted by the author

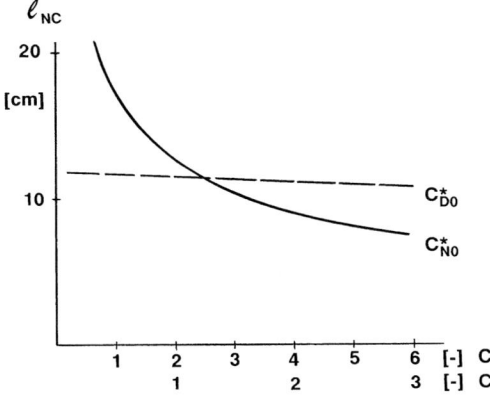

Fig. 3-26. Relationship between the length of the narrow channel (ℓ_{NC}) and bullet parameters.
C^*_{D0}: Drag coefficient relative to effective surface in contact with the medium.
C^*_{N0}: Corresponding lateral force coefficient.

(KNEUBUEHL) yielded a mean narrow channel length of 85 mm for the same bullet at the same distance.

One obtains the same length from the equations above if the (plausible) value of 2.3 is used for the constant C^*_{N0}. As we have seen, the other constants are of secondary importance.

Relationship between narrow channel length on the one hand and the length of the bullet and its lateral moment of inertia on the other. For a given angle of incidence, the point at which the bullet starts to yaw in a dense medium depends largely on the ratio of bullet length to lateral moment of inertia (ℓ_g/J_q). This ratio is shown in Table 3-4, together with the corresponding narrow channel length, calculated using the system of equations described, with a constant angle of incidence.

It is now possible to write an equation for the relationship between the quotient ℓ_g/J_q (calculated from the bullet characteristics) and the calculated length of the NC ℓ_{NC} for the bullet concerned. At a range of 30 m:

$$(3.2{:}17) \qquad \ell_{NC} = 31.6 \cdot \left(\frac{J_q}{\ell_g}\right)^{0.212}, \qquad \text{where } r = 0.98. \qquad [m]$$

The first thing that becomes apparent is that the smaller-calibre bullets, Nos 3 and 4, produce a shorter narrow channel than the 7.62 mm bullets. This is due to the ratio ℓ_g/J_q being higher for the smaller-calibre bullets. Generally speaking, short bullets and bullets with a high lateral moment of inertia produce long narrow channels.

When, during the Vietnam War, the US forces used their M 16 assault rifle with the corresponding ammunition, and saw that the wounds it caused were considerably more severe than with "conventional" bullets, the reaction was to blame the small calibre as such for the grim wounding potential of the round. That this conclusion was mistaken is clear from the behaviour of newer small-calibre bullets (e.g. No. 5), which behave significantly better in terms of narrow channel and deformation than the old 5.56 mm. It is possible to influence the length of the narrow channel significantly by modifying the parameters of the bullet (e.g. its moment of inertia) and the twist length of the weapon. Or, to put it differently, within certain limits it is possible to produce small-calibre bullets with acceptable behaviour.

3.2.3.4 The effect of the sectional density of a bullet on the shape of the temporary cavity

One might ask what happens, in terms of physics, if the velocity of a projectile is increased while keeping E_0 constant (which implies reducing mass). In order to eliminate any effects linked to the angle of incidence, we shall initially just look at spheres.

As mentioned above, energy transferred per cm E'_{ab} is given by:

$$(3.2{:}18) \qquad E'_{ab} = \rho \cdot C_D \cdot \frac{A}{m} \cdot E(s) = \rho \cdot C_D \cdot \frac{1}{q} \cdot E(s). \qquad [\text{J/cm}]$$

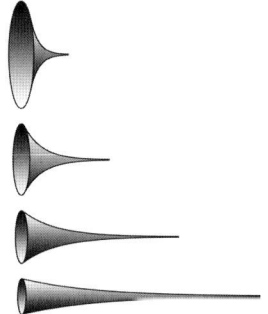

Fig. 3-27. Relationship between sectional density of a bullet and shape of temporary cavity. Lowest sectional density at top, highest at bottom. (Schematic, after SELLIER 1982).

If E(s) remains constant, then E'_{ab} depends only on $1/q$ and C_D. However, q is proportional to k, which means that as the calibre of the sphere decreases the sectional density will also decrease and hence the energy transferred E'_{ab} will increase. However, if E'_{ab} increases, a smaller sphere will lose more energy than a larger one, which means that penetration depth will increase. If we assume that even the largest of the spheres used comes to rest inside the block – i.e. it transfers all of its energy to the block, which implies that E_{ab} is the same for every sphere – then the volume of the cavity will also be the same for every sphere, as $V = \mu \cdot E_{ab}$. However, a larger value of E'_{ab} implies a greater cavity diameter, as we saw earlier.

In summary, therefore, smaller spheres produce a wound channel that is shorter but has a larger radius of destruction. This is shown schematically in Fig. 3-27 (after SELLIER 1982). The effect of high levels of energy being transferred in the case of small spheres fired at very high velocities is accentuated by the increase in C_D as the velocity of the spheres approaches the speed of sound in water or media with a high water content, such as gelatine and tissue (approx. 1500 m/s).

At these very high velocities a further effect occurs, which also tends to shorten the wound channel. The pressure on the face of the sphere (the stagnation pressure) is proportional to velocity squared. At about 2800 m/s, the stagnation pressure exceeds the resistance of steel, and the sphere fragments in the tissue. The fragments now form separate projectiles, and their sectional densities are lower than that of the sphere. As a result, they are subject to much higher deceleration. The fragments therefore transfer all their energy over a very short penetration distance.

CHARTERS and CHARTERS (1976) experimented with steel spheres 6.35 mm and 3.18 mm in diameter (1/4" and 1/8" respectively). The spheres were fired at velocities of between 750 m/s and 5510 m/s. The authors confirmed the theoretical predictions outlined above; if spheres with the same energy but of different mass are fired at a target, the smaller spheres will produce shallower cavities at lesser penetration depths but with larger openings. At a v_a of 2850 m/s and above, the steel spheres fragment. As mentioned above, the resulting fragments achieve only limited penetration depths (see Table 3-5).

Table 3-5. Temporary cavity as a function of E_a. (After CHARTERS and CHARTERS 1976)

Projectile ⌀ [mm]	v_a [m/s]	E_a [J]	Max. cavity ⌀ [cm]	Max. cavity Sec. Area [cm²]	Max. cavity Vol. [cm³]	Penetration depth [cm]	Condition of projectile
6.35	750	295	7.1	39	423	26.6	Intact
3.18	1940	234		No data		15.6	Intact
3.18	2090	279	8.6	58	561	18.0	Intact
3.18	2850	518	9.8	75	488	5.2	Fragmented
3.18	2880	529		No data		8.2	Fragmented
3.18	5510	1936	14.3	160	1051	7.4	Fragmented

We can draw the following conclusions from Table 3-5: If the sphere remains intact, then despite the values of E_a being approximately the same, a smaller sphere penetrates less than a larger one. See Rows 1 and 3. This is clear from Eqn 3.2:18. At a velocity of 2850 m/s – if not before – the resistance of the steel is exceeded. The sphere fractures and penetration depth decreases substantially, for the reasons mentioned above.

The authors saw the curve of C_D (see 2.1.4.4, Eqn 2.1:64) as a function of the Mach number Ma. According to DUBIN (1974), the value of C_D in a medium ranges from approx. 0.38 where Ma < 0.8 to 1.0 where Ma ≥ 1. The authors conclude from this that when the velocity of the sphere exceeds the speed of sound the temporary cavities "suddenly" become substantially shallower (see Eqn 3.2:18 regarding the relationship between energy transferred on the one hand and q and C_D on the other). However, this statement is not supported by the measurements published (Table 3-5). The results for the fragmented spheres cannot be compared with those for intact spheres, as the conditions are qualitatively quite different; the sectional density decreases as a result of the fragmentation (the total surface area is greater) and, in accordance with Eqn 3.2:18, more energy is transferred, causing the penetration depth to decrease considerably. As a result, it is completely impossible to compare the penetration depths of the intact spheres with those that have undergone fragmentation.

The authors' (unconvincing) conclusions prompted FACKLER et al. (1986) to conduct further experiments. Their approach was as follows: If one were to compare the volume of the first part of the temporary cavity – from the entry wound to a point approximately 10 calibres into the target (V_{entr}) – with the total volume of the temporary cavity (V_{tot}), then the quotient $Q = V_{entr}/V_{tot}$ should change significantly when the velocity of the sphere exceeds the speed of sound, as the sharp increase in C_D at this point (Ma = 1) would also cause E'_{ab} to change suddenly, as E'_{ab} is proportional to C_D. The experiments showed, however, that the function Q v_a produced a smooth curve to within the tolerances of experimental scatter. Table 3-6 presents the principal results of the experiments.

However, the negative results of the experiments conducted by FACKLER et al. (i.e. no excessively large cavity diameter at supersonic velocities) does not of itself constitute adequate proof that the

Table 3-6. Results of experiments with 6 mm steel spheres (from FACKLER et al. 1986)

v_a	E_a	V_{tot}	Q [a]	Penetration depth	Diameter [b]
[m/s]	[J]	[cm³]	[-]	[cm]	[mm]
392	960	330	0.20	41.0	6.00
537	1124	371	0.23	43.0	6.00
741	1320	890	0.09	48.0	6.00
918	1470	900	0.09	52.0	6.00
963	1505	1140	0.09	50.5	6.14
1107	1614	1175	0.09	49.0	6.19
1486	1870	1980	0.06	47.0	6.24
1675	1985	1690	0.13	30.5	6.90

[a] $Q = V_{ent}/V_{tot}$ (see text). [b] Diameter of sphere after experiment.

phenomenon does not exist, for the following reason:

Substituting $\rho = 1000$ kg/m³, $q = 15.6$ kg/m² (for steel spheres 3.18 mm in diameter), $C_D = 1$ and $E_a = 234$ J into Eqn 3.2:18 shows that after only 1 cm (just under 2 calibres) the sphere has lost 150 J, which means that it has slowed to below the speed of sound in less than 1 cm (at the speed of sound, a 3.18 mm steel sphere possesses about 140 J of energy). The authors considered the cavity from the point of penetration up to a depth of 10 calibres, and used that to determine V_{entr} empirically. However, most of that volume is created while the sphere is travelling at subsonic speeds. Any effects attributable to the sphere moving at supersonic speeds would be hidden by the excessively large volume taken into consideration (corresponding to a depth of 10 calibres). One must therefore use a volume corresponding to a smaller penetration depth (approx. 2 calibres, according to rough calculations) to calculate the quotient Q (see above), in order to obtain reliable results.

It is clear from these results that as the sectional density decreases, the initial diameter of the temporary cavity increases and the penetration depth decreases. Any sudden increase in the amount of energy transferred around Ma = 1 is of little significance for the effectiveness of normal bullets, as the velocity of such a bullet within a dense medium is lower than the speed of sound (Ma < 0.7).

In summary, one can say that the four parameters listed below are required in order to describe the effect of a bullet, and that the extent and form of the temporary cavity – factors that depend on the energy transferred – are the most important:

1. $E_{ab}(s)$ – Energy transferred as a function of penetration depth (proportional to the total volume of the temporary cavity up to penetration depth s).
2. $E'_{ab}(s)$ – The gradient of the energy transfer function, which corresponds to the local cross-sectional area of the temporary cavity at s.
3. The condition of the bullet during penetration (intact, deformed or fragmented).
4. Penetration depth.

3.2.4 The effect of bullet design on behaviour

3.2.4.1 Categories of bullet

In this section, we shall discuss the changes that a bullet can undergo when penetrating and passing through tissue, ballistic soap and gelatine. We have already looked at the general behaviour of deforming and fragmentation bullets in 3.2.2.1 (rifles) and 3.2.2.2 (handguns).

The factors that determine the nature and degree of the changes that a bullet undergoes are its design, its impact energy and certain characteristics of the target medium.

As far as changes in shape are concerned, we can divide bullets into three categories:

- shape-stable;
- deforming;
- fragmenting.

Generally, we distinguish between deformation and fragmentation by looking at the mass of the largest fragment remaining after impact. If the mass of this fragment is equal to at least 90% of the original nominal mass of the bullet, we speak of deformation rather than fragmentation.

As regards their design, bullets can be divided into the following categories (see also 2.2.2.1, Fig. 2-9 and the overview of bullet types and names in A.5):

- solid bullets;
- full metal-jacketed bullets (FMJ) – lead core, with jacket covering all but the tail;
- semi-jacketed bullets (SJ) – lead core exposed at tip;
- hollow-point bullets (HP) – cavity in tip;
- solid bullets – bullets made of a single material and designed to deform on impact.

However, many modern bullets – especially those designed for hunting – cannot be assigned to one of these groups (see Fig. 2-24 for examples).

3.2.4.2 Deformation and fragmentation; general points

Until some point in the second half of the 20th Century, it was generally possible to predict the behaviour of a bullet from its design, as long as it hit the target at design velocity. One could assume that semi-jacketed and hollow-point bullets would deform or fragment on impact with soft tissue or equivalent simulants. There are, however, good reasons for designing bullets of these types that are not intended to deform, or for which deformation would in fact be undesirable.

If the most important factor is accuracy, then it is the tail of the bullet that must be manufactured to close tolerances, more than the tip. The jacket is manufactured by drawing, and in such a case it is useful if the tail rather than the tip can be drawn. However, this requires that the lead core be introduced via the tip, which results in a bullet with an open tip, with the lead core exposed to a greater or lesser extent. The Sierra Matchking is a typical example of this type of bullet.

Certain types of sport shooting (such as dynamic and silhouette shooting) involve firing at targets made of hard materials. For reasons of safety, it is important to minimize the risk of ricochets. Solid lead and semi-jacketed bullets with exposed lead tips deform much more on hitting hard targets than do full metal-jacketed bullets, and hence are significantly less likely to ricochet. The risk of jacket fragments flying back towards the shooter is also reduced.

Today, appropriate choice of copper alloy for the jacket, lead alloy for the core and shape of bullet make it possible to manufacture semi-jacketed and hollow-point bullets which, at the normal impact energies for the calibre concerned, either do not deform at all or else deform so little that the cross-sectional area does not increase. This means that it is now virtually impossible to predict whether a bullet will deform in a given target medium purely on the basis of its design.

With deforming bullets, the increase in cross-sectional area depends not only on bullet design but also – to a very large extent – on the impact velocity v_a. Impact velocity is often ignored when discussing the behaviour of bullets.

Examples: When a 38 Spl HP bullet fired from a long-barrelled revolver hits a soft target, it deforms as required. The same bullet, with the same cartridge, but fired from a short-barrelled revolver, hardly deforms at all on hitting the same target, as it is travelling much more slowly.

A hunting bullet that deforms as required at normal velocity penetrates the body without deforming if it is fired into the air and falls tip-down. Its velocity under these conditions is approximately 120 m/s – terminal velocity.

We can therefore assign two velocity thresholds to every deforming bullet: the first (v_{D50}) is the velocity at which the increase in cross-sectional area is only half as great as before and the second (v_{D0}) is that below which the bullet does not deform at all. The first threshold could limit the practical range for hunting. For fragmentation bullets, there is a further threshold velocity: that below which no fragmentation takes place.

In a soft, dense medium, deformation is affected not only by impact velocity but also by the time for which the bullet tip is subjected to pressure. Under certain circumstances, a bullet can pass through a thin layer of material so quickly that there is no time for the work required for deformation.

Example: A wax candle that pierces a thin board at high speed undergoes virtually no deformation, as the impact time is very short and the force-time integral hence very small. Any attempt to push the candle through the board slowly would completely deform it.

The design of the bullet and the materials of which it is made both play a decisive role. Solid bullets made of copper alloys deform significantly faster than do conventional semi-jacketed or hollow-tip bullets with lead cores.

3.2.4.3 Experimental results

Rifles. Under certain conditions, jacketed spitzer bullets for rifles – especially for small calibres – can deform or even disintegrate. As long as this type of bullet passes through the medium normally – i.e. tip first – the favourable ballistic shape of the tip endures that the forces on the bullet remain relatively small. If the angle of incidence increases, however, the bullet will present a larger surface area to the medium. If the bullet is perpendicular to its direction of travel, a large force (up to a few tens of kilonewtons) will be acting on a narrow meridian surface along the side of the bullet. Depending on the impact velocity and the design of the bullet, this pressure can compress the bullet, squeezing lead out of the rear, and the bullet may break at the cannelure. If the forces are sufficiently large, the jacket will be ripped open. This behaviour only occurs at short range, however (i.e. up to around 50 m).

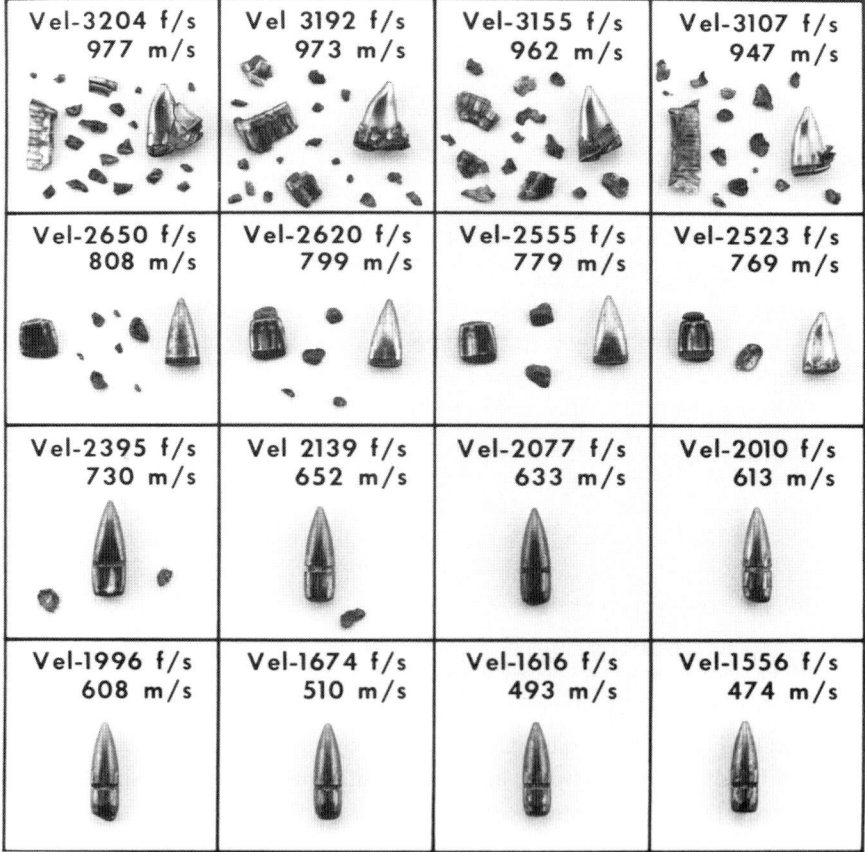

Fig. 3-28. Deformation/fragmentation of a 5.56 × 45 (223 Rem.) M193 bullet as a function of impact velocity (with kind permission of FACKLER).

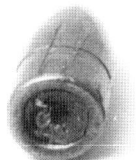  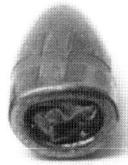  

Fig. 3-29. Deformation of the tail of a bullet in gelatine as a function of impact velocity (KNEUBUEHL). *From left to right:* 600 m/s, 630 m/s, 670 m/s, 690 m/s and 710 m/s (7.62 × 39 bullet).

At low impact velocities (i.e. less than approx. 500-600 m/s) the bullet remains undeformed and intact, even after turning through 90°.

Fig. 3-28 contains an instructive presentation of the results of bullet disintegration from FACKLER et al. (1988b). The points made above are illustrated very clearly by the state of the bullets and bullet fragments (5.56 × 45 M193).

It is clear from the photos that crushing is much more pronounced in the rear, cylindrical part of the bullet than in the leading part, the ogive. This is primarily because the tip is stiffer and because the tail is either open or else sealed simply by a thin brass disc.

It is interesting to note that deformation depends not on the calibre or form of the bullet but merely on its design (full metal-jacketed, lead core). Similar deformation is observed with 7.62 × 39 and 7.62 × 51 bullets at the same range of speeds (see Figs 3-29 and 3-30). The velocity at which changes in shape begin (approx. 600 m/s) remains approximately the same. In forensics, this behaviour can be used to estimate the impact velocity. Bullets of the type under investigation are fired into soap or gelatine repeatedly, adjusting the velocity until the deformation of the test bullet matches that of the bullet from the crime scene. This is then the approximate impact velocity (FACKLER).

Handguns. The velocity of full metal-jacketed bullets fired from handguns is generally substantially lower than the threshold of 500-600 m/s mentioned above. As a result, the bullet undergoes no change in shape when fired into a soft medium. However, deforming bullets deform in accordance with velocity and penetration depth.

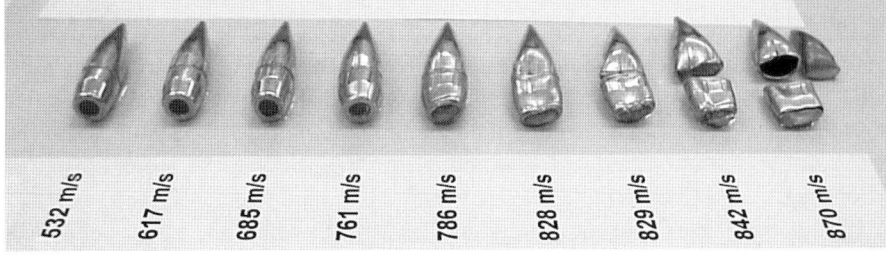

Fig. 3-30. Deformation/fragmentation of a British 7.62 mm L2A2 NATO bullet in water as a function of impact velocity (with kind permission of Lucien HAAG).

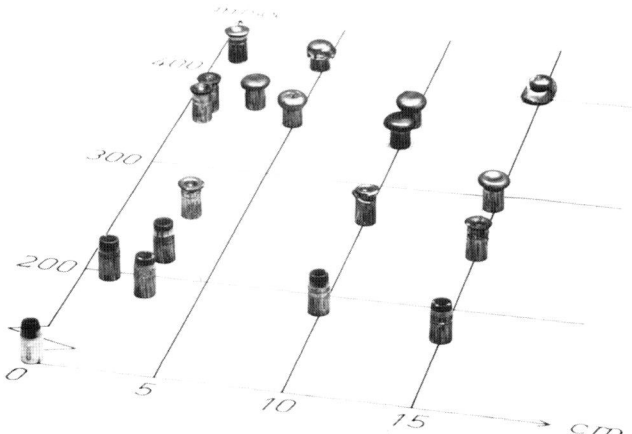

Fig. 3-31. Deformation of an SJHP bullet (NORMA No. 689, 38) after passing through 20% gelatine, as a function of the thickness of the gelatine sheet (x-axis) and velocity (y-axis).

Fig. 3-32. Bullets from Fig. 3-31. Left: an unused bullet.

This is illustrated by the results of systematic experiments (SELLIER 1982) that involved firing NORMA SJHP bullets into 20% gelatine blocks of various thicknesses at various velocities. The results are shown in Figs 3-31 and 3-32 (see also RAGSDALE and SOHN 1988, for instance). It is apparent that for a constant impact velocity, deformation of the bullet becomes increasingly marked as the thickness of the layer through which it passes increases (up to a certain limit), as long as impact velocity does not fall below 200 m/s. Below that velocity, this particular type of bullet does not deform.

Handguns clearly illustrate the greater tendency of solid copper-alloy bullets to deform, by comparison with conventional lead-cored deforming bullets. Solid copper bullets achieve maximum effectiveness (and hence maximum deformation) after penetrating just 2 to 4 cm, whereas this only occurs after 4 to 6 cm with lead-cored bullets (see Fig. 3-33).

The first point of maximum energy transfer occurs somewhat earlier with non-deforming full metal-jacketed bullets (greater energy for almost the same sectional density).

Deformation and penetration depth are also related to velocity in the case of solid bullets. As impact velocity decreases, so does the diameter of the bullet tip after impact, until a point is reached at which the bullet remains shape-stable. With the 9 mm Luger Action 4 bullet, for example, this point comes at around 230 m/s (KNEUBUEHL and GLARDON, 2010, see Figs 3-34 and 3-35).

It is interesting to note the change in the depth to which the bullet penetrates ballistic soap. This remains virtually constant at velocities of 400 to 460 m/s (tip diameter of between 12.5 and 11 mm) but increases as the velocity drops further,

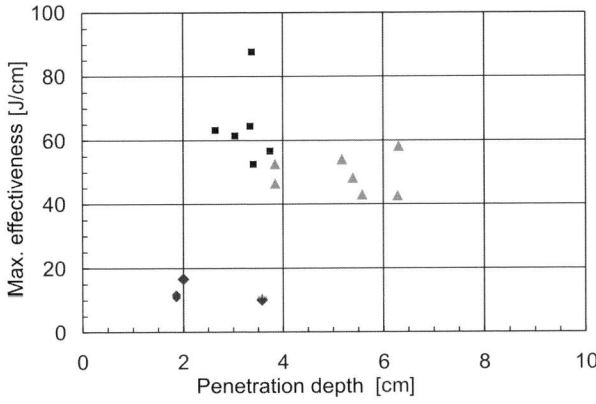

Fig. 3-33. Position and magnitude of maximum effectiveness.
■ Solid bullets.
▲ Lead-cored bullets.
◆ Full metal-jacketed bullets.
(KNEUBUEHL 1999a).

reaching a maximum at approx. 270 m/s (see Fig. 3-34) before starting to decrease again.

The increased penetration depth is accounted for by the increase in sectional density that accompanies the decrease in diameter. This factor clearly has a greater influence than does the falling energy level.

3.2.5 Patterns in bullet wounds to bones

If a bone is hit by a bullet of which the calibre is less than the diameter of the bone, fracture is caused primarily by hydraulic pressure in the bone marrow. In the absence of bone marrow (e.g. in the case of the shoulder blade), the bullet

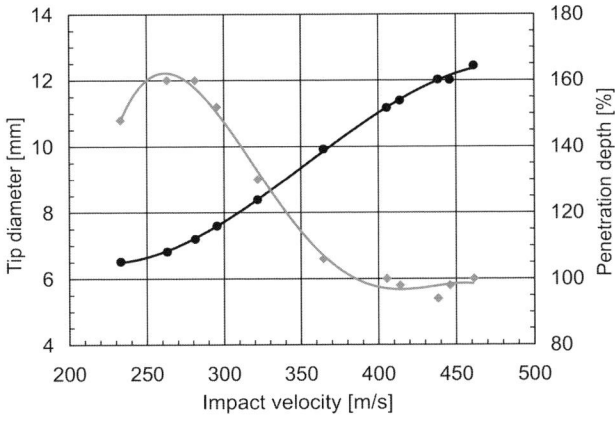

Fig. 3-34. Tip diameter of an Action 4 bullet after impact, as a function of impact velocity (black line) and corresponding penetration depths in ballistic soap (grey line).

233   281   322   365   414   446 m/s

Fig. 3-35. A number of bullets from the experiment in Fig. 3-34, with the corresponding velocities.

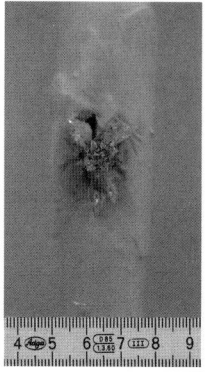 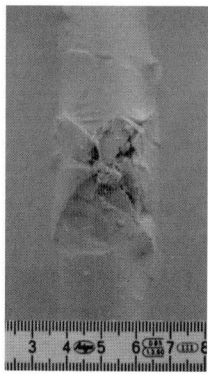

Fig. 3-36. Entry wound side of plastic bone. *Left:* 9 mm Luger full metal-jacketed bullet (impact energy approx. 500 J). *Right:* 7.62 mm NATO full metal-jacketed bullet (impact energy approx. 3 200 J). A 5 cm covering of gelatine was used in both cases.

generally leaves a hole, with little fissuration. A fracture is therefore to be expected when sufficient pressure can be created in the marrow, i.e. when the projectile can transfer sufficient energy.

The amount of energy transferred depends on both the total available energy and the time for which the bullet remains inside of the bone. This time is inversely proportional to the velocity of the bullet.

If a relatively slow full metal-jacketed bullet from a handgun impacts the bone in stable flight, it may cause as much damage as a substantially faster stable rifle bullet carrying significantly more energy. Indeed, it may even cause more damage than the rifle bullet. See Fig. 3-36.

If the narrow channel of a rifle bullet passes through a hollow bone, the bullet will lose comparatively little energy (some 200 J at an impact energy of approx. 3000 J) and will undergo no deviation. The narrow channel will be somewhat shorter, but any decrease in the size of the temporary cavity will be virtually undetectable (see Fig. 3-37).

Regardless of bullet type, the damage done to the bone by the pressure build-up in the marrow corresponds approximately to the dimensions of the temporary cavity (see Fig. 3-38). The pressure inside the bone will propel bone fragments in the direction of the shot and in the opposite direction, with fragments entering the temporary cavity. Bone fragments possess too little energy to create their own wound channels outside the temporary cavity – they do *not* act as secondary pro-

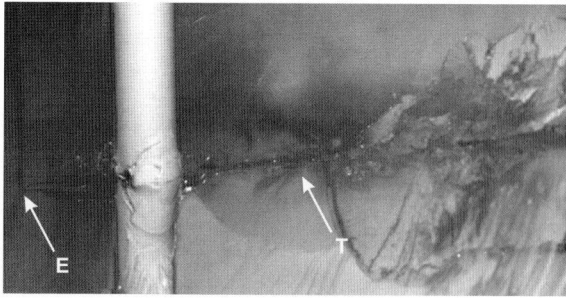

Fig. 3-37. A bone pierced by a full metal-jacketed rifle bullet. The length of the narrow channel is almost identical to that observed in the absence of bone (bullet fired from left).
E: Bullet entry. T: Start of temporary cavity.

 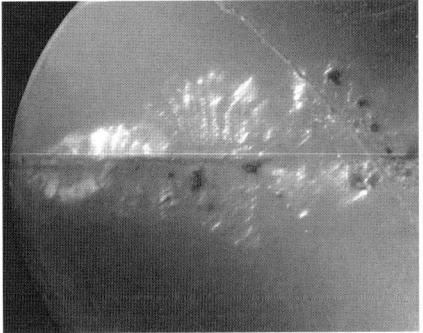

Fig. 3-38. Bone damage corresponding to the extent of the temporary cavity. Semi-jacketed rifle bullet. Shot from front left. E: Bullet entry.

Fig. 3-39. Wound channel in gelatine after bullet has passed through bone (model skull). All bone fragments lie within the area previously occupied by the temporary cavity.

jectiles. After the temporary cavity collapses, the particles are always found within the damaged tissue (see Fig. 3-39).

3.2.6 Bullet temperature and sterility

3.2.6.1 Historical background

The following discussion is based on the work of THORESBY and DARLOW (1967).

It is still a widely-held belief that a bullet is sterilized by being fired, and remains sterile until it reaches its target. SPENCER (1908) stated that most bullet wounds were initially aseptic. OGLIVIE (1944) assumes that bullets are generally sterile. SLESINGER (1943) speaks of high velocity bullets as having a clean surface, sterilized by heat. But as early as 1892, LAGARDE had stated in his work that if a bullet is not sterile before firing, it will not be sterile after firing. He based his conclusions on the work of German army doctor VON BECK, who had shown that the maximum temperature of a bullet was approx. 110°C and that bullets contaminated with septic blood infected the wound channels of animals. LAGARDE (1895) developed his conclusions further and WOODRUFF (1898) introduced into wound ballistics the term "cavitation," which comes from maritime technology (see 3.2.1.1), partly to explain the suction effect. In his textbook, LAGARDE (1914) emphasizes that a bullet wound is never bacteriologically clean. In experiments using both animals and gelatine, DZIEMIAN and HERGET (1950) showed that barium sulphate powder and various dyes could be detected at both entry and exit wounds, as a result of the temporary cavity. ZIPERMAN (1961) came to the same conclusion. In his work on tetanus in wounds, and on the basis of work he had conducted previously (1943), MACLENNAN (1962) concluded that contaminated clothing was the primary source of infection in the case of bullet wounds.

3.2.6.2 Bullet temperature

There are two relevant questions regarding the risk of infection from a bullet:

1. What is the maximum temperature that a bullet can reach between leaving the cartridge and entering the body?
2. Is a bullet sterile on impact, or not?

To some extent, the answer to the second question depends on the answer to the first. If, for instance, the maximum temperature at the surface of a bullet is very high, one can assume that the bullet will be sterilized by this high temperature.

On its path through the barrel, the air and the target body, a bullet is subjected to a number of factors that influence its temperature. The different sources of heat do not affect the bullet equally. Two sources will tend to raise the temperature of the bullets, and their effects are additive.

At the moment of firing, the base of the bullet is exposed to the hot combustion gases. Shortly after it starts to move, the bullet is pressed into the lands of the barrel, which also generates heat.

The resistance due to friction as the bullet moves down the barrel is small by comparison with the engraving resistance.

Quantitatively speaking, the situation is as follows: LAMPEL and SEITZ (1983) concluded that in the case of a small-calibre bullet with a hard jacket, passing down a long barrel, some 7% of the total energy from the propellant is used to heat the bullet, through friction in the barrel and heating of the base of the bullet through direct contact with the combustion gases. With a larger-calibre bullet, a shorter barrel and a thinner or softer jacket, considerably less energy is converted into heat. Approximately one third of the energy delivered by the propellant is transferred to the bullet in the form of kinetic energy. See Table 2-11.

The increase in temperature that a bullet undergoes as the result of a quantity of heat Q being applied to its surface can be estimated using the following formula:

(3.2:19) $$\Delta T = \frac{Q}{c_p \cdot m} \quad [K]$$

where c_p is the specific heat capacity, m the mass and ΔT the change in temperature.

If Q, c_p and m are known, it is therefore possible to calculate the temperature change ΔT. For a fully-jacketed 9 mm Luger bullet with 0.35 g of propellant, for instance, the total propellant energy E_{tot} is 1450 J. According to LAMPEL and SEITZ, 7% of this (101.4 J) is used to heat the bullet. The steel jacket of the bullet ("Schweizer Ordonnanz") weighs about 1.85 g and the specific heat capacity of steel is 0.5 J/(g·deg). Substituting these values into Eqn 3.2:19 yields an estimated temperature increase of approximately 110 °C, raising the temperature of the bullet's surface to approximately 130 °C.

On leaving the barrel, the bullet is exposed to drag, with a friction component of approx. 10 %. From the energy that the bullet loses during its trajectory we can therefore calculate the amount of heat transferred to the bullet. However, the air flowing past the bullet is considerably cooler than the surface of the bullet, which means that heat will be lost to convection. As far as we are aware, no research has been conducted into whether temperature increases or decreases during flight.

MARTY et al. (1994b) used infrared thermography to measure skin temperature at the entry wound in the case of a shot from close range. As part of the same study, they also succeeded in measuring the temperature of a bullet in flight. The surface temperature of the full metal-jacketed 9 mm Luger bullet was 152 °C, which is somewhat higher than one would expect from the calculation above. Given the limited distance involved, this difference cannot be ascribed to drag. It is more likely that the hot gases that surrounded the bullet after it left the barrel (see Fig. 2-47) had an additional heating effect.

The situation is broadly similar with rifle bullets. LAMPEL and SEITZ (1983) state that jacket temperatures of up to approximately 170°C are possible. If one bears in mind that at normal ranges these temperatures are only present for a fraction of a second, it is perfectly possible for bacteria to survive for the duration of a bullet's trajectory, and hence for infection to take place, especially if the bullet remains in the wound.

3.2.6.3 Bullets contaminated with bacteria

Experiments carried out by JOURNÉE et al. (1930) showed that a bullet contaminated by bacteria was capable of transferring those bacteria to the target.

Equipment used: 1. Gewehr 98 rifle, 7.9 mm, spitzer bullet with steel jacket. 2. 7.65 Browning pistol. For some experiments the bullets were fired as issued, for others the tip was machined down and a cup 4 mm in diameter was inserted. The bullet was sterilized in a Bunsen burner flame and 2 to 3 mm of a Bacterium prodigiosum culture were deposited on it, either 5 mm from the tip (but still on the ogive) or in the cup. The target consisted of a 7 cm layer of cotton wool, with a 4 mm steel plate behind. This was wrapped in three layers of filter paper and the assembly was sterilized. The rifle was fired from a distance of 7.6 m, the pistol from 2 m. After firing, the targets were checked for bacteria by creating a culture and by examining them under a microscope. The bacteria were found to have been transferred to the target in a virulent state by the bullet.

THORESBY and DARLOW (1967) conducted similar experiments using blocks of gelatine. Here again, Bacterium prodigiosum was used.

This type of bacterium is often used for experiments, as it forms clearly visible red colonies in a culture medium at incubation temperatures of under 30°C. The bacteria are heat labile, and do not survive long in the laboratory. The bullets used (22 Hornet, m = 2.9 g, v_0 = 490 m/s) were sterilized, and approximately ten of these heat-sensitive organisms were deposited on the head of each. The gelatine block was found to have been contaminated along the entire length of the wound channel.

WOLF et al. (1978) fired solid lead and copper jacketed 38 bullets contaminated with Staphylococcus aureus into sterile sand. The bacteria were transferred to the sand.

In point of fact, the question as to whether a bullet is sterile or contaminated with bacteria is of secondary importance. Even if the bullet were to be sterile on impact, the wound would become infected either by the skin carried into the wound (skin is never sterile) or by fibres from clothing.

TIAN et al. (1982) conducted experiments on this. They fired 5.56 mm bullets weighing 3.5 g into both hind legs of dogs. The bacteria used were Serratia marcescens (=Bacterium prodigiosum). Pieces of soft cloth measuring 6 × 6 cm were soaked in the bacteria. In the first series of experiments, the prepared piece of cloth was placed in front of the site of the entry wound before firing. After the experiment, bacteria were found all along the wound channel, including the exit wound. This could have been because the bullet had carried contaminated fibres into the wound channel, or because the bullet itself had become contaminated on passing through the cloth. In the second series of experiments the cloth was placed over the site of the exit wound. The exit wound was contaminated in all cases. The middle of the wound channel was contaminated in almost all cases, and in the majority of cases the entry wound was also contaminated, which clearly showed that bacteria had been sucked in when the temporary cavity was created, rather than being carried in on the bullet.

3.2.6.4 Burns due to bullets

The information above regarding the temperature of a bullet indicates that it is not possible for a bullet to burn skin on impact or in passing through the skin, especially when one considers that bullet and skin are in contact for well under a millisecond.

Nevertheless, one sometimes hears doctors unfamiliar with ballistics state that they have found a black ("charred") ring around the entry wound, and that this is the result of contact with the hot bullet. What is interpreted as a blackened zone caused by charring of the skin is in fact the bullet wipe that occurs when a bullet penetrates skin without first passing through clothing. When this occurs, the black gunshot residue on the bullet is wiped off on the edge of the wound, forming a black ring that could look like charring.

The question is sometimes raised of whether the residual rotational energy of a bullet that has come to rest in the body is capable of causing heat damage to the surrounding tissue. The answer is no. Rotational energy accounts for less than 0.5% of the total kinetic energy of a bullet. With a 7.62 mm NATO bullet, for instance, this equates to only 9 J (see Table 2-11).

3.3 Simulants

3.3.1 General

In the context of wound ballistics, a simulant is a material that reacts to bullets in a manner similar to tissue as regards elasticity, capacity to absorb energy, strength, etc. Simulants serve to model the *physical ballistic* aspect of wound ballistics.

The density ρ of the simulant must be similar to that of tissue. Soap and gelatine have densities very similar to that of muscle ($\rho = 1.06$ g/cm^3). Water would be suitable in terms of density, but not as far as its other properties are concerned.

There are good reasons to use simulants for wound ballistics experiments rather than animals or cadavers. Ethical concerns are the most important. A further argument against using animals is that tissue is not homogenous, which means that no two wound channels are identical. It is virtually impossible to establish patterns, and a large number of experiments are required in order to draw statistically valid conclusions. In addition to which, it is by no means certain that one can extrapolate the results to humans.

By contrast, experiments with simulants are reproducible and it is possible to establish patterns by changing parameters in a systematic fashion. It is possible to validate the simulant by comparing the results of experiments with cases from real life, of which forensics and war surgery provide all too many examples.

One further point should be borne in mind when using simulants. Their physical characteristics (plasticity, viscosity, elasticity, etc.) are homogenous along the entire length of the wound channel, whereas the body consists of different types of tissue, with different characteristics. Interestingly, this has little effect on the physical ballistic behaviour, particularly the energy transferred, unless the bullet is moving slowly. However, it is of great importance when interpreting the results from a medical or biological point of view.

One can make a broad distinction between hard tissue and soft. By hard tissue we mean bone, while soft tissue consists of skin, muscle, internal organs such as the liver and kidneys, and brain. The different types of soft tissue have quite different ballistic characteristics. Muscle, for instance, is fairly elastic and tear-resistant. The kidneys are less so and the liver is especially prone to tearing when pierced by a bullet; being well supplied with blood, it bleeds profusely. From a ballistics point of view, the brain can be compared to water or gelatine.

Simulants must meet the following requirements:

1. Experiments must be reproducible and must always produce the same results under the same conditions.
2. The dynamic motion of the bullet and the reaction of the medium to it (e.g. deformation) must correspond closely to the situation of a bullet in tissue.
3. The values of the physical parameters along the wound channel (deceleration, force and timing) must be close to those encountered in real life.

Cost, availability and ease of use are also important. Archival (e.g. for teaching purposes) may be relevant, and gelatine is completely unsuitable for this purpose, whereas soap is ideal.

Not all media meet all of the above requirements. For instance, soap is completely incapable of showing the dynamics of the temporary cavity but is well suited to quantitative studies, such as the measurement of energy transfer.

3.3.2 Gelatine

3.3.2.1 Characteristics and fabrication

Gelatine is a protein produced by submitting collagen to an irreversible process that renders it water-soluble. It is usually made of skin or bone. The raw materials and the manufacturing process influence such characteristics as molecular weight, isoelectric point, transmittance, viscosity and gel strength. The last two parameters are of particular relevance to the use of gelatine for modelling. While gelatine batches or mixtures are homogenous, various types of gelatine are produced in the manufacturing process, as is the case with petrochemicals, and the manufacturer assembles the required product from these various types.

Gelatine is obtained by treatment with acid or alkali, followed by extraction using hot water. Gelatine is amphoteric; it reacts to an electrical field like a kation in an acidic solution and like an anion in an alkaline solution. At a particular pH value, the reaction of the solution is neutral. This is known as the isoelectric point. The pH value of the isoelectric point depends largely on the treatment to which the collagen has been subjected. If it has been treated with acid, the isoelectric point lies between pH 7 and 9 (Type A gelatine), whereas treatment with alkali produces a Type B gelatine, with an isoelectric point somewhere between pH 4.7 and 5.4. Gelatine has a relative molecular mass of between 15,000 and 250,000 D (Dalton), but this factor can be influenced by the manufacturing process.

The gel strength of the gelatine (which is of importance for its strength) is measured in Bloom.

The Bloom is a measure of the strength of a gel and is defined as the mass of a cylindrical probe with a diameter of 12.7 mm that is required to deflect the surface of the gel 4 mm. This test is carried out on a sample of gel with a concentration of $6\frac{2}{3}$ % at a temperature of 10 ± 0.1 °C.

Gelatine is obtainable at strengths of 50 to 300 Bloom. That used for ballistic experiments is generally Type A, 250 to 300 Bloom. The strength of gelatine also depends on temperature and concentration, though to a lesser extent.

Gelatine is a polypeptide. The discovery that the configuration of a polypeptide chain is greatly stabilized by the formation of hydrogen bonds between C=O and NH groups was of great significance. The more hydrogen bonds there are, the more stable the structure. Existing hydrogen bonds are broken at higher temperatures during the manufacturing process, or when the material is dissolved. When the temperature of the material falls below the gelling temperature, the hydrogen bonds re-form and the gelatine sets. In order to obtain reproducible results, it is important to standardize the production of gelatine blocks and to avoid both high temperatures (over 60 °C) and extended exposure to heat, as heat causes gelatine to deteriorate. Thermal deterioration affects such parameters as gel strength and viscosity. In turn, this may affect the strength of the gelatine block.

3.3.2.2 Fabrication of gelatine blocks; preparation for experiments

Gelatine is used to determine energy transfer and to observe the movement of the bullet about its own axis. It therefore makes sense to standardize the gelatine used for ballistic purposes, in order to ensure that results from different laboratories are comparable.

One suitable gelatine is "Type Ballistic 3 Photographic Grade" manufactured by GELITA AG, 69412 Ebersbach, Germany (gel strength 245 to 275 Bloom, viscosity 3.4 to 4.6 mPa·s, pH 4.7 to 5.7).

There are various international standards concerning the ratio of water to gelatine and the temperature at which to conduct experiments. FACKLER and MALINOWSKI (1988b) introduced 10% gelatine at 4 °C as a standard, while NATO specifies 20 % at a temperature of 10°C and RUAG Ammotec, Fürth, also use 20 % gelatine, but at a temperature of 15 °C. Gelatine to FACKLER's specifications is used in the USA and in a number of ballistics laboratories in Europe. The technical guidelines of the police technical institute at the German police college in Münster specify the use of RUAG Ammotec gelatine.

There are also a number of different standards regarding the size of the blocks used in experiments. The German guidelines mentioned above specify for handguns 20×20×30 cm and for rifles 25×25×30 cm, but dimensions of 15×15×30 cm, 25×25×40 cm (KNEUBUEHL) and 25×25×50 cm (FACKLER) are also common.

The process of producing gelatine blocks consists of a number of stages:

- Stir the gelatine granules into cold water, ensuring that all granules are moistened. Stop stirring as soon as all the water has been absorbed, to prevent the formation of bubbles.
- Allow to swell for 1 to 2 hours.
- Heat the mixture to 55 °C in a bain marie or heated container, without stirring.
- Once the mixture has reached the specified temperature, stir to an even consistency. A preservative (such as thymol dissolved in ethanol) can be added during the stirring phase.
- Allow air to rise to the surface by letting the mixture stand at a temperature of 45 °C to 55 °C. Remove the froth that has formed on the surface.
- Pour the homogenous solution into moulds while still hot.
- Allow the gelatine to set in the mould overnight, in a fridge.
- Before use, store the gelatine blocks for at least 18 hours at the temperature specified for the experiment. Use within three to four days.

The various standards contain detailed descriptions of the fabrication process.

As gelatine is produced from biological material, there will be inconsistencies. Manufacturers even out these inconsistencies by mixing several types of gelatine, in a manner analogous to the practice in a refinery, ensuring that the user receives gelatine with the required characteristics. Nevertheless, there will be variations between batches, because of the maximum accuracy that can be achieved with the measuring methods used. For instance, the 3 σ value for a Bloom measurement for a 260 g Bloom gelatine is 12 g Bloom. If comparative measurements are to be carried out, it is therefore best to use the same batch for the whole series.

3.3.2.3 Evaluating gelatine experiments

Crack length method. When a bullet passes through a gelatine block it creates a temporary cavity. Because of the elasticity of the gelatine, the cavity collapses, as it does in muscle. In gelatine, the cavity leaves cracks that radiate from the wound channel. Studies (conducted mainly by KNAPPWORST, Dynamit Nobel AG) have shown crack length to be proportional to the volume of the temporary cavity and hence to the energy that the bullet transfers to the block per cm, E'_{ab}. KNAPPWORST was able to demonstrate that under constant conditions we can write

$$(3.3:1) \qquad \sum r_i = c \cdot (E'_{ab})_i \qquad [\text{cm}]$$

where Σr_i is the sum of all crack lengths in a given cross-section i of the block and c is a constant, of which the value has to be determined.

If we measure Σr_i for a series of adjacent sections (e.g. every 2.5 cm) and plot these values against penetration depth, we obtain a graph as shown in Fig. 3-40.

The choice of section thickness – 2.5 cm in this example – is arbitrary. Ideally, the sections should be as thin as possible, but thinner sections mean expending more effort on analysis. A thickness of 2.5 cm has proved to be a reasonable compromise between precision and effort.

If the bullet has passed right through the block, then the whole block will exhibit these radial cracks. The total length of all cracks (i.e. not just the total length for a given section) is equal to the area under the curve (the grey area in Fig. 3-40). For a section thickness Δs, and taking account of Eqn 3.3:1:

$$(3.3:2) \qquad \sum_0^s \left(\sum r_i \cdot \Delta s\right) = c \cdot \sum_0^s (E'_{ab} \cdot \Delta s) \ .$$

The left-hand side of the equation corresponds to the area under the curve, while

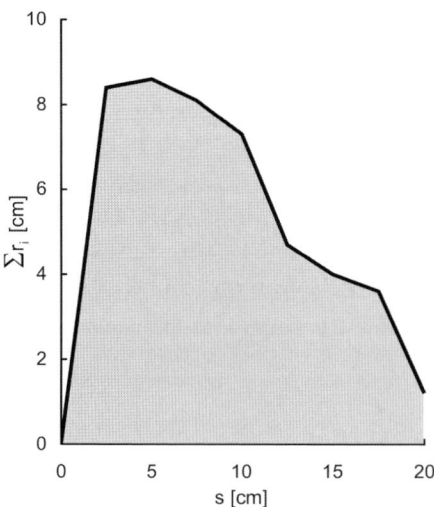

Fig. 3-40. Analysis of shots fired into gelatine blocks, using KNAPPWORST's crack length method. See text for details.

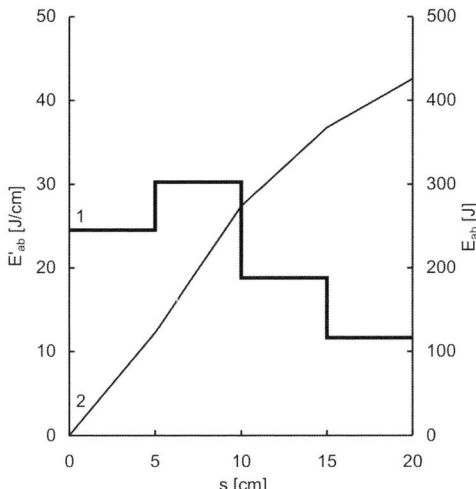

Fig. 3-41. Energy transfer (E_{ab}) and energy transfer per unit distance (E'_{ab}) in gelatine as a function of distance travelled, for an SJHP bullet (after KNAPPWORST). *Curve 1*: E'_{ab} in 5 cm sections, *Curve 2*: E_{ab} (total E'_{ab}).

the right-hand side (without the constant c) is the total energy transferred to the block, E_{ab}.

E_{ab} is the difference between energy on impact and energy on exit, which can be calculated from velocities v_a and v_e. In practice, a bullet exiting the block at an angle may take gelatine with it, and the energy of this gelatine can affect the value of E_{ab}.

Because we know the left-hand side of Eqn 3.3:2 (sum of all crack lengths) and the right-hand side (total energy transferred, E_{ab}), we can now calculate the constant c and hence use Eqn 3.3:1 to calculate the total energy transferred:

$$(3.3:3) \qquad (E'_{ab})_i = \frac{1}{c} \cdot \sum r_i \ . \qquad [J/cm]$$

Using Eqn 3.3:3, we can now calibrate the y axis of Fig. 3-40 for E'_{ab} [J/cm] rather than for total crack length (see Fig. 3-41, left-hand y axis).

This graphic shows not only E'_{ab} (which KNAPPWORST terms E_{ab}/s), on the left-hand y axis, but also the total energy transferred up to the point under consideration (Curve 2, right-hand y axis).

FACKLER's wound profile. This method can be seen as complementary to the method of analysing experiments with gelatine described above. FACKLER includes four elements of the effect of the bullet on gelatine (representing tissue):

– Penetration depth.
– Deformation and fragmentation (if any).
– Size (diameter) of the temporary cavity.
– Size (diameter) of the permanent wound channel.

142   3 General wound ballistics

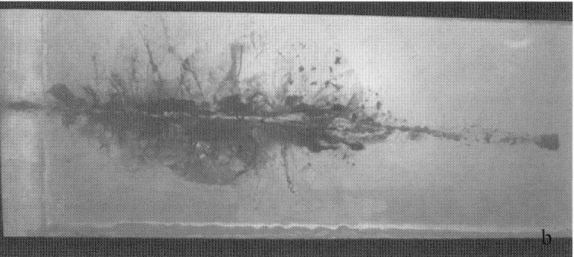

Fig. 3-42.   Gelatine block after firing. **a**. In direction of shot. The cracks are clearly visible, **b.** from the side. Here, the "pockets" formed by the cracks are visible.

The sizes of the temporary cavity and of the permanent wound channel can be determined fairly clearly from the cracks in the gelatine (Fig. 3-42 a, b). This is used to derive the graphics in Figs 3-43 a, b and in Fig. 3-44 a, b). FACKLER (FACKLER and MALINOWSKI, 1985) calls the description of all these effects the "wound profile." He included the entire wound channel in all instances.

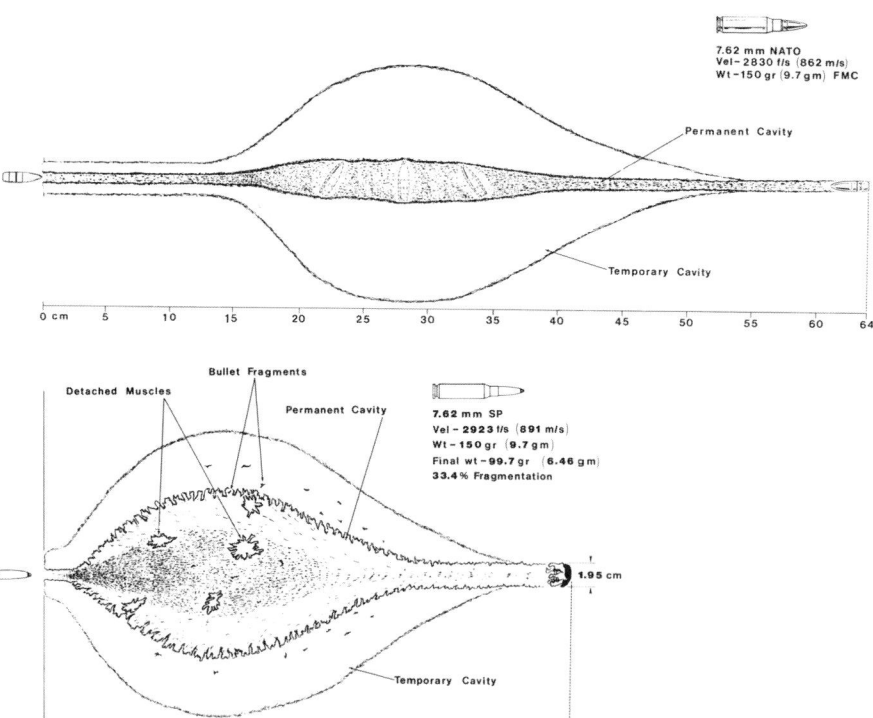

Fig. 3-43 a, b.   FACKLER's wound profile (1988). The illustrations are self-explanatory. The cartridge used is shown in the top right-hand corner, together with its ballistic data. (Continued on next page).

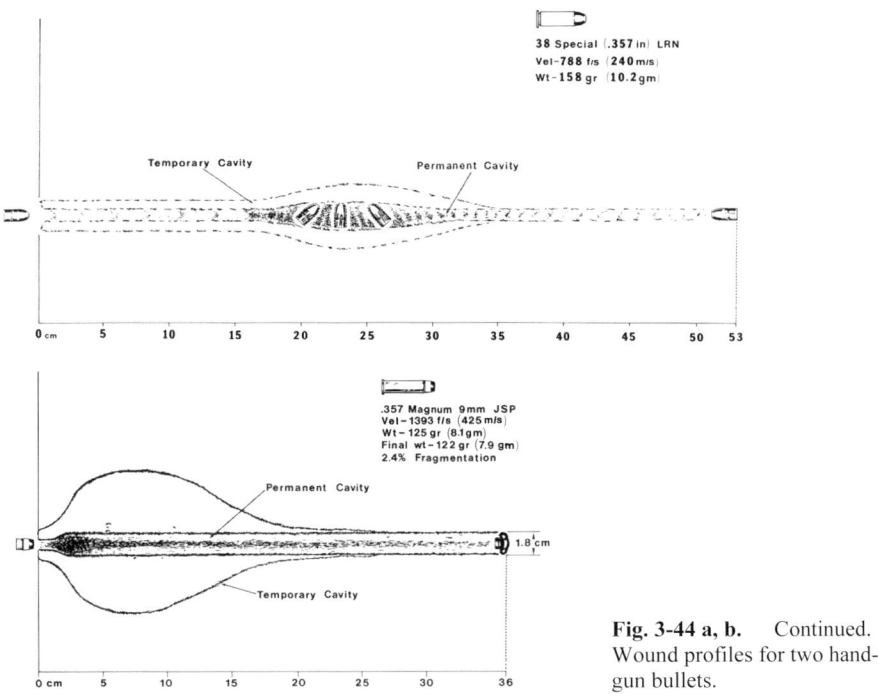

Fig. 3-44 a, b. Continued. Wound profiles for two handgun bullets.

3.3.3 Glycerine soap (ballistic soap)

3.3.3.1 Characteristics and fabrication

Conventional soaps are mixtures of solid, water-soluble sodium salts of various higher fatty acids. The only soaps that can be used for wound ballistics experiments are glycerine soaps (moulded, transparent soaps). Producing this type of soap is relatively complex, but by mixing the constituent fats and oils with care, it is possible to maintain consistency over periods of many years, which is an important precondition for accuracy.

The main raw materials are:

- coconut oil (lauric acid with small percentages of other acids), tallow, castor oil (ricinoleic acid) and stearin (palmitic and stearic acid).

Other constituents:

- glycerine, sucrose, ethylene glycol, ethanol, water and sodium hydroxide.

The fatty acids are transformed into soap by boiling in sodium hydroxide. The fatty acid content of ballistic soaps varies between 39% and 43%, but water and alcohol are also present in varying proportions. It is possible to modify the density and softness of a soap by varying its components. This allows us to achieve the penetration depth required for a particular calibration shot. It is therefore perfectly possible to standardize ballistic soap.

Depending on its composition, the density of new soap (1 to 3 weeks old) lies between 1060 and 1100 kg/m³ at room temperature. The mechanical properties are of secondary importance for the uses to be described here. Soap has a very low limit of elasticity (0.5 N/mm²), which is an indicator of its plastic behaviour. The important parameters in connection with the motion of a bullet are the speed of sound in the medium and its viscosity. Both are closely related to temperature, and both decrease as temperature increases (CORTHÉSY 1985, 1987, see Fig. 3-45).

The deformation of soap due to the passage of a bullet is almost entirely plastic, and the wound channel only collapses very slightly. This gives us a direct image of the temporary cavity, on which we can readily take measurements.

3.3.3.2 Ageing

Soap can be stored for comparatively long periods, but some of its properties do change over time. Because the components that evaporate fastest during drying are alcohol and water, the soap becomes denser over time (rising to 1150 kg/m³ after approx. four years). This is accompanied by an increase in the speed of sound in the medium (CORTHÉSY 1985, see Fig. 3-46).

As the physical properties of soap only change slowly, it can be used for up to six months with no loss of accuracy, depending on storage conditions. The behaviour of the bullet (movement, deformation, fragmentation) remains the same. The only difference is that the cavity volumes will be slightly smaller with older soap, because the bullet encounters more resistance when penetrating the drier outer layer. It is possible to delay the onset of this effect by removing that layer.

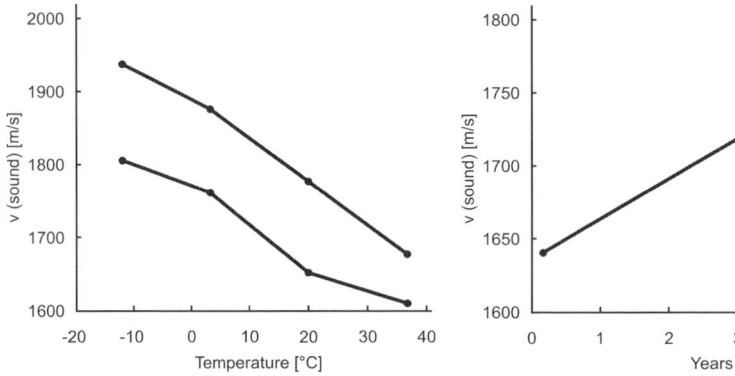

Fig. 3-45. Speed of sound (c) in soap as a function of the temperature of the soap.

Fig. 3-46. Speed of sound (c) in soap as a function of the age of the soap.

3.3.3.3 *Evaluating soap experiments*

As with gelatine, a linear relationship between the volume of the temporary cavity and the energy transferred has been demonstrated empirically (see also 2.3.7.3, MARTEL's Theory). In order to measure the energy, we simply need to measure the volume of the channel.

The easiest method is to slice the wound channel along its axis and then take photographs. The diameter can then be measured on the photograph as a function of penetration depth. As the cross-section of the wound channel is circular (except at the end), it is possible to perform a fairly accurate calculation of volume as a function of distance.

The channel cross-section becomes elliptical towards the end, because the bullet is moving more slowly, but if one takes the mean diameter for this section, the error will be small.

If the channel is sufficiently large, one simple method of measuring volume is to fill it with water a section at a time. The shape of the cross-section is then irrelevant as far as measuring the volume is concerned. However, it is also possible to cut the block into a sufficient number of slices (as for the crack length method when using gelatine), to measure the cross-sectional areas and to obtain the total volume by adding them together:

$$(3.3:4) \qquad V = \sum_i \frac{A_i + A_{i+1}}{2} \cdot d_i, \qquad [cm^3]$$

where A_i is the cross-sectional area at the start of the ith slice [cm^2] and d_i is the thickness of the same slice [cm]. Very small volumes, such as those left by small, low-energy fragments, can be calculated from the penetration depth and the cross-sectional area of the start of the channel. From Eqn 3.2:5, it is reasonable to assume that the boundary of the volume can be represented by an exponential function. Volume is therefore given by the following equation:

$$(3.3:5) \qquad V = \tfrac{1}{4} \cdot \pi \cdot \ell_s \cdot \frac{d_1^2 - d_2^2}{\ln(d_1^2) - \ln(d_2^2)}, \qquad [cm^3]$$

where ℓ_s is the penetration depth, d_1 the diameter at the start of the channel and d_2 the diameter at the end.

Another point to bear in mind when measuring volume is that the volume in the vicinity of the entry and exit points will be larger than in real life, as the bullet will displace material towards the surface of the block as it penetrates, and again on exit, enlarging the channel. To obtain reproducible results, it is important that none of the sides of the block swell out, as this would allow the volume to increase in an uncontrolled manner.

The equation below gives the relationship between volume and energy transferred:

(3.3:6) $V = \mu \cdot E_{ab}$.   [cm³]

The proportionality factor μ remains the same for energy levels of up to more than 2500 J, *regardless of the type of bullet*. Fig. 3-47 shows the results of experiments conducted with a wide range of projectiles. Calculating μ by means of linear regression yields a value of 0.182 cm³/J (with a correlation coefficient r of 0.98). With new soap, one can therefore expect values of 0.16 to 0.2 cm³/J. As the soap ages, this will fall to around 0.12 cm³/J.

3.3.3.4 Using soap to conduct measurements

As there is a relationship between energy transferred and the volume of the cavity, soap can also be used to determine velocity and energy. We can derive the following directly from Eqn 3.3:6:

(3.3:7) $E_{ab} = \dfrac{1}{\mu} \cdot V$.   [J]

If fresh glycerine soap is used, mixed so as to allow the same penetration depth as 10% gelatine at 4°C, we can substitute a value of 5.5 to 6 J/cm³ for 1/μ.

We can increase the accuracy of the measurements by firing calibration shots of known energy transfer into the soap used. Measuring the volumes created allows us to determine the actual proportionality factor 1/μ. Experience shows that the values lie between 5.0 and 6.25 J/cm³. If the

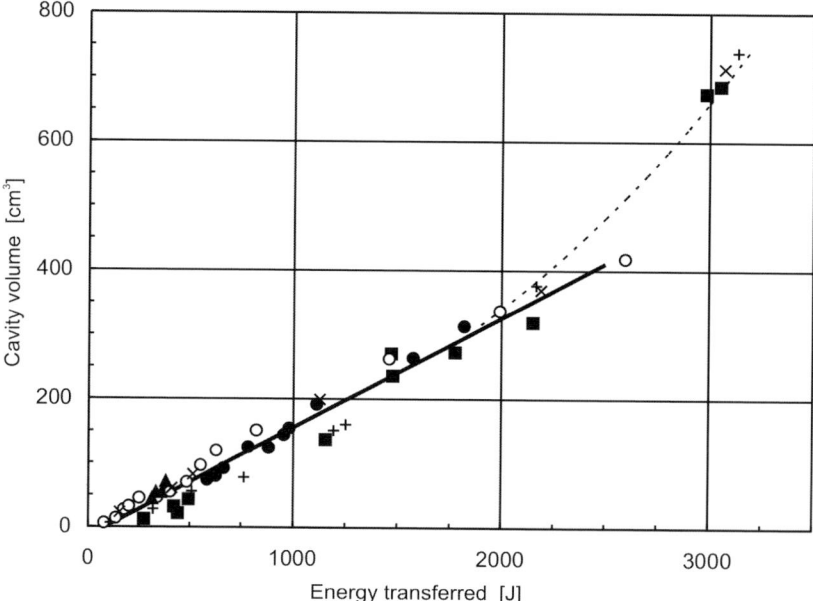

Fig. 3-47. Relationship between the energy transferred to the soap E_{ab} and the volume of the cavity created. Irrespective of the type of projectile, a linear relationship exists between the two parameters up to energy levels of at least 2500 J, with a high degree of correlation (KNEUBUEHL).

projectile transfers large quantities of energy at an early stage, we may observe lower values (around 4).

This method is particularly useful where more usual methods of measuring velocity cannot be used, e.g. for fragments from fragmentation shells, fragments from the rear of glass surfaces, fragmented bullets or shotgun pellets, or when determining the residual energy of a bullet that has passed through an object.

The soap blocks are so arranged as to catch the relevant projectiles. From the volume, it is possible to use Eqn 3.3:7 to calculate the energy transferred and hence the impact energy. From the impact energy and the mass of the projectile we can then calculate the impact velocity (KNEUBUEHL 1990a).

3.3.4 Comparison between soap and gelatine

3.3.4.1 General

A number of studies have demonstrated that gelatine and soap are both suitable simulants for wound ballistics. However, some of the properties of the two media differ quite substantially. This raises the question as to which material one should use for which purposes. In the next section, we shall examine the advantages and disadvantages of soap and gelatine, together with the differences in behaviour between the two media.

3.3.4.2 Availability, handling and measuring techniques

Gelatine is used in a wide variety of applications (the food industry, medicines, for photographic film, etc.) It is therefore readily obtainable, in a range of qualities. It is possible to store gelatine powder for relatively long periods if it is kept dry. One can produce gelatine blocks oneself, although they can only be stored for a short period.

In order to use soap, one needs access to a soap manufacturer prepared to produce soap of a specific composition and constant quality. However, as soap can be kept for several months, good stock management will ensure a ready supply.

Soap is far easier to handle, and experiments with soap are easier to analyse. It is also less sensitive to temperature, which makes the temperature requirements for experiments less stringent. The measurements of volume that must be conducted in order to determine energy transferred are easier and more pleasant than the crack length method generally used with gelatine.

Gelatine is more suitable for observing movement. It is transparent to visible light and hence allows the use of optical methods (film and photography, see Fig. 3-48). Most types of soap are opaque (as is tissue), which means that X-ray photography or electromagnetic methods must be used. The elasticity of gelatine means that it only permits indirect measurements of the volume of the temporary

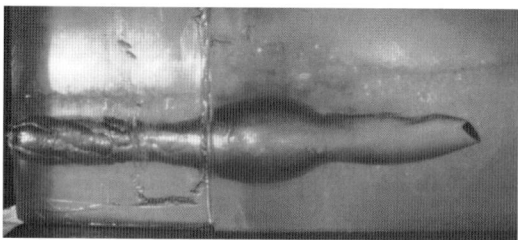

Fig. 3-48. A bullet passing through gelatine. The orientation of the bullet and the dynamics of the temporary cavity are clearly visible.

cavity (via crack lengths), whereas the plastic behaviour of soap makes it possible to measure the cavity directly. Each simulant has its advantages and disadvantages, and each has its role to play in wound ballistics experiments.

Gelatine should not be melted down and re-used, as this would affect reproducibility. As gelatine consists of natural materials (water and protein) it can be disposed of easily. Soap can be returned to the supplier, where it can be melted down and additional raw materials can be added, allowing the soap to be re-used.

3.3.4.3 Reaction to bullets

The dynamic behaviour of a bullet is very similar in the two media. If the appropriate composition is selected, the penetration depths are also very similar (see Fig. 3-49 a-d) and bullet deformation is almost identical (see Fig. 3-50 a-d).

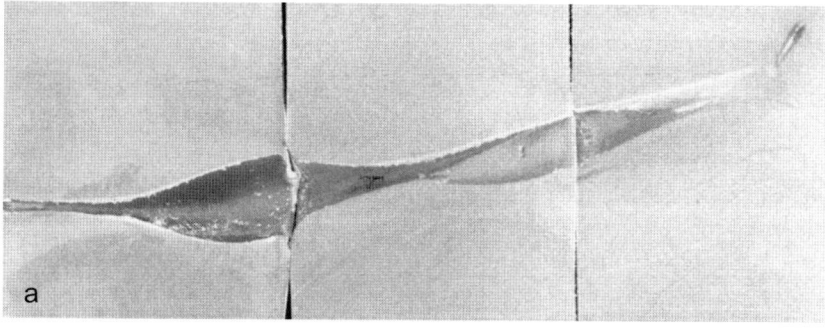

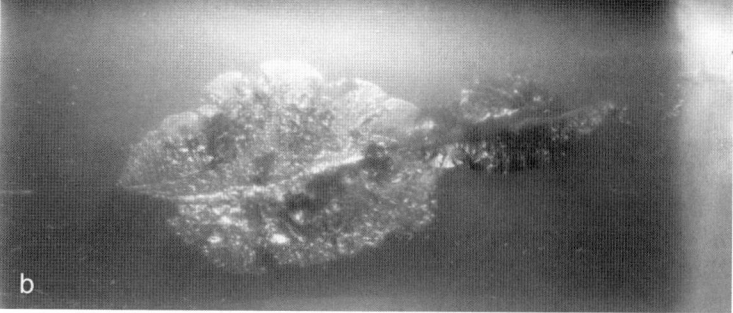

Fig. 3-49 a, b. Comparison between glycerine soap (a) and gelatine (b): 5.45 x 39 (Kalashnikov).

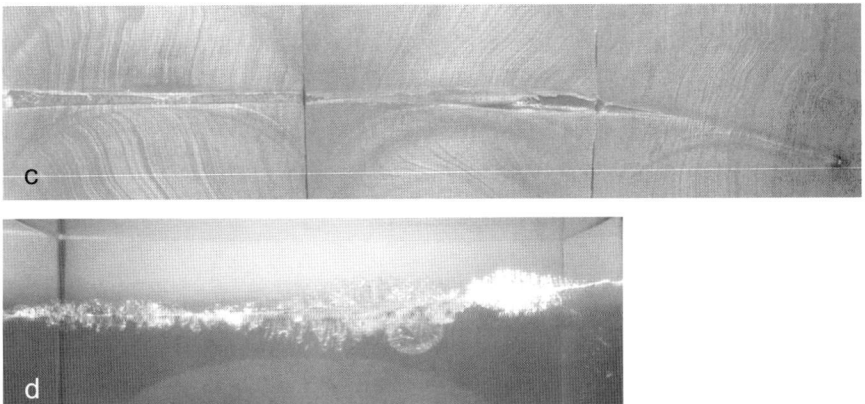

Fig. 3-49 c, d. Comparison between glycerine soap (c) and gelatine (d): 9 mm Luger ("Schweizer Ordonnanz").

This indicates that the forces – and hence the deceleration values – must be very similar, as JANZON and FACKLER have confirmed. There is therefore no physical reason to prefer one simulant over the other.

At lower velocities (when a large proportion of the bullet surface is in contact with the medium), certain bullets travel significantly further in soap than in gelatine. This may be because surface friction (which is the determining factor in such cases) is greater in gelatine than in soap, which

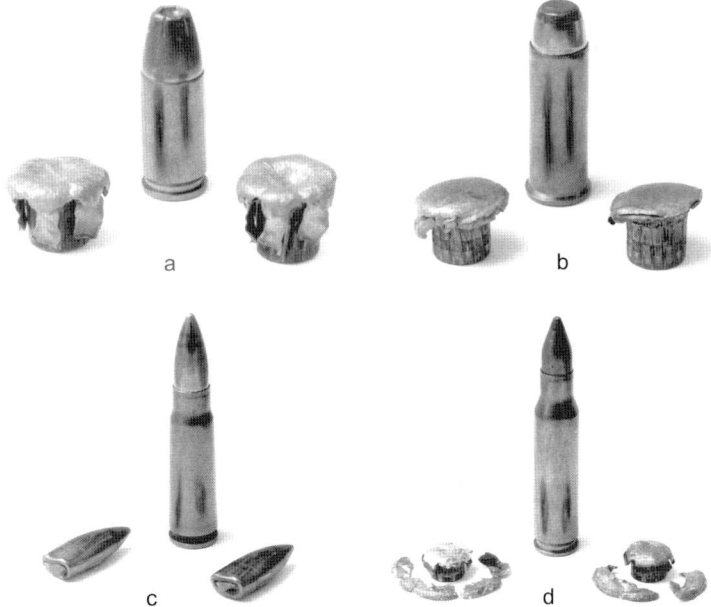

Fig. 3-50. Similarity of deformation in soap (left) and gelatine (right). **a.** 9 mm Luger, Winchester Subsonic. **b.** 44 Rem. Mag. SJFP. **c.** 7.62 × 39, jacketed spitzer. **d.** 308 Win. SJCone.

150   3 General wound ballistics

Table 3-7.   Advantages and disadvantages of each simulant

	Soap	Gelatine
Handling	+	–
Measuring volume	+	–
Recording movement	–	+
Measuring fragment energy	+	–
Price	–	+

has a high fat content. As this effect only arises at low velocities and great penetration depths, it is virtually irrelevant to wound ballistics studies.

Soap is unsuitable for experiments involving projectiles that have low impact velocities (arrows, crossbow bolts, etc.) or jets of gas or fluid (alarm pistols, rocket motors, etc.) Such "projectiles" penetrate significantly less, if at all.

3.3.4.4 Which simulant for which purpose?

Table 3-7 only shows those criteria for which there are clear differences. If we bear in mind that it is possible to make gelatine easier to handle by using it at lower temperatures (e.g. 4°C), the advantages and disadvantages of the two simulants more or less balance out.

The choice of simulant should therefore depend on the purpose of the experiment. Standardisation of the simulants used for wound ballistics is desirable, and should cover both materials. Standardization should also ensure that a given bullet penetrates both media to approximately the same depth under the same impact conditions. Furthermore, this penetration depth should correspond to that of the

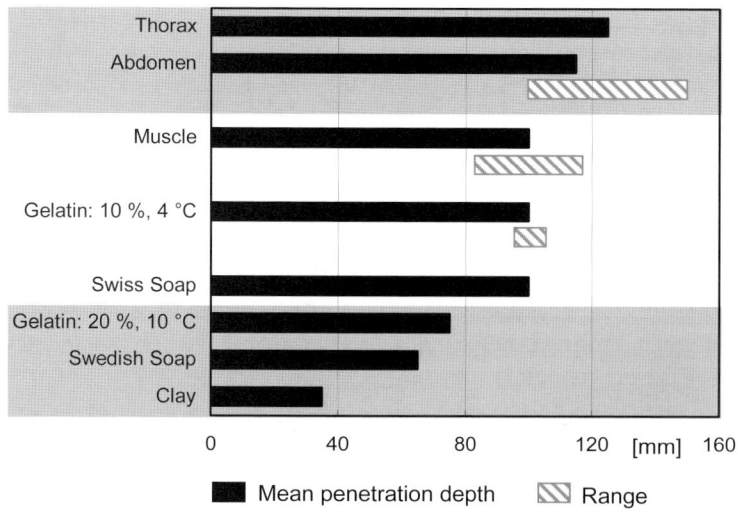

Fig. 3-51.   Penetration depth of a steel sphere in various media (after FACKLER).

same bullet in muscle. According to FACKLER, it is possible to achieve equivalence between muscle and simulant using 10 % gelatine at 4 °C. By choosing the composition appropriately, it is also possible to obtain soap that performs in a similar manner (see Fig. 3-51).

3.3.4.5 Connection between the analysis methods

As both gelatine and soap are suitable for measuring energy transfer, it must be possible to convert between the two standard evaluation methods (crack length for gelatine and volume measurement for soap).

The formal relationship between μ and c is based on the dimensions of the two variables (see Eqns 3.3:3 and 3.3:7). μ is measured in cm^3/J, while c is measured in cm^2/J. Therefore:

(3.3:8) $$c \propto \mu^{2/3}.$$ $[cm^2/J]$

We can therefore compare experiments conducted with gelatine and soap (if we apply a constant factor).

3.3.5 Bone

3.3.5.1 General

We have already discussed the advantages of using simulants in 3.3.1. The same arguments apply to other types of tissue, such as bone. The effects on the bullet wound and the course of the wound channel of the bullet striking a bone are of interest to both surgeons and forensic specialists. To date, however, experiments appear only to have been conducted on animal bones (see 4.2.1.3 and 4.3.2.6 and the literature cited in those sections). It is only possible to carry out systematic and reproducible studies of the effects of bullets on bone if the bone is standardized, as in the case of muscle. Bearing in mind the conditions that must be fulfilled by a simulant for soft tissue (see 3.3.1), the requirements for a bone simulant, other than reproducibility, are as follows:

- Similar deceleration to that which occurs when a bullet passes through a bone;
- Similar threshold velocity for penetration;
- Similar fracture behaviour and similar propagation speed for the fracture lines.

The most suitable material for simulating bone in experiments with bullets is that used for the artificial bones employed in the training of surgeons. This consists of a three-layer polyurethane structure, in which a relatively soft interior (similar to cancellous bone) is enclosed by a harder outer layer. Using this material, it is possible to represent both heavy and light bone structures. In order to simulate the behaviour of bone fragments, the plastic bone is covered with a thin layer of latex, to simulate the periosteum.

The natural form of the bone is of secondary importance when carrying out systematic studies of the effects of bullets. Building artificial bones with lifelike shapes would affect reproducibility, as it would be more difficult to achieve identical impact conditions over multiple shots. Using simple geometric forms (e.g. hollow cylinders to simulate hollow bones or a sphere to represent the skull) makes it possible to maintain the same relative positioning of wound channel and bone over an entire series of shots. This is essential when conducting parameter studies, for instance.

However, it may sometimes be necessary to use accurate reproductions of bone shapes when conducting reconstructions of specific bullet wounds. This is perfectly possible, as entire skeletons of suitable artificial bone are available (source: Synbone AG, CH-7208 Malans, Switzerland).

3.3.5.2 Hollow bones

Simulations of hollow bones are of particular importance for the study of bullet wounds to the limbs. The best form to use is a hollow cylinder, which can be produced with different wall thicknesses. In order to simulate bone marrow, the hollow space is filled with gelatine and the two ends sealed so that they are pressure-tight (KNEUBUEHL and THALI 2003).

Without the gelatine, a bullet passing through the bone will not generate hydraulic pressure. It will perforate the bone without causing any of the typical fractures, such as a butterfly fracture.

If bullets are fired directly at plastic bones prepared in this manner, it is possible to measure both the velocity and energy lost by the bullet and the velocity of the

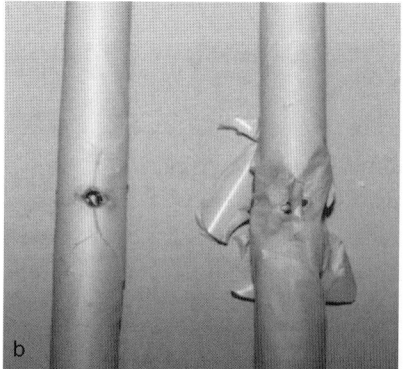

Fig. 3-52. Comparison between the effect of full metal-jacketed and deforming bullets on model bones embedded in gelatine. The extent of bone damage corresponds approximately to the diameter of the temporary cavity. **a.** Rifle. **b.** Handgun.

Fig. 3-53. Model skull shot through with a lead ball fired from a muzzle-loading musket. The aim here was to determine approximate bullet velocity in the case of skeletons recovered during archaeological investigations, by comparing the fracture to the real skull with the fracture observed on the model.

bone fragments produced. If the bone is cast into a gelatine block, one can study the dynamic processes that occur when a bullet hits a bone (e.g. the degree of destruction associated with the temporary cavity, see Fig. 3-52 a, b; the positions of bone fragments in the wound channel and the behaviour of a bullet that ricochets off a bone).

3.3.5.3 Modelling the head

It is also possible to construct a model of the head using artificial bone. The most obvious shape to use is a hollow sphere, with which it is possible to simulate any angle of attack. The sphere is filled with gelatine, to simulate the brain. Materials to simulate muscle and skin can be added on the outside if required (THALI et al. 2002a, 2002b, 2003d). Models like this can be used to study many different aspects of bullet wounds to the head (see Fig. 3-53).

3.3.6   Other simulants

It is possible to simulate other biological structures using plastics that react in a realistic manner to a bullet. These structures include skin and blood vessels.

Skin. It is unlikely that any simulant would simulate skin in such a way as to meet all forensic requirements. One material that has proved successful, however, consists of a layer of silicon in conjunction with synthetic fibres (synthetic suede) (THALI et al. 2002a, 2002b, see Fig. 3-54). For simple experiments one can just soak the synthetic suede in wax.

Blood vessels. One can conduct systematic studies of blood vessels using plastic tubes that react to bullets in a manner similar to that of real blood vessels. These have been evaluated by comparing them with the aortas of pigs (see Fig. 3-55) which apparently are fairly similar to those of humans (SCHANTZ 1979). To ensure that the bullet generated hydraulic pressure, both the artificial blood vessels and the real blood vessels used for comparison purposes were filled with "film blood"

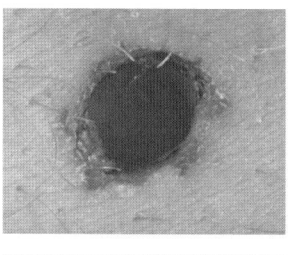
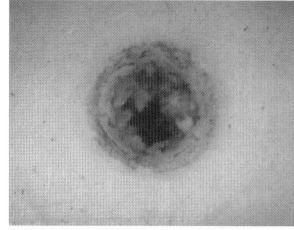

Fig. 3-54. *Left:* Real bullet hole (filled with blood). *Right:* Bullet hole in artificial skin.

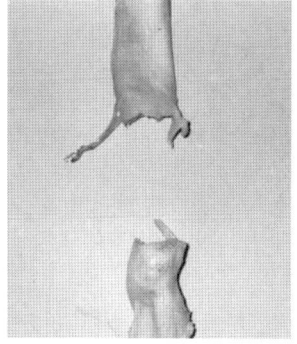
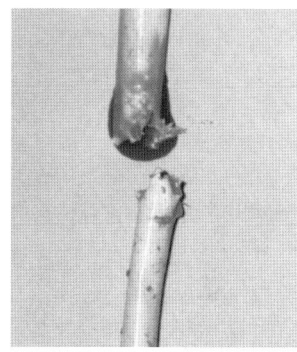

Fig. 3-55. *Left:* Real blood vessel severed by a bullet. *Right:* Artificial blood vessel severed in the same manner.

and closed off at both ends, with the contents placed under slight pressure.

The basis of comparison was energy density for the direct hit (perforation) and effectiveness for indirect damage (failure of the blood vessel due to excessive stretching resulting from the pressure generated by the temporary cavity). The plastic tubes were considered acceptable as simulants if their threshold values were similar to those of natural blood vessels.

The various simulants can be combined to simulate entire structures, with which one can study the formation of a bullet wound repeatedly, making controlled changes to the parameters. This is particularly useful in reconstructions of crimes involving firearms.

3.4 Other approaches to simulation

3.4.1 Experiments on animals and cadavers

3.4.1.1 Animals

SCHANTZ, a member of a Swedish research team, submitted a contribution on the selection of animals for the study of bullet wounds to the 3rd International Symposium on Wound Ballistics (1979). We shall look at the more important veterinary aspects of his work, which contains numerous bibliographical references.

General. The problem with all animal experiments is that of extrapolating the results to humans. The aim is, of course, to learn about the reaction to bullets of humans, not of the animals used for the experiments. One therefore needs to select an animal that is as similar to humans as possible. The greater the difference between the animal and humans, the less accurate the extrapolation (e.g. cat $\Rightarrow$ human, chimpanzee $\Rightarrow$ human).

Every animal has different characteristics, in terms of anatomy, physiology, behaviour, etc.

Simple examples from hunting show how differently animals react to bullets. For instance, a deer can run for up to 100 m after its heart has been completely destroyed by a bullet. By contrast, rabbits are extremely sensitive; a few shotgun pellets are fatal. Pellets are often found in the muscle of a dead rabbit, and not in vital organs (heart or brain), which means one must assume that a genuine shock effect is at work.

If one wishes to study the effect of a bullet along the full length of the wound channel (e.g. the mass of debridement), the animal must be of sufficient size, which excludes rabbits and cats.

Before any experiments are carried out, one must ask:

– How closely do the organs of the animal correspond to those of humans?
– Is it possible to obtain wound channels of sufficient length?

Regardless of which animal is chosen, what effects are to be studied and which organs are involved, the projectile has to penetrate the skin. However, the skin of many animals has physical properties quite different to those of humans. This is of particular importance if we wish to study the behaviour of soft-tipped bullets. The strength of the skin will have a significant effect on how the tip of the bullet deforms on contact with it, and hence on the behaviour of the bullet in the body.

During a study on the reaction of skin when the muzzle is in contact (as is generally the case in suicides) initial experiments involved firing shots at the skull of a horse. High-speed video revealed that the skin around the entry wound bulged out slightly but remained intact, except at the actual entry hole. The same experiments under the same conditions, using the skull of a human cadaver, revealed a pronounced outwards movement of the skin, which caused radial tears and undermining of the skin. Pigs also have considerably tougher skin than humans. The skin of goats and sheep comes closest.

Horses and cattle. Journée already used horse cadavers, as we saw in 3.1.2. Both horses and cattle have large muscles, in which it is possible to create long wound channels. The size of the animals means that experiments require considerable effort. Keeping these animals anaesthetized for long periods (over 30 minutes) requires complex equipment. As ruminants, cattle have a complex digestive system. This can lead to problems with the animals inhaling solid matter. The skin of both animals is thick and rough (see above regarding the role of the skin). Observing the animals over a longer period after the shot is difficult, whether they are unconscious or not. The overall expenditure (for the purchase and care of the animals and for the experiments themselves) is very high.

Sheep and goats. While these animals are also ruminants, a general anaesthetic is not nearly as dangerous as it is for cattle. Furthermore, one can use the same equipment and drugs as for humans. Both animals have fairly thin, supple skin, and their skin comes closest to that of humans. The muscles of the limbs are slender, which means that larger breeds are needed in order obtain longer wound channels.

Dogs. For many years, dogs have been the animal of choice for experimental surgery. They are available in a wide range of sizes and forms, easily obtainable and highly suitable for laboratory purposes. They can also be trained for post-operative treatment and observation. Their peripheral arteries and veins are easily accessible, and one can readily insert a cannula through their skin. As with sheep and goats, it is possible to use the same anaesthetic equipment as for humans, including tracheal tubes. It is important to monitor the circulation carefully, but dogs can safely undergo long periods of inhalation anaesthesia. Sufficient blood should be obtained and stored to replace blood loss, if needed. However, SCHANTZ reports that experiments which involve firing bullets at "man's best friend' encounter considerable opposition.

Pigs. Pigs are readily obtainable in any size required. It is possible to create long wound channels even in animals of medium size (30 kg), but animals of this weight are still young, and their bones are not yet fully developed (e.g. the epiphyses are still open). One should bear this point in mind when carrying out experiments involving bones.

Because adult pigs of typical breeds are heavy, miniature varieties have been bred, with an adult weight of 40 kg to 60 kg. However, such breeds are expensive, and they have the reputation of being susceptible to disease.

It is more difficult to collect blood from pigs and to anaesthetize them, as the only easily accessible peripheral veins are those in the outer part of the auricle. Pigs have a funnel-shaped oral cavity and the larynx and trachea are at an angle to each other. Furthermore, the larynx is particularly prone to spasm. As a result, endotracheal intubation is difficult and requires trained personnel. Premedication with atropine and the use of muscle relaxants alleviate the problem.

The skin of most breeds is not pigmented, and has little hair, making it easy to observe and analyse bullet wounds.

Despite their disadvantages by comparison with dogs, pigs are the animals best suited to wound ballistics experiments (SCHANTZ).

3.4.1.2 Cadavers

One can use cadavers to study mechanical phenomena in the human body as long as one observes certain rules.

Following death, the physical characteristics of tissue change over time. The cadavers used should therefore be as fresh as possible. Body parts such as bones are often preserved for later use, which almost always affects the properties of the bone. We shall look at this aspect in 4.2.1.3.

With soft tissue (muscle, fat, liver, etc.) the temperature at which the experiments are carried out plays a major role. In general, cadavers are preserved in cold storage and are generally only just above freezing. If one were to conduct physical measurements on cadavers at such temperatures, e.g. the penetration depth of a sphere, one would obtain results very different from those obtained with a living person.

By definition, it is impossible to study the physiological reaction of a person to a bullet using a cadaver.

3.4.1.3 Cell cultures

In recent years, cell cultures have been successfully used for certain studies in order to avoid experiments on animals. The pharmaceutical industry was the first to use cell cultures for this purpose. SUNESON et al. (1987) were the first to apply this method to the field of wound ballistics, in their work on the effects of shock waves in cells. We shall be looking at these experiments and results in 4.3.2.2 (Shock waves).

3.4.2 Physical/mathematical models

3.4.2.1 General

There are advantages to first representing a real object using a physical model and then constructing a mathematical model. The most obvious advantage is that this approach makes it possible to conduct parameter studies of unlimited scope, simply and cheaply. However, mathematical models – especially those intended to model the human body – face almost insoluble problems. For instance, it is extremely difficult, if not impossible, to obtain the necessary physical data regarding tissue in vivo, especially for extremely high forces acting over extremely short periods. The properties of a material under such conditions often differ substantially from the properties of the same material under quasistatic conditions. Finite element models of the type described by WIND et al. (1988) are hence unlikely to enter widespread use.

In order not to forego the advantages of a model, one must make do with simpler representations of biological structures, which means accepting that the results will also be simpler.

If one is to make practical use of models, the datasets required to calculate the physical processes must be available and the model must have been validated for the field of application against known, real-life cases. It is possible to generate the datasets from simulations of the types described in 3.3 and 3.4, but non-biological simulants are particularly suitable, on account of their availability, reproducibility and usability. Validation requires real-life cases for which one knows the ballistic data of the projectile and has a sufficiently exact description of the damage done to the biological structure.

3.4.2.2 SELLIER's velocity profiles

SELLIER's velocity profile (SELLIER 1982) is a one-dimensional model, in which the decrease in velocity is calculated for each layer of tissue (each layer being assumed to be homogenous) using the equations derived in 3.2.3.2 and 4.2.1. From Eqn 3.2:11 we have:

(3.4:1) $\qquad v_e = v_a \cdot e^{-\Re \cdot d}$, [m/s]

or, for $\Delta v = v_a - v_e$:

(3.4:2) $\qquad \Delta v = v_a \cdot (1 - e^{-\Re \cdot d})$. [m/s]

where v_a is the impact or entry velocity, v_e is the velocity after passing through a thickness d and $\Re$ is the retardation coefficient, which is dependent on the type of tissue.

It is therefore possible to calculate Δv for each layer of tissue, with the exit velocity for one layer constituting the entry velocity for the next. From this, we can create a velocity profile, as shown in Fig. 3-56. As this model is one-dimensional – the bullet only moves in one spatial dimension – deviations (e.g. as a result of impact with a bone) are not taken into account, the bullet always strikes a layer at right-angles and the forces on the bullet act only along the line of movement.

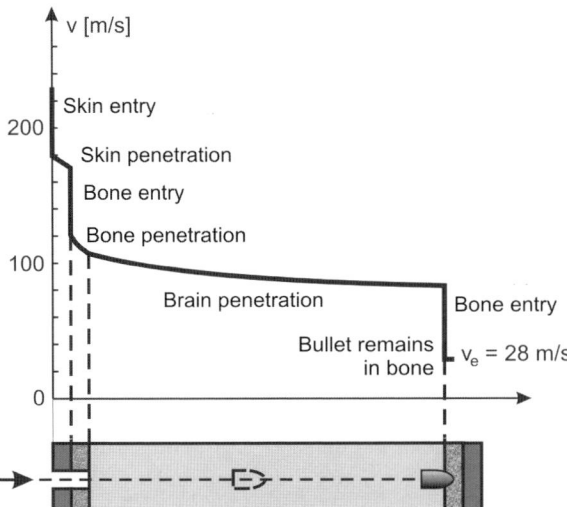

Fig. 3-56. One-dimensional velocity profile of a bullet in the calibre 6.35 browning, passing through a head (after SELLIER 1982).

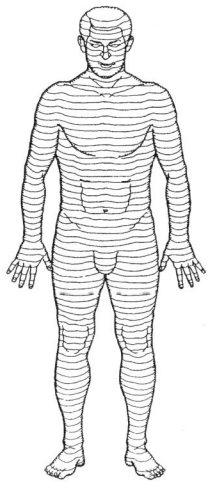
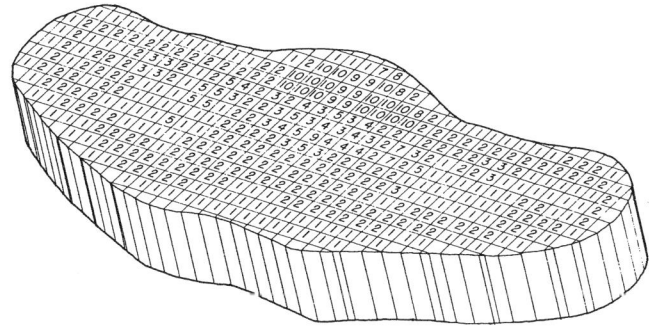

Fig. 3-57. Original version of Computer Man. **a.** Left: the body is divided into 25 mm layers. **b.** Right: the layers are divided into cells measuring 5 mm × 5 mm. The layer shown is a section through the body at shoulder height. An "Injury Criteria Component Vulnerability Number" was assigned to each cell (from BRUCHEY W. J. et al. 1983).

This velocity profile can be useful when, for instance, a forensic specialist is asked whether a bullet would produce an exit wound, given a specific wound channel.

3.4.2.3 Computer Man

In 1975, the US Department of Justice's National Institute of Justice (NIJ) published a study on the effectiveness of handguns for the police. The study had been conducted against the background of a long-running controversy in the US regarding the most effective handgun calibre.

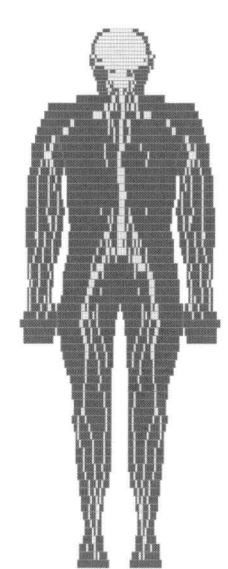

The study was based on a large number of shots using gelatine and a large number of virtual shots fired at a computer model of the human body, all of which were brought together and subjected to statistical analysis. The model was dubbed "Computer Man."

This is a three-dimensional model of the human body, divided into 25 mm horizontal layers (see Fig. 3-57). Each layer was divided into cells measuring 5 mm × 5 mm. This gave a total of over 150,000 cells. A team of doctors then looked at whether a bullet wound to a given cell would produce an instantaneous incapacitating injury (Fig. 3-57, right). Simulated shots were fired at the model and for each wound channel the vulnerability number was determined for every point. This procedure was repeated (taking account of shooter accuracy) until the mean values stabilized (at around 10,000 shots). This gave a "Vulnerability Index Function," which was then used to evaluate the weapon and ammunition

Fig. 3-58. Computer Man. Major blood vessels.

concerned, resulting in a relative incapacitation index, or RII. See 4.1.2.3.

Computer Man was later used in the military field, in modified form, to assess the effectiveness of fragmentation munitions. In this later version, each cell was linked to an organ, making it possible to represent the internal structures of the body, at least approximately (see Fig. 3-58). This model consisted of 167 layers and 124,000 cells, and distinguished between 280 types of tissue.

Fragment wound analysis also underwent modification, adding the work of SPERAZZA and KOKINAKIS (1965) and a large volume of data from experiments with animals to the existing gelatine experiment data (BELLAMY and ZAJTCHUK 1991a).

3.4.2.4 The "Verwundungsmodell Schütze" (VeMo-S)

Towards the end of the 1990s, the German Federal Office of Defence Technology and Procurement commissioned a model with similar aims. The original task was to verify the well-known probability of incapacitation figures that SPERRAZZA and KOKINAKIS (1965) had put forward. Over time, this developed into a computer model of the human body that offered a number of interesting possibilities.

The formulas and datasets proposed by SPERRAZZA and KOKINAKIS result in incapacitation probabilities that are clearly too high when compared with real cases and intuitive reasoning, particularly with small fragments.

Unlike Computer Man, Verwundungsmodell Schütze ("rifleman wound model," VeMo-S), rather than modelling a particular person, approximated biological structures using geometrical shapes (see Fig. 3-59). The positions and dimensions of these structures correspond to human anatomy. The concept is based on the idea that, given the wide variation between one person and another, and the wide range of threats (calibres, projectiles, fragments, etc.) and wound channels, there

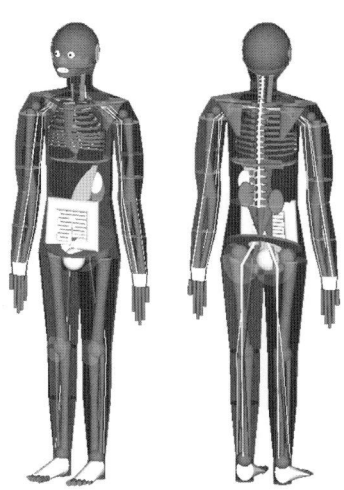

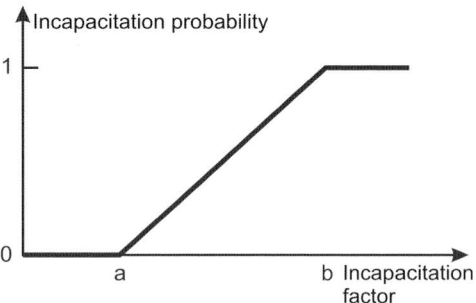

Fig. 3-60. (Top) Basic principle of the incapacitation function. a: Lower threshold. b: Upper threshold.

Fig. 3-59. (Left). Structure of the human body in VeMo-S

was no point in modelling a single object very precisely – that indeed, this was counter-productive.

VeMo-S models approximately 400 structures. A time-dependent incapacitation factor relevant to wound ballistics was assigned to every structure (e.g. effectiveness or energy density). For each structure, lower and upper thresholds were defined for each incapacitation factor. Below the lower threshold, no incapacitation would occur, while incapacitation was certain if the upper threshold were exceeded (see Fig. 3-60). The incapacitation factors therefore fitted into a mathematical framework subject to the laws of fuzzy logic (KNEUBUEHL 2000). This made it possible to reflect time factors, multiple hits and various levels of partial incapacitation. The thresholds were defined by W. TITIUS, a surgeon with exceptional experience in the field of bullet wounds.

The ballistic data of the bullet (energy, energy density and effectiveness) were calculated continuously along a geometrical wound channel and the degree to which the structure had been destroyed was determined from this. The ballistic data were obtained by conducting experiments using the simulants discussed in 3.3. As a result, the model was not restricted to fragments, but could perform calculations for any projectile, as long as the necessary ballistics data were available.

VeMo-S was validated against real cases from war surgery and forensics (GRÄMIGER 2010). If the required ballistic data are available, the wound channel can be calculated using VeMo-S and compared with the actual wound channel. If the two correspond, the incapacitation function is confirmed. If not, the function must be adjusted.

The role of VeMo-S is not restricted to the calculation of incapacitation probabilities; the program can also be used in forensics and in crime reconstructions. The program can calculate velocity profiles (see 3.4.2.2) for any wound channel, making it possible to estimate the exit velocity of a bullet and hence determine hazard levels or the range of a bullet on exit.

The program can also assess the effectiveness of body armour, and this assessment can include the effect of the energy remaining after a projectile has penetrated the protective layer.

4 Wound ballistics of bullets and fragments

B. P. KNEUBUEHL

4.1 The effectiveness of bullets

4.1.1 Effectiveness versus effect

4.1.1.1 Definitions

If we are to talk about the "effect" of a bullet, we have to start by defining our terms. Here, when we talk about the *effect* of a shot, we mean the reactions that the shot provokes in the human body. Consequently, an effect is always linked to a unique event.

For instance, an airgun pellet with a muzzle energy of 7 J that penetrates the spine and damages the spinal cord has a much greater effect than a 44 Rem. Mag. hollow-tip with an energy of 1500 J that grazes the upper arm. This means that statements like "A hollow-point bullet has a greater effect than a full metal-jacket bullet" are meaningless.

Clearly, the effect of a bullet depends on a wide variety of factors. What is clear is that effect cannot be defined (and most certainly cannot be measured) purely in terms of the physical or design characteristics of a bullet, such as mass, velocity, energy and deformability.

However, there is no denying that the physical and design characteristics of a bullet can have a decisive influence on its effect. If we consider these factors in isolation, we can establish the "potential effect," which we shall call the "effectiveness" of a bullet.

Effectiveness and effect are, of course, related to one another. A highly effective bullet, if it hits a certain part of the body, will very probably have a significant effect. Conversely, a less effective bullet will probably have a lesser effect.

4.1.1.2 Factors that contribute to the effect of a bullet

The above comments show that the effect of a bullet depends not just on its effectiveness but also, to a large degree, on the point of impact and on the path of the wound channel in the body. The physical and mental state of the victim are also important, and these two factors are usually underestimated when assessing the effect of a bullet – if they are taken into account at all.

Determination to achieve an objective can considerably reduce the immediate effect of a bullet. Under other circumstances, the mere sight of a weapon can be sufficient to prevent an attack. In the latter case an effect has been achieved even though there is no point of impact, and the effectiveness of the bullet is irrelevant.

The *effect* of a bullet on a living being therefore depends on the following factors:

- the effectiveness of the bullet;
- the point of impact and the path of the wound channel in the body;
- the physical and mental state of the victim.

The point of impact also reflects the decision taken by the shooter, and his or her mental state.

Analysis of cases in which a firearm failed to achieve the required effect shows that in many instances failure was caused by psychological stress on the firer leading to an inappropriate decision, rather than to any lack of effectiveness in the bullet.

Unless a bullet passes through certain parts of the brain or spinal cord, damaging vital areas, its effect depends not only on the purely mechanical effect (and the path of the wound channel) but also, to a very great degree, on the victim. His or her emotional and psychological state are decisive. Drugs, alcohol and similar also influence the effect of a bullet. So while *effectiveness* can be determined in a general fashion on the basis of physical parameters, *effect* depends entirely on the circumstances. There are numerous examples of this.

The influence of the mind on the effect of a bullet can best be illustrated by a few examples. Victims of bullets can be divided into three groups, as Hatcher (1927) has pointed out.

1. The victim is a harmless passer-by. He is fired at by chance (accidentally), having done nothing to provoke an attack. Surprise will play a major role. The effect of being hit by a bullet is generally very significant.

To take an extreme case, one could imagine an old lady standing next to an alarm pistol when it is fired. If she faints, a dramatic effect has been achieved despite the absence of any injury.

2. The second group consists of those who expect to be, such as soldiers in combat or police officers in a firefight with criminals. If a person in this group is hit, the effect will generally be less marked than in the case of the first.
3. The last group consists of persons whose consciousness is severely limited by extreme excitement, fanaticism or concentration on a combat situation – the "blinker effect."

Examples include a bank-robber who emerges from a bank with a hostage and finds himself confronted by a large police presence, a soldier storming an enemy position or a man shot during a pub brawl.

Experience shows that persons in the third group, if hit in a non-vital part of the body, will initially be quite unaware that they have been hit. Only when the excitement has died down will the presence of blood draw their attention to the wound. Even injuries which eventually lead to the death of the victim (such as rupture of the aorta or the heart) may not necessarily incapacitate the victim immediately, if vital areas of the brain are not damaged.

During their colonial wars, the British often found that even several fatal hits to major blood vessels from a large-calibre revolver did not stop an attacker. The victim was able to continue his charge and slit his enemy's throat before finally bleeding to death.

In one case known to the author, a policeman suffered a mortal wound to the aorta from a 45 Auto pistol bullet fired by a terrorist. Despite his injury, he was still able to empty his magazine at the attacker.

These examples show the degree to which the psychological element influences the effect of a bullet. It is impossible to predict this in advance, just as it is impossible to predict the point of impact or the course of the wound canal. The *effect* of a bullet can only be determined after the shot has been fired.

The only factor contributing to the effect of a bullet that can be determined physically and influenced by design is the *effectiveness*. Effectiveness depends primarily on the impact energy of the bullet (which constitutes its total potential effect) and on its ability to transfer that energy along the course of the wound channel.

4.1.2 Measures of effectiveness

4.1.2.1 Historical background

Ever since the end of the 19th Century, attempts have been made to determine the effectiveness of a bullet, using measurements or calculations. The first attempts were undertaken by doctors who had conducted a close study of bullet wounds. In his books published at the end of the 19th Century, the famous surgeon and first recipient of the Nobel Prize for Medicine, Prof. Theodor KOCHER of Bern University, assumed that what he called the "living force" of the bullet, which today we would term its "kinetic energy," played a decisive role in determining the effect.

Not long after, the effectiveness of a bullet was seen to be related to its ability to transfer energy to tissue along the course of the wound channel. In 1908, C. G. SPENCER, Professor of War Surgery at the Royal Army Medical College, London, published a book with the title "*Gunshot Wounds*". In the chapter entitled "The Wounding Power of Bullets," he described in detail how the "wounding capacity" of a bullet depends upon:

1. Its (kinetic) energy.
2. The ease with which it can convert its energy into work following impact.

SPENCER related the second point to the cross-sectional area of the bullet, the degree to which it tended to deform or fragment and its strength.

Seen from today's perspective, SPENCER's points are correct. Unfortunately, his conclusions were overtaken in the first decade of the 20th Century by the erroneous theory that effectiveness meant a bullet being able to "stop" an attacker, or even knock him down.

4.1.2.2 The "stopping power" fallacy

The idea that a bullet can knock an enemy down has passed into the literature and into everyday language through such terms as "stopping power." The assumption is that a bullet is capable of physically stopping a person who is in the process of carrying out an attack.

This (erroneous) understanding of how bullets work is supported by film scenes in which a person hit by a bullet is knocked flying.

In terms of physics, the transfer of movement from the bullet to the body is governed by the laws of collisions, which are based on the basic principle of conservation of momentum (see 2.1.3.5). The movement transferred from the bullet to the body is maximized if the bullet remains in the body. This is then a case of perfectly inelastic collision. If m_G is the mass of the bullet and its velocity is v_G, then its momentum before impact is given by:

$$(4.1:1) \qquad p_G = m_G \cdot v_G \,. \qquad [\text{N·s}]$$

If m_K is the mass of the body struck by the bullet and v_K is its velocity, then:

$$(4.1:2) \qquad p_K = (m_G + m_K) \cdot v_K \,. \qquad [\text{N·s}]$$

Bearing in mind the principle of conservation of momentum ($p_G = p_K$), v_K is given by:

$$(4.1:3) \qquad v_K = \frac{m_G}{m_G + m_K} \cdot v_G \,. \qquad [\text{m/s}]$$

Even if we assume a very heavy bullet, and also assume that it remains in the body (and hence transfers all of its momentum to the body), the "throwback velocity" is completely irrelevant by comparison with the normal movement of a pedestrian, and certainly by comparison with the velocity of an enemy during an attack (v > 2 m/s). See also KARGER and KNEUBUEHL (1996).

Even the rotation about the point of contact between the soles of the feet and the ground that acts upon the upper body when hit by a bullet (the force tending to knock the victim over) is so slight as to be unnoticeable. Table 4-1 shows the throwback and rotational velocities that typical bullets might cause.

The theory has been proven empirically. Alex Jason, an American forensic ballistician, allowed himself to be shot with a 308 Win. assault rifle while wearing body armour and standing on one leg. The impact of the bullet did not even cause him to wobble. The throwback velocity was approximately 10 cm/s. The experiment is documented on video.

Table 4-1. Throwback velocity (v_K) and rotational velocity (v) (m_K = 80 kg)

Calibre	m_G [g]	v_G [m/s]	E [J]	p_G [N·s]	v_K [cm/s]	v [s^{-1}]
9 mm Luger	8.0	350	490	2.8	3.5	0.009
45 Auto	14.9	260	505	3.9	4.9	0.012
44 Rem. Mag.	15.6	440	1510	6.9	8.6	0.022
308 Winchester	9.5	830	3270	7.9	9.9	0.025
12/70 slug	31.5	400	2400	12.6	15.8	0.038

4.1.2.3 Traditional measures of effectiveness

"Stopping power" and "relative stopping power." In 1927, HATCHER published his book "Pistols and Revolvers and Their Use," in which he defined the term *stopping power* (StP). He started with the assumption that a bullet requires a certain amount of energy to penetrate the body to sufficient depth (about 25 cm). If the bullet passes through the body, then only that part of the energy has an effect that is transferred to the body. For a given energy, a bullet with a large cross-section transfers more energy than a bullet of smaller calibre. The StP must therefore be proportional to the cross-sectional area A. In a similar manner, HATCHER included the effect of the shape of the bullet tip, via a form factor f. Round-nose bullets were assigned a smaller figure than blunt bullets, as they transfer less energy. His formula was as follows:

(4.1:4) $$StP = E \cdot A \cdot f .$$ "energy formula".

In HATCHER's original publication, energy E was expressed in foot-pounds and A in square inches, with the factor f depending on the shape of the bullet tip (the "form factor"). A 9 mm Luger full metal-jacketed round-nose bullet (E = 370 ft-lbs, A = 0.099 sq.in., f = 0.9) therefore has an StP of 33.0. For metric units, a conversion factor of 0.1144 is used.

The form factors in Table 4-2 appear to have been chosen in a somewhat arbitrary fashion, on the principle of "the blunter the bullet, the greater the effect." For instance, there is no reason why a full metal-jacketed round-nose bullet should be exactly 10% less effective than a lead round-nose bullet, even allowing for deformation of the softer bullet.

Table 4-2. Form factors for StP (stopping power) and RSP (relative stopping power)

Shape of bullet tip	Form factor
Full metal-jacketed round-nose	0.9
Lead round-nose	1.0
Semi-Wadcutter	1.1
Wadcutter	1.25
Semi-jacketed [a]	1.25 – 1.35

[a] Dependent on the degree to which the bullet mushrooms.

In the new edition of his book in 1935, HATCHER replaced energy by momentum I, which was supposed to better represent the effect of the bullet. He termed this measure of effectiveness the "relative stopping power" (RSP) by contrast with the StP and defined it as follows:

(4.1:5) $RSP = 1000 \cdot I \cdot A \cdot f$.   "momentum formula"

In the original, momentum is expressed in "pounds velocity" and the area is once again given in square inches. As the figures for I and A are very small, the right-hand side of Eqn 4.1:5 is multiplied by 1000, to yield more convenient numbers for the RSP. If I and A are expressed in metric units, a conversion factor of 0.0179 is required.

In the case of a 9 mm Luger cartridge, for instance: If I = 2.8 N·s, A = 0.64 cm² and f = 0.9, we obtain an RSP of 28.9.

HATCHER preferred the momentum formula, as it produced results that corresponded more closely to those of LAGARDE's empirical work (see 3.1.2).

WEIGEL's measure of effectiveness. (Effectiveness of a bullet as a function of the size of the channel in wood). WEIGEL, a well-known German ballistician, assumed effectiveness to be proportional to the volume of the channel in wood (WEIGEL 1975). He was hence the first person to use a simulant to study the interaction between bullet and target.

WEIGEL measured the depth to which various bullets penetrated fir, and derived an empirical formula from the results. Multiplying the penetration depth by the cross-sectional area gives the theoretical channel volume, which he took as indicating the effectiveness. His equation was:

(4.1:6) $W_H = 0.00024 \cdot m \cdot v^{1.5}$.   [kg·(m/s)$^{1.5}$]

It is interesting to note that if we disregard the constants, WEIGEL's measure of effectiveness W_H corresponds to the geometric mean M of momentum I and energy E:

(4.1:7) $W_H \propto \sqrt{I \cdot E} \propto \sqrt{m \cdot v \cdot m \cdot v^2} = \sqrt{m^2 \cdot v^3} = m \cdot v^{1.5}$.

Tables 4-3 and 4-5 list values of W_H for a number of bullets.

Because the volume of the geometrical channel in the wood is often substantially less than the volume of the temporary cavity, and because its relationship to penetration depth is different, W_H is only of limited use as a measure of the effectiveness of a bullet. Nonetheless, it is better than either of the HATCHER formulas.

SELLIER's measure of effectiveness. SELLIER proposed the pain caused by the temporary cavity as a measure of effectiveness. Of the two possible geometrical dimensions – volume and internal surface area – he selected volume, on the grounds that, as we have seen in Chapter 3, internal surface area is directly related to work done, and hence also to damage caused.

The principle is therefore: effectiveness is proportional to the energy that the bullet is able to transfer to the tissue, i.e. the potential damage. This energy is proportional to the volume of the temporary cavity and hence is also related to the amount of damaged tissue, which naturally depends on the nature of the tissue.

The thought process is as follows: The volume of the temporary cavity is given by the following equation:

(4.1:8) $$V = \mu \cdot E_{ab} \cdot \qquad [cm^3]$$

We have already derived the residual energy of a bullet that has penetrated a distance s (see 3.2.3.2 and Eqn 3.2:15):

(4.1:9) $$E(s) = E_a \cdot e^{-2 \cdot \Re \cdot s}, \qquad [J]$$

with retardation coefficient $\Re$:

(4.1:10) $$\Re = \frac{\rho}{2} \cdot \frac{C_D}{q} \cdot \qquad [m^{-1}]$$

If two bullets have the same energy and similar form, but different sectional densities q, the bullet with the smaller value of q will have the greater retardation coefficient. It will therefore transfer more energy and hence create a larger temporary cavity. That bullet is therefore considered to be the more effective of the two. From this, SELLIER developed a simple measure of effectiveness W_{TH}. Taking the entire wound channel into consideration, he made this measure proportional to impact energy and inversely proportional to sectional density:

(4.1:11) $$W_{TH} \propto \frac{E_a}{q} \cdot$$

If we substitute the basic variables mass (m), velocity (v) and the cross-sectional area of the bullet (A) for E_a and q, assume a proportionality factor of 2 and divide v by 1000, to obtain convenient numbers, we obtain:

(4.1:12) $$W_{TH} = A \cdot \left(\frac{v}{1000}\right)^2,$$

(with A in mm² and v in m/s). Table 4-3 lists this value for a number of calibres, alongside the WEIGEL and HATCHER criteria.

Determining effectiveness using the relative incapacitation index. The measures of effectiveness discussed so far can all be determined purely from the bullet's physical/ballistic data (mass and velocity) and arbitrary constants. The channel in a body (or simulant) is ignored.

The oldest and probably most well-known means of including the form and development of the temporary cavity in the process of determining effectiveness is based on the vulnerability index (VI). See Fig. 4-1. Depending on the position of the wound channel and on the organs damaged by the bullet as it passes along this channel, VI will vary as a function of penetration depth.

Table 4-3. Effectiveness of handgun cartridges, measured in accordance with four criteria (W_H, W_{TH}, StP and RSP)

Cartridge	m [g]	v [m/s]	E [J]	A [mm²]	ℓ [a] [cm]	W_H	W_{TH}	StP	RSP
22 short	1.85	285	75	24.6	8.2	2.0	2.0	1.9	2.4
22 L.R.	2.55	330	142	24.6	14.4	3.6	2.7	3.5	3.3
6.35 Browning	3.25	230	88	31.7	8.4	2.7	1.7	2.8	4.2
7.63 Mauser	5.5	440	535	46.0	25.4	11.7	8.9	25	20
7.65 Browning	4.8	300	216	46.0	12.6	5.8	4.1	10	12
9 mm Browning, short	6.0	265	215	63.6	9.5	6.1	4.5	14	19
9 mm Luger	8.0	350	490	63.6	17.2	11.0	7.8	31	32
38 Spl	10.2	265	330	63.6	14.5	9.3	4.5	24	35
357 Mag.	10.2	450	1010	63.6	36.0	23.0	13.0	74	60
44 Rem. Mag.	16.0	475	1800	98.5	41.7	41.3	22.3	182	135
45 Auto	16.0	260	516	99.4	14.9	14.8	6.7	15	59

[a] Penetration in fir for solid lead bullets: theoretical maximum values.

The VI and the cross-section of the temporary cavity as a function of distance travelled s are used to calculate the relative incapacitation index, RII, using the equation below.

$$(4.1:13) \qquad RII = \sum_{s=0}^{s=s_{max}} \pi \cdot r^2(s) \cdot VI(s) \cdot \Delta s ,$$

s : distance travelled
s_{max} : maximum distance (if s > 22 cm, s_{max} = 22 cm)
VI(s): value of VI at s
r(s) : radius of the temporary cavity at s
Δs : size of summation increments (the thickness of the "slices" into which the temporary cavity is divided).

The VI/s curves in Fig. 4-1 are the result of computer simulation using Computer Man) see 3.4.2.3 and the Monte Carlo method (approx. 10,000 shots per curve).

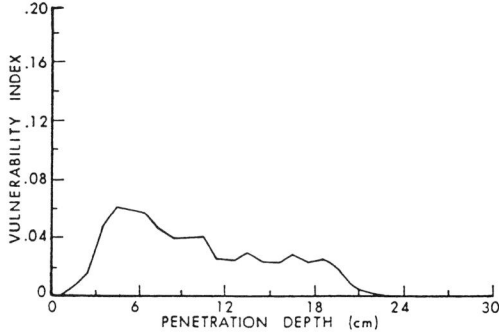

Fig. 4-1. Vulnerability index for a 9 mm Luger Silvertip bullet produced by W-W. Energy transfer and anatomy have been taken into account. The curve is therefore valid only for one direction of fire.

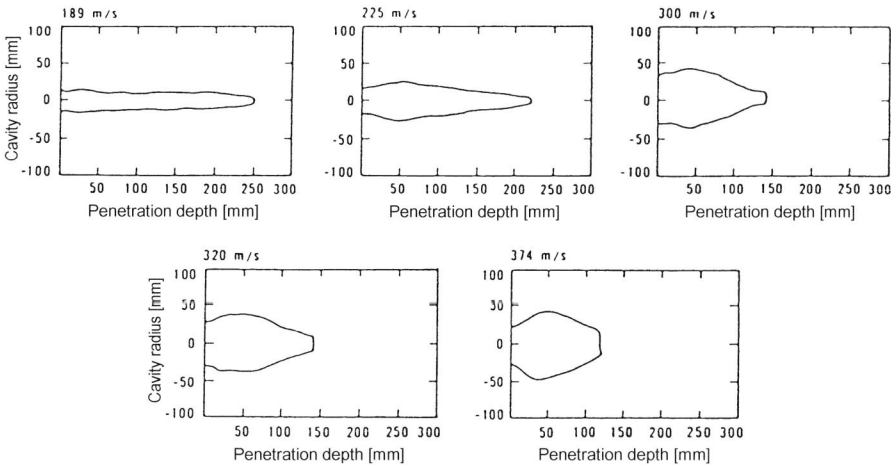

Fig. 4-2. Form and size of the temporary cavity as a function of impact velocity (top left). Bullet as in Fig. 4-1.

As we have seen, the degree to which a fragmentation bullet fragments depends on velocity and position. If one fires such a bullet into gelatine or soap, varying impact velocity v_a, one obtains temporary cavities of differing forms (see Fig. 4-2). For each of these cavities, we can use Eqn 4.1:13 to calculate the RII as a function of velocity. For a given calibre, we can therefore determine the velocity that yields maximum effectiveness. See Fig. 4-3: curve for Silvertip 9 mm Luger bullet manufactured by Winchester Western (m = 5.5 g).

Power Index Rating. Because of the extensive experiments and calculations required to determine the RII, and because the RII is unfavourable to certain calibres popular in the USA, new experiments were conducted, in an effort to describe the effectiveness of handgun bullets using easily-obtained data.

American author MATUNAS (1984) published an effectiveness formula to calculate what he called a "power index rating"(PIR). The formula was as follows:

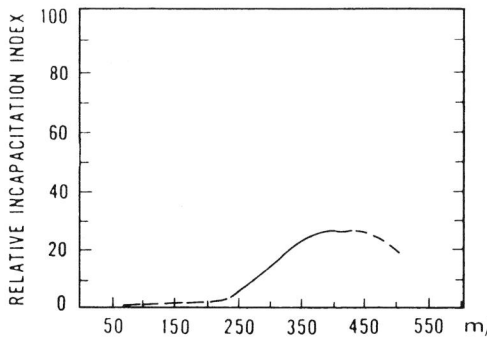

Fig. 4-3. RII (relative incapacitation index) calculated from Figs 4-1 and 4-2, as a function of impact velocity.

(4.1:14) $$\text{PIR} = \frac{v_a^2 \cdot \text{ET} \cdot B}{12111} \cdot D\ .$$ British/U.S. system

Where: v_a is impact velocity in ft/s, ET the "energy transfer value" from Table 4-4 a, B the mass of the bullet in grains and D a "diameter value" from Table 4-4 b.

Using the figure of 12111 in the numerator, the author standardized what he saw as a satisfactory bullet (38 Spl with PbHP bullet) to 100. He used this as the basis of an assessment scale (see Table 4-4 c).

The PIR is also based on impact energy. In metric units, substituting energy direct, the formula for PIR is as follows:

(4.1:14a) $$\text{PIR} = 27.4 \cdot E \cdot \text{ET} \cdot D\ .$$ E in Joule

The two additional factors ET and D introduce a substantial measure of subjectivity into the value of PIR, although these values could be derived from the author's experience. Why, for instance, should calibres of 10.15 to 11.41 mm automatically be 10% more effective than those of 8.88 to 10.14 mm? The same applies to ET.

Table 4-4 a. Energy transfer value (ET) for PIR

ET	Description
0.01	Bullets of which the cross-section increases on impact
0.0085	Shape-stable bullets of which the face accounts for at least 60% of the cross-section
0.0075	All other shape-stable bullets

Table 4-4 b. Diameter value (D) for PIR

D	Calibre range [mm]	Calibre range [in]
0.80	5.05–6.33	(.200–.249)
0.85	6.34–7.60	(.250–.299)
0.90	7.61–8.87	(.300–.349)
1.00	8.88–10.14	(.350–.399)
1.10	10.15–11.41	(.400–.449)
1.15	11.42–12.69	(.450–.499)

Table 4-4 c. Assessment table for values of PIR (MATUNAS)

PIR	Effectiveness
< 24	Unusable
25–54	Usable only if particular areas of the body are hit
55–94	Usable to some extent, but often unsatisfactory in practice
95–150	Ideal
151–200	Very effective
> 200	Too powerful

"Knockout Value". The "Knockout Value" (KO) formulated by TAYLOR (1948) is another momentum-based effectiveness formula. TAYLOR was a British big-game hunter. His formula for assessing hunting ammunition was as follows:

(4.1:15) $\qquad \text{KO} = m \cdot v \cdot k$.

The above formula uses mass in pounds, velocity in feet per second and calibre in inches. In order to produce the same figures when using SI units (grammes, metres per second and millimetres respectively), one must apply a conversion factor of 0.000285 to the right-hand side.

Unlike HATCHER's RSP, this formula uses calibre rather than the face area A. As a result, this plays a lesser role than in the RSP formula. The KO formula takes no account of deformation, which indicates that it may well have been intended only for solid and full metal-jacketed bullets of the types used for big-game hunting. We include this formula purely for the sake of completeness.

CARANTA and LEGRAIN's experiments using clay as a simulant. Like wood, clay is heterogeneous, and its behaviour depends to a very large extent on its composition and moisture content. It is significantly denser than that of soft biological tissue and its flow behaviour when struck by a bullet is totally different. As a result, clay can only be used for purposes of comparison, and even then only if the density, granulometry, water content, plasticity, etc. of the blocks of clay are sufficiently similar. Reproducing experiments with clay at a later date, or at another location, is likely to be very difficult. Nevertheless, clay is still used on occasion to assess the effectiveness of bullets.

CARANTA and LEGRAIN (1993) conducted an extensive set of experiments. The reason for their study was the famous controversy as to which is the more effective handgun bullet – the 9 mm Luger or the 45 Auto.

A propos this "famous controversy": The question as to whether the 9 mm Luger or the 45 Auto is the more effective pistol cartridge is an old one, but one which continues to be discussed today. The two cartridges have approximately the same energy, but the differences in calibre and mass give them different characteristics, rendering the one more suitable for certain roles and the other more suitable for others. The debate has given rise to numerous studies.

CARANTA and LEGRAIN confirmed a military criterion, whereby a bullet is "effective" if it passes through a 25 mm piece of fir. For close-quarter defensive weapons they increased this value (arbitrarily) to 50 mm.

The cavities that the bullets caused in clay were measured up to a depth of 15 cm. Although both authors noted major differences, they decided that a cavity volume of 0.62 l was a representative value for the better of the two calibres and set a minimum volume of 0.60 l as the criteria for classifying a handgun cartridge as "effective." The connection between this and the criterion of a bullet passing through a fir plank of a given thickness is not entirely clear.

The authors conducted a large number of tests on what was unfortunately a highly dubious basis, using 10 different weapons and a variety of ammunition types. They recorded the results with great thoroughness, measuring volume, longitudinal cross-section and the diameter at various points along the cavity. Table 4-5 shows examples of their results.

Comparison between the various effectiveness criteria. To conclude our discussion of the common bullet effectiveness criteria, Table 4-5 summarizes the rating that the different criteria assign to a number of typical handgun bullets (KNEUBUEHL, 1990b). We have been careful only to compare values that are indeed comparable. In the case of the effectiveness criteria derived from experiments, differences in velocity between the experiments to compare should not make more than 10 m/s.

Table 4-6 summarizes the various rankings, with 1 representing the bullet rated most effective according to the criterion concerned.

These figures are highly informative, in that they clearly show that it is wellnigh impossible by means of simple numbers to measure the effectiveness of rifle or handgun ammunition objectively and in a manner that is generally valid. As the table shows, the 38 Spl LRN and lead wadcutter are at the bottom of the rankings, regardless of which effectiveness criterion is used, Conversely, three bullets take first place, with seven in second place. One bullet takes first place or last place, depending on the criterion chosen, while another is to be found anywhere between second and tenth place. The smallest difference between best and worst place for a given bullet is 4.

This is the clearest possible illustration of how disparate the assessment of bullet effectiveness can be if simple criteria are applied. It also shows how much subjectivity is involved in the formulas.

MARSHALL and SANOW's "street results" MARSHALL and SANOW (1992) addressed the question of bullet effectiveness from the point of view of practical experience. They called their work a "definitive study" and believed they had found the secret of producing a definitive assessment of effectiveness, using data from real-life cases. MARSHALL spent a huge amount of time analysing over 1800 firefights ("street results"), looking exclusively at first hits to the body or head. A shot was deemed to have been effective if it caused instantaneous incapacitation, or prevented the victim from fleeing more than 3 m. For each calibre and type of bullet, he calculated the ratio of effective hits (according to these criteria) to the total number of cases analysed. This percentage figure was then taken as the effectiveness of the bullet ("mean effect" would have been a more precise term).

At first sight, this definition of effect, drawn from "real life," would appear to solve the old problem. However, the validity of the results is beset by a number of statistical and other pitfalls.

The numbers the authors give to describe effect are in fact simple frequencies, just like RII. Statistically speaking, the two criteria are analogous.

4.1 The effectiveness of bullets 175

Table 4-5. Comparison between the various effectiveness criteria

Calibre	No.	Bullet	M [Ns]	E [J]	RSP	StP	W_H	PIR	C/L	RII	W_{TH}
9 mm Luger	1	FMJ	2.58	415	26.5	14.0	10.8	85	0.69	11.3	3.30
	2	SJHP	2.77	511	39.5	23.9	12.4	140	1.63	28.2	7.80
38 Spl	3	LRN	2.54	316	29.0	11.8	9.4	65	0.37	4.8	1.97
	4	SJHP	2.19	388	31.2	18.2	9.7	106	1.72	28.9	7.18
	5	LSWC	2.57	324	32.2	13.3	9.5	76	1.17	6.7	2.02
	6	LWC	2.13	235	30.4	11.0	7.4	64	0.59	9.0	2.16
357 Mag.	7	LRN	3.58	628	40.8	23.5	15.7	129	0.70	21.0	3.92
	8	SJHP	3.40	714	48.5	33.4	16.3	196	1.69	40.8	10.10
	9	SJHP	3.50	600	49.9	28.1	15.2	165	1.45	22.3	6.74
45 Auto	10	FMJ	3.92	515	64.6	27.8	14.9	117	–	4.3	3.53
	11	SJHP	3.58	492	81.9	36.9	13.9	148	1.98	18.0	6.95

Table 4-6. Rankings

Calibre	No.	m [g]	v_0 [m/s]	I	E	RSP	StP	W_H	PIR	C/L	RII	W_{TH}
9 mm Luger	1	8.0	322	7	7	11	8	7	8	8	7	8
	2	7.5	369	6	5	6	5	6	4	4	3	2
38 Spl	3	10.2	249	9	10	10	10	10	10	10	10	11
	4	6.2	354	10	8	8	7	8	7	2	2	3
	5	10.2	252	8	9	7	9	9	9	6	9	10
	6	9.7	220	11	11	9	11	11	11	9	8	9
357 Mag.	7	10.2	351	2	2	5	6	2	5	7	5	6
	8	8.1	420	5	1	4	2	1	1	3	1	1
	9	10.2	343	4	3	3	3	3	2	5	4	5
45 Auto	10	14.9	263	1	4	2	4	4	6	–	11	7
	11	13.0	275	3	6	1	1	5	3	1	6	4

Abbreviations used in Tables 4-5 and 4-6
m   mass of bullet
I   momentum
RSP "Relative Stopping Power" (Hatcher)
W_H Weigel's formula (wood model)
RII „Relative Incapacitation Index"
C/L Clay (Caranta and Legrain)
v_0 impact velocity
E   energy
StP "Stopping Power" (Hatcher)
PIR "Power Index Rating" (Matunas)
W_{TH} Sellier's formula

For instance, if Cartridge A has a "street result" value of 88% and Cartridge B a value of 76%, one may conclude that, over a large number of incidents, Cartridge A will achieve a few more successes than Cartridge B. Whether Cartridge A will actually prove to be the better type in the next incident (which could be one's last), cannot be deduced from these figures.

The number of cases analysed varies enormously from one cartridge type to another. Some percentages were calculated on the basis of just five cases (!), while others are based on 400. It is misleading to compare two types of cartridge on the basis of such widely different frequencies.

The data used to assess the effect of the cartridge are derived from the effect observed in real life. In theory, there can be no objection to this. However, a large number of parameters are missing – exact point of impact, condition of the weapon, mental state of the victim, etc. – which introduces considerable scatter. This in turn makes it very difficult to draw comparisons and renders any predictions highly unreliable.

MACPHERSON's "Wound Trauma Incapacitation" In his book "Bullet Penetration" (1994) MACPHERSON introduces a criterion for assessing the effectiveness of a bullet that can be determined using gelatine. His method is based on the damage done to the tissue, i.e. the effect of the bullet. He therefore excludes energy as a criterion because, for instance, a bullet fired into water transfers energy to the water, yet does not damage it.

In the case of tissue, the damage caused depends to a large extent on the structure of the tissue and its physical characteristics, and on the extent to which it can absorb the energy from the bullet and convert it into work. For his "Wound Trauma Incapacitation" (WTI) MACPHERSON therefore chose the volume of the gelatine damaged by the bullet.

It is virtually impossible to determine this volume empirically to a sufficient degree of accuracy. Methods of measuring the effect of a bullet by measuring the volume of tissue damaged (debridement) had already been proposed (BERLIN et al. 1976, BERLIN et al. 1977), but were rejected because the results varied too widely from one person to another. The same result is to be expected when using gelatine.

MACPHERSON determined this volume analytically, using empirical constants (coefficient of resistance and bullet constant) obtained from experiments with gelatine. Despite his detailed instructions, it would appear to be relatively difficult to repeat the process in other laboratories in such a way as to produce comparable results.

4.1.2.4 Summary and conclusions

We can divide the effectiveness criteria discussed in 4.1.2.3 into three categories:

1. Criteria based on the momentum of the bullet.
2. Criteria based on the energy of the bullet.
3. Statistics-based criteria.

The momentum criteria include HATCHER's "Relative Stopping Power" (RSP) (1935) and TAYLOR's "Knockout Value" (KO). As mentioned earlier, the momentum of a bullet has very little to do with its possible effect in the body. It is therefore impossible to apply momentum-based criteria to bullets that penetrate the body.

The energy-based criteria include HATCHER's "Stopping Power" (StP) (1927), MATUNAS' "Power Index Rating" (PIR) and the measures of effectiveness proposed by WEIGEL (W_H) and SELLIER (W_{TH}). All these criteria include the impact energy of the bullet and hence the maximum energy that the bullet could transfer to the body. However, the effectiveness of a bullet depends not on the energy available to it, but on the energy that it is capable of transferring to the tissue. Each formula therefore requires additional variables that take account of the behaviour of the bullet in tissue. In the cases of StP and PIR, these are coefficients chosen by the authors, which naturally include a high degree of subjectivity (e.g. a preference for larger calibres). W_H uses a formula that approximates the volume of the wound channel in wood, while W_{TH} uses theoretical considerations regarding energy transfer to introduce the sectional density of the bullet (taking account of deformation). CARANTA and LEGRAIN (C/L) use the volume of the cavity created by the bullet, which is proportional to the energy transferred. Unfortunately, they chose clay, which is a very unsuitable medium.

None of these criteria is capable of representing the local potential energy transfer, which is what counts as far as effectiveness is concerned. If we disregard those that include subjective adjustments and those based upon cavity volumes in unrepresentative materials, the only measure of effectiveness remaining is that of SELLIER, W_{TH}. However, even this value can only be taken as a rough indication, as it does not tell us the amount of energy transferred at a given point.

The statistical methods – "Relative Incapacitation Index" (RII), "Street Results" and "Wound Trauma Incapacitation" (WTI) all aim to record the effect of the bullet. RII is based on the simulation of bullet wounds using Computer Man, which in turn is based on the results of animal experiments. The "Street Results" are based on statistics from a few thousand real-life firefights. WTI is calculated with the aid of parameters derived from experiments with gelatine.

All three methods use the *effect* of the bullet on the human body – real or simulated. The RII and "Street Results" also take account of elements of the effect that are independent of the bullet. RII takes account of the point of impact and the path of the wound channel, but not of the physical or mental state of the victim. The "Street Results" are based on actual effect, but are subject to systematic and random errors (selection of cases, definition of "stop," missing ballistic data, etc.).

WTI, despite claiming to describe effect (i.e. the destruction of tissue), takes no account of the direction or length of the wound channel, which means that it is only valid for gelatine.

Effect criteria ignore the fact that the *energy that a bullet transfers to a medium* cannot be compared with the *energy that the medium absorbs* or the *energy converted into work*. The energy *transferred* as a function of penetration depth is virtually the same in all accepted simulants, and in biological tissue, and can hence be used as a measure of the effectiveness of the bullet (its *potential* effect), independent of the medium and of the path of the wound channel. The energy converted into work, which is what does the actual damage, depends very much on the tissue and its physical characteristics. It therefore cannot be seen purely as a characteristic of the bullet.

4.1.3 Determining the effectiveness of a bullet

4.1.3.1 Definition of effectiveness

From 4.1.1.1, we see that the wounding potential of a bullet – its effectiveness – has to be based on the mechanical work that the bullet does on tissue. This work is equal to the energy transferred from the bullet to the tissue. Because the amount of energy transferred can vary significantly along the wound channel, it would appear appropriate to define the effectiveness of a bullet as the energy transferred per unit distance, as a function of penetration depth. The unit of measure is hence energy per unit distance, with J/cm yielding a suitable order of magnitude. For a distance Δs, Eqn 3.2:5 (see 3.2.2.5) gives:

$$(4.1\!:\!16) \qquad \frac{\Delta E}{\Delta s} \approx E'_{ab}(s) = -2 \cdot \Re(s) \cdot E(s) \; . \qquad [\text{J/cm}]$$

According to this definition, effectiveness (in mathematical terms) is the gradient $E'_{ab}(s)$ of the energy transfer function $E = E(s)$ for $E(0) = 0$ and $E(s_{max}) = E_A$, where s_{max} is the depth at which the bullet comes to rest and E_A is the impact energy.

The energy transferred at a point s along the path Δs is hence proportional to the energy remaining at that point $E(s)$ and the retardation coefficient $\Re$. $\Re$ depends on the density of the tissue and on the resistance coefficient C_D, and is inversely proportional to the instantaneous sectional density of the bullet at s (remaining mass divided by instantaneous cross-sectional area in the direction of travel). See Eqn 3.2:4a.

$$(4.1\!:\!17) \qquad \Re(s) = \frac{C_D \cdot \rho}{2 \cdot q(s)} \; . \qquad [\text{m}^{-1}]$$

If a bullet deforms inside the medium, or undergoes yaw, sectional density decreases and the retardation coefficient increases. According to Eqn 4.1:16, the amount of energy transferred will also increase.

The effectiveness of the bullet is independent of the path of the wound channel and of other circumstances that influence the effect of the bullet, such as psychological factors. Effectiveness is one of the characteristics of a particular bullet. This implies that the effectiveness of a bullet can only be measured using predefined, suitable simulants (as described in 3.3). This is an essential precondition for comparing the effectiveness of one bullet with that of another.

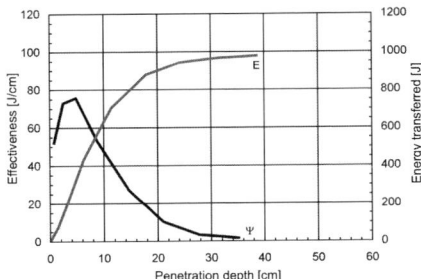
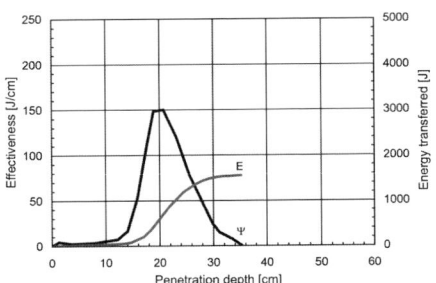

Fig. 4-4. Effectiveness 'I' (black line) and energy transferred E (grey line) for a 357 Mag. semi-jacketed bullet.

Fig. 4-5. Effectiveness 'I' (black line) and energy transferred E (grey line) for a 5.56 × 45 full metal-jacketed.

4.1.3.2 Measuring effectiveness

MARTEL's law (see 2.3.7.3) comes into play when determining effectiveness empirically. If a bullet penetrates a deformable medium and displaces material, the volume created is proportional to the energy used. This is true not only for glycerine soap (which undergoes plastic deformation) but also for gelatine, in which only a temporary cavity is created. Determining the effectiveness of a bullet is therefore a matter of determining the volume of the cavity that the bullet creates in a simulant.

We have already described in detail (in 3.3) the methods used for gelatine and glycerine soap. Figures 4-4 and 4-5 show the effectiveness functions for two bullets, by way of example.

For many years, the ballistic laboratories of Dynamit Nobel AG have been determining the effectiveness functions of bullets using gelatine (GAWLIK and KNAPPWORST 1972, and KNAPPWORST 1976). Methods using soap have been developed at the ballistics laboratory of the Swiss ministry of defence (KNEUBUEHL 1999a). The results of the two methods correspond closely.

4.1.4 Military effectiveness criteria

4.1.4.1 Definitions of effectiveness

For military purposes, the effectiveness of a projectile is generally defined according to one of the following principles:

1. A projectile (be it a bullet or a fragment) is "effective" if it has enough kinetic energy to pass through a specified thickness of a specified material.
2. A level of kinetic energy is specified for the projectile that is held to be just sufficient to incapacitate a person.

3. The probability is specified of a projectile with a given energy incapacitating a person.

Regarding the *first principle*: It is very easy to verify that a projectile meets effectiveness criteria based on a projectile passing through a hard material. As a result, such criteria are popular and are in widespread use. In general, common materials are used that can be defined reasonably well and are easy to obtain.

The materials used are generally pine or fir boards with thicknesses ranging from 20 to 40 mm, and steel or aluminium sheet with thicknesses ranging from 1 to 3 mm. Wood is less suitable, as it has a heterogeneous structure and is not easy to describe, on account of such factors as moisture content.

In Germany, for instance, a projectile is deemed effective against persons if it passes through a sheet of galvanized steel sheeting 1.5 mm thick (St 37, DIN 17100). A 7 mm steel sphere requires 80 J to pass through such a sheet. Naturally, the energy required varies with the diameter of the sphere.

The disadvantage of determining effectiveness in this manner is that whether or not a projectile passes through a hard material depends primarily on the energy density of the projectile, whereas effectiveness depends on the energy transferred. As a result, this method can lead to totally erroneous decisions.

Principle 2: The amount of kinetic energy a projectile needs, in order to put a soldier out of combat is generally termed the *casualty criterion*.

RHONE (1896) formulated the first effectiveness criterion, according to which the transfer of 80 J of energy was sufficient to cause incapacitation. GURNEY (1944) was of the opinion that this value applies to projectiles weighing between 50 mg (fragments) and 30 g. Over the years, therefore, various States have laid down energy thresholds of this type (see Table 4-7).

The large difference between the French and Russian values (1:6) indicates that this criterion is not very useful. FINCK (1965) stated that it would appear difficult and unrealistic to lay down a standard of this type. Many other factors, such as the form of the bullet tip, the path of the wound channel, etc., have a major effect on the energy required to cause incapacitation.

It is worth repeating what was said about the *effect* of a bullet in 4.1.1: The effect depends not only on the energy transferred (the *effectiveness*) but also, to a large extent, on the physical and mental state of the victim. A highly motivated fighter (e.g. in the Pacific theatre during the Second World War, see CHURCHILL,

Table 4-7. Energy thresholds laid down by States

State	Threshold [J]	Note
France	40	
Germany	80	RHONE, 1896
USA	80	GURNEY, 1944
Switzerland	150	Until ca. 1975, later 80 J.
Russia	240	

"The Second World War") or the native defending his land against the white invader (i.e. the colonial wars) requires a more effective projectile for a given effect than does the unmotivated soldier. The path of the wound channel in the body is also highly significant. For a given energy transfer, a bullet passing through the neck will have a very different effect to that of the same bullet passing through the soft tissue of the thigh.

Principle 3: this is the subject of the next section.

4.1.4.2 Probability of incapacitation

General. The probability that a person will be incapacitated due to the effectiveness of a bullet or a fragment is determined not by a single energy value, but by a number of parameters. These are:

1. The ballistic properties of the projectile, generally the impact energy $\frac{1}{2} \cdot m \cdot v^2$ or a more general function of m and v: $m^{\alpha} \cdot v^{\beta}$.
2. The point of impact and the path of the wound channel in the body.
3. The activity on which the soldier is engaged (attack, defence, etc.). Psychological and other factors come into play here.

Military wound ballistics literature express such probability functions as P(I|H), where I = incapacitation and H = hit. P(I|H) is hence the probability of incapacitation occurring if a soldier is hit.

P(I|H) = 0.2 therefore means that if we consider a large number of soldiers, 20% will be incapacitated, but that 1 - P(I|H) = 0.8 (80%) will still be capable of fighting. This figure takes no account of the fate of the individual; P(I|H) is a statistical figure, which is only relevant when taken as the mean over a large number of persons.

It is interesting to consider the form a P(I|H) function should take in principle, if it is to take account of energy and other factors. For very small amounts of energy E_a, P(I|H) = 0. As E_a increases, so does P(I|H), tending towards 1 as E_a increases further. A further increase in E_a produces no further increase in P(I|H), leading only to what in military terms would be dubbed "overkill." The function must therefore yield the following:

for $E_a \to 0$ $\Rightarrow$ P(I|H) $\to$ 0 and

for $E_a \to \infty$ $\Rightarrow$ P(I|H) $\to$ 1.

A large number of functions fulfil these requirements. The simplest and most widely-used belong to the following two classes of exponential function:

(4.1:18) $\quad y = 1 - \exp(-a \cdot x + b)^n$,

(4.1:19) $\quad y = \dfrac{1}{1 + \exp(-a \cdot x + b)^n}$.

For small values of exp(–a·x + b)^n (y close to 1) the two equations are related as follows:

(4.1:20) $$\frac{1}{1+z} \approx 1 - z, \qquad \text{if } |z| \ll 1.$$

Under these conditions, the results of the two equations are practically identical.

Incapacitation formulas. In addition to impact energy E_a (or the mass and velocity of the projectile), the two equations include the following parameters:

1 The probability of a particular area of the body being hit, depending on the soldier's activity at the time.
2 The probability of incapacitation for a given wound channel in a given area of the body.

The first of these probabilities can only be determined from combat experience. BEEBE and DE BAKEY (1952) give the values listed in Table 4-8, for instance. It is interesting to note that these figures remain virtually constant over different wars, even though the percentage of wounds caused by different projectiles (fragments vs. bullets) is very different in some cases.

The second type of probability is determined separately for each area of the body (head and neck, chest, abdomen, arms and legs). The P(I|H) for a person as whole is hence the sum of the individual P(I|H) values, multiplied (weighted) by the probability of a particular area of the body i sustaining a hit (p_i). See Table 4-8 for examples.

We can therefore write:

(4.1:21) $$P(I|H) = \sum_{i=1}^{5} P(I|H)_i \cdot p_i. \qquad [-]$$

While values for p_i are known from statistics or combat experience, or can be determined without too much work, determining P(I|H) demands considerable resources in terms of money, personnel and time, quite apart from the incredible number of animals required (over 10,000). As a result, the values and parameters determined empirically are almost always secret, to ensure that these hard-won values do not fall into the hands of a potential enemy.

Table 4-8. Hit distribution for US infantrymen in the Second World War

Area of body	Body surface area [%]	Hits [%]
Head and neck	12	21
Chest	16	13
Abdomen	11	8
Arms	22	23
Legs	39	35

To determine P(I|H), it is necessary to carry out test shots for each part of the body i, for a specific bullet, at a number of different impact velocities v_a. This then yields the length and diameter of the wound channel for a standing person. An experienced doctor can then determine the probability of incapacity if a particular organ (or part of an organ) is hit. See also 3.4.2.3 and 3.4.2.4.

It is not possible to determine the wound ballistic effect on a person without comparing the results with real wounds. The well-known formulas drawn up by SPERRAZZA and ALLEN in 1956 (SPERRAZZA 1962, SPERRAZZA and KOKINAKIS 1965) used the results of animal experiments and involved firing bullets into body parts of goats. The decrease in velocity in various organs was then measured. In addition, shots were fired into living animals, and their behaviour was observed. After the animals had been killed they were dissected and the findings analysed from a pathological point of view, in order to draw conclusions regarding the effects of bullet wounds on humans. It is easy to see how much work went into determining the parameters for the P(I|H) function.

We shall now look at a number of equations for P(I|H), as proposed by various authors and States.

SPERRAZZA and ALLEN produced the following formula, which applies to spherical and cuboid fragments:

(4.1:22) $\qquad P(I|H) = 1 - \exp[-a \cdot (m \cdot v_a^{3/2} - b)^n]$. $\qquad$ [-]

The values of a, b and n were kept secret for many years. These parameters allow the formula to take account of the activity the soldier is undertaking when hit. It is interesting to note that in this instance P(I|H) depends not upon v_a^2 (i.e. energy) but upon $v_a^{3/2}$, i.e. the geometric mean of the energy and momentum of the bullet. Eqn 4.1:22 was not sufficiently general, however (e.g. it did not describe the wound ballistic effects of dart-shaped projectiles), and so attempts were made to find a better (in the sense of more general) equation.

DZIEMIAN (1958) produced an equation that related P(I|H) and the energy $E_{1-15\,cm}$ transferred to a gelatine block 15 cm in length. The block length was chosen on the assumption that a wound channel 15 cm in length will reach any vital organ, regardless of the direction from which the projectile comes, and that deeper penetration will achieve no further effect. It was also assumed that a projectile that penetrates no further than 1 cm will not damage any tissue that would incapacitate the person (except in the case of the hand).

DZIEMAN's equation is as follows:

(4.1:23) $\qquad P(I|H) = \dfrac{1}{1 + \exp[-(a + b \cdot \log E_{1-15\,cm})]}$. $\qquad$ [-]

Constants a and b should be the same regardless of the projectile type – fragment or bullet.

Initially, energy $E_{(1-15\,cm)}$ was determined using high-speed film of shots into gelatine. This technique is very expensive and time-consuming, however. Attempts were therefore made to find simpler methods. The tried-and-trusted ballistic pendulum was chosen, although this approach does only allow one to measure the energy ΔKE transferred to the gelatine block.

Using energy transfer ΔKE yielded the following incapacitation formula:

(4.1:24) $\qquad P(I|H) = 1 - \exp(-a \cdot \Delta KE)^n$. [-]

The significance of the parameters a and n has already been discussed above.

The NATO wounding criterion, formulated in 1975 by STURDIVAN on the basis of secret research, uses *expected kinetic energy* (EKE), which is calculated as follows:

(4.1:25) $\qquad EKE = \int p(x) \cdot F(x) \cdot dx$. [J]

where p(x) is the probability of the bullet still being in the body at point x and F(x) is the decelerating force acting on the bullet at point x. This force is measured from centimetre to centimetre. The integral for F(x)·dx is simply the energy transferred to the body. However, energy can only be transferred if the bullet is still in the body at x. This is reflected in the inclusion of p(x). For a given individual, p(x) is 1 if the bullet is still in the body and zero if it has left the body. This is known as a rectangular function. However, if one considers a population of individuals, p(x) does not drop from 1 to 0 instantaneously: the value of p falls gradually, on account of the scatter in the dimensions of the human body. Fig. 4-6 shows probability curves (p(x)) for bullet track length in various parts of the body.

In practice, x-ray flash photography was used to measure the decrease in the velocity of the bullet in the 30 cm gelatine block from centimetre to centimetre, and the velocity/distance curve extrapolated out to 45 cm. EKE was than calculated using the discretized form of Eqn 4.1:25.

STURDIVAN's equation is as follows:

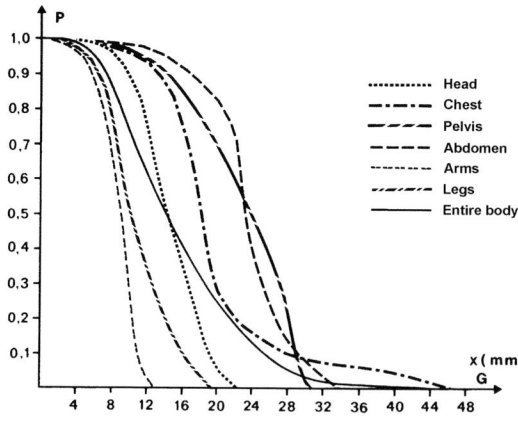

Fig. 4-6. Probability curves p(x) for bullet track length in various parts of the body as a function of penetration depth x. The solid curve is the mean of the other two curves, and refers to the body as a whole.

$$(4.1:26) \qquad P(I|H) = \frac{1}{1 + \alpha \cdot \left(\dfrac{EKE}{\gamma} - 1\right)^{-\beta}}, \qquad [-]$$

where α, β and γ are once again (secret) constants that serve to adapt the curve to the circumstances. At a later stage, the equation was improved (or so it was claimed) by replacing the numerator with a constant L, of which the value was something other than 1.

In summary, it would appear that a lot of "playing with numbers" is going on here, in connection with all the expressions for P(I|H) cited above. This could well be linked to the excessive and uncritical use of the computer to analyse the results. Indeed, this is hinted at by the large number of terms used to describe one thing – the incapacitation of a soldier as a function of impact energy. It also shows that there is considerable variation in the reaction of a biological system to trauma, although "integrating" over a large number of individuals can compensate for this to some extent.

One should therefore take a critical view of these expressions. Ultimately, they are simply a mathematical description of the situation. They do not establish any physical or pathological relationship. At best, they show that for low impact energies the effect (i.e. incapacitation) is negligible ($P(I|H) \approx 0$), whereas for very high energies ("overkill") a virtually 100% effect is to be expected ($P(I|H) \approx 1$).

Table 4-9 lists the most significant of the effectiveness criteria proposed in the USA over the years.

Table 4-9. Summary of the US effectiveness criteria

Author	Criterion	Year
BENTON	Bullet passes through fir plank	1867
RHONE	Kinetic energy: 80 J	1896
ZUCKERMAN	$m^{0.4} \cdot v$ (bullet passes through)	1942
SPERRAZZA and ALLEN	$m \cdot v^{3/2}$	1956
DZIEMIAN	$E_{1-15\,cm}$, energy transfer in 15 cm of gelatine	1960
STURDIVAN	Expected kinetic energy (EKE)	1975

4.2 Wound ballistics of handgun bullets

4.2.1 Penetration depth of handgun bullets and ability to pass through gelatine, soap, muscle and bone

4.2.1.1 General

The material in this section applies primarily to bullets fired from handguns (i.e. pistols and revolvers), and such bullets generally carry relatively little energy. They are therefore significantly less effective than rifle bullets, which carry many times more energy. The direct effect of the bullet is hence more important than the effects of the temporary cavity (indirect damage). Two factors that are of more importance in connection with handgun bullets are penetration depth and the minimum velocity required for the bullet to penetrate and pass through various materials. These factors are of secondary importance when considering rifle bullets, as a rifle bullet will always have more than enough energy to pass right through the body, except at the end of its trajectory or in the case of a ricochet.

In the next few sections, we shall look at equations linking penetration depth to bullet characteristics and ballistic parameters (mass, form factor and velocity) and to the properties of the target. In all cases, we shall attempt to establish physical relationships between the variables.

Penetration depth s can be seen as the "braking distance" of the bullet. We are therefore looking at the movement of an object as it undergoes deceleration, which can be described using the equations of motion (see 2.1.3.6). The determining factor is the characteristic deceleration resulting from the force acting on the bullet. This is generally given by the following equations:

(4.2:1a) $\qquad a = -c_0$, $\qquad$ [m/s^2]

(4.2:1b) $\qquad a = -c_1 \cdot v$, $\qquad$ [m/s^2]

(4.2:1c) $\qquad a = -c_2 \cdot v^2$. $\qquad$ [m/s^2]

In Eqn 4.2:1a, deceleration is independent of velocity. This equation is mainly used for materials held together by cohesive forces, such as bone, concrete and thin steel.

In Eqn 4.2:1b, deceleration is proportional to velocity. This equation is valid for viscous fluids (the Navier-Stokes equations) and is only relevant at very low velocities.

Eqn 4.2:1c is a square law. This equation applies when deceleration is determined primarily by the density of the medium, but not by its cohesive force. This category of medium includes air, water and predominantly aqueous media, such as gelatine, the liver, muscle, etc. In media such as muscle, the decelerating affect of mass inertia is far more important than that of the strength of the fibres.

If we substitute deceleration from Eqns 4.2:1a-c into the general equation of motion for distance and integrate, we obtain the following equations for penetration depth as a function of velocity:

(4.2:2a) $\quad \frac{1}{2} \cdot v^2 = -c_0 \cdot s + C \Rightarrow s \propto v^2 - v_{ad}^2$,

(4.2:2b) $\quad v = -c_1 \cdot s + C \Rightarrow s \propto v - v_{ad}$,

(4.2:2c) $\quad \ln v = -c_2 \cdot s + C \Rightarrow s \propto \ln \frac{v}{v_{ad}}$.

The integration constant C is chosen such that when $s = 0$, $v = v_{ad}$, the entry velocity (the velocity of the bullet just after it penetrates the outer layer, e.g. the skin).

The properties of the bullet, the ballistic parameters and the characteristics of the target medium are reflected in c_0, c_1 and c_2.

4.2.1.2 Penetration depth in gelatine, soap and muscle

If the composition corresponds to certain criteria, these three media can be grouped together, as the equations for penetration depth as a function of velocity all have the same form – only the constants differ.

Minimum energy density and velocity. One notable characteristic of skin, gelatine, soap and bone is that a bullet cannot penetrate these materials if it is moving at less than a certain velocity, v_{gr}. This velocity depends on the sectional density of the bullet. A bullet moving at less than v_{gr} rebounds from the surface of the target. In the case of skin it will leave a bruise, unless it is moving very slowly.

For most purposes, v_{gr} is sufficiently accurate. In his work on the minimum velocity for projectiles from airguns, MISSLIWETZ (1987) distinguishes between a number of different thresholds.

He defines a projectile as having 'bounced off' if it has struck the surface and rebounded, regardless of whether it has left a visible wound. The author defines a projectile as "protruding" if the front is embedded in the epidermis, with the projectile still visible. The corresponding velocity is v_{St}. A penetrating projectile is one that has passed through the epidermis and the dermis and has come to rest in the subcutis or deeper in the muscle. The corresponding velocity is v_t. Such differences are of course only of relevance in the case of projectiles moving at very low velocities.

The impact energy is not the deciding factor as regards penetration. For instance, a 3.2 mm steel sphere (m = 0.13 g) requires a velocity of approximately 50 m/s in order to just penetrate the skin (E = 0.16 J). A 4.4 mm lead sphere (m = 0.5 g) travelling at approximately 48 m/s (E = 0.6 J, i.e. almost four times as high) does not penetrate, however. JOURNÉE (1907) experimented with 11.3 mm lead spheres (m = 8.5 g), because shells filled with shrapnel, i.e. spheres of this calibre, were in common use against soldiers in the open at the time. These spheres caused bruising at velocities of 46 m/s (E ≈ 10 J). At 70 m/s, these spheres penetrated muscle (E ≈ 20 J).

As we can see from JOURNÉE's measurements, the impact energy is not a suitable indicator of the ability of a bullet to penetrate the skin. What counts is the energy *density*, E'. E' is the energy per unit contact area A. E' = E/A, and for stable flight A = ¼·π·k².

There is a certain energy density beyond which skin cannot prevent a bullet from penetrating. This is the threshold energy density, E'$_{gr}$. If this energy density is exceeded, the skin will tear and the projectile will penetrate. The energy transferred to the skin as the bullet passes through it is significantly less than the threshold energy required by the threshold energy density and the impact surface A of the bullet (see 2.3.7.4). We can therefore write:

(4.2:3) $\qquad E_{ds} = E_a - E_{ad} < E_{gr} = E'_{gr} \cdot A$. [J]

where E_a is the impact energy, E_{ds} the energy used in passing through the skin and E_{ad} the residual energy of the bullet.

In the discussion below, "threshold" velocity or energy is the velocity or energy at which the bullet is just unable to penetrate the skin.

Naturally, the skin is not equally thick or equally firmly supported at all points of the body. We can therefore expect the threshold energy density to vary. In a comprehensive study on a "non-lethal" 12-bore projectile, BIR et al. (2005b) determined the threshold energy density for various parts of the body. As one would expect, there was some overlap between the energy density thresholds for penetration and non-penetration, together with considerable scatter. In such cases, it is wise to use the mean energy density, i.e. the value at which one can expect penetration to occur in 50% of cases. Table 4-10 lists the results of the experiments carried out by BIR et al.

The study also confirmed that at the threshold energy density (from SELLIER's work) of 0.1 J/mm² one could indeed assume, with a sufficiently high degree of

Table 4-10. 50% energy density thresholds for various parts of the body, in J/mm² (after BIR, STEWART et al. 2005)

Area of body	E'$_{gr}$ (50%)	Scatter
Sternum	0.329	0.018
Rib, front	0.240	0.017
between ribs, front	0.333	0.036
Liver	0.399	0.029
Abdomen (at level of navel)	0.343	0.028
Shoulder blade	0.506	0.053
Rib, rear	0.527	0.110
Buttock	0.381	0.077
Femur, proximal	0.261	0.097
Femur, distal	0.281	0.084

certainty, that a projectile would not penetrate the skin on any part of the body except the eyes.

It is not possible to measure energy directly. In practice, therefore, it is usual to state the velocity of the bullet, and the criterion that a particular bullet must satisfy in order to penetrate skin is often given as a threshold velocity. We can calculate this velocity from the threshold energy density E'gr, using the following formula:

$$(4.2:4) \qquad v_{gr} = \sqrt{\frac{2000 \cdot E'_{gr}}{q}}, \qquad [m/s]$$

with threshold energy density E'_{gr} in J/mm^2 and the sectional density of the bullet q in g/mm^2.

Substituting the general threshold value for E'_{gr} of 0.1 J/mm^2 mentioned above, we obtain:

$$(4.2:5) \qquad v_{gr} = \frac{14.1}{\sqrt{q}} . \qquad [m/s]$$

This means that v_{qr} is not constant; it depends on the sectional density q of the bullet. As q increases, v_{gr} decreases.

SPERRAZZA and KOKINAKIS (1968) use the following equation for v_{gr} as a function of q (in g/mm^2):

$$(4.2:6) \qquad v_{gr} = \frac{1.25}{q} + 22 . \qquad [m/s]$$

This equation, of which the constants were determined empirically, can be seen as a linear approximation of the function in 4.2:4 as regards 1/q. The results do not significantly contradict the threshold of 0.1 J/mm^2 mentioned above, especially at the sectional densities normally encountered in the case of handgun bullets.

Other than the work of BIR et al. mentioned above, the literature yields little experimental work in this area. JOURNÉE was the first to investigate threshold velocities, as mentioned above. MATTOO (1974) obtained fairly similar results, 80 years later. DI MAIO et al. (1982) and SELLIER (1976) carried out further work. MISSLIWETZ (1987) studied projectiles fired from airguns.

While airgun pellets are of less importance in the field of wound ballistics, it is still worth noting the results of the measurements made by MISSLIWETZ, as they show that v_{gr} and v_{ad} (or v_{st} and v_t in this author's work) vary substantially between individuals and between different parts of the same individual's body. These results also illustrate the effect of tip shape.

As we have already seen in the results from BIR et al., (see Table 4-10), v_{gr} is greatly influenced by the thickness of the skin and the supporting layer beneath it. This is also clear from the results obtained by the various authors. These are shown in Tables 4-11 and 4-12 a and b.

TAUSCH et al. (1978) conducted experiments using 4 mm, 9 mm and 11.4 mm (45 calibre) bullets with different tip shapes. Table 4-13 summarizes the results.

190   4 Wound ballistics of bullets and fragments

Table 4-11. Threshold energy density and velocity (measured values)

Type of projectile	Mass	Surface area	q	v_{gr}	E'_{gr}	Author
	[g]	[mm^2]	[g/mm^2]	[m/s]	[J/mm^2]	
11.25 mm lead sphere	8.4	99.4	0.0855	70	0.210	JOURNÉE (1907)
8.5 mm lead sphere	4.5	56.7	0.0793	71	0.200	MATTOO ET AL. (1969)
177 Diabolo	0.53	15.9	0.0333	101	0.170	DI MAIO ET AL. (1982)
22 Diabolo	1.07	24.5	0.0434	75	0.122	DI MAIO ET AL. (1982)
38 round nose	7.3	73.2	0.1147	58	0.193	DI MAIO ET AL. (1982)
(Theoretical) [a]			0.02 - 0.16		0.10	SELLIER (1976)

[a] On the basis of measurements conducted by SPERRAZZA and KOKINAKIS (1968).

Table 4-12 a. Results of firing various projectiles from an airgun into the thigh of an adult (after MISSLIWETZ 1987)

Type of projectile	Calibre	v_{st}(min)	v_{st} [a]	v_{st}(max)	v_t(min)	v_t [a]	v_t(max)
	[mm]	[m/s]	[m/s]	[m/s]	[m/s]	[m/s]	[m/s]
Pointed bullet (KS)	4.5	50	99.0	118	72	109.0	135
Lead sphere	4.5	56	99.8	130	72	110.4	135
Diabolo (FK)	4.5	73	130.3	167	79	135.6	167
Hollow-point bullet	4.5	74	123.5	149	75	133.2	172
Brass sphere	4.0	66	109.9	132	83	120.6	145
Steel sphere	4.0	67	114.1	151	91	126.1	157
Glass sphere	4.0	119	180.1	238	145	197.8	229

[a] Mean values

Table 4-12 b. Threshold velocity and energy density, calculated from the results shown in Table 4-12 a

Type of projectile	Calibre	Mass	Surface area	q	v_{gr} [a b]	E''_{gr} [a]
	[mm]	[g]	[mm^2]	[g/mm^2]	[m/s]	[J/mm^2]
Pointed bullet (KS)	4.5	0.56	15.9	0.035	99.0	0.173
Lead sphere	4.5	0.54	15.9	0.034	99.8	0.169
Diabolo (FK)	4.5	0.49	15.9	0.031	130.3	0.262
Hollow-point bullet	4.5	0.44	15.9	0.028	123.5	0.211
Brass sphere	4.0	0.31	12.6	0.025	109.9	0.149
Steel sphere	4.0	0.26	12.6	0.021	114.1	0.135
Glass sphere	4.0	0.08	12.6	0.0064	180.1	0.103

[a] The values for children were significantly lower (E'_{gr}: 0.085 to 0.12 J/mm^2, see original work for details).
[b] In accordance with the definition of threshold velocity, the values of v_{st} from Table 4-12 a were used to determine the threshold energy density.

Table 4-13. Determining v_{gr}: Results of measurements conducted by TAUSCH et al. (1978)

Type of projectile	k [mm]	Mass [g]	q [g/mm²]	v_{gr} [m/s]	E'_{gr} [a] [J/mm²]
Lead sphere	9.0	5.3	0.083	68.7	0.20
Round-nose	9.0	6.2	0.097	66.2	0.21
Round-nose	9.0	10.6	0.166	41.8	0.15
Flattened cone	9.0	7.9	0.123	54.5	0.18
Pointed cone	9.0	7.9	0.123	57.9	0.21
45 lead sphere	11.4	9.0	0.087	56.7	0.14
45 lead round-nose	11.4	11.7	0.143	37.0	0.10
Lead sphere	4.0	0.47	0.032	68.7	0.23
					E'_{gr} [b]
6.35 Browning	6.35	3.3	0.104	81.3	0.109
7.65 Browning	7.65	4.8	0.104	70.5	0.119
9 mm Browning, short	9.0	6.0	0.094	63.9	0.123
9 mm Luger	9.0	8.0	0.125	55.3	0.122
9 mm lead round-nose	9.0	10.6	0.166	47.0	0.117

[a] Values calculated by us (based on Eqn 4.2:4).
[b] Values calculated by the authors, using their equations.

TAUSCH et al. propose the following equations (adapted to match the values measured) to describe the mathematical relationship between sectional density q (or mass m) and threshold velocity v_{gr}:

(4.2:7a) $\qquad v_{gr} = 277.7 \cdot e^{-4.82 \cdot \sqrt{q}}$, [m/s]

(4.2:7b) $\qquad v_{gr} = 162.1 \cdot e^{-0.38 \cdot \sqrt{m}}$. [m/s]

These equations may well describe the values obtained by these authors better than, for instance, Eqn 4.2:5, but they do not reveal any physical/ballistic relationship. It would be quite impossible to extrapolate to other conditions, and nothing is added to the sum of knowledge regarding internal relationships and laws, which is what one should be aiming to achieve.

Closer analysis of the penetration process reveals the following: At $v_a < v_{gr}$, the projectile does not penetrate. Its energy is transferred to the surface of the target and the target itself, in various forms (elastic, kinetic, heat, deformation, etc.). If v_a is somewhat greater than v_{gr}, the projectile destroys the surface of the target and starts the process of penetrating the target, at a residual velocity v_{ad}. In accordance with Eqn 4.2:3, v_{ad} is generally greater than the difference between v_a and v_{gr}. This is because the opening created in the surface prevents the energy of the bullet being distributed over a larger area. As a result, the bullet cannot transfer the same amount of energy to the surface of the target as it would if it did not penetrate. The energy remaining allows the bullet to penetrate to a depth s_{ad}.

We can observe this behaviour for all thin layers of material with a certain degree of elasticity (such as skin, metals, gelatine and cloth). Shots fired into textiles, for instance, yielded a v_{gr} of about 80 m/s. Bullets with an impact velocity of 90 m/s passed through the cloth, and the residual velocity v_{ad} measured behind the cloth was 60 m/s (KNEUBUEHL 1992).

It is therefore impossible, for certain materials, to find a bullet that penetrates and comes to rest at a depth between 0 and s_{ad}.

If the impact velocity v_a is greater than v_{gr}, but less than the velocity required to pass through the target medium, the bullet will come to rest in the target. If the bullet does pass through, its velocity will drop progressively and predictably. A wound channel will be created, and damage will occur. In the last part of the channel, the bullet is moving so slowly that almost its entire surface is in contact with the medium. At such velocities, frictional force is greater than inertial force, and the Navier-Stokes equations apply (Eqns 4.2:1b and 4.2:2b). The distance over which these equations are valid is generally so short that it is of virtually no significance when compared with the channel as a whole, unless the impact velocity is low (e.g. in the case of a projectile from an airgun).

In many cases, the bullet does not come to rest at the end of the channel. This was described by NENNSTIEL (1990) and was confirmed by KNEUBUEHL and TROESCH (1992, unpublished) using high-speed video. In such cases, the forces created by the vacuum in the temporary cavity are greater than the frictional forces acting on the surface of the bullet.

This behaviour is confirmed by forensic experience. The penetrating power of lead round-nose 22 and 38 Spl bullets is such that they can just pass through the trunk, as long as the wound channel only passes through soft body parts. A bullet that hits the skin at a velocity less than v_{gr} will not penetrate it, because of the elasticity of the skin. The bullets are held back by the skin on the opposite side. In many cases, one finds the bullet just under the skin, where it is easily detected by palpation.

Equations of penetration depth in gelatine, soap and muscle. In order to describe the entire penetration process mathematically, it is necessary to introduce two further characteristic velocities, together with the corresponding energies.

- The decrease in the velocity of the bullet as it passes through the surface of the medium, $v_{ds} = v_a - v_{ad}$.
- The velocity of the bullet immediately before it comes to rest in the medium, v_{stk}.

The following two equations apply:

(4.2:8) $\qquad v_{ds} < v_{gr} \quad \text{und} \quad v_{stk} < v_{gr}$. $\qquad$ [m/s]

For the deceleration of a bullet that penetrates a dense medium at typical velocity, we use Eqn 4.2:1c (from 4.2.1.1), as the cohesive forces are negligible by comparison with inertia force. According to Eqn 3.2:4, we must set the value of the coefficient c_2 equal to the retardation coefficient $\mathfrak{R}$. By integrating the corresponding equation of moment, we obtain an equation analogous to Eqn 4.2:2c for penetration depth as a function of instantaneous velocity:

(4.2:9) $\quad s = -\dfrac{1}{\Re} \cdot \ln(v) + C$. $\quad$ [m]

But we still have to determine the value of C. If the bullet has penetrated the surface but has not yet moved further into the medium (i.e. if s = 0) then its velocity is still:

$$v_{ad} = v_a - v_{ds} .$$

C is hence given by:

$$C = \dfrac{1}{\Re} \cdot \ln(v_a - v_{ds}) . \qquad [m]$$

and therefore:

(4.2:10) $\quad \begin{cases} s = \dfrac{1}{\Re} \cdot \ln\left(\dfrac{v_a - v_{ds}}{v}\right) & \text{falls } v > v_{gr} \\ s = 0 & \text{falls } v \leq v_{gr} \end{cases}$ $\quad$ [m]

A sphere that strikes the target at a velocity $v_a > v_{gr}$ therefore covers a distance:

(4.2:11) $\quad s = \dfrac{1}{\Re} \cdot \ln\left(\dfrac{v_a - v_{ds}}{v_{stk}}\right),$ $\quad$ [m]

before coming to rest in the target. For some calculations, it is useful to express velocity as a function of penetration depth. Solving Eqn 4.2:11 for v_a we obtain:

(4.2:12) $\quad v_a = v_{stk} \cdot e^{\Re \cdot s} + v_{ds}$.

If $v_a \leq v_{gr}$, then $s_a = 0$ (Eqn 4.2:10) and from Eqn 4.2:12:

(4.2:13) $\quad v_{gr} \geq v_{stk} + v_{ds}$, $\quad$ [m/s]

This equation for the characteristic velocities would also appear to be plausible from a physics point of view. If we substitute pairs of measured values for s_a and v_a in Eqn 4.2:12, we can calculate v_{ds} and v_{stk} by regression.

KNEUBUEHL measured the impact velocity and penetration depth for 4.5 mm lead spheres in gelatine ($C_D = 2$, NENNSTIEL 1990). Analysing the results of measurements using the method described produced values of 28.7 m/s for v_{ds} and 14.0 m/s for v_{stk}, with a high degree of correlation (r = 0.941). Spheres with impact velocities of less than 44 m/s did not penetrate the gelatine.

If the sphere exits the block, we can calculate its residual velocity with the following equation (derived from Eqn 4.2:12):

(4.2:14) $\quad v_{rst} = (v_a - v_{ds}) \cdot e^{-\Re \cdot \ell}$. $\quad$ [m/s]

Table 4-14. Values for C_D from the literature

Material	C_D	Author(s)	publicated
Water	0.30	HARVEY et al.	1962
Water	0.29	HOERNER	1965
Gelatine 10°	0.375	DUBIN	1974
Gelatine 24°	0.34	HARVEY et al.	1962
Gelatine M < 1	< 0.4	CHARTERS and CHARTERS	1976
Gelatine M ≥ 1	1.0	CHARTERS and CHARTERS	1976
Soap	0.33	SCEPANOVIC	1979
Muscle (cat)	0.44	HARVEY et al.	1962
Muscle (pig)	0.36 a	JANZON et al.	1979

[a] Tolerance: +0.003 / -0.008

The corresponding energy values can readily be calculated from the velocities, as long as the bullet loses no mass as it passes through the block.

The derived equations are only valid if the retardation coefficient remains constant. This implies that the sectional density remains constant (i.e. the cross-section of the face does not change) and that the drag coefficient C_D does not change. The drag coefficient for soap, gelatine and animal tissue has been measured many times. Table 4-14 gives a number of values from the literature.

All authors (with the exception of CHARTERS and CHARTERS 1976) assume a constant value for C_D across the range of velocities at which measurements are made. JANZON (in RYBECK and JANZON 1976) notes that the value of C_D for muscle differs little at 1500 m/s from the value of this coefficient at 1000 m/s (0.36). The author felt that this was surprising, as one knows from experience that the speed of sound c (approx. 1500 m/s in gelatine) leads to a sharp increase in C_D.

JANZON's explanation is that a sphere striking the target at a velocity greater than the speed of sound in the medium concerned decelerates to below the speed of sound over a distance equal to just a few times its own diameter. The value of C_D is calculated by taking the mean over a relatively long distance (which depends on the sampling rate used for the measurements of position or velocity), and high values of C_D over a short distance have little influence on the mean.

NENNSTIEL (1990) measured the value of C_D for 7 mm steel and tungsten spheres at velocities of 150 to 1500 m/s. At velocities below 150 m/s, the author noted that as velocity decreased, the value of C_D rose sharply. The increase as the velocity approached c was not observed, for the reason mentioned above.

The high values of C_D at low velocities can be explained by the fact that the entire surface of the bullet comes into contact with the medium at these velocities, with the high drag coefficients that are to be expected in a highly viscous medium.

PETERS (1990b) introduced an extension to the conventional approach (Eqn 4.2:1c) to calculating the motion of a bullet. His deceleration model includes a second component, related to velocity in a linear fashion (see Eqn 3.2:6 in 3.2.2.5).

Another factor must be borne in mind: while the muscles of humans and animals display similar levels of resistance, the skin of animals such as pigs and horses is considerably more resistant than that of humans. A bullet penetrating the skin of an animal uses up more energy than one which penetrates human skin. As a result, a bullet undergoes a greater loss of velocity (v_{ds}) on passing through the skin of an animal. According to Eqn 4.2:8, the threshold velocity will therefore increase. The above points apply mainly to handgun bullets moving at velocities close to v_{gr}. This difference can safely be ignored in the case of rifle bullets, as is obvious when one considers the energy levels involved.

4.2.1.3 Penetration capacity in bone

Threshold velocity and energy. For bone, as for skin, there is a threshold velocity v_{gr} below which a bullet will not penetrate it. For bone, v_{gr} is approximately 60 m/s. This is value given by JOURNÉE and, more recently, by HUELKE et al. (1968a,b) and HARGER and HUELKE (1970). The literature establishes no relationship between v_{gr} and the sectional density of the bullet q.

It is possible to calculate the threshold energy E_{gr} and threshold energy density $E'_{gr} = E_{gr}/A$ at which the bullet will just fail to penetrate bone. The values for certain types of ammunition are as follows:

6.35 Browning	$E_{gr} =$	6.0 J	$E'_{gr} =$ 0.19 J/mm^2 ,
7.65 Browning		8.7 J	0.19 J/mm^2 ,
9 mm Luger		14.0 J	0.22 J/mm^2 .

On *penetrating* a bone, a bullet of larger calibre loses more energy.

Penetration depth and penetration capacity, theory and practice. Two relationships are of particular interest as regards the effect of a bullet on a bone:

1. Penetration depth as a function of velocity.
2. The energy lost by the bullet as it passes through a given bone thickness d.

Regarding 1: GRUNDFEST (1945) conducted numerous tests that involved firing steel spheres with diameters of 3.18, 4.74 and 6.35 mm ($^1/_8$, $^3/_{16}$ and $^1/_4$") into the bones of cows. In metric units, the equation relating penetration depth s to velocity derived from his experiments is:

(4.2:15) $\qquad s = 0.863 \cdot 10^{-4} \cdot k^2 \cdot (v_a - v_{gr})^2$, [mm]

where k is the diameter of the sphere in mm, v_a is the impact velocity in m/s and $v_{gr} \approx 60$ m/s.

As it stands, this equation is valid only for steel spheres and cannot be applied to practical situations, as bullets generally consist of lead (apart from the jacket). They also have a greater volume and mass than spheres of the same calibre, especially if those spheres are made of steel, not lead.

For instance, a steel sphere 9 mm in diameter weighs approximately 3 g, whereas a 9 mm Luger bullet weighs 8 g. This means that if one were to substitute k = 9 mm into the equation, the value of s obtained would be valid only for a steel sphere of that radius, not for a Luger bullet. One has to multiply the value of s obtained by the quotient of the mass of the Luger bullet and that of the steel sphere of the same calibre, i.e. $^8/_3$, to obtain the correct value for a Luger bullet.

The equation therefore requires modification to render it suitable for general use. k^2 is simply the expression m/k multiplied by a constant factor. Substituting the variables m and k and using the density of steel, we can derive the following relationship from Eqn 4.2:15:

(4.2:16) $\qquad s \;=\; 0.21 \cdot 10^{-2} \cdot \dfrac{m}{k} \cdot (v_a - 60)^2$. [mm]

As the structure of human bones differs from that of cow bones (e.g. as regards the distribution of cortical bone and diploë) it is clear that this equation requires further correction before it can be applied to human bones. Indeed, tests (SELLIER and KNÜPLING 1969) have shown that the a value of 0.21 for the constant is too low for human bone (i.e. cow bones are harder). The correct value for a bullet with an approximately spherical head is around 0.4.

The penetration depth of lead round-nose bullets is less than that calculated using Eqn 4.2:16, as their calibre increases on impact and as they pass through the bone. The calculated penetration depth is also smaller in the case of wadcutter bullets, because of their flat tips. Overall, the calculated penetration depth should only be seen as an estimate, as the precise value will depend on the composition of the bone and, in particular, on the ratio of solid to soft matter (see below for further factors).

Regarding 2: HUELKE et al. (HUELKE and DARLING 1964, HUELKE et al. 1967, HUELKE et al. 1968a,b, HARGER and HUELKE 1970) have conducted comprehensive empirical studies on the energy lost by a bullet ΔE as it passes through human bone, as a function of impact velocity v_a or impact energy E_a. They used steel spheres with diameters of 6.35 mm (0.25"; m = 1.04 g) and 10.3 mm (0.406"; m = 4.48 g). They presented the results of their experiments in empirical form. Impact velocity and energy transferred to the bone could in all cases be approximated using quadratic equations, with only a small degree of error.

For 6.35 mm steel spheres fired into the distal end of the femur, for instance, the equation was as follows (using values from HUELKE et al. 1967):

(4.2:17) $\qquad \Delta E \;=\; 0.65 \;+\; 0.0077 \cdot v_a \;+\; 0.0000127 \cdot v_a^2$, [ft-lbs]

(velocity in ft/s).

Although HUELKE did not interpret this quadratic relationship between impact velocity and energy loss, even in his later works, one immediately thinks of a linear relationship between impact energy and energy transfer, and such a relationship does indeed appear logical from a physics point of view.

From the results set out in this literature (particularly HUELKE et al. 1968a), one can derive the following equations (in metric units):

For 6.35 mm steel spheres:

(4.2:18) $\quad \Delta E = 6.6 + 0.416 \cdot E_a$, $\quad (r = 0.993)$, $\quad$ [J]

and for 10.3 mm steel spheres:

(4.2:19) $\quad \Delta E = 14.9 + 0.262 \cdot E_a$, $\quad (r = 0.988)$. $\quad$ [J]

However, these equations require some refinement. If one looks more closely at the measurement results (Fig. 4-7), it becomes apparent that they are better described by two straight lines: The first line has its origin at the intersection of the axes and has a gradient of 1. This is followed at higher values of E_a by a second line, of which the gradient is significantly less than 1.

The meaning of the straight line with a gradient of 1 originating at the intersection of the axes is clear. In this area, $\Delta E = E_a$, i.e. all the energy is used up in the bone. At velocities below v_{gr} this is because the bullet rebounds. At velocities above v_{gr} because the bullet comes to rest in the bone. The line joining just the measurement points for the higher values of E_a is less steep than the line that joins all of them. As a side effect, the projection of the line onto the y axis increases, as one can easily see on the graph.

If the measurement values are divided up as per this diagram, we obtain the following equations for the less steep part, in place of Eqns 4.2:18 and 4.2:19:

(4.2:18a) $\quad \Delta E = 8.57 + 0.402 \cdot E_a$, $\quad (r = 0.993)$, $\quad$ [J]

(4.2:19a) $\quad \Delta E = 17.9 + 0.252 \cdot E_a$, $\quad (r = 0.990)$. $\quad$ [J]

The constant part can now be interpreted as follows: The bullet has to separate the piece of bone directly in front of it from the rest. To achieve this, work must be done. The amount of work required is determined by the shear strength of the bone (50 N/mm^2, according to VON GIERKE 1968), the generated surface of the detached bone fragment and the distance over which the bullet displaces the fragment.

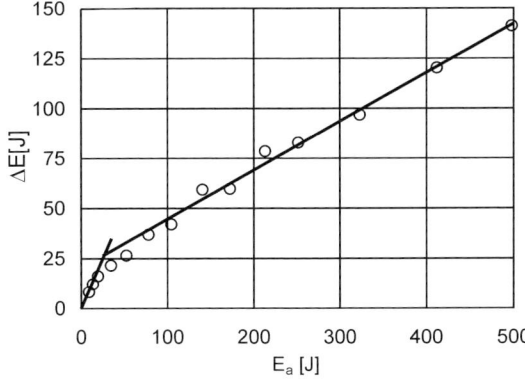

Fig. 4-7. Energy lost by a 10.3 mm steel sphere as a result of passing through a bone, as a function of impact energy (after HUELKE et al.).

Once the fragment has been punched out of the bone, the bullet will accelerate it in accordance with the law of collisions. The velocity imparted to the fragment is proportional to the impact velocity of the bullet (with the corresponding mass ratio as proportionality factor). From this it follows that the energy transferred from the bullet to the bone fragment is proportional to the impact energy (or, more precisely, to the energy the bullet still has after it has separated the fragment from the bone). This explains the linearity of Eqns 4.2:18a and 4.2:19a.

In these equations, energy loss ΔE depends only upon Ea, in other words on the combination of m and v. It does not depend on m and v as such. The material of which the sphere is made (steel, brass, lead, tungsten alloy, etc.), should therefore make no difference, as long as it the diameter remains constant, as the only factor determining the damage done by the sphere, and hence the energy consumed, is the size of the sphere, or its cross-sectional area.

This does not apply to the damage done to the bone by the temporary cavity or the hydrodynamic pressure in the bone marrow. For both of these types of damage, the bullet's v_a plays an important role.

Indeed, measurements carried out by HUELKE et al. showed that spheres of the same calibre but different masses (e.g. steel or tungsten alloy) lost the same amount of energy ΔE if the impact energy E_a remained constant.

For the impact energy and calibre to be the same despite the mass being different, the heavier spheres must have a lower impact velocity v_a. The sectional density q must also be greater.

The influence of calcium content and treatment of the bone on the energy lost by the bullet. Clear, statistically valid differences in energy loss are observed when the bones used in experiments are ranked according to their calcium content. HUELKE et al. (1968b) divided them into three categories: normal, mildly osteoporotic and osteoporotic. They observed clear differences in energy lost between the three categories, as was to be expected (see Fig. 4-8 a, b).The differences are significant, with the projectiles losing up to 40% less energy in osteoporotic bones than in normal bones.

In order to be able to extrapolate the results of tests on cadavers to living persons, it is important to know whether the storage or treatment of the bone has an effect on its reaction to a projectile. The above authors conducted further experiments to investigate this point. They fired into both fresh, untreated bones and embalmed bones, and observed no difference in energy loss.

The temporary cavity and bone damage. A hollow bone can be seen as a fluid-filled hollow organ with a rigid wall. One can therefore assume that phenomena occur similar to those observed when a bullet passes through the head, heart or full bladder, as long as the bullet is moving at sufficient velocity and certain other conditions are met.

Damage to bone away from the actual track of the bullet is caused by the pressure generated in the marrow by the temporary cavity, and this pressure acts in all

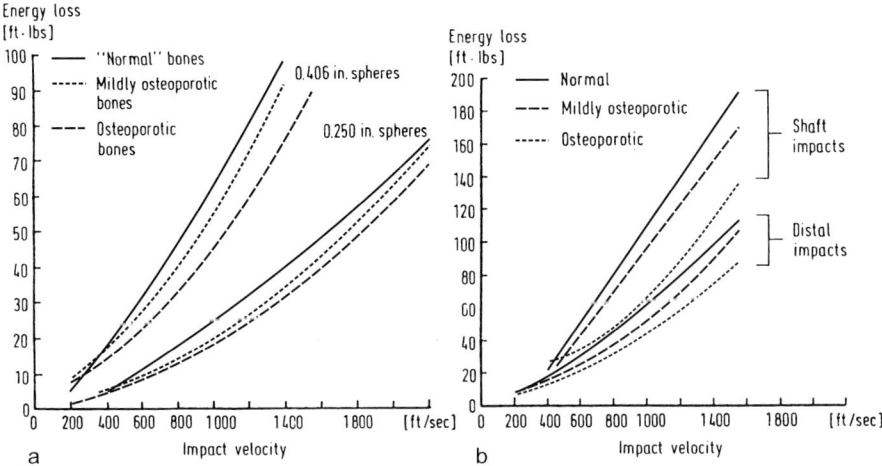

Fig. 4-8. a. Effect of bone condition on the relationship between impact velocity and energy loss (after HUELKE et al. 1968a).
b. Effect of various bone structures on the relationship between impact velocity and energy loss (after HUELKE et al. 1968b).

directions. As a result, bone splinters will be found in the wound channel not only in the direction in which the bullet was moving but also in the opposite direction. One can expect bone damage to extend beyond the actual track of the bullet. The diameter of the cross-sectional area within which bone damage occurs will correspond approximately to that within which soft tissue is damaged (see Fig. 4-9 and also Fig. 3-19 c).

Experiments by HUELKE et al. (1968a,b) have shown that the extent of the cavity created in bone depends not only on v_a but also on the calibre of the bullet. The findings were as follows:

With 10.3 mm spheres, temporary cavities were observed at values of E_a of approximately 73 J and above (v_a approx. 180 m/s). In the case of 6.35 mm spheres, cavities were already observed at energies of 30 J (240 m/s). The 10.3 mm sphere loses 36 J, the 6.35 mm sphere 21 J. The cavities caused by the 10.3 mm spheres were so large at energies E_a of approx. 200 J and above ($v_a \approx 300$ m/s) that the femoral condyle was almost separated from the diaphysis. In the case of the

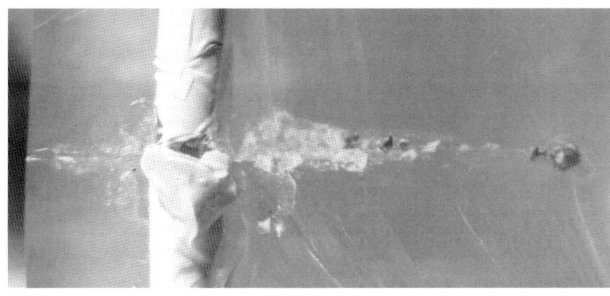

Fig. 4-9. Effect of firing a projectile at a plastic bone embedded in gelatine. The extent of bone damage corresponds approximately to the diameter of the cavity in soap.

6.35 mm spheres, this occurred at an E_a of 130 J ($v_a \approx 500$ m/s). The energy required was similar for the two calibres (60 J to 70 J).

At velocities below those cited above, a piece of bone the size of the calibre is punched out, causing a "drill hole fracture." Closer observation reveals that even in the case of non-deforming projectiles such as steel spheres the diameter of the hole is almost always slightly less than that of the projectile (BERG 1955). This is because, as mentioned in Chapter 3, the projectile displaces the surrounding tissue radially, and this includes the bone. The bone is temporararily pushed apart to create a gap of which the diameter corresponds to the calibre of the projectile as it passes through, with the two parts moving back towards each other afterwards. This means that a non-deforming projectile that has created a hole in a bone cannot be passed through that hole afterwards. Naturally, this is even more the case for projectiles that deform as they pass through a bone, as deformation continues until the projectile has exited the bone.

4.2.1.4 Threshold velocities for eyes

The eyes account for only 2 ‰ of the frontal surface of the body, so the probability of their being hit is small. In view of their vulnerability, however, eye injuries are highly significant, as they often lead to blindness. The risk is particularly high during hunting, if shot is used. Similar injuries can occur in the workplace, e.g. when working with metal. It is therefore not surprising that numerous studies have addressed the problem of eye injuries, especially the question of the threshold velocity for penetration, v_{gr}. We shall now present values for v_{gr} from various studies in tabular form.

Table 4-15 presents values of v_{gr} for spheres. The original study (STEWART 1961) gives diameter d, mass m and v_{gr}. We have calculated the values of cross-sectional area A, sectional density q and E'_{gr}. STEWART's measurements give a mean value for v_{gr} of 0.06 ± 0.015 J/mm². This is significantly lower than the value for skin, which is what one would expect.

Table 4-15. Threshold velocities for eyes (rabbit). With one exception, a steel sphere was used

d [mm]	A [mm²]	m [g]	q [g/mm²]	v_{gr} [m/s]	E'_{gr} [J/mm²]	Author(s)
1.00	0.79	0.004	0.0052	146	0.056	STEWART (1961)
2.36	4.4	0.054	0.0123	108	0.072	STEWART (1961)
3.20	8.0	0.135	0.0168	100	0.084	STEWART (1961)
4.37	15.0	0.341	0.0227	77	0.068	WILLIAMS and STEWART (1964)
4.37	15.0[1]	0.513	0.0342	65	0.072	WILLIAMS and STEWART (1964)
6.40	32.2	1.037	0.0321	47	0.035	STEWART (1961)

[1] Lead sphere

Table 4-16. Threshold velocities for eyes (rabbit). Steel spheres (STEWART 1961)

a [mm]	A [mm^2]	A$_m$ [mm^2]	M [g]	q [g/mm^2]	v$_{gr}$ [m/s]	E'$_{gr}$ [J/mm^2]
2.60	6.8	10.2	0.136	0.0135	63	0.027
3.27	10.7	16.1	0.272	0.0169	38	0.012
5.11	26.1	39.2	1.037	0.0265	36	0.017
8.11	65.8	98.7	4.147	0.0420	29	0.018[1]
12.3	147.6	221.4	14.58	0.0639	22	0.015[1]

[1] We have swapped these two values, as there was clearly a mistake in the original (inconsistency between the two values).

Table 4-16 presents values of v$_{gr}$ for steel cubes. The author states the mass of the cubes and the measured values of v$_{gr}$. We started by calculating the length of the edges of the cubes from the mass. Unlike the sphere, the face surface of a cube is not constant – it depends on the orientation of the cube at the moment of impact. One therefore takes the mean projected area A$_m$. For a convex body this equals one quarter of the surface, so for a cube A$_m$ = $^6/_4 \cdot a^2$.

It is easy to see why E'$_{gr}$ for a cube (0.018 ± 0.006 J/mm^2) is lower than that for a sphere. The probability of the cube striking the target with an edge or corner is fairly high. This results in a high local force (stress concentration) and hence a low value of E'$_{gr}$.

4.2.2 Characteristics of handgun bullets

4.2.2.1 General

A handgun bullet for self-defence purposes has to fulfil a number of requirements. On the one hand, it should have sufficient effect on the human body to cause the other person to abandon their intended course of action (attack or flight) without causing unnecessary and life-threatening injury. The residual energy of the bullet, should it exit the body, must be sufficiently low as not to endanger any bystanders. On the other hand, the bullet is required to pass through hard materials (such as steel and glass) and still have enough energy to achieve an adequate effect in the body of the other person.

According to Eqns 4.1:16 and 4.1:17, effectiveness in a person (transfer of energy) is inversely proportional to the sectional density q of the bullet (i.e. a high level of effectiveness requires a low value of q), whereas according to Eqn 2.3:9, the ability to penetrate hard materials is directly proportional to q (i.e. q needs to be large). This is the third ballistic paradox (KNEUBUEHL 2004, p. 27 et seq.). Penetrating hard targets and achieving a high degree of effectiveness in a human target both demand high impact energy, which is at odds with the principle of proportionality. From the above it is clear that the requirements levelled at a handgun bullet are mutually exclusive in physical terms, which means that no design can meet all of them.

The contradictions between the various physical characteristics also account for the huge number of handgun bullet types and calibres. There are well over 60 types of bullet for Luger 9 mm ammunition, for example, and new types are arriving on the market all the time.

4.2.2.2 Bullets with good penetration properties

The higher the sectional density of a bullet, the better its penetration power, as we have seen above. This means that the bullet should be heavy, while its cross-sectional area should be small. This implies that bullets which deform easily, causing their cross-sectional area to increase, are not suitable for such purposes.

Typical bullets with the above characteristics include the classic full metal-jacketed round-nose bullets, lead-cored bullets with reinforced jackets (e.g. the Swedish Skarp bullet, see Fig. 4-10), bullets with a hard (steel or tungsten alloy) core or solid steel bullets (such as the Alpha, with its cage, see Fig. 4-11).

Bullets in this category produce long channels, and the diameters of the temporary cavities and of the permanent wound channels are small. If no vital areas are hit, such as major blood vessels, brain or spinal cord, the probability of survival is high and the risk of permanent damage low.

Tracer ammunition constitutes a special sub-category among full metal-jacketed bullets. The tail of a tracer bullet contains an illuminating charge. This burns with no external source of oxygen. If the bullet does not exit the body, burns may occur inside the wound channel.

4.2.2.3 Bullets designed for maximum effectiveness

General. We can derive the following simple formula for the effectiveness of a bullet (its wounding potential) from the equations for effectiveness (Eqns 4.1:16 and 4.1:17):

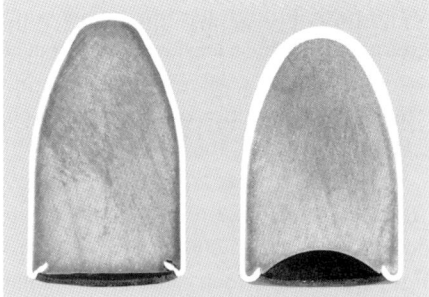

Fig. 4-10. 9 mm Luger Skarp reinforced jacket bullet (right) compared with a normal full metal-jacketed bullet (left).

Fig. 4-11. 9 mm Luger Alpha bullet. Solid steel bullet with plastic cage.

Fig. 4-12. French solid bullets, 9 mm Luger and 38 Spl: Left to right: Arcane, 4.6 g, THV, Version 1, 2.6 g. THV, Version 2, 3 g. Both with hollow tails. (THV = Très Haute Vitesse = very high speed).

(4.2:20) $$E'_{ab}(s) \propto \frac{E(s)}{q(s)} \quad . \qquad [J/cm]$$

Effectiveness at a given point along the bullet track is hence directly proportional to the energy remaining at that point E(s) and inversely proportional to the sectional density at that same point.

With the exception of unusually large calibre rounds (such as the 44 Rem. Mag., 454 Casull or 50 A.E.), the energy delivered by a handgun cartridge is very limited. From Eqn 4.2:20, it therefore follows that bullets designed for maximum effectiveness must have a small sectional density. Since the sectional density is equal to the mass of the bullet per unit area of cross-sectional area perpendicular to the direction of movement, an effective bullet has either to be light or else to have a large cross-sectional area. These are therefore the two parameters of interest to designers of high-effectiveness handgun bullets.

Low-mass bullets. Because most handguns are fairly high-calibre weapons, and because the barrel needs to have a certain length in order to guide the bullet, the only two ways of producing a light bullet are to use a light metal or to incorporate a hollow space in the tail. If the mass of the bullet is kept low, then the muzzle velocity will have to be high, in order to ensure that the empty case is ejected and the working parts re-cocked. It often happens, however, that the required muzzle momentum is not achieved, which can lead to the weapon jamming. Furthermore, a very fast-burning propellant will be needed, and the peak pressure such a propellant generates can easily exceed the maximum permissible pressure. For physical reasons, therefore, low-mass bullets have never been widely used.

This type of bullet is generally made of aluminium or copper alloys. They include the French Arcane and THV bullets produced by SFM (in two versions). See Fig. 4-12. The THV bullet achieves a velocity of approx. 600 m/s when fired from a pistol with a 10 cm barrel. Its channel in soap is similar to that caused by a steel fragment with the same energy (see Fig. 4-13).

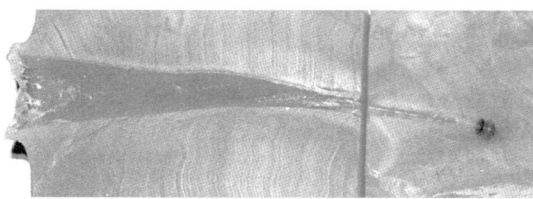

Fig. 4-13. Channel in soap, THV, Version 2. Sectional density remains approximately constant, so energy transfer decreases exponentially.

The limited penetration depth by comparison with a full metal-jacketed bullet is a positive feature in a bullet for police use. If the bullet exits the body at all, it will have very little energy and hence present virtually no danger to a person standing behind the target. However, such bullets have lethal effects, as the diameter of the temporary cavity is already very large at the entry wound and just beyond. This causes serious damage and heavy bleeding in organs that are vulnerable to tearing, such as the liver, which lies just under the abdominal wall, whereas elastic tissue such as muscle and lungs are less affected.

Cartridges with low-mass bullets included the plastic training (PT) cartridges produced by the former Dynamit Nobel AG (now RUAG). The plastic bullet used in this cartridges weighs only 0.42 g. The muzzle velocity is approx. 1000 m/s and the muzzle energy approx. 200 J. At close range, these bullets cause large but shallow entry wounds (2 to 4 cm, see Fig. 3-27, top). At approx. 30 m, the bullet has just enough energy to penetrate the eye.

Bullets of which the surface area increases. The surface area of a bullet can increase either immediately after it leaves the barrel or on impact with the target. These are the only two points at which the high forces are present that are needed in order to deform a bullet that is designed to deform.

The bullet can be deformed by the high gas pressure at the muzzle or by the high penetration resistance offered by the density of the target.

The large cross-sectional area is only needed once the bullet strikes the target. If the increase in cross-sectional area occurs at the muzzle, this has a negative effect, in that the increased drag will cause the bullet to lose more energy along its trajectory, energy that is no longer available when the bullet strikes the target. As a result, most bullets intended to achieve high effectiveness (deformation and fragmentation bullets) are so designed as to increase their cross-sectional area on impact with the target.

Deforming bullets lose virtually no mass as their cross-section increases. Deformation can start only once the bullet is in contact with the target. The deformation process then takes a certain amount of time, depending on the material used and the design of the bullet. As a result, the diameter of both the temporary and

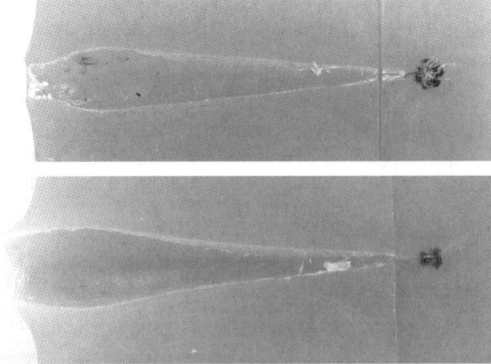

Fig. 4-14. Channels in soap. Top: lead-core deforming bullet (Remington Golden Saber). Bottom: solid bullet (MEN QD P.E.P.). The QD P.E.P. deforms noticeably sooner.

the permanent cavity increases to a maximum after 2 to 6 cm. Thereafter, the sectional density of the bullet remains constant, causing the diameter of the cavities to decrease steadily, and approximately exponentially. See Fig. 4-14. The large area of the leading surface of the bullet results in shoulder stabilization, as described in 2.3.5.4. The bullet therefore does not yaw and leaves a virtually straight track.

There are two basic types of deforming bullet. The first type consists of semi-jacketed and semi-jacketed hollow-point bullets, in which a lead core is partly covered by a harder metal (steel, aluminium or copper alloy). The tip of the lead core is exposed, and in the case of hollow-point bullets an axial cavity is drilled into it. This type of bullet is mainly manufactured and used in the USA. Typical examples include Remington's Golden Saber, (see Fig. 4-14), Federal's Hydra Shok and Gold Dot and Winchester's Subsonic, Silvertip and Ranger.

The other category consists of the solid deforming bullets. These are made of a single material (generally a copper alloy) and also have cavity at the tip, generally closed off by a piece of plastic or metal. This type is produced and used almost exclusively in Europe. Most of these belong to the first generation range of police bullets such as Dynamit Nobel AG/RUAG Action 1, MEN Quick Defense or to the second generation, such as RUAG Action 4 and Action 5, MEN QD P.E.P. (see Fig. 4-14) and Hirtenberger EMB.

Fragmentation bullets. If a bullet breaks up into two or more parts, the masses of the individual parts will be smaller than that of the original bullet, and the larger the number of fragments, the smaller the mass of each. At the same time, the distribution of the fragments means that the effective cross-sectional area (in the direction of travel) will be larger. Together, these two factors lead to a huge reduction in the sectional density and hence (in accordance with Eqn 4.2:20) a correspondingly greater wounding potential.

Fragmentation is achieved in one of the following ways:

- Structure as for a deforming bullet, but so designed as to disintegrates on impact as a result of the forces applies. A sufficiently high level of energy is required to ensure fragmentation, so these bullets are generally found in larger calibres such as the 357 Mag. and 44 Rem. Mag. (see Fig. 4-15).
- Explosive charge built into the body of the bullet. The charge detonates on impact, causing the projectile to fragment. There are very few handgun bullets of

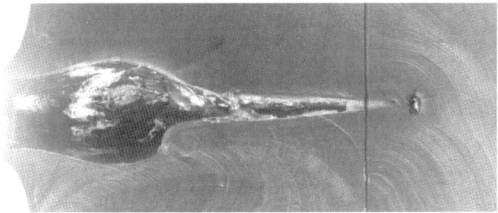

Fig. 4-15. Channel of a 44 Rem. Mag. fragmentation bullet in soap. Note the large fragment.

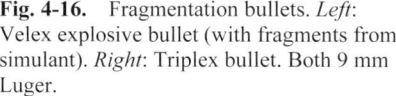
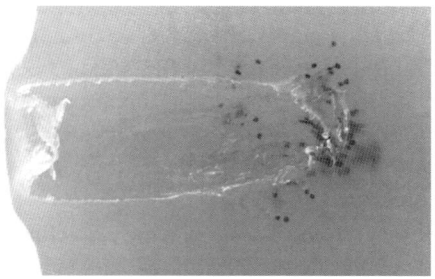

Fig. 4-16. Fragmentation bullets. *Left*: Velex explosive bullet (with fragments from simulant). *Right*: Triplex bullet. Both 9 mm Luger.

Fig. 4-17. Channel of a 9 mm Luger Glaser Safety Slug. Penetration depth: 11.8 cm (KNEUBUEHL).

this type. One of the few examples is the Velex bullet that was produced for a number of common calibres. The charge consisted of a primer cap in the tip of the bullet that incorporated a small quantity of explosive (see Fig. 4-16, left).

- Bullets with secondary projectiles. These bullets include a number of smaller projectiles, which are released either immediately after the bullet leaves the barrel or on impact with the target.

Colt has developed the Salvo Squeeze Bore system. The end of the barrel is tapered conically on the inside. The projectile consists of three pointed conical bullets, each with a hollow in the base, which fit over one another and are held together by a thin plastic jacket (Triplex bullet, Fig. 4-16, right). These are compressed and separated in the conical barrel, such that three projectile reach the target.

Cartridges containing shot are available for a number of common calibres, such as 38 Spl The diameter of the shot in such cartridges does not exceed approx. 3 mm. The shot is contained in a plastic capsule which bursts in the barrel, releasing the shot. The individual pellets carry only a small amount of energy. Nevertheless, their energy density is fairly high – they are capable of penetrating unprotected skin out to a significant distance. Penetration depth and wounding potential are both limited, however.

With the 2 mm pellet diameter typical for a 38 Spl cartridge and a v_0 of 280 m/s, a pellet carries approx. 1.8 J of energy. Such a pellet can penetrate unprotected skin at ranges of up to approx. 50 m, but without causing major damage. The same pellet is capable of penetrating the eye at ranges of up to 70 m.

Bullets that release their secondary projectiles after impact include the Glaser Safety Slug, which is available in 9 mm Luger, 38 Spl and 357 Mag. This bullet consists of a copper capsule containing approx. 350 pellets with a diameter of 1.5 mm. After penetrating the body, the capsule bursts and the pellets are scattered around the geometric wound channel (see Fig. 4-17). Because of the low sectional density, the wound channel is wide but comparatively shallow.

Bullets that discharge a small number of secondary projectiles on impact include the MagSafe Razor Ammo (also known as Rhino Ammo), Swat and Defender (containing a small number of large pellets) and High Safety Ammo (HSA). The HSA contains six pointed steel darts. The tip of the bullet is closed off by a rubber sphere. On impact, this sphere is forced between the darts, pushing them outwards.

Triton (USA) produce a fragmentation bullet known as the Quik-Shok. This is a conventional semi-jacketed hollow-tip bullet, but with the lead core pre-fragmented into three or six parts. As the bullet penetrates and mushrooms, it fragments into the jacket and the three or six lead fragments.

It has never been proven that a handgun bullet that transfers a large amount of energy early on is more likely to stop an attacker. It is quite common for the victim to feel no pain whatsoever for several minutes after being hit. Throughout that time, he or she remains fully operational. This being the case, there are no good reasons for using fragmentation bullets. They make treatment of the victim more difficult, and are therefore questionable from a humanitarian point of view.

4.2.2.4 Unconventional bullet designs

Deforming full metal-jacketed bullets. There have been a number of attempts to design full metal-jacketed bullet that deform on penetrating soft tissue.

The aim is clearly to achieve a high degree of effectiveness, while conforming to international law. However, the conventions prohibit bullets that "... expand or flatten easily in the human body ..." and only give semi-jacketed bullets as an example of the type of bullet that is prohibited. As a result, there is no good reason for developing such bullets.

The British CBXX prototype bullet has a core made of a gelatinous material. On penetrating a simulant, the tip of the bullet mushrooms slightly and its cross-section increases to a moderate extent, but without the jacket opening.

More recently, Federal has developed the Expanding Full Metal Jacketed (EFMJ) bullet. This bullet is available in 9 mm Luger, 40 S & W and 45 Auto. The tip has a plastic core. In a soft simulant, the tip flattens, causing the jacket to split in a predetermined pattern. The cylindrical part of the bullet has a normal lead core, providing the bullet with typical mass for its size (see Fig. 4-18).

Fig. 4-18. Federal Expanding Full Metal Jacket, 9 mm Luger. Deformation in a simulant.

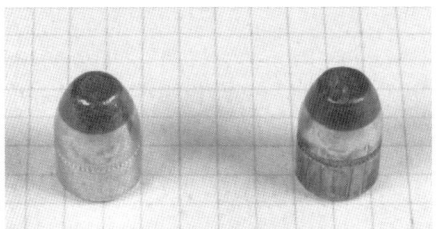 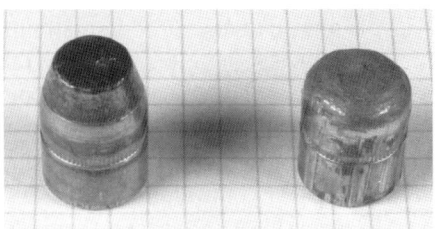

Fig. 4-19. Fiocchi non-deforming semi-jacketed bullet, 38 Spl *Left*: Before firing. *Right*: after firing into a simulant.

Fig. 4-20. Winchester semi-jacketed bullet, 454 Casull. *Left*: Before firing. *Right*: after firing into a simulant. The bullet has only undergone slight mushrooming at the tip. Its diameter has not increased.

Non-deforming semi-jacketed bullets. It is generally assumed that semi-jacketed bullets mushroom on impact with soft tissue or equivalent simulants, causing their cross-section to increase. There are, however, good reasons for designing bullets of these types that are not intended to deform, or for which deformation would in fact be undesirable.

If the most important factor is accuracy, then it is the tail of the bullet that must be manufactured to close tolerances, more than the tip. The jacket is manufactured by drawing, and in such a case it is useful if the tail rather than the tip can be drawn. However, this requires that the lead core be introduced via the tip, which results in a bullet with an open tip. The Sierra Matchking rifle bullet is a typical example.

Certain types of sport shooting (such as dynamic and silhouette shooting) involve firing at targets made of hard materials. For reasons of safety, it is important to minimize the risk of ricochets. Solid lead and semi-jacketed bullets deform or fragment much more on hitting hard targets than do full metal-jacketed bullets, and hence are significantly less likely to ricochet.

Appropriate choice of copper alloy for the jacket and lead alloy for the core make it possible to manufacture semi-jacketed handgun bullets which, even at high impact energies, either do not mushroom at all or else do so to such a limited extent that the cross-sectional area does not increase (see Figs 4-19 and 4-20).

4.2.3 Gas and fluid jets as projectiles

4.2.3.1 General

A jet consists of a mass flow moving at a certain velocity. If it strikes a medium, then a certain mass will rebound at this velocity within a certain time limit. The jet therefore possesses kinetic energy and a certain energy density, which is related to the diameter of the jet (see 2.1.5). This would seem to indicate a degree of similarity with a projectile.

One major difference between a projectile and a jet is the length of time for which they have an effect. While the impact or penetration process of a bullet lasts only for a fraction of a millisecond, a jet can continue to act for several milliseconds or even hundredths of a second. This leads one to suspect that the penetration capacity of a jet – and possibly that of a projectile – is determined not by its energy density, but by its energy-flux density, i.e. the energy density acting per unit time.

This has no implications for the way in which we have been considering the penetration capacity of a bullet so far. As the time over which a bullet acts is always of the same order of magnitude, we can safely ignore the time aspect.

However, this might explain certain phenomena that occur when a bullet is fired into body armour or other protective materials. Elastic or plastic layers allow for a longer stopping distance. This extends the exposure time, reducing the energy-flux density. That would be an explanation for the reduced penetration observed when the victim is wearing soft layers under body armour.

In order to cause open skin wounds, jets have to produce energy-flux densities which, in a very short exposure time (a matter of milliseconds) create energy densities in excess of the threshold energy density for the penetration of skin (0.1 J/mm^2).

4.2.3.2 Liquid jets

Injuries due to liquid jets can occur if a hydraulic hose is punctured, or in connection with a high-pressure cleaner, for instance. As the medium is approximately 1000 times as dense as a gas, even relatively low velocities can cause sufficiently high energy-flux densities.

For instance, liquid will escape from a pipe under a pressure of 10 bar at a velocity of approx. 50 m/s, which produces an energy-flux density of 0.05 J/(mm^2·ms). An exposure time of as little as 2 ms (which corresponds to a jet length of 10 cm) is enough to yield a critical energy density.

4.2.3.3 Gas jets

Gas jets are created by such devices as rocket motors. High energy-flux densities can occur in the vicinity of such motors, the magnitude depending on the size of the motor. However, the energy-flux density decreases with the third power of distance (see Eqns 2.1:84 and 2.1:85), so open wounds can only occur within a relatively short distance.

The velocity of a gas jet, however, only decreases linearly with distance, so there is a danger of injury due to the effect of high stagnation pressure and of open wounds at a greater distance. As a result, the danger zone behind jet engines is determined by stagnation pressure and not by energy-flux density.

The gas velocity 10 cm behind an anti-tank rocket is approximately 1800 m/s. Fig. 4-21 shows the pattern of energy-flux density. Open wounds can occur at up to 40 cm, if the skin is exposed to the jet for 2 to 3 ms (STAUFFER, KNEUBUEHL 1992).

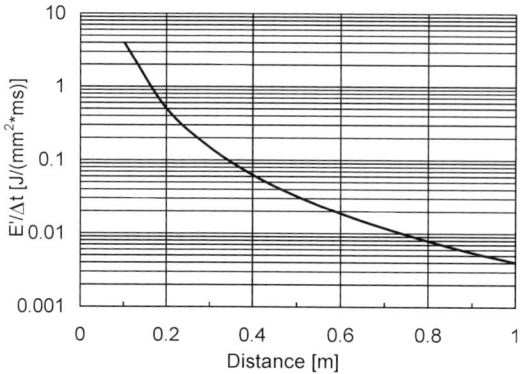

Fig. 4-21. Energy-flux density ($E'/\Delta t$) in the gas jet of an unguided anti-tank missile as a function of distance (KNEUBUEHL).

4.2.3.4 The effects of gas jets in the case of gas and alarm pistols

The literature describes numerous cases in which gas jets from alarm pistols have caused serious or even fatal injuries. See the comprehensive work by ROTHSCHILD (1999) for details. The design of the weapon (especially the length of the barrel) and the mass of propellant are the decisive factors.

The main factor determining the danger presented by a gas or alarm pistol is the velocity/distance function in the free jet. From the fluid velocity we can determine the energy-flux density, and from that we can determine the relevant parameter, which is the energy density at the point of impact for a given exposure time.

The propagation velocity of the gas jet can be determined in a number of ways (KNEUBUEHL 1998):
    It is possible to carry out direct measurements using multiple shadowgraphs with very short exposure times (of the order of 1 μs). The velocity can then be calculated directly from the distance covered between two successive images (indicated using a reference scale) and the time taken for the jet to cover that distance, which equates to the time between the two images.
    The momentum of the gas can be determined using a ballistic pendulum. As the mass of the gas is known (as it is equal to the mass of the propellant) we can calculate the fluid velocity at the muzzle directly.
    A third option is to measure the pressure in the gas jet at various distances from the muzzle. Velocity can then be calculated using Bernoulli's equation (Eqn 2.1:61a).

KNEUBUEHL and ROTHSCHILD conducted extensive ballistic studies on various alarm pistols using all three measuring techniques (KNEUBUEHL 1998, ROTHSCHILD 1999). Figs 4-22 and 4-23 show the velocities measured for two revolvers with different barrel lengths.

The energy-flux density can be calculated from the measured velocities, using Eqn 2.1:85. The exposure time (which is required in order to calculate energy density) is obtained from the pressure/distance recordings.

As the pressure and the velocity are related to each other (in accordance with Bernoulli's equation), the pressure/time curve also indicates velocity over time and hence energy over time.

4.2 Wound ballistics of handgun bullets

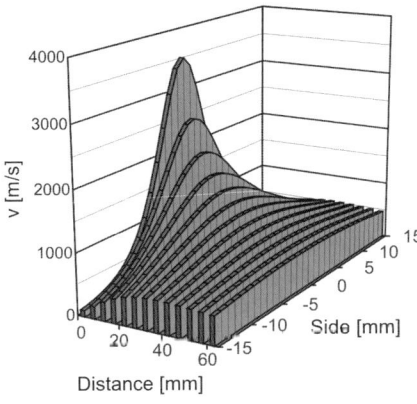

Fig. 4-22. Velocity as a function of distance in the gas jet produced by a 380 cartridge. Barrel length: 105 mm.

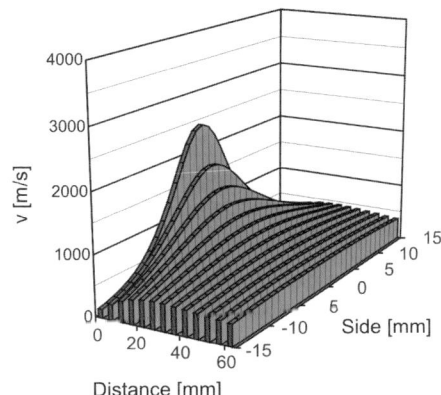

Fig. 4-23. Velocity as a function of distance in the gas jet produced by a 380 cartridge. Barrel length: 140 mm.

Table 4-17 shows energy densities for a number of typical alarm pistol calibres with different barrel lengths. The data in that table allow us to draw the following conclusions:

– Long-barrelled weapons produce significantly higher energy densities than do weapons with short barrels. The difference is more pronounced in the case of the pistols (8 mm and 9 mm P.A.) than of the revolvers (380). In the case of the revolvers, the size of the gap between cylinder and barrel and the associated loss of pressure play a major role, as does the mass of the gas that escapes at that point.

– The 380 produces significantly higher energy densities than the 8 mm. This is readily explained by the much higher mass of propellant that must be used

Table 4-17. Energy densities of a number of alarm pistols as a function of distance from the muzzle

Calibre	Type of weapon	Barrel length	Energy density in Joule/mm^2 at various distances		
		[mm]	0 mm	5 mm	10 mm
8 mm	Pistol	57	0.57	0.20	0.09
		56	0.46	0.16	0.08
		53	0.51	0.18	0.08
		125	0.27	0.10	0.05
380	Revolver	90	0.61	0.22	0.10
		105	0.75	0.27	0.13
		105	0.56	0.21	0.10
		140	0.24	0.09	0.04
9 mm P.A.	Pistol	68	0.50	0.19	0.09

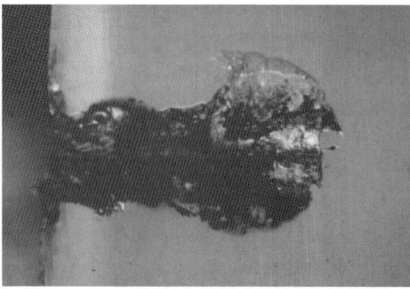

Fig. 4-24. Channel created by a 380 SP blank cartridge fired into gelatine with the muzzle touching. Short barrel. Penetration depth: 5 cm.

in these cartridges to compensate for the losses at the gap between cylinder and barrel.
- At a distance of 10 mm, the energy densities of these weapons are already at the threshold for penetrating skin.

In terms of effectiveness, gas and alarm pistols can be divided into two groups on the basis of the measurements. The short-barrelled weapons produce energy densities of between 0.45 and 0.75 J/mm^2 at the muzzle. This is certainly enough to produce open wounds in unprotected skin and to allow the gas jet to penetrate a certain distance into the underlying tissue (see Fig. 4-24). Even at a distance of 5 mm from the muzzle, the energy density lies between 0.2 and 0.3 J/mm^2.

The long-barrelled weapons (pistols over 120 mm, revolvers over 135 mm) produce significantly lower energy densities at the muzzle, although 0.25 J/mm^2 is still above the threshold for skin. However, given the uncertainty regarding the duration of exposure to the gas jet, it is not absolutely certain that serious injury would occur, even if the muzzle were in contact with the skin.

4.3 Wound ballistics of rifle bullets

4.3.1 Introduction

The barrel of a rifle is between twice and six times as long as that of a handguns. Furthermore, the volume of a rifle cartridge is significantly greater (see 2.2.2.2) as is the mass of the propellant. As a result, projectiles fired from rifles (with the exception of rimfire and airgun ammunition) carry many times the energy of a typical handgun bullet. Given that rifle and handgun bullets have similar masses, this extra energy can only come from a higher muzzle velocity (see Eqn 2.1:20), and indeed, rifle bullets leave the barrel at some 700 to 1000 m/s, whereas typical handgun muzzle velocities lie between 200 and 500 m/s.

We have seen in 3.2.3.2 and 4.1.3 that the wounding potential (i.e. the effectiveness) of a bullet depends in part on the energy that the bullet is carrying. Rifle bullets are therefore *substantially more effective* than handgun bullets. However, there is no difference of principle between the behaviour of a rifle bullet and that of a handgun bullet in wound ballistics terms.

There is currently no evidence for the existence of a velocity or energy threshold for conventional designs of handgun or rifle bullets above which different wound ballistics phenomena would occur. As impact energy increases, the mechanical effect on biological tissue increases steadily, as long as the bullet remains shape stable at the velocities under consideration.

However, there is a velocity threshold at which the behaviour of the bullet changes. Full metal-jacketed bullets remain shape stable on impacting gelatine or water at velocities of up to 500 or 600 m/s. At higher velocities, they begin to deform and fragment (see Figs 3-30 and 3-31). It is only possible to recover a bullet undamaged from water if it has struck it at a velocity below this threshold.

There is no justification for often-used differentiation between "low-velocity" and "high-velocity" bullet wounds. This distinction is, at best, another form of differentiation between wounds caused by handguns and rifles.

However, rifle bullets do cause certain effects that are almost undetectable in the case of handgun bullets. Most of these effects are related to the larger amount of energy transferred. We shall discuss these below.

4.3.2 Remote effects

4.3.2.1 General

By *remote effects* we mean those effects that are related to the bullet wound but manifest themselves somewhere other than in the wound channel or the temporary cavity. The agent of such effects may be mechanical, i.e. the bullet, or biological (physiological or pathological).

On impact, the *bullet* causes shock waves, which are transmitted through the medium. The radial displacement of tissue as the tip of the bullet passes through it leads to the creation of the temporary cavity, which in turn causes pressure waves that propagate through the tissue.

It is important to distinguish between *shock waves* on the one hand and *pressure waves* (or, more precisely, pressure changes) on the other. They differ in their propagation velocity and duration, and in the energy they carry. The duration of a shock wave is measured in microseconds, whereas that of a pressure change is measured in milliseconds. The shock wave passes right through the body very shortly after the bullet enters it, whereas the temporary cavity and accompanying pressure change arise only after the bullet has left the body.

We must also make a careful distinction between primary and secondary biological effects. By primary effects, we mean those associated directly with the bullet and its effects, i.e. the wound itself, damage caused when the temporary cavity is created, the effects of shock waves on tissue, the stimulation of nerves and its consequences, the effects of pressure waves on blood vessels, etc. The secondary effects are the consequences of the primary changes, which occur minutes or hours later. We know that the wound causes a large number of physiological changes in the body, blood loss leads to a redistribution of the blood, hormones are released and pathological changes (*secondary lesions*) occur in the damaged tissue. All of these secondary changes are medically relevant and are important for the understanding of the physiological and pathological processes that occur following trauma (see for instance SCHANTZ 1982). However, these secondary processes belong not to the domain of physical wound ballistics, but to that of the medical and biological consequences of bullet wounds. Given the aims of the present book, we shall examine the primary effects.

4.3.2.2 Shock waves

Physics of shock waves. A shock wave is a particular kind of sound wave. Sound waves are longitudinal waves, in which particles of the material concerned oscillate about a neutral position. They propagate through the material at a velocity c that is related to temperature.

The propagation speed of small-amplitude waves is known as the *speed of sound* (see 2.1.4.3). The speed of sound in gelatine, soap and water at 20 °C lies between 1500 and 1600 m/s (see A.2.1 for other values).

However, the velocity of sound waves depends not only on temperature but also on pressure. As pressure increases, so does velocity. In turn, pressure depends on compressibility, which is low in the case of a fluid. The wave does not move any material from one place to another, but it can transmit energy.

At small amplitudes the oscillations are harmonic, i.e. the positive half of the wave (overpressure) is equal in magnitude to the negative half (underpressure) – the only difference is in the sign. However, if amplitude approaches or exceeds the pressure at rest, then while the positive pressure can increase indefinitely, the negative pressure cannot fall below a certain limit. Because the speed of sound increases with amplitude, this means that the positive half-wave moves faster than the negative. This results in an increasingly steep wavefront. If the wavefront becomes so steep that pressure changes from minimum to maximum virtually instantaneously, the resulting wave is known as a shock wave. A shock wave is therefore a form of acoustic wave caused by a sudden, single, intensive excitation. After a certain time, or after the wave has covered a certain distance, loss of energy or velocity will cause it to become an acoustic wave packet of small amplitude.

The danger of a shock wave (in biological terms) lies in its large pressure gradients, i.e. in the sudden changes between positive and negative pressure with relation to pressure at rest.

A shock wave that starts from a point source propagates spherically. Its amplitude decreases with $1/r^2$, where r is distance from the source. If the source is linear, the wave propagates in the form of a cylinder, and amplitude decreases with 1/r. A bullet penetrating a medium creates a wave that is neither spherical nor precisely cylindrical. The exponent of r therefore lies between 1 and 2.

We can calculate the pressure at r by introducing a term r_0, the point of maximum pressure p_{max} (e.g. the surface of the projectile). If the source of the shock wave is spherical, r can be taken as equal to the radius of the sphere. We therefore obtain:

(4.3:1) $\qquad p(r) = p_{max} \cdot \left(\dfrac{r_0}{r}\right)^n$, [Pa]

with n = 1 for a cylindrical wave and n = 2 for a spherical wave.

Unlike a sinusoidal sound wave, with only one frequency, a shock wave is a unique event of very short duration with an extremely steep front, its amplitude p decaying approximately exponentially thereafter. Its half-life period τ is less than half a millisecond.

(4.3:2) $\qquad p = p_{max} \cdot e^{-\frac{t}{\tau}}$, $\qquad \tau < 0.5$ ms . [Pa]

After a given period > 0.5 ms, amplitude decays more gradually, with a graph of the waveform showing a "tail" (Fig. 4-25 a). This tail is of no significance to the present discussion.

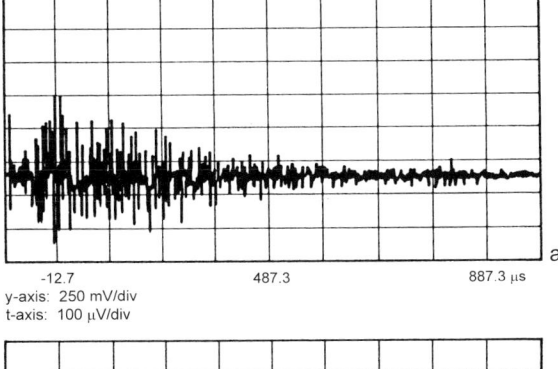

-12.7   487.3   887.3 µs
y-axis: 250 mV/div
t-axis: 100 µV/div

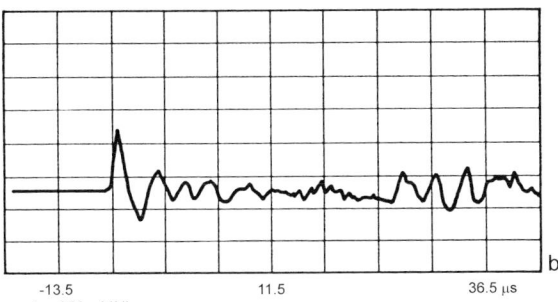

-13.5   11.5   36.5 µs
y-axis: 250 mV/div
t-axis: 5 µV/div

Fig. 4-25. Shock wave as a function of time:
a. Timebase 100 µs. The shockwave is right at the beginning of the pressure graph.
b. Timebase 5 µs. The shockwave at the beginning is clearly visible. It reaches a maximum after approx. 0.5 µs, followed by a negative "tail."

A steep wavefront always indicates that a mixture of frequencies is present (Fourier analysis). The steeper the wavefront, the higher the frequencies making up the mixture. Conversely, the presence of high frequencies is an indicator of a steep wavefront.

The decisive factor is the steep wave front mentioned above, with amplitude reaching a maximum in less than a microsecond. This means that pressure rises from zero to a maximum in less than a millimetre. This distance corresponds to the dimensions of biological structures (cells), and one could therefore expect cellular reactions to occur.

One parameter of relevance to the propagation of a shock wave is the characteristic wave impedance Z of the medium. This is defined as:

(4.3:3) $\qquad Z = \rho \cdot c$, $\qquad\qquad$ [kg/(m²·s)]

where ρ = density and c = the speed of sound in the medium.

If a shock wave passes through a homogenous medium, its amplitude will decrease in accordance with Eqn 4.3:1. If the medium is heterogenous, its amplitude will also be reduced by reflection at the interfaces. The magnitude of the reflections depends on the reflection coefficient. This is defined as:

(4.3:4) $\qquad \beta = \dfrac{Z_1 - Z_2}{Z_1 + Z_2}$, $\qquad\qquad$ [-]

Negative values for β indicate that the wave has been reflected with its phase reversed, i.e. positive pressure has become negative and vice versa.

Bone is the tissue type with the highest characteristic wave impedance and therefore causes reflections with a high amplitude. Lung tissue has the lowest impedance and, because of the air it contains, the greatest damping effect.

Shock waves are used in medicine, to break up kidney stones, in a process known as *shock-wave lithotripsy*. The waves are focused on an area some 10 mm in diameter, within which they produce a pressure of 1000 to 1500 bar followed by an underpressure of around 300 bar. Here also, we can assume that it is not the overpressure that destroys the kidney stone, but the underpressure that follows it, creating tensile stress. This is confirmed by slow-motion video.

The interesting aspect from a wound ballistics point of view is the amplitude of a shock wave propagated through a dense medium by a bullet. Using Eqn 2.1:62 (see 2.1.4.4) the stagnation pressure at the tip of the bullet is given by:

(4.3:5) $\qquad p_{max} = C_p \cdot \tfrac{1}{2} \cdot \rho \cdot v^2$. $\qquad\qquad$ [Pa]

where ρ = the density of the medium [kg/m³].

For instance, substituting the density of water (1000 kg/m³) for ρ and 0.3 for the pressure coefficient C_p of the bullet (see Table 4-14), we obtain:

(4.3:6) $\qquad p_{max} = 150 \cdot v^2$. $\qquad\qquad$ [Pa]

If we know the exponent n from Eqn 4.3:1 (1 < n < 2), we can calculate pressure amplitude p(r) at a distance r from the bullet track.

SUNESON et al. (1989) fired a 6 mm steel sphere into water at a velocity of 1200 m/s. At a distance of 500 mm from the bullet track, they recorded a pressure of 2.12 bar. The stagnation pressure calculated for such a sphere is 2160 bar. From this data, we can calculate the value of the exponent n. Using Eqn 4.3:1, we obtain a value of approximately $^4/_3$.

Shock waves in the human body. A shock wave initiated in a homogenous medium, or applied to such a medium, results in a defined waveform with a sharp (positive) wavefront followed by a negative phase. Its duration is very short – of the order of microseconds. Because of the inhomogeneity of the body, this simple wave undergoes significant reflection and scatter at the interfaces. It is transformed into a "burst" of longer duration, with a statistically irregular frequency and amplitude distribution. If one expands the timebase far enough, one can just about identify the sharp front of the original shock wave at the start of the burst.

SUNESON et al. (1987, 1989, 1990a–c) carried out extensive studies on shock waves in the body, with the primary aim of determining whether such waves cause changes in cells (see 4.3.2.3).

Equipment used: Young pigs weighing 19 and 28.5 kg were used. Pressure transducers were inserted into the brain, the abdomen and the muscle of the right thigh. 6 mm steel spheres weighing 0.88 g were fired into the left thigh at an impact velocity of approx. 1500 m/s. The wound channel measured 11 cm in one test group and just under 8 cm in another. The calculated energy transferred was 770 J in one case and 630 J in the other.

From the difference in the times at which signals were recorded in the right thigh and in the brain, and the distance between the two, the speed of sound was calculated as lying between 1444 and 1450 m/s.

The peak pressure p_{max} in the right thigh was 8.78 and 8.62 bar, while the difference between maximum and minimum was 14.0 and 9.74 bar. The bursts were 700 to 900 μs long, the frequency range extended from a few hundred kilohertz to over 1 MHz and the rise times were around 400 ns. This clearly shows that we are dealing with shock waves (see Fig. 4-25 a, b). The entire (initial) shock wave complex lasted for only approx. 3.5 μs.

After a delay (which corresponded to the transit time), a similar waveform was recorded in the upper abdomen. In general, the upper frequency limit lay at about 500 kHz – the upper frequencies had decreased, and hence so had the steepness of the wave front. p_{max} was 3.80 bar and the peak-to-peak value was 6.60 bar.

Later still (after approx 300 to 350 μs), the wave reached the brain. The maximum frequency was now only about 200 kHz, the oscillations lasted for 500 to 700 μs, p_{max} was 1.46 bar and the peak-to-peak value was 2.68 bar.

To sum up, therefore: the projectile causes a shock wave, which propagates at approximately the speed of sound (slightly faster at first). As distance from the point of excitation increases, the amplitude of the wave decreases because of reflections and damping and it becomes less steep, as evidenced by the drop in the maximum frequency present in the frequency mix. Finally, the shock wave becomes a simple mix of acoustic frequencies.

4.3.2.3 Biological/pathological consequences of shock waves

General. This areas has recently become the object of renewed interest, especially in connection with "non-lethal" weapons. HARVEY et al. carried out studies on tissue damage due to shock waves as early as 1947.

One of the authors' conclusions was that blood cells in Ringer's solution that were subjected to a large shock wave suffered no damage (i.e. no haemolysis). Suspended frog hearts likewise exhibited no changes. The amplitudes of the shock waves lay in the range 50 to 100 bar, measured with a tourmaline pressure transducer. Overall, the authors maintained that shock waves caused no damage to tissue, although in view of the manner in which they conducted their experiments it would be more correct to say that they observed no macroscopic damage. The authors did not conduct any histological investigations.

Excitation of nerves by shock waves. It is the change in pressure that can provoke a nervous reaction. A static pressure of over 100 bar does not induce any excitation, nor does it impair the functioning of the nerve, as was demonstrated by physiologist EBBECKE (Bonn) several decades ago.

As part of a study by WEHNER and SELLIER (1981) on the effects of bullets striking persons wearing body armour, experiments were also conducted to investigate the possible excitation of nerves by shock waves.

The experiment is shown in Fig. 4-26. A steel block contains a channel 4 cm in diameter, filled with 40% Ringer gelatine. The nerve (taken from a bullfrog) lies perpendicular to the channel and is connected to stimulating and recording electrodes outside the block. The right-hand side of the channel is closed off by a pressure transducer with a high natural frequency (Kistler 603 B), while the left-hand side is sealed by a metallic membrane. When a projectile strikes the membrane, it produces an almost plane shock wave, which passes via the nerve as it moves towards the right-hand end of the channel. The amplitude of this wave is then measured by the pressure transducer. It is possible to vary the amplitude of the shock wave by altering the impact energy of the sphere.

Fig. 4-27 shows a typical recording. To test the apparatus, the nerve is first stimulated using a 4 V square-wave pulse (Curve b). The nerve reacts by producing an electrical signal that moves along the nerve and can be detected at the recording electrode (Curve a). This is the *compound action potential*. A shock wave

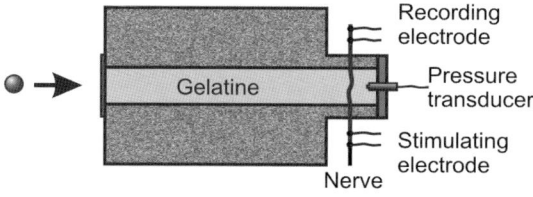

Fig. 4-26. Experiment to detect stimulation of a nerve by a shock wave.

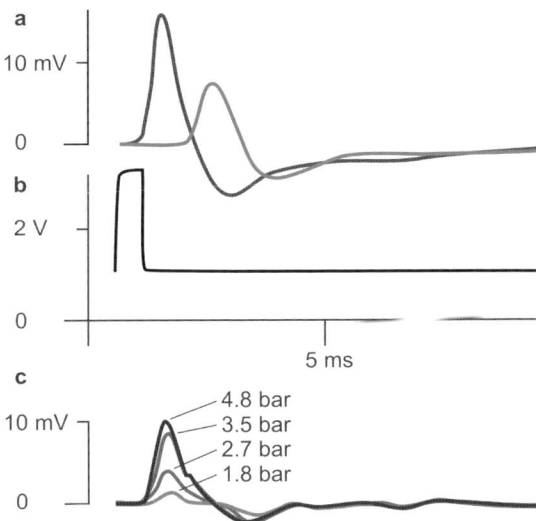

Fig. 4-27. Results of exciting a nerve using a shock wave:
a. The reaction of the nerve to mechanical stimulus (left) and electrical stimulus (right) takes the same form.
b. Square-wave impulse used to excite the nerve. The time axes of a and b are identical, and share a common zero.
c. Results of mechanical stimulation. The y-axis shows the magnitude of the reaction of the nerve to stimulus (in mV). E.g. 4.8 bar produces approx. 10 mV.

is then produced which also induces an electrical stimulus in the nerve, as shown in Curve b. The nerve reacts in the same manner to electrical stimulus as to mechanical.

Below about 0.75 bar, the nerve used for the experiment did not produce a compound action potential. As one would expect, the amplitude of the compound action potential increased with the amplitude of the shock wave (see Fig. 4-26). Saturation was reached at about 5 bar. Higher pressures led to no increase in compound action potential.

These experiments show that a shock wave of sufficient amplitude can excite a nerve. In other words, the nerve reacts to the wave as if it had undergone physiological stimulation. The effector organ on which the nerve acts is therefore incapable of distinguishing between an endogenous and an exogenous electrical nerve impulse, i.e. one from within the body or from outside.

The eye is a good example of the effect of mechanically stimulating a nerve. A sharp blow to the eye causes the victim to "see stars" because of the stimulation undergone by the cells in the retina, which in principle are sensitive to light.

The purpose of these experiments was to determine whether a shock wave could cause death by shock. "Death by shock" does not have a clear medical definition. The term is used in a number of different senses. Here we shall define "death by shock" (if indeed it exists) as follows: death following a non-life-threatening minor injury that did not cause significant bleeding, from which it may be concluded that the victim suffered sudden cardiac arrest shortly after the injury occurred. Theoretically such an effect is possible, in terms of the mechanism by which heart and circulation are controlled.

The mechanism is as follows: At the point where the carotid arteries divide (i.e. where the Arteriae carotis communis splits into the A.carotis interna and externa) there is a pressure transducer, the carotid sinus, which takes the form of a plexus (the Glomus caroticum). This measures the blood pressure in this vessel and passes the result to a control system that adjusts the pressure. An increase in blood pressure leads, via a branch of the vagus nerve, to a drop in heart rate and vice versa. A large pulse applied to the pressure transducer or the nerve that leads from it causes the nerve to transmit an electrical signal to the heart. The result is a false "Pressure Increasing" message, which leads to a reduction in heart rate (this is known as the carotid sinus reflex). If the impulse were powerful enough this could, in theory, lead to cardiac arrest.

The carotid sinus syndrome is a known phenomenon in internal medicine and neurology. Pressure on the corresponding point in the neck causes a temporary drop in blood pressure, leading to dizziness (due to a shortage of blood in the brain) and pallor.

One argument against such an attempt to explain death by shock in this manner is that when the bullet passes through the body the shock wave will stimulate not only the vagus nerve (the parasympathetic nervous system) but also the sympathetic nervous system, if the amplitude of the wave is sufficiently great, and the effects of the two would cancel each other out. It could of course be that at a certain, low amplitude (higher than the threshold at which the nerve is excited) the two systems could be stimulated to different degrees, as propagation of the shock wave is damped proportionally to 1/r and undergoes reflection from inhomogeneous structures. At present, this is still a theory. No experimental confirmation exists. The problem should be borne in mind, however.

In one case dealt with by the Institute of Forensic Medicine of the University of Berne (WYLER et al., unpublished) it appeared that an instance of "death by shock" had been discovered. An elderly woman was hit in the soft parts of the neck by a shotgun blast at close range. She clearly died instantly, although the shot had failed to damage large blood vessels or to perforate the cranial cavity or vertebral canal. On the other hand, the brain did show evidence of "coup contrecoup."

Ballistic reconstruction involving shooting tests showed that when pellets of this nature pass through the neck in a very short time (approx. 0.2 ms) the amount of energy transferred to the neck is very large, corresponding to a fall from a height of 1.7 m, but with the energy transferred 50 times as fast. The shotgun pellets striking the soft parts of the neck applied kinetic energy eccentrically, and the resulting rotational acceleration of the brain caused instantaneous death.

A further theoretically possible manner in which death by shock could result from shooting – stimulation of the pressure receptors in the carotid arteries – will be discussed in 4.3.2.4. See also "Changes in EEG as a result of shock waves" later in this section.

Cell damage caused by shock waves, and histological changes to tissue. As we have already seen, the amplitude of the shock wave rises from zero to a maximum over a very short distance, causing a large pressure gradient, and this distance corresponds approximately to the size of a cell, i.e. about 10 μm. One would therefore suspect that such a gradient might cause damage to cells. A number of authors have mentioned this possibility (e.g. BERLIN et al. 1979 and SELLIER 1983). What is certain is that it is not the pressure itself that causes damage, but rather the rapid increase in pressure or the rapid drop in pressure that follows it, as a biological structure can only be damaged by shear or stretching, not by compression.

Logically, the degree of cell damage will depend on the amplitude of the shock wave. In other words, a shock wave with a small amplitude will cause only reversible interference to the functioning of the cell, whereas a large amplitude will also destroy cells, depriving the organism of the corresponding function. Furthermore, there will probably be a threshold below which there is no visible reaction on the part of the cell. This would be analogous to the threshold below which a shock wave fails to stimulate a nerve.

SUNESON et al. (1987, 1989, 1990a–c) measured the pressure in the thigh, abdomen and brain of a pig when a projectile was fired into its left thigh (see 4.3.2.3). These experiments also involved histological studies aimed at identifying any modifications or damage to cells caused by shock waves. We shall discuss these below. Inevitably, our discussion will involve a number of medical terms. For those unfamiliar with medical topics, the essential finding was as follows: depending on the intensity (i.e. the amplitude) of a shock wave, it caused either reversible interference to the functioning of a cell or irreversible changes to that cell. However, no massive cell destruction was observed at the shock wave amplitudes produced in these experiments (of the order of 1 to 10 bar).

Results: Macroscopic examination revealed scattered petechial (minuscule, punctiform) bleeding on the left n. ischiadicus, but no visible destruction of tissue. This effect could be the result of the temporary cavity, with the projectile having missed the nerve itself. The right n. ischiadicus and n. phrenicus were unaffected.

Examination under an optical microscope revealed the myelin sheaths of the above-mentioned nerves to be undamaged, with no sign of deformation or shear. The axons, however, displayed varying degrees of deformation of the myelin sheaths, myelin invagination and, in some cases, voids between the axolemma and the myelin. The frequency and extent of the changes were more marked in the case of the animals killed 48 hours after the experiment and those examined immediately after. One must therefore assume that these were not direct, primary results of the shock wave but rather secondary effects. However, we still need to identify the primary mechanism by which damage is caused.

If the blood-brain and blood-nerve barriers become permeable, this is a clear sign of cell damage.

It is possible to test for this by injecting a dye, such as Evans Blue. Only if the barrier has become permeable will the organs (in this case the brain or the nerves) take on the colour. The dye is injected intravenously 10 minutes before the shot is fired.

The left N. ischiadicus was strongly coloured, which was to be expected in view of the other indications of significant damage (petechial bleeding, etc.). The N. ischiadicus on the other side showed slight coloration, visible to a varying degree. As we have seen above (in 4.3.2.2), the largest shock wave amplitudes were recorded in the right hind leg, through which the N. ischiadicus passes, as this is the closest point to the wound channel, apart from the left hind leg. No coloration was observed in the brain or in the left or right N. phrenicus.

SUNESON et al. also studied the reaction of cell cultures to shock waves. This technique was at that time unknown in the field of wound ballistics, although it had been applied to pharmaceutical research. It has two advantages over experiments involving animals: not only is it considerably less expensive, but more importantly it uses no animals, which is highly satisfactory from an ethical point of view.

The apparatus used by SUNESON et al. was as follows: A reinforced rubber pipe 100 cm long, with an internal diameter of 15 cm and walls 0.3 cm thick was set up vertically. It was sealed at the bottom by a convex acrylic glass disc placed with the convex side inwards, so as to scatter the shock wave. Two openings approximately 5 cm in diameter were made in the walls of the tube about 20 cm from the upper end. These openings were situated diametrically opposite each other and were sealed by thin plastic membranes. The tubes were filled with bubble-free water at a temperature of 37°C.

6 mm steel spheres weighing 0.88 g were fired through the two plastic membranes at a velocity of approx. 1,200 m/s. The calculated energy transferred to the water was approximately 500 J.

The plate with the cell cultures was placed 50 cm below the planned track of the projectile. This distance was selected because it corresponded to the distance between the left hind leg and the brain in the animals used; the idea was to achieve the same mixture of frequencies. A pressure transducer was installed under the plate and its cable routed through the acrylic glass disc.

The cell cultures consisted of dorsal root ganglions from 17-day-old rat embryos. The plate was enclosed in a surgical glove containing cell culture medium at a temperature of 37°C. This enclosure allowed the shock wave to pass with virtually no reflection.

The maximum shock wave amplitude at the cell culture was approx. 2.1 bar. The frequencies in the wave ranged from 0 Hz to 250 kHz.

Results: Immediately after the shot, examination with an optical microscope revealed no pathological changes in the cells. However, a number of the larger neurones showed more intensive coloration than did those of the controls. After six hours, most of the neurones displayed signs of damage: their cores had shrunk and some had become pyknotic, they also displayed vacuoles and irregularities in the structure of the cytoplasm. The Schwann cells and fibroblasts showed similar changes after six hours, albeit to a lesser degree.

The Evans Blue colour test mentioned above was negative immediately after the shot was fired, with rare exceptions. This corresponds with the results for the controls. After six hours, virtually all the neurones were coloured, indicating serious functional failure of the cell membranes.

Interestingly, extensive cell damage was noted at impact velocities of approx. 1,500 m/s (i.e. 1.25 times the velocity used, producing waves with an amplitude 1.5 times larger). Some of the cells had become mechanically separated from their support.

The authors also measured the oxygen consumption of the (living) cells before and after the shock wave. Oxygen consumption is a good indicator of whether the cells are functioning correctly. Under the above-mentioned experimental conditions and ballistic values (velocity, energy transfer and shock wave amplitude), oxygen consumption fell to 20% of the previous level after the shot was fired.

It is clear that the shock waves damaged the cells. What is not clear is the mechanism by which this happened. Immediately after being subjected to shock waves, the cells show no change. Change only becomes apparent later. The damage is therefore not caused directly by the wave, but by some (non-specific) reaction by the cell to a cause of which we are currently ignorant.

The question as to which physical characteristic of the shock wave is capable of damaging the cell remains to be answered. The candidates are: the rise in pressure, the pressure gradient (from maximum positive amplitude to minimum negative) or just the negative pressure. We discussed this point to some extent in the previous section, "Excitation of nerves by shock waves." From that discussion it is clear that the only possible causes are the pressure gradient (which effectively means shearing of the cells) or the negative pressure alone.

Changes to the EEG as a result of shock waves. The experiments conducted by SUNESON et al. (see 4.3.2.2) showed that even when a projectile was fired into the hind leg of a pig, pressure changes of approx. 1.5 bar occurred in the brain. Waves of this nature could have physiological or psychological effects. In the case of non-fatal hits, they could even cause incapacitation or other mental effects.

GÖRANSSON et al. (1988) studied this problem, observing the changes in the EEG of a pig (m ≈ 25 kg) when a shot was fired into the hind leg. Obviously, these studies address only one aspect of the problem.

They found that the amplitude of the EEG curve dropped to 50% of normal for about 30 s after the shot. In three out of nine pigs, however, no such effect was observed. None of the animals underwent a change in blood pressure or ECG during the period in which the EEG was observed, but six of the nine animals stopped breathing for up to 45 s. Control experiments confirmed that sound waves or concussion from the shot had no influence on the EEG.

4.3.2.4 Pressure changes in blood vessels

In 3.2.3.1, we say that the creation and collapse of the temporary cavity causes changes in the pressure acting radially about the wound channel. We also saw that this process, which requires a certain amount of space, presses and stretches the surrounding tissue, with these effects becoming less marked as distance from the wound channel increases. It is therefore clear that the blood vessels and nerves running along the tissue subjected to these loads will also be compressed and stretched. A pressure wave is created in the blood vessel where the pressure is applied, and is propagated in the blood vessel in accordance with the laws of haemodynamics. As a result, pressure changes occur in blood vessels at points remote from the temporary cavity.

One point should be borne in mind: in terms of physical behaviour, the pressure change resulting from the sudden compression of the blood vessels can be seen as virtually equivalent to a pulse wave. The velocity v_p of the pulse wave is approximately 4 m/s in the aorta of a young person. In peripheral vessels, v_p is approximately 7 to 12 m/s. Velocity v_p increases with age. This indicates that v_p is related to the elasticity of the blood vessels and their radius. In general:

$$(4.3:7) \qquad v_p = \sqrt{\frac{Y \cdot h}{2 \cdot \rho_f \cdot r}} \, , \qquad [m/s]$$

where Y is the Young's modulus of the wall material [N/mm^2], ρ_f the density of the fluid (blood, in this instance) and r the radius of the blood vessel.

As the Young's modulus increases (i.e. the elasticity and radius of the blood vessel decrease), pulse velocity v_p increases. This should not be confused with the mean flow velocity v_f of the blood. When blood pressure is at a maximum, v_f in the aorta is approximately 0.6 m/s, which is substantially slower than v_p. The amplitude of the pulse wave undergoes exponential damping. After a distance x, the amplitude A_0 has decreased to:

$$(4.3:8) \qquad A = A_0 \cdot e^{-k \cdot x} \, , \qquad [N/m^2]$$

where $k \approx 1$.

As part of a study on the magnitude of the traumatic effects on persons wearing body armour, SELLIER and WEHNER (unpublished) measured the pressure in the aortas of cadavers when bullets strike the outside surface of body armour and deform the chest wall without penetrating. They measured pressures in the aorta of between 0.1 bar (22 L.R.) and 0.5 bar (45 Auto). See Fig. 4-28. In comparing these values, one should bear in mind that the lungs, being filled with air, prevent the amplitude from rising further.

4.3.2.5 The effects of pressure pulses on blood vessels

Blood vessels can be damaged in either of two ways. They can suffer internal damage through excessive stretching (tearing of the intima or damage to the muscles of the blood vessels) without any direct blood loss occurring (this damage can be demonstrated histologically, see RICH 1968b, RICH et al. 1967 and AMATO et

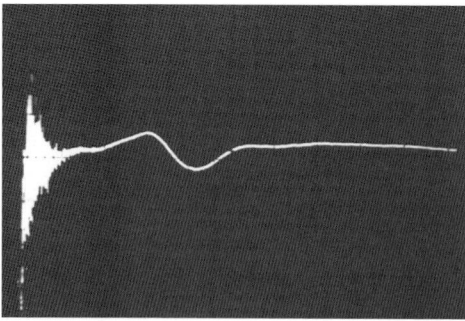

Fig. 4-28. Pressure in the aorta following after a bullet strikes a chest protected by body armour. Bullet: 45 Auto. Oscillations at the start of the curve due to the impact. Pressure amplitude.

al. 1970, 1971). The blood vessel can also tear, allowing blood to escape. This occurs when the vessel is subject to a sufficiently high load suddenly, in the area to the side of the temporary cavity.

The forces acting on the blood vessel depend both on the distance from the geometrical wound channel and on the maximum volume of the temporary cavity. In turn, this volume is linked directly to the energy E_{ab}' transferred from the projectile to the tissue in the area of the shortest distance from the blood vessel.

To obtain an idea of the amount of energy that must be transferred in order to tear a blood vessel, KNEUBUEHL conducted experiments using simulated blood vessels (see 3.3.6).

The plastic tubes were embedded in gelatine at various depths and various types and calibres of bullet were fired into the gelatine in such a way as to pass by the "blood vessels" without touching them. Using this approach, it was possible to generate energy transfer levels of between 30 and 200 J/cm adjacent to the blood vessel.

Fig. 4-29 presents the results. At effectiveness levels of less than about 100 J/cm, the blood vessels were only destroyed by direct contact with the bullet. As the distance between the bullet track and the blood vessel increased, there was a clear increase in the effectiveness E_{ab}' required in order to tear the blood vessel. Fig. 4-29 shows the approximate threshold (dashed line).

4.3.2.6 Bone fractures at locations remote from the wound channel

If a bullet hits a bone directly and its velocity/energy is sufficiently high, the bone will break. We have discussed the quantitative aspects of this in 4.2.

But even if the bullet does not touch the bone a fracture may occur if it passes close by. This is due to the impact on the bone of muscle displaced by the change in pressure that is induced by the temporary cavity. Such fractures occur if the impacting mass bends the bone so far that the maximum permissible tensile stress for bone σ_B (approx. 90 N/mm^2) is exceeded on the opposite side of the bone to that on which the force is applied. For this movement to occur, the stress must act on the bone for a sufficient length of time.

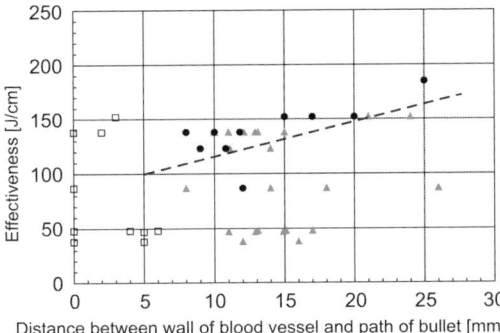

Fig. 4-29. Tearing of blood vessels as a result of the pressure pulse generated by a temporary cavity as a bullet passes to one side of a blood vessel.
- Blood vessel torn
- Blood vessel not torn
- Direct hit (distance from centre of track to blood vessel smaller than radius of bullet).

It has been suggested that shock waves (as described in 4.3.2.2) can cause fractures. However, one can readily show that the duration of a shock wave (a few microseconds) and the energy it carries would not suffice to bend a bone past the maximum permissible stress limit. SELLIER carried out calculations on this:

"If one considers the measured or calculated amplitude of the shock wave acting on the bone, together with the energy density that the shock wave can produce, there is room for considerable doubt as to whether a shock wave is capable of breaking a bone". (SELLIER and KNEUBUEHL 2001)

During experiments intended to demonstrate bone fractures due to shock waves, it is extremely difficult to separate the shock wave from the pressure pulse chronologically.

4.3.3 Wound ballistic characteristics of rifle bullets

4.3.3.1 Bullets designed for military use

In recent years, larger calibres have come into use for military rifles and machine guns, in addition to the existing calibres (5.5 mm to 8 mm). These are particularly popular for use with long-range sniper rifles. Common calibres include the 338 Lapua Magnum (8.4 mm) and the 12.7 × 99 (50 Browning). Full metal-jacketed bullets in these calibres behave in basically the same manner as smaller-calibre bullets, both in soft media and on impact with bone (see 3.2.2).

Two differences are of relevance, however. The first is the significantly higher muzzle energy (of between approx. 6,800 and 16,000 J), which ensures that such bullets achieve a high degree of effectiveness, despite their higher sectional densities (see Eqn 3.2:18).

For cartridges and bullets of similar geometry, sectional density increases linearly with calibre. Muzzle energy, however, increases in proportion to the cube of the calibre (on account of the increase in volume of the cartridge case).

The second is that these larger bullets have higher moments of inertia, which delays the onset of yaw. They therefore produce a longer narrow channel (see Figs 4-30 and 4-31).

The Nammo (Raufoss) 12.7 × 99 Multipurpose bullet is the subject of controversy, in terms both of wound ballistics and of international law. This projectile

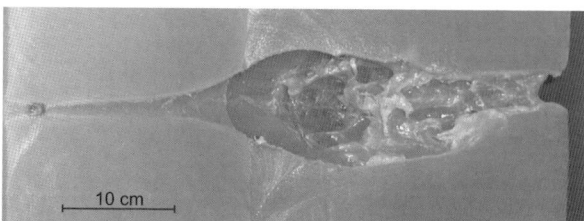

Fig. 4-30. Channel of a 338 Lapua Mag. full metal-jacketed hard-core bullet in soap.

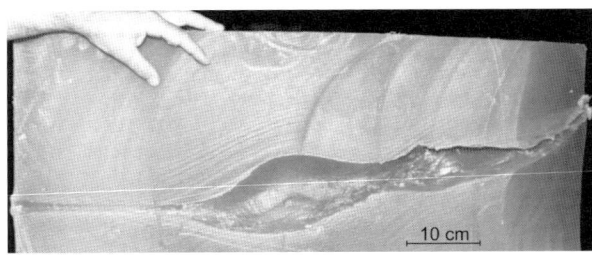

Fig. 4-31. Channel of a 12.7 × 99 full metal-jacketed bullet in soap.

was developed for use against hard targets. In addition to a tungsten alloy core, it includes an incendiary compound and explosive (see Fig. 4-32). The latter is triggered on impact, exclusively by the friction and heat resulting from deformation.

As it is also possible to fire this bullet from sniper rifles, one must expect persons to be hit, and it is perfectly possible for the explosive to detonate when this happens (MOOSBERG 2003). In such a case, the bullet would possess an exceptionally high wounding potential, bringing it into conflict with international law (specifically, the St Petersburg Declaration; see 7.3.2.2).

However, bullets exist that contravene international law, but of which the effectiveness corresponds to that of accepted full metal-jacketed bullets. For ballistics reasons, precision bullets are crimped shut at the tip, rather than the base. This leaves a small hollow in the tip (see Fig. 4-33 a), which is contrary to the Hague Convention of 1899. However, the diameter of the hollow is generally so small that no pressure can be built up on the lead core that would cause the bullet to mushroom. There is no visible difference between the wound channel of such a bullet and that of a full metal-jacketed bullet (see Fig. 4-33 b). However, smaller-calibre (e.g. 5.56 mm) versions of the same type of bullet are capable of deforming and of creating wound channels similar to those of similar hunting bullets. The only way to be certain as to how a bullet will behave is to conduct experiments.

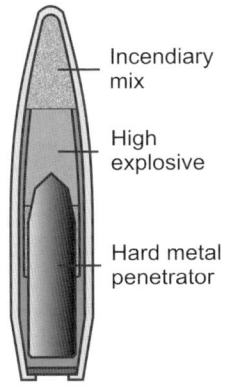

Fig. 4-32. Structure of the "multipurpose" bullet (from Wikipedia).

Tracer bullets are widely used by the armed forces. The pyrotechnic charge in the base of the

 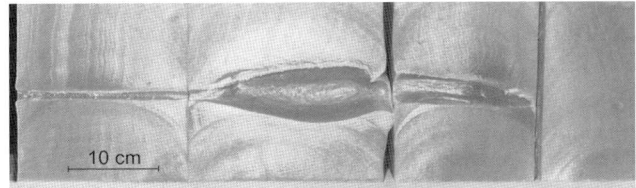

Fig. 4-33. a. *Left*: Tip of a Sierra Matchking 308 Winchester bullet (7.62 mm NATO). b. *Right*: Channel created in soap.

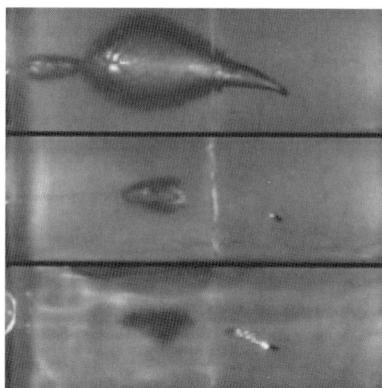

Fig. 4-34. High speed video recording of a 5.56 × 45 tracer bullet in gelatine. *Picture 1*: 3.75 ms after impact. Pre-ignition phase. *Picture 2*: 7.5 ms later. The tracer is burning, the temporary cavity is in the final phase of collapse. *Picture 3*: 65 ms later, the tracer is fully ignited. Total duration of burning: 0.62 s (frame rate: 2000 f/s).

bullet burns for approx. 1 s and indicates the trajectory and point of impact to the firer. It is of relevance to wound ballistics that the charge can burn with no external source of oxygen and therefore carries on burning until the charge is exhausted even if it comes to rest in the body (see Fig. 4-34). If this occurs, severe burns are to be expected in the vicinity of the bullet.

4.3.3.2 Hunting bullets

The requirements for a hunting bullet are just as contradictory as those for a handgun bullet (see 4.2.2.1). On the one hand, as much energy as possible should be transferred to the animal so that it dies as quickly and painlessly as possible. On the other hand, the bullet should be capable of causing an exit wound so that the animal loses enough blood to enable the hunter to track it if it escapes.

The design principle applied in order to meet these two contradictory requirements has remained the same since the beginning of the 20th Century: a tip that deforms slightly (or fragments) and a compact, relatively heavy rear part, which ensures that an exit wound is created. A large number of designs have been produced along these lines, all of which are similar from a wound ballistics point of view. Depending on the design of the bullet, energy transfer increases for the first 3 to 10 cm of the wound channel (the deformation or fragmentation phase), decreasing (very approximately exponentially) thereafter, as sectional density remains the same from this point on. Maximum effectiveness is determined by the energy available and by the increase in cross-section. Fig. 4-35 shows channels in soap for four different bullet designs, a, b and d deliver an impact energy of approx. 3300 J, while c delivers approx. 3500 J. As with handgun bullets, it is apparent that the bullet made of copper alloy (d) deforms substantially faster than those of conventional jacket/core design. The laws of physics dictate that greater penetration depth can be achieved only at the expense of reduced energy transfer.

As we saw in 3.2.4.2 in our discussion concerning full metal-jacket and handgun bullets, the deformation of a bullet depends to a high degree on its impact

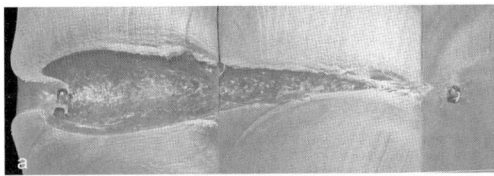
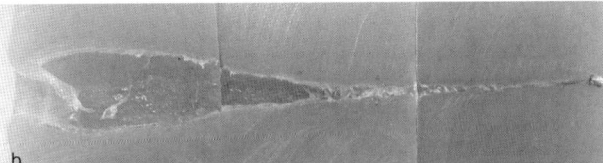
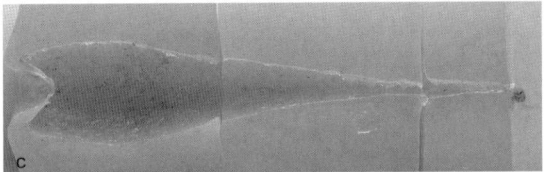
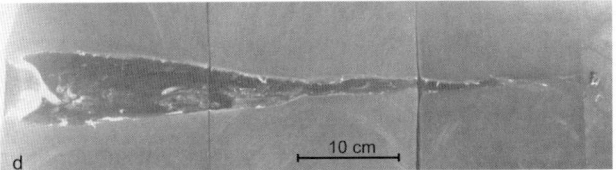

Fig. 4-35. Channels created by various hunting bullets:
a. 308 Win. RWS cone-point bullet.
b. 7 × 64 Brenneke TIG
c. 30-06 Brenneke TOG,
d. 308 Win. Brenneke TAG.

velocity/energy. As far as hunting bullets are concerned, we can take the RWS cone-point bullet as an example. In the vicinity of the muzzle (at approx. 830 m/s) this bullet produces the channel shown in Fig. 4-35 a. At a lower velocity (100 m/s lower, at a range of 100 m), the bullet penetrates slightly further, creating a cavity with a smaller diameter (Fig. 4-36 a). At approx. 645 m/s (at a range of 200 m), deformation is already delayed substantially, and after the bullet has penetrated to a depth of approx. 20 cm, the channel expands for a second time, probably because the orientation of the now-deformed bullet has changed (see Fig. 4-36 b). At an impact velocity of 603 m/s (at a range of 250 m), the bullet no longer deforms at all, creating a channel like that of a very stable full metal-jacketed bullet (see Fig. 4-36 c and cf. Figs 3-28, 3-29 and 3-30). The effect of velocity on deformation must be taken into account in a hunting context.

4.3.3.3 Shot and slugs

Shotguns – weapons with smooth, unrifled barrels – are in principle designed to fire cartridges containing shot (see 2.2.2.2). These cartridges contain lead or soft iron pellets, all of the same diameter, their number depending on the capacity of the cartridge case. Pellets are available in sizes ranging from 1 mm to 9 mm. Typical diameters lie between 2 mm and 4 mm.

230 4 Wound ballistics of bullets and fragments

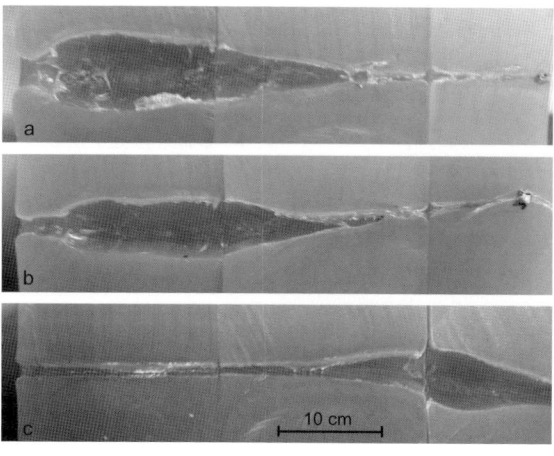

Fig. 4-36. Relationship between channel and impact speed, taking the RWS cone-point bullet as an example:
a. v = 729 m/s, d = approx. 100 m
b. v = 644 m/s, d = approx. 200 m
c. v = 603 m/s, d = approx. 250 m.

Shot.[1] From a wound ballistics point of view, an individual pellet can be treated as if it were a preformed spherical fragment (see 3.2.2.3). The diameter of the wound channel decreases exponentially and the greatest damage potential is immediately after entry. Penetration depth depends on impact velocity and sectional density, and hence on pellet diameter. This parameter can be estimated to a good degree of accuracy (see 4.4.2.4).

At very short ranges (less than about 5 m) the shot sheaf is still quite compact and can cause a large entry wound, surrounded by wounds caused by individual pellets (see Fig. 4-37). The diameter of the scatter can be used to estimate the range. In doing so, however, one must take care to use the same ammunition and barrel type (cylindrical or choke), as these factors affect scatter considerably (see Fig. 4-38).

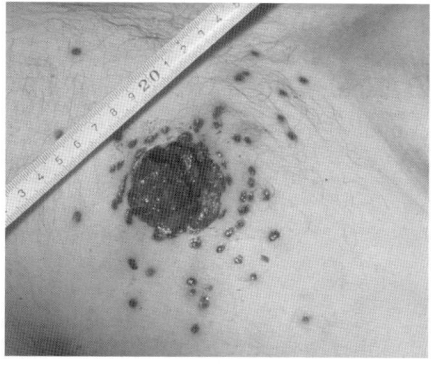

Fig. 4-37. Shotgun wound to the chest under the left armpit, close range.

---

[1]) For a comprehensive discussion of shotgun wounds, see Gunshot Wounds (V. J. M. DI MAIO, 1999). See bibliography.

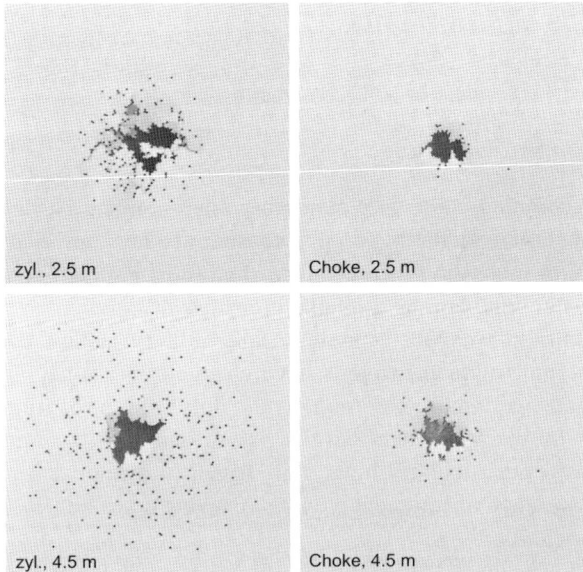

Fig. 4-38. Distribution of 2.41 mm shotgun pellets at short ranges:
Left: Cylindrical barrel.
Right: Full choke.
The squares measure 25 cm by 25 cm.

Shotgun slugs. The *slug* was developed in order to give a hunter armed with a shotgun the advantages of a single, heavy projectile. For reasons of interior ballistics, the mass of a slug must be similar to that of a shot load or less. This, together with the large calibre of a shotgun barrel, means that a slug has a relatively small sectional density.

The sectional density of a 12/70 Brenneke slug (31.5 g) is only 0.121 g/mm^2, which is less than that of a 9 mm Luger bullet (0.125 g/mm^2).

The combination of high muzzle energy (2750 J in the case of a 12/70) and low sectional density give shotgun slugs a high degree of effectiveness. These projectiles (which are generally blunt) generally fly stable in a dense medium due to shoulder stabilization, giving a straight, tapered wound channel similar to that of a fast, heavy fragment (see Fig. 4-39). Other slugs, such as the Balle Blondeau, give very similar results.

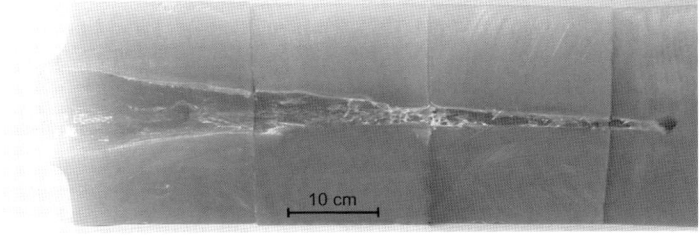

Fig. 4-39. Channel of a 12/70 Brenneke slug in soap. Diameter of entry wound: approx. 10 cm. Total penetration: 68 cm.

4.4 Wound ballistics of fragments

4.4.1 General

4.4.1.1 Frequency of fragment wounds

Fragments are still the most common cause of wounds in armed combat. At one time, they were virtually unknown in the field of forensics. Today, however, fragment wounds are becoming increasingly frequent in the civilian sphere as a result of attacks or accidents involving bombs and hand grenades.

As a result of developments in weapon technology and military tactics, the relative percentages of bullet and fragment wounds have changed over time. In the 19th Century, bullet wounds were more common, as the only sources of fragments were grenades filled with black powder, and the risk of being hit by fragments from one of these was not particularly high. This began to change in the First World War, with the percentage of fragment wounds reaching a peak of 90% in the Korean War.

It was the high risk of fragment injuries that led to the invention of the flak jacket during that war.

In the 1960s (during conflicts such as Borneo and Vietnam), bullets once again accounted for the majority of injuries. This may be related to the nature of jungle warfare. The latest data from the International Committee of the Red Cross (ICRC) indicate that fragments are once again by far the most common cause of injuries.

One should always examine wound statistics critically, as they are often subject to unintentional, systematic forms of error. The figures generally reflect the numbers of casualties treated rather than the numbers of soldiers injured. Because fragment wounds are much less likely to be lethal than are bullet wounds, one can not draw conclusions regarding percentages of bullet and fragment wounds from the numbers of cases treated.

GANZONI (1975) includes an excellent overview of the percentages of wounds accounted for by different causes in armed combat over a period of 100 years (see Table 4-18).

4.4.1.2 Wounds caused by fragments and similar projectiles

Fragment wounds have a clear pattern (see 3.2.2.3). The most striking characteristic is that the wound is largest at the point of entry, becoming progressively smaller as the projectile penetrates the body.

This is because fragments are often roughly spherical in form, in that all points of the surface are at about the same distance from the centre of gravity. Fragments adopt a position perpendicular to their direction of travel very rapidly after impact, as they have no stabilizing mechanism, and pass through tissue oscillating about this position.

Table 4-18. Percentages of wounds caused by bullets and fragments in conventional warfare (from GANZONI 1975)

		Bullets	Fragments	Source
Austro-Prussian War, 1866	Prussians	79	16	BERNDT (1897)
	Austrians	90	3	BERNDT (1897)
Franco-German War, 1870-1871	Germans	70	25	BERNDT (1897)
	French	94	5	BERNDT (1897)
First World War (British)		39	61	MATHESON (1968)
Second World War (British)		10	85	MATHESON (1968)
Korea 1950-1951 (USA)		7	92	MATHESON (1968)
Borneo 1963-1965 (British)		90	?	MATHESON (1968)
Vietnam 1965-1966 (USA)		52	44	MATHESON (1968)
Vietnam 1967 (USA)		49	50	RICH (1967)
Vietnam 1971 (USA)		40	?	BYERLY (1971)

The difference between the total and 100% is accounted for by other wounding mechanisms.

Sectional density remains approximately constant, whether the projectile is spherical or elongated. This means that the retardation constant of Eqn 3.1:1 can also be assumed as constant. That being so, the energy of the projectile decreases exponentially, causing the volume of the temporary cavity to do likewise.

There are special cases of stabilized fragments. A spin-stabilized shell, especially in the range of 20 to 40 mm, may release its base on detonation in the form of a single fragment subject to the full rotation of the shell. This fragment may fly like a discus for some considerable distance, often in the opposite direction to the direction of travel. If such a fragment enters the body, it causes an entry wound in the shape of a slit and maintains its pre-impact orientation.

Handgun and rifle bullets that have been caused to tumble by striking an object behave in a manner very similar to that of a heavy fragment when they strike the body. They turn perpendicular to their direction of travel immediately on penetration and move through the body in that position, with sectional density remaining virtually constant. Ricochets therefore cause injuries very similar to those of heavy, high-energy fragments.

This also applies to deforming bullets and fragmentation bullets that are unable to deform on impact because of the large angle of incidence and likewise adopt a position perpendicular to their direction of travel. There is no difference between their behaviour and that of a full metal-jacketed bullet that has ricocheted.

If a deforming bullet undergoes deformation as a result of passing through a body and then enters a second body, it will behave like a fragment and cause the corresponding type of wound channel, assuming it still has sufficient energy.

4.4.2 Equations of motion and energy for fragments

4.4.2.1 Hypotheses

For the purposes of this discussion, we shall use the following hypotheses and assumptions, which are based on empirical experience:

1. Fragments (and projectiles in general) behave in the human body as they do in the simulants described in 3.3.
2. The damage potential of a projectile at a given position corresponds to the energy transferred at that position.
3. The energy transferred at a position x is proportional to the energy of the projectile at that position:

(4.4:1) $$\frac{dE}{dx} = -\Re \cdot E(x) \,. \qquad [J/m]$$

4. The volume created in a glycerine soap is symmetrical about its longitudinal axis and is proportional to the energy transferred:

(4.4:2) $$dV(x) = -\frac{1}{\beta} \cdot dE(x) \,. \qquad [m^3]$$

By integration we obtain the following equation for the total volume from the entry wound to a position x:

(4.4:3) $$V(x) = \frac{1}{\beta} \cdot [E_a - E(x)] \,, \qquad [m^3]$$

where E_a is the impact energy and $E(x)$ the energy remaining at position x.

$\Re$ is the retardation coefficient [1/m]) and β is a material constant of the simulant [J/cm^3].

4.4.2.2 The geometrical form of the wound channel

From Eqns 4.4:1 and 4.4:3 we obtain an inhomogeneous differential equation with the following solution as our means of determining the volume:

(4.4:4) $$V(x) = \frac{1}{\beta} \cdot E_a \cdot \left(1 - e^{-\Re \cdot x}\right) \,. \qquad [m^3]$$

As long as the wound channel is symmetrical about its longitudinal axis, we obtain the following equation for the total volume from the entry wound to a position x:

(4.4:5) $$V(x) = \frac{\pi}{4} \cdot \int_0^x [d(\xi)]^2 \cdot d\xi \,. \qquad [m^3]$$

Equating Eqns 4.4:4 and 4.4:5 and differentiating, we obtain the following expression for the diameter d of the wound channel:

(4.4:6) $$d(x) = 2 \cdot \sqrt{\frac{\Re}{\beta \cdot \pi} \cdot E_a} \cdot e^{-\frac{1}{2} \cdot \Re \cdot x} \ .$$ [m]

This is hence an exponential function with negative exponents, which corresponds to the results of experiments (the form of the channel in soap). See Fig. 3-15.

4.4.2.3 Equation of energy and motion

From Eqn 4.4:1, we obtain the following value for the energy remaining after the projectile has penetrated to a depth x:

(4.4:7) $$E(x) = E_a \cdot e^{-\Re \cdot x} ,$$ [J]

and hence for velocity we can write:

(4.4:8) $$v(x) = v_a \cdot e^{-\frac{1}{2} \cdot \Re \cdot x} \ .$$ [m/s]

Differentiating this function and equating with a fluid dynamics approach to flow resistance yields the following:

(4.4:9) $$\frac{dv(x)}{dx} = -\tfrac{1}{2} \cdot \Re \cdot v_a \cdot e^{-\frac{1}{2} \cdot \Re \cdot x} = -C_D \cdot \frac{\rho}{2} \cdot v(x) \cdot \frac{1}{q} \ .$$ [s^{-1}]

Where C_D is a dimensionless drag coefficient, ρ is the mean density of the medium and q the sectional density of the projectile.

From Equations 4.4:8 and 4.4:9 we obtain the following equation for the retardation coefficient $\Re$:

(4.4:10) $$\Re = \frac{C_D \cdot \rho}{q} \ .$$ [m^{-1}]

4.4.2.4 Entry wound diameter and penetration depth

If we take x = 0 in Eqn 4.4:6, we obtain an equation for the diameter of the entry wound, which can be presented as follows with the aid of Eqn 4.4:10:

(4.4:11) $$d_0 = 2 \cdot \sqrt{\frac{C_D \cdot \rho}{\beta \cdot \pi}} \cdot \sqrt{\frac{E_a}{q}} = \lambda_1 \cdot \sqrt{\frac{E_a}{q}} \ .$$ [m]

This equation also defines the proportionality number λ_1 [s/m].

The diameter of the entry wound, which is correlated with the size of the surface wound, is therefore proportional to the square root of the impact energy and the reciprocal of the sectional density.

The model of an exponential decrease in energy obviously requires an unlimited penetration depth on the part of the fragment. To bring the model into conformity with the real-life situation, a residual energy E_{stk} is included, which in physical terms can be interpreted as the heat energy released at the moment the fragment comes to rest. If ℓ is the penetration depth of the fragment, and taking account of Eqn 4.4:8:

(4.4:12) $$E_{stk} = E(\ell) = E_a \cdot e^{-\Re \cdot \ell} \,. \qquad [J]$$

Using Eqn 4.4:10, the penetration depth is given by:

(4.4:13) $$\ell = \frac{q}{C_D \cdot \rho} \cdot \log \frac{E_a}{E_{stk}} = \frac{q}{\lambda_2} \cdot \log \frac{E_a}{E_{stk}} \,. \qquad [m]$$

This equation also defines the proportionality number λ_2.

Penetration depth is therefore proportional to the sectional density and to the logarithm of the impact energy.

From Equations 4.4:11 and 4.4:12 we obtain the following equation for the relationship between λ_1 and λ_1:

(4.4:14) $$\lambda_2 = \frac{\pi}{4} \cdot \beta \cdot \lambda_1^2 \,. \qquad [kg/m^3]$$

4.4.3 Experimental verification of the models

4.4.3.1 Method

From Hypothesis 1 (4.4.2.1) we can verify the models using soap and gelatine. This involves determining the mass and sectional density of the fragment and measuring its impact energy and penetration depth and the diameter of the entry hole.

In calculating the sectional density of a spheroid fragment we take its mean face area which, according to an approximation from geometry, corresponds to ¼ of the total surface. The mean face area of a cube is hence $1.5 \cdot b^2$ (where b is the length of one edge) and that of the sphere $\pi/4 \, k^2$ (where k is the diameter of the sphere).

Using the relationships already derived, we balance the measured values using regression. The correlation coefficient also gives an indication as to the reliability of the model chosen.

4.4.3.2 Entry wound diameter

To penetrate a medium, a projectile needs a certain minimum energy. If it strikes the surface of the medium with less energy than this minimum, it will not penetrate. This threshold energy E_{ad} does not appear in the model above, as the model

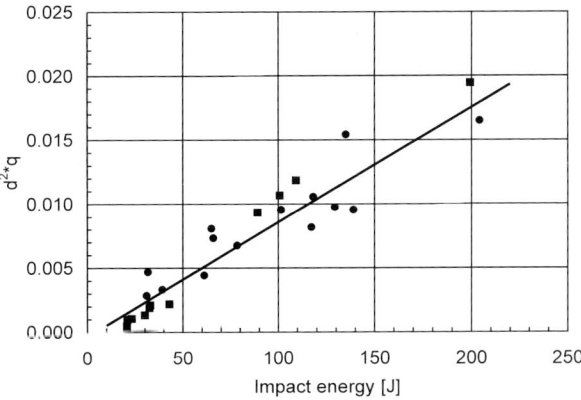

Fig. 4-40. Relationship between impact energy, sectional density and the diameter of the entry hole (for steel and tungsten spheres and cubes).

applies only to projectiles that have penetrated the medium. We must therefore add E_{ad} to Eqn 4.4:11 as follows:

$$(4.4:15) \qquad d_0 = \lambda_1 \cdot \sqrt{\frac{E_a - E_{ad}}{q}} \; . \qquad [m]$$

From this equation we can derive the following linear regression:

$$(4.4:16) \qquad d_0^2 \cdot q = \lambda_1^2 \cdot E_a - \lambda_1^2 \cdot E_{ad} \; . \qquad [kg]$$

If we substitute pairs of values for E_a and d_0 then we obtain λ_1 from the gradient and E_{ad} (with the known value of λ_1) from the axis.

Analysis of 26 shots using spheroid and cuboid fragments of various masses (0.13 to 1.04 g) and materials (steel and tungsten) yielded a value of 0.00946 s/m for λ_1 and 4.15 J for E_{ad}. The correlation coefficient was 0.944. Fig. 4-40 shows the individual values and the best-fit line.

For a wide range of fragment shapes, masses and materials, the following formula allows us to estimate the diameter of the entry hole:

$$(4.4:17) \qquad d_0 = \frac{0.95}{\sqrt{q}} \cdot \sqrt{E_a - 4.15} \; . \qquad [cm]$$

The relationship set out in Eqn 4.4:16 also applies to fragments with higher levels of energy. Older measurement results obtained with higher energy levels were used for the best-fit exercise in Fig. 4-41, with correlation being surprisingly good.

Fitting the line to the full set of measurement results does, however, give somewhat different values in the lower energy range.

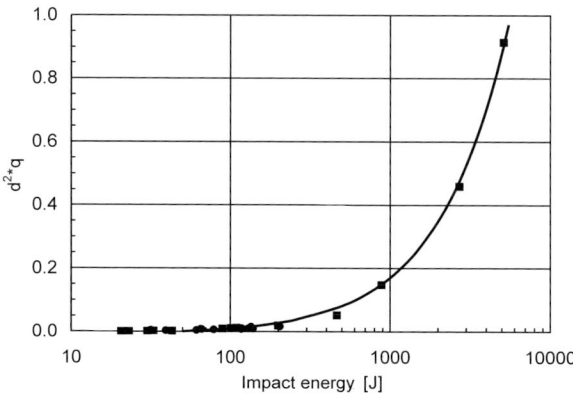

Fig. 4-41. As in Fig. 4-40, but range extended to approx. 5000 J.

4.4.3.3 Penetration depth

Rearranging Eqn 4.4:12 we obtain:

$$(4.4\!:\!18) \qquad \log E_a = \lambda_2 \cdot \frac{\ell}{q} \cdot \log E_{stk} \,. \qquad [-]$$

A best-fit line obtained using this equation pairs of ℓ/q and E_a values from measurements reveals an almost linear relationship between E_{stk} and E_a. As λ_2 and λ_1 are also related (see Eqn 4.4:14), we can also write:

$$(4.4\!:\!19) \qquad E_{stk} = \gamma \cdot E_a + \delta = E_a \cdot e^{-\lambda_2 \cdot \frac{\ell}{q}} \,. \qquad [J]$$

This gives us an equation for penetration depth that includes just impact energy, sectional density and the parameters:

$$(4.4\!:\!20) \qquad \ell = -\frac{q}{\lambda_2} \cdot \log\!\left(\gamma + \frac{\delta}{E_a}\right). \qquad [m]$$

From Eqns 4.4:8 and 4.4:10, the residual energy of a fragment that has penetrated a distance x is given by:

$$(4.4\!:\!21) \qquad E(x) = E_a \cdot e^{-\frac{\lambda_2}{q} \cdot x} \,. \qquad [J]$$

Analysis of the experiments conducted by JEANQUARTIER (1996a) yields the values shown in Table 4-19.

Table 4-19. Coefficients for (4.4:18) to (4.4:21)

Parameter	Unit	Spheroid	Cuboid
λ_2	[kg/m³]	377.8	390.0
γ	[-]	0.011	0.050
δ	[J]	1.03	1.33

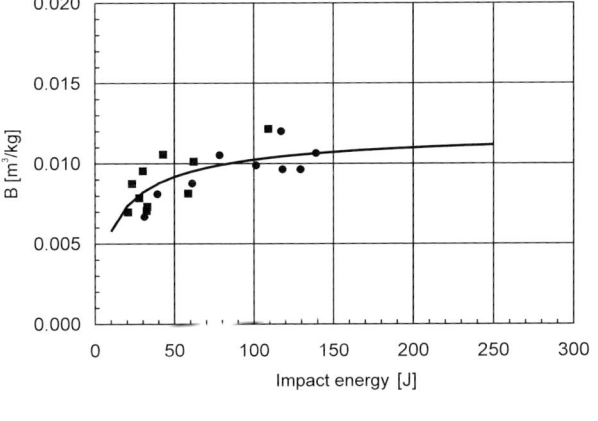

Fig. 4-42. Penetration depth/sectional density (indicated as B) as a function of impact energy. Spheroid fragments.

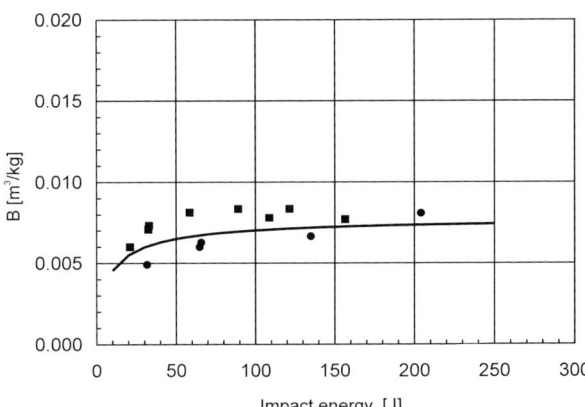

Fig. 4-43. Penetration depth/sectional density (indicated as B) as a function of impact energy. Cuboid fragments.

Figs 4-42 and 4-43 show the corresponding best-fit curves for the measurements (penetration depth as a function of sectional density).

4.4.3.4 Comparison with other studies

If we compare the values from Eqn 4.4:20 and Table 4-19 with the results of gelatine experiments we obtain sufficiently close correlation for the purposes under discussion. A test using 6.35 mm steel spheres (see CHARTERS 1976) yields a difference of 7%, for instance.

FACKLER et al. (1986) conducted experiments using 6 mm steel spheres at higher energy levels (400 to 1000 J). Here, however, the difference between the calculated and measured values is greater. This is mainly because the best-fit calculation above is based on points that lie relatively close, which means that extrapolation is always critical. However, if we summarize the results from FACKLER et al. (1986) and JEANQUARTIER (1996a), they can be fitted onto a single curve using the model above to a high degree of correlation ($r = 0.92$), with comparable results ($\lambda_2 = 395$, $E_{stk} = 1.4$ J). The lower value for residual energy could be linked to the use of gelatine.

In the same study, FACKLER also describes experiments using short cylinders, which clearly confirm the hypothesis of a maximum penetration depth (in accordance with Fig. 4-43) at increasing impact energy.

4.4.3.5 Applications

A number of practical applications arise from these relationships, both for forensics and for military purposes.

As fragments generally remain in the body, it is possible to estimate impact energy from the diameter of the entry wound, the penetration depth and the mass of the fragment (and hence its sectional density) using Eqns 4.4:15 and 4.4:20. This in turn makes it possible to reconstruct the trajectory and hence to determine the approximate area from which a fragment came, as long as the shell or grenade type is known. Where a fragment has passed through the body, it is possible to make an estimate of the impact energy using the diameters of the entry and exit wounds and the length of the wound channel, although such estimates are subject to a high degree of uncertainty.

For military purposes, one can build these analytical equations into vulnerability models; these can be used for effectiveness and hazard studies, for instance.

4.5 "Non-lethal" projectiles

4.5.1 General

When the police and special military units are maintaining law and order, they often require a means of placing a criminal hors de combat as rapidly as possible (ideally instantaneously) from a certain distance. Serious damage to health and life-threatening injury are to be avoided as far as possible. These so-called "non-lethal" or "less-than-lethal" means use chemicals (such as tear gas or pepper spray), electromagnetic technology (electromagnetic fields, lasers or bursts of high-frequency electricity) or mechanical force (pressure pulses, acoustic shock or kinetic energy).

Wound-ballistic issues arise almost exclusively in connection with the use of kinetic energy. We shall therefore restrict our discussion to projectiles that act directly on the body by means of kinetic energy. Serious physical injury can be excluded only if the projectile does not penetrate the body, or else penetrates only by a very few centimetres. This means that either the sectional density (i.e. mass and impact surface) and impact energy must be such that the penetration threshold is not crossed, or the penetration depth does not exceed the safety limit (see Eqn 4.4:20).

4.5.2 Projectile design

4.5.2.1 Projectiles with low sectional density

It is only possible to reduce sectional density by either reducing the mass of the projectile or increasing its calibre without increasing the mass to compensate. A very low projectile mass (or sectional density) allows a higher muzzle velocity. However, this does mean accepting a large decrease in velocity and energy over the trajectory of the projectile.

One extreme case, in ballistics terms, is the bullet used in 9 mm Luger PT (plastic training ammunition), which weighs only 0.39 g. It should be pointed out, however, that this ammunition is not intended for a "non-lethal" role. Despite its muzzle velocity of approx. 1000 m/s (energy: 195 J, energy density: 3.06 J/mm^2) this bullet is no longer able to penetrate the skin at ranges of over about 30 m, and does not penetrate the eye at ranges over 36 m.

The large drop in energy over the trajectory means that the bullet must have a fairly high muzzle energy if it is to be usable at a given range. As a result, the energy density of the bullet over the first part of its trajectory is many times higher than the "non-penetration" threshold. At the same time, the bullet becomes ineffective a short distance after its energy density has dropped below that threshold. The minimum safe distance and the maximum effective distance are very close together in most cases, requiring the user to estimate range very accurately.

The effectiveness of a bullet with low sectional density depends to a large extent on the clothing worn by the target. Even a single layer of cloth significantly increases the minimum energy required to create an open wound. Heavy winter clothing can easily increase the minimum energy required several times over.

Because of this, some "non-lethal" ammunition is produced with two or more muzzle energy levels for use with targets wearing different amounts of clothing.

These characteristics are common to all light projectiles. Nor is it possible to change them by increasing the calibre. However, larger calibres to transfer significantly more energy to the body, if they reach the threshold mentioned, on account of their larger surface area. They are therefore held to be more effective.

Most projectiles of low sectional density are made of rubber or plastic. Calibres range from 18 mm to almost 60 mm. Examples include the 35 mm MR-35 Punch made by Manurhin (Fig. 4-44, KNEUBUEHL 1993, SCHYMA and SCHYMA 1997b). Table 4-20 summarizes the ballistic data for a number of typical projectiles.

4.5.2.2 Expanding bullets

These are bullets so designed that the cross-sectional area expands after the bullet leaves the muzzle, thereby reducing the sectional density. If this happens immediately after the bullet leaves the muzzle, it behaves as described at 4.5.2.1. If deformation only occurs on impact, the sectional density can be kept relatively

Fig. 4-44. MR-35 Punch, 35 mm self-defence weapon. The weapon is made mainly out of plastic and has a five-round magazine.

high. This reduces the amount of energy lost in flight and increases the effective range.

It does take a certain amount of time for a bullet do deform after impact, and during this time it is already penetrating the body.

Table 4-20. Ballistic data for a number of "non-lethal" projectiles of low sectional density

Type			12-bore, sphere Fig. 4-45		12-bore, sabot		MR-35 Punch Fig. 4-46		Flash Ball Fig. 4-47	
∅	[mm]		18		18		35		44	
Mass	[g]		4.6		5.8		21.5		30	
q	[g/mm^2]		0.0181		0.0228		0.0224		0.0197	
			E [J]	E' [J/mm^2]	E [J]	E' [J/mm^2]	E [J]	E' [J/mm^2]	E [J]	E' [J/mm^2]
Distance	[m]	0	243	0.955	42	0.164	187	0.195	198	0.130
		10	156	0.611	34	0.134	154	0.160	160	0.105
		20	110	0.434	28	0.109	128	0.133	129	0.085
		30	82	0.324	23	0.089	106	0.110	105	0.069
		40	63	0.247	19	0.073	88	0.091	85	0.056
		50	49	0.192	15	0.060	73	0.076	69	0.046

Key: q: Sectional density   E: Energy   E': Energy density

Fig. 4-45. Rubber sphere, 12-bore. Diameter approx. 18 mm.

Fig. 4-46. Projectile of the MR-35 «Punch», rubber hollow sphere, Diameter 35 mm.

Fig. 4-47. 44 mm Flash Ball, cartridge and solid rubber projectile.

Fig. 4-48. Example of an expanding bullet (SCHIRNEKER).
Left: 38 Spl bullet, unfired, seen from behind.
Right: Bullet after undergoing expansion due to muzzle pressure. (By kind permission of the manufacturer).

It is possible to achieve an increase in cross-sectional area after the bullet has left the muzzle by any of the following means:

1. Expansion due to gas pressure at the muzzle;
2. Expansion due to centrifugal force or drag;
3. Expansion on impact.

Option 1. Bullets of this type are hollow, with predetermined breaking lines parallel to the axis, or have slits in the sides. After the bullet leaves the muzzle, it is forced apart by the muzzle gas pressure inside it, which increases its cross-section and reduces its sectional density. Because it expands shortly after leaving the muzzle, the section density of the bullet is low (and drag is therefore high) over the entire distance from the muzzle to the point of impact. The effective range of such bullets is therefore limited to about 10 m.

The SCHIRNEKER PbHB 38 is an example of such a bullet. It includes six slits and expands on leaving the muzzle, forming a six-armed star with a diameter of approx. 65 mm (see Fig. 4-48).

Option 2. A small bag filled with shot is inserted into the cartridge in place of a bullet. The bag unfolds at a certain distance after leaving the barrel. The bag takes on the form of a discus, with a substantially higher surface area and hence a lower sectional density. However, the discus does sometimes strike the target edge first.

The 38 Spl Short Stop cartridge can be fired from any 38 Spl or 357 Magnum revolver. The bag weighs 4.5 g. Once it has fully deployed, it forms a disc approximately 2.5 cm in diameter (see Fig. 4-49). Muzzle velocity is approximately 320 m/s, but the velocity of the projectile falls rapidly on account of the high sectional density, to approx. 185 m/s at 10 m and 105 m/s at 20 m.

Fig. 4-49. Short Stop 38 cartridge.
Left: Unfolded bag containing shot. *Right:* Unfired cartridge consisting of a red plastic case containing the bag of shot, rolled up.

A similar 12-bore cartridge used to be available, under the name of "Bean Bag." The square bag containing the shot weighed 40 g and measured approx. 45×45 mm once deployed.

The effective "non-lethal" range is small. When the projectile has just left the barrel and has not yet deployed (and its sectional density is still high) it is capable of penetrating the body, with life-threatening results. At greater ranges (i.e. over 10 m) it has lost so much velocity that it can no longer achieve the intended effect.

Option 3. This type of projectile generally consists of a soft, deformable case, filled with granules or a powder. The idea is that the projectile flattens on impact, increasing its surface area.

Examples include the 56 mm Bliniz projectile. This consists of a balloon filled with flour and weighing 83 g. It has a muzzle velocity of 60 m/s, and hence a muzzle energy of 150 J. Despite its relatively high energy, the projectile does not penetrate unprotected skin, as it expands to a diameter of approx. 100 mm on impact with a soft surface.

Another design uses a fragile plastic case filled with a heavy metal powder (e.g. bismuth or tungsten). The case bursts on impact with the body, releasing the powder and hence drastically reducing the sectional density. Examples of such designs include the 17.3 mm projectile designed for the FN 303 air rifle manufactured by FN Herstal (see Fig. 4-50) and Simunition's CQP 9 mm Luger bullet. Because of the time it takes for these projectiles to expand, they penetrate the skin before opening fully, and hence distribute the powder under the skin.

4.5.2.3 Rubber shot

Rubber and plastic shot cartridges are available in 12-bore shotgun calibre and for launchers in the range 40 mm to 60 mm (see Figs 4-51 and 4-52). However, the two types of cartridge are used differently. The shotgun-calibre cartridges are intended to affect a single person, with several small hits, while the launchers are intended for use against groups, with several persons being hit by the contents of one cartridge.

The rubber of plastic projectiles are equally diverse. Cartridges in shotgun calibres generally contain 10 to 15 spheres. Their diameter generally lies between 7 mm and 8 mm and their mass between 0.25 g and 1 g. Muzzle velocities also vary, but generally lie between 250 m/s and 350 m/s. Table 4-21 summarizes the ballistic data for two designs.

Fig. 4-50. 17.3 mm projectile designed for the FN 303 air rifle. The plastic casing is filled with bismuth granules. The tail incorporates a container filled with a glycol, to which a dye or Capsaicin can be added.

4.5 "Non-lethal" projectiles   245

Fig. 4-51. 12-bore rubber shot cartridge.

Fig. 4-52. Rubber shot cartridge for 56 mm grenade launcher.

The rubber spheres in the cartridges for large-calibre launchers are often the same as those used as single projectiles in shotgun cartridges (see Fig. 4-45). However, the muzzle velocity is generally much lower (around 100 m/s). At these values, the energy density E' is already lower than the threshold of 0.1 J/mm² at the muzzle. In addition to spheres, prismatic rubber bullets are also available for law-enforcement purposes, with a mass of 10 g and a muzzle velocity of 70 m/s. The energy density of such projectiles depends to a large extent on its orientation at the time of impact.

Table 4-21. Ballistic data for various types of rubber shot

Weapon		12/70 shotgun Fig. 4-51		12/70 shotgun		56 mm launcher Fig. 4-52		57 mm launcher	
⌀	[mm]	7.2		8		18		18[a]	
Mass	[g]	0.25		1.0		4.5		10	
q	[g/mm²]	0.0181		0.0199		0.0224		0.0385	
		E	E'	E	E'	E	E'	E	E'
	[m]	[J]	[J/mm²]	[J]	[J/mm²]	[J]	[J/mm²]	[J]	[J/mm²]
Distance	0	13.2	0.324	26.5	0.526	24.8	0.097	24.5	0.094
	10	4.4	0.107	20.0	0.398	19.6	0.077	15.8	0.060
	20	2.1	0.050	15.5	0.309	15.5	0.061	10.2	0.039
	30	1.0	0.025	12.2	0.243	12.3	0.048	6.6	0.025
	40	0.5	0.013	9.8	0.194	9.8	0.038	4.3	0.017
	50	0.3	0.007	7.8	0.156	7.8	0.031	2.9	0.011

Key: q: Sectional density   E: Energy   E': Energy density
[a] Prism. Calculation assumes end-on impact.

4.5.2.4 Special projectiles for handguns

In addition to the shot cartridges mentioned in 4.2.2.3, which are sometimes (and often wrongly) described as "non-lethal," special ammunition is available for handguns that is often classed as "non-lethal."

Simunition produces the FX, 9 mm Luger and 38 Spl/357 Mag. cartridge for training that involves two-way firing (see Fig. 4-53). The 8 mm plastic bullets used weigh 0.5 g or 0.65 g. They contain a dye and are designed to burst on impact. They have a muzzle energy of approx. 5.6 J and an energy density of 0.11 J/mm^2. Protective clothing is therefore essential. The training cartridges are fired from normal weapons, and to avoid the risk of confusion between this ammunition and ball, a conversion kit (with a special barrel and working parts) are required in order to fire the training cartridges. The CQT cartridge (see Fig. 4-53) permits training using targets without danger to surrounding structures. The same conversion kit is used. At 17 J (and an energy density of 0.34 J/mm^2), the energy of this projectile is substantially higher than that of the FX. This ammunition is therefore classed as "less lethal."

Fig. 4-53. *Left:* FX, *right:* CQT

4.5.3 Wound ballistics of "non-lethal" projectiles

4.5.3.1 Penetrating projectiles

General. We saw in 4.1 that the effect of a bullet depends not just on its effectiveness but also, to a large degree, on the point of impact and on the path of the wound channel in the body. This means that no projectile (with a calibre of up to 13 mm, or in shotgun calibres) is "lethal," because it can always hit a person in such a way that they survive. If, on the other hand, we assume that the projectile creates the most efficient wound channel in terms of effect achieved, then we must class almost every projectile as "lethal," because even an airgun pellet with an energy of just a few Joule can cause a fatal injury.

Clearly, therefore, it only makes sense to speak of projectiles with a low probability of lethality, the exact probability remaining unclear. That probability must be determined by those who bear responsibility for the use of the weapon system in question.

Given the role for which they are intended, these so-called "non-lethal" projectiles must not only have no more than a very low probability of killing anyone, but must also cause no serious injury, and in particular no permanent injury. This effectively means that there must only be a very low probability of their doing so.

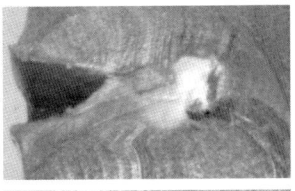
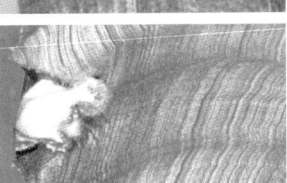

Fig. 4-54. A Bean Bag "non-lethal" 12-bore projectile. The bag containing lead shot weighs 40 g. At 5 m, it penetrated approximately 8 cm into soap (top), whereas at 7 m it penetrated only approximately 3.5 cm (bottom).

As a consequence, the penetration depth of such projectiles must not exceed a very few centimetres.

Projectiles for handguns and shotguns. The wound ballistic behaviour of projectiles with a low sectional density and those that expand before impact depends to a large extent on the distance to the target. Because their sectional density is low, these projectiles lose a larger amount of energy along their trajectories, and this has to be compensated for by relatively high muzzle energies. At short ranges, the projectile may penetrate too far, while a few metres further on it has too little energy to be effective. Fig. 4-54 shows the penetration depths for the first model of the 12-bore Bean Bag projectile at ranges of 5 m and 7 m. The Short Stop 38 cartridge behaves in a similar manner (see Fig. 4-49).

HUBBS and KLINGER (2004) report a fatal shooting in which a 22-year-old man was hit in the breast by a 12-bore Bean Bag projectile at a range of 6.4 m.

Compact projectiles such as rubber spheres (see Fig. 4-45 and Table 4-20, Column 1) often have such high energy densities that they cannot possibly be classed as "non-lethal." At short range, these projectiles are well capable of causing fatal wounds (CHOWANIEC et al. 2008).

Rubber shot is available in a range of forms. The energy density of the spheres contained in one type of cartridge drops to the threshold level at 10 m (Table 4-21, Column 1), whereas the energy density of another type (Column 2) is 1.5 times as high at 5 times this distance. At short distances, therefore, projectiles from this type of ammunition are therefore capable of penetrating to a substantial depth, and the risk of death is correspondingly high (CHOWANIEC et al. 2008).

Large-calibre projectiles. Projectiles designed for large-calibre launchers (30 mm to 60 mm) are significantly less likely to penetrate the body than shotgun projectiles, as their energy density is generally fairly small, because of their larger surface area. However, weapon systems with high muzzle energies (approx. 200 J) are capable of breaking the skin and penetrating the body at close range, or at least of causing large crush injuries/lacerations. Examples include the MR-35 Punch;

see Fig. 4-44 and Table 4-20, Column 3. The energy of such projectiles is in any case so great that they can cause serious injury even without penetrating the body. We shall discuss this point further in the next section.

4.5.3.2 Non-penetrating projectiles

General. In principle, the effect of non-penetrating projectiles on the human body belongs to the field of the physics of blunt force rather than to that of wound ballistics. However, it is worth mentioning a few points regarding these projectiles in the context of "non-lethal" projectiles and their potential effects.

One can assume that the severity of a wound caused by such a projectile will depend mainly on the impact energy. Energy density, the thickness of the body wall (skin, fat, muscle and/or bone) and clothing will all play a role. Relatively little is known about thresholds, partly because a projectile acts over a much shorter timeframe than do other sources of blunt force.

Known thresholds. A study carried out by the US Army Land Warfare Laboratory (JONES 2000) yields the following orders of magnitude:

- Impact energies of between 40 and 120 J can cause dangerous wounds (bruises, abrasions, broken ribs, concussion, blindness and damage to organs near the surface, such as the liver).
- At energies over 120 J, severe damage is to be expected, such as severe crush injuries/lacerations, skull fractures, tearing of kidneys or the heart and heavy bleeding.

BIR and STEWART (2005) carried out experiments with the aim of determining the threshold for penetration of the skin. Their results indicate the energy densities necessary to cause secondary injuries. However, their values are derived from experiments using fin-stabilized 12-bore rubber projectiles. Calculation of the corresponding energies gives the following values:

Fracture of a rib:	28 to 156 J
Fracture of the breastbone:	94 J
Fracture of the shoulder blade:	146 J
Tearing of the liver:	104 J
Penetration of a lung:	153 J
Damage to muscle:	127 to 135 J

Table 4-22 contains further data regarding the relationship between certain projectiles (including balls and other projectiles used in sport), their energy and the injuries they have caused or could cause. The data is taken from a number of sources.

STURDIVAN's equation. STURDIVAN was commissioned by the National Institute of Law Enforcement and Criminal Justice (NILECJ) to develop a formula for esti-

Table 4-22. Wounding potential of various projectiles, balls, pucks and similar

Projectile	Energy [J]	Wounding potential	Source
Fragment, approx. 10 g	5	Minimum mass required to cause serious injury	(1)
Falling golfball	15	Not lethal	(2)
38 mm projectile, 4.5 kg	50-70	Minimum required to cause concussion, fractured skull, fractures of hollow bones or internal injuries if victim hit in trunk	(3)
38 mm rubber or plastic projectile	250-300	Significant risk of serious injury, little risk of death	(4)
Football	165	Fractured nasal bone, concussion	(5)
Hockey ball	160	Fracture	(5)
Baseball	150	Can be fatal	(2)
Nouss (puck-like projectile used in Swiss sport of hornussen)	62	Fractured facial bones, black eye, crush injury/laceration	(6)
Tennis ball	60	Minimal risk	(2)
Squash ball	20-30	Severe black eye	(5)

(1) NATO Paper AC/225 (Panel VII) N/19, dated 10 October 1969.
(2) US Summary Report No. 1192-01(03)ER on "Tearspot Non-Hazardous 40 mm Incapacitating Agent CS Munition" (Contract DAAA 15-68-C-0078) CD/L 37553.
(3) Edgewood Arsenal Technical Reports Nos 4251 and 4319.
(4) TCD/TP 4020/612/72 dated 21 March 1972.
(5) Details regarding the various sports found on Internet.
(6) KNEUBUEHL (2001). Hornussen is a traditional Swiss team sport.

mating the lethality probability of blunt force applied to the thorax over the heart or the lungs (U.S. CONGRESS, OFFICE OF TECHNOLOGY ASSESSMENT 1992). Converted to metric units, and taking energy as an independent variable, the equation is as follows (cf. Eqn 4.1:26):

(4.5:1) $$P(L) = \frac{1}{1 + \exp\left[39.9192 - 3.597 \cdot \ln\left(\frac{E}{m_K^{1/3} \cdot d_H \cdot d}\right)\right]} \cdot \quad [-]$$

Where: E = impact energy, m_K = body mass of victim, d_H = thickness of body wall, d = calibre of projectile (in m, kg and s).

Using this equation in combination with exterior ballistics calculations, it is possible to present the probability of death resulting from use of a non-penetrating "non-lethal" projectile, as a function of distance. Fig. 4-55 contains curves of this type for three of the four projectiles listed in Table 4-20. The fourth projectile yields very low values. Such a diagram clearly shows the extent to which the description "non-lethal" does or does not apply to a particular projectile.

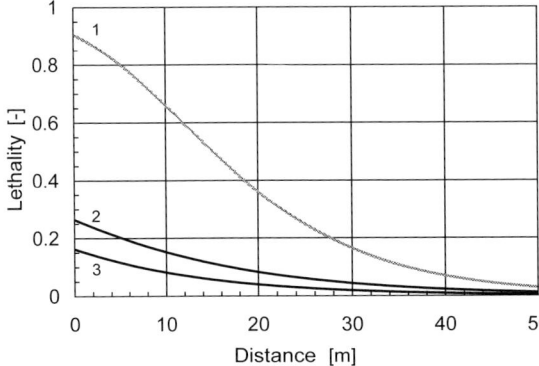

Fig. 4-55. Probability of a "non-lethal" projectile from Table 4-20 proving lethal (STURDIVAN (from Eqn 4.5:1)):
1: 12-bore rubber sphere;
2: 35 mm MR-35 Punch;
3: 44 mm "Flash-Ball".

4.5.4 Dangerosity of projectiles

4.5.4.1 Criteria of dangerosity

Unfortunately, one often hears it said that fragments and projectiles that are not effective are not dangerous either. But even if a projectile fails to meet any of the effectiveness criteria mentioned, it should not be classed as "not dangerous." This means that in order to describe the dangerosity of projectiles we require specific criteria based on the effects they may have on the human body.

In the case of rifle bullets, handgun bullets and fragments, one could adopt the following criteria, for instance: A projectile (be it a bullet or a fragment) is non-hazardous if:

A   It is highly probable that it will not cause any injury to the unprotected human body. This criterion applies to civilians not involved in an incident.

B   It is highly probable that it will cause only minor injury (requiring no more than out-patients treatment). This criterion applies to persons who, in the execution of their duties, are involved in an incident in which firearms may be used.

In order to satisfy Criterion A, the projectile may not penetrate the body and may not cause subcutaneous injury (bruising). As a consequence, its energy density may not exceed the limits for skin or eyes.

According to BIR et al. (2005) the threshold for skin (0.1 J/mm^2) that has been mentioned at a number of points is a valid practical danger threshold (probability of penetration < 5 %). By contrast, the limits for the eye (0.06 J/mm^2 for spheres and 0.02 J/mm^2 for cubes) are mean values. A projectile may only be classed "non-hazardous" if it carries substantially less energy than these limits – approximately 50%.

Experience with Softair guns has shown that spheres with an energy density of just 0.025 J/mm^2 are capable of causing irreversible eye injuries.

A projectile can still satisfy Criterion B if it causes minor injury. This means that a projectile may exceed the threshold of 0.1 J/mm^2, as long as its energy level is so low that penetration does not exceed approx. 2 cm. This means, however, that the maximum permissible energy is just a few Joule. It depends primarily on sectional density and can be estimated using Eqn 4.4:20, taking 0.02 m for ℓ.

4.5.4.2 Determining hazard areas

The question of the danger area associated with fragmentation ammunition and bombs is often raised when assessing the safety of shooting ranges and the danger to one's own side when such weapons are used.

The following points regarding fragmentation grenades apply mutatis mutandis to ricochets and stray bullets near shooting ranges.

The energy and energy density of a fragment can be determined as a function of distance travelled using exterior ballistics calculations based on initial velocity (the vector sum of exit velocity and the velocity of the grenade or other explosive device at the time of detonation).

The exit velocity of a fragment is virtually independent of its mass. Exit velocity depends exclusively on the ratio of fragment mass to mass of explosive (Gurney equations, see 2.3.6.1). However, light and heavy fragments lose quite different amounts of speed along their trajectories. It would therefore seem logical to include fragment mass when determining dangerosity. We can present energy as a function of distance for various fragment masses in graphical form (see Fig. 4-56). This presentation allows us to draw threshold curves for different dangerosity criteria.

Fig. 4-56 shows curves for the mean threshold energy density of skin (0.1 J/mm^2) and the eye (0.03 J/mm^2) (Criterion A). This graph also shows the threshold curve for Criterion B, i.e. 2 cm penetration in soap (dashed line). For very light fragments, this criterion leads to substantially smaller hazard distances.

This type of diagram (which has to be drawn separately for each device) allows us to derive hazard distance as a function of fragment mass.

For instance, a 0.1 g cuboid fragment with an initial velocity of 1800 m/s is dangerous at up to approx. 55 m according to the 0.1 J/mm^2 criterion but up to approx. 72 m according to the 0.03 J/mm^2 criterion(see Fig. 4-56).

If we also know the distribution of fragment masses for a device (from explosion tests) then for a given surface (e.g. a standing person) we can calculate the probability of a dangerous hit as a function of distance from the point of detonation.

4.5.4.3 Danger area for persons wearing protective equipment

In a similar manner, we can calculate danger areas for persons wearing protective equipment (helmet, visor and protective clothing), together with the probability of penetration. In order to carry out these calculations, we must first determine experimentally the minimum energy density (or energy) for penetration as a function of fragment mass. This involves determining the mean energy density required to penetrate the equipment, together with the corresponding standard deviation.

Whether or not a projectile will penetrate a protective material (other than ceramics and glass) depends primarily on the energy density of the projectile. In the case of ceramics and similar, vitreous materials, the impact energy is the deciding factor.

Very light fragments constitute an exception to this rule, as they can pass through the weave of woven protective materials at low energy levels.

Hazard probability as a function of distance from the point of detonation helps wearers of protective equipment (e.g. those responsible for disarming bombs and mines) to assess risk and hence also to plan their work.

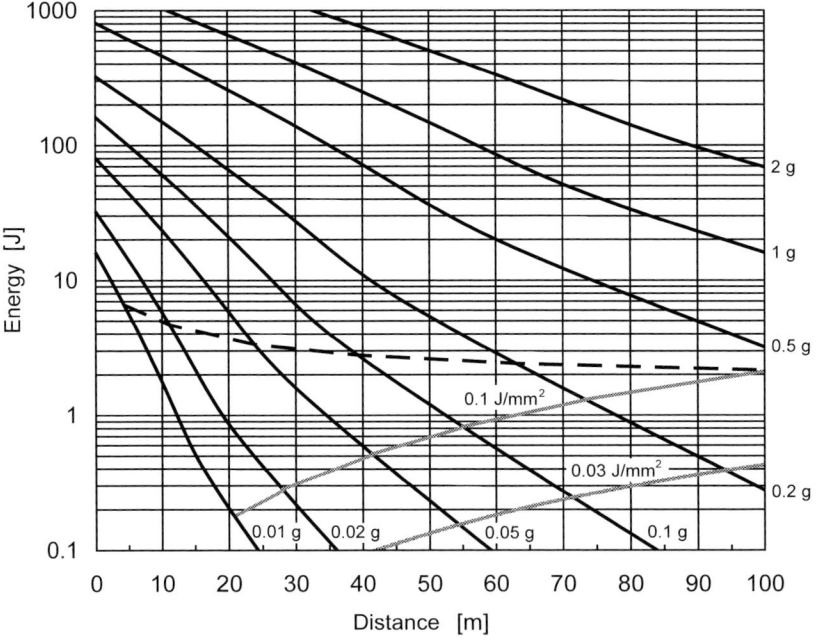

Fig. 4-56. Distance/energy diagram for a cuboid fragment with an initial velocity of 1800 m/s. This graph shows curves of the energy density required to penetrate skin and of that required to cause eye damage (Criterion A, grey lines). This graph also shows the threshold curve for Criterion B, i.e. 2 cm penetration in soap (dashed line).

5 Wound ballistics and forensic medicine

5.1 Conventional forensic medicine
M. A. ROTHSCHILD

5.1.1 General

Analysing gunshot wounds is one of the many roles of forensic medicine. The preceding chapters have shown clearly that the multiplicity of physical parameters and effects renders forensic analysis of such wounds quite complex. The short duration of the processes involved (a bullet wound occurs within a few milliseconds) plays a decisive role here.

Rather than discussing bullet wounds systematically, in the manner of a forensic textbook, we shall build on the material contained in Chapters 2, 3 and 4, taking a wound ballistics approach. The questions that guided the drafting of this chapter were: How can wound ballistics help the forensic pathologist produce his report? What can the forensic pathologist learn from wound ballistics?

5.1.2 Crime-scene investigation

5.1.2.1 Bullet damage at the crime scene

The great advantage of performing a forensic examination on a body at the crime scene (or at the location where the body is found) is that a good number of details relevant to the pathologist's report will be available on the spot. The forensic pathologist's first task will be to gain an overview of the situation at the scene and obtain some basic information from the police to guide his or her work. The more shots were fired, the more weapons were used and the more persons were killed or injured, the more complex and difficult will be the investigation and reconstruction process.

A visual inspection of the scene by the forensic pathologist will give a first indication as to whether a bullet may have lost a relevant percentage of its energy before hitting the target. This may be indicated by damage to materials through which the bullet has passed, such as windows, doors or furniture. Ricochet marks on the floor or walls, for instance, are also an important indicator of modified bullet behaviour. A bullet that has passed through another medium or ricocheted off an object before hitting the body will not only strike the body at a lower velocity but will also in many cases have undergone deformation and have struck the body at a greater angle of incidence. This will result in an increased impact area and hence in a reduced sectional density. As a result, the entry wound may be irregular and the narrow channel significantly shorter than usual, with a disproportionately large temporary cavity.

If a bullet passes through a person lying on or leaning against a solid surface, it may rebound into the wound channel via the original exit wound (RESCHELEIT et al. 2001). It is therefore particularly important to make a note of marks indicating that a bullet has rebounded from a wall or the floor. Damage of this sort in combination with bloody contact marks or smearing is a particularly strong indicator of such a phenomenon. This constitutes grounds for extreme caution when identifying a wound as an entry or exit wound (see 5.1.3).

When considering the posture of the victim at the time of impact, it is often wrongly assumed that he or she was standing upright, in a fairly static position. The situation in the autopsy room, where the body is lying on its back on a table, makes it all the more tempting to assume this. However, if the bullet has passed through the abdomen from front to rear, the victim may have been standing upright, seated, kneeling, jumping or even prone. Even if the wound channel(s) is or are correctly recorded at autopsy, this information alone does not allow for reliable reconstruction of the shot(s); further information is required regarding the situation at the scene. The position of the wound channel, of blood traces at the scene and of the point at which the bullet first struck another object after passing through the victim generally make it possible to narrow down the position of the victim and of the shooter at the moment of the shot to a reasonable degree. If more than one bullet has passed through the body, it may be especially useful for the forensic pathologist to revisit the scene following the autopsy.

5.1.2.2 Examination of the body at the scene

A visual inspection of the body at the scene will give the forensic pathologist an initial impression and overview. The significant factors are the position of the body relative to its surroundings, its posture and the state of clothing.

Blood trickles can indicate the posture that the victim adopted immediately after being struck by the bullet. Traces of blood on the soles of the feet or shoes indicate that the victim may have been capable of walking through blood on the floor after suffering a wound that caused bleeding.

It is essential to ensure that no projectiles are lost when turning the body over or placing it on a stretcher. It is not unusual for a bullet to reach the end of the wound channel with enough energy to exit the body but not to pass through all layers of clothing. Such a bullet may slip out of the clothing when the body is moved and may possibly be missed. This applies in particular to examinations conducted outdoors. Examination of a body outdoors should be kept to a minimum, and consideration should be given to placing the body on plastic sheeting.

Depending on the situation at the scene, it may be advisable to put individual items of clothing to one side before the body is moved, in order to protect any traces of gunshot residues.

5.1.2.3 Bloodstain pattern analysis

How much blood is to be expected at the scene depends on a number of factors. The entire system – in particularly the ammunition used – will determine the morphology of the entry wound and of any exit wound. Clothing and, above all, the body parts/structures concerned, have a major effect on visible blood loss.

In the case of contact shots and close range shots, backspatter (involving both blood and tissue) is a well-known phenomenon, and one of relevance to the reconstruction process. In particular, KARGER et al. (1996, 1997) and FACKLER (cited in BEVEL and GARDNER 2002) established that this is not a direct consequence of the action of the bullet, i.e. that the bullet does not expel this blood directly. High-speed photography showed that the blood was not ejected from the entry and exit wounds until a certain time after the bullet had emerged from the exit wound. The acceleration of this backspatter is due to a combination of factors: Muzzle gases flowing out of the powder gas cavity, together with the collapse of the temporary cavity, eject blood and tissue via the entry wound. Backspatter may therefore be particularly marked in the case of close range shots to the head, as the flat bone under the skin favours the creation of a gas cavity.

In extensive experiments that involved firing 9 mm Luger bullets into calf heads, KARGER et al. (1996, 1997) noted that in the case of contact and close range shots blood emerged from the skin approximately 0.7 to 4 ms after impact and that the initial velocity of the blood lay between 10 and 100 m/s. Most of this backspatter was found within 50 cm of the entry wound, distributed in the form of a hemisphere. If suitable surfaces are present, it is not unusual to find clouds of very fine blood droplets. These may occur both at the entry and the exit wound, as the direct result of the action of the bullet, taking the form of an aerosol of minute blood droplets (diameter under 0.1 mm). The very small size of the droplets means that they do not travel very far (less than 50 cm). The larger backspatter droplets (larger than 0.5 mm) are far less numerous and can be projected up to 1.2 or 1.5 m further. This can give an indication of the distance between the part of the body concerned and the surface affected and, under certain circumstances, the position of the body (KARGER et al. 1996, JAMES and SUTTON 1998, BEVEL and GARDNER 2002).

Other types of tissue may be included in backspatter, such as skin, hair, subcutaneous fatty tissue, muscle and bone fragments. These may leave traces at greater distances. The wide variation in the findings with regard to backspatter are reason to be careful not to over-interpret their meaning. Ultimately, backspatter on an object only tells us that the object must have been near the entry wound at the time.

Backspatter is of particular value in determining who was holding the weapon at the time the shot was fired (YEN et al. 2003). Together with gunshot residues (which are not always a reliable indicator), backspatter may give a first indication at the scene. In the case of suicides, backspatter could be found on the victim's hand in up to one third of instances, the exact percentage depends, varying from one investigator to another (BETZ et al. 1995). In the vast majority of cases, the front part of the barrel is even closer to the entry wound than the hand (except in cases where the other hand was used to steady the barrel). As a result, backspatter was found even more frequently on and inside the barrel (BRÜNING 1934, STONE 1987, 1992).

The bloodstain pattern from the exit wound is generally more symmetrical than the pattern caused by backspatter from the entry wound. Whether this has to do with the position of the temporary cavity is currently still unclear.

A further phenomenon may occur when several bullet wounds lie close together. If a bullet strikes the body close to an existing bullet hole, the stretching and shearing of tissue that accompanies the creation of the temporary cavity may compress the blood-filled wound channel and eject the blood it contains. This often results in large spots of blood being found a short distance from the source.

From the above points, it will be clear that it is impossible to predict the nature and intensity of bleeding from entry and/or exit wounds. As DIRNHOFER is supposed to have said: *"Every shot is different."* The type of ammunition used affects the morphology of the external bullet wound and the creation of the temporary wound cavity. When a projectile strikes an unclothed part of the body, it initially ejects skin particles in the form of a spray. A depression is created in the wake of the bullet, with a vacuum forming in the wound channel as the bullet penetrates further. This draws tissue and blood into the temporary cavity as it is created, behind the bullet. A bullet that strikes the body in stable flight and undergoes only minimal deformation is unlikely to cause significant direct bleeding. On the other hand, a bullet with a lower sectional density will probably cause more intensive backspatter, as it will cause a larger entry wound.

From 4.4.2.4, we know that the diameter of an entry wound is inversely proportional to the square root of the sectional density (as a first approximation).

If the entry wound is in the vicinity of a body cavity, such as the chest or abdominal cavity, one should not necessarily expect to find any significant visible bleeding, as most of the blood leaving the damaged blood vessels can collect in those cavities. Whether the blood exits the body via a bullet wound under the influence of gravity or remains mainly in the body will depend on the final position of the victim.

Bleeding is likely to be heavier if the bullet passes through the body than if it remains inside. Even so, the ammunition used, the type of bullet, its residual energy, its orientation and its shape when it leaves the body will all have an effect, as will the body part/structure affected and the clothing worn. If the bullet leaves the body, one can expect to see bleeding. If the exit wound lies in the narrow channel, one will generally observe less direct bleeding than if the bullet leaves the body when the temporary wound channel is at its maximum diameter.

Contact shots to the head result in particular phenomena. In such cases, blood and tissue backspatter is often to be found not only on the outside of the weapon or on the hand used, but also in the barrel. The cuffs of suspects' clothing should therefore be examined for traces of the victim's blood and tissue.

5.1.3 Morphology of entry and exit wounds

5.1.3.1 Entry wounds

The morphology of the entry wound is of particular importance in assessing bullet wounds. This element provides indicators regarding the distance between muzzle and victim, the angle of impact and any special characteristics of the bullet. Even in the case of lightly-clothed body surfaces, the morphology of the entry wound will provide a wealth of information.

An entry wound caused by a full metal-jacketed bullet has a number of typical characteristics (concentrically from the centre outwards): *1*. A central skin defect surrounded. *2*. A circular bullet wipe. *3*. Contusion ring. *4*. Margin of distension.

1. **Central skin defect**, caused by direct contact with the bullet. By contrast with the exit wound, it is not usually possible to make the edges of the entry wound match up. The bullet crushes the underlying skin tissue on impact, and drags it far into the wound channel (GROßE-PERDEKAMP et al. 2005). A hole (often circular) is created in the tissue. As the tissue is accelerated away from the bullet, however, the skin is stretched briefly to a diameter greater than that of the bullet as it passes through, before contracting again due to its own elasticity after the bullet has passed. As a result, the diameter of the opening is generally somewhat smaller than that of the bullet (STRASSMANN 1885). Entry wounds in particularly hard areas of skin (such as the hands and the soles of the feet) may be particularly small (POLLAK 1980). The flatter the angle of attack, the more elliptical the bullet hole.

2. **Bullet wipe**. This is caused by matter being wiped off the surface of the bullet as it enters the skin (STRASSMANN 1885).

As it passes along the barrel, the bullet picks up oil, together with propellant and explosive from previous shots. As it penetrates the skin, it wipes these deposits off onto the entry hole (HABERDA 1919). In addition, propellant overtakes the bullet via the rifling as it passes down the barrel, and this too is deposited on the bullet.

The result is a dark, shiny ring around the entry wound, generally about 1 to 3 mm wide. The flatter the angle of attack, the more elliptical the ring, as in the case of the entry wound. The external diameter of the ring gives us some idea as to the calibre of the bullet. It does not indicate the exact calibre, as the interaction between the surface of the projectile and the surface of the skin on impact is highly dynamic. In the case of a corpse, there will also be the effect of drying out, which will cause the skin to shrink, and with it the entry wound and related effects.

3. **Contusion ring**. The appearance of this ring gives the impression that it was caused by the bullet invaginating the skin and then abrading it, hence the name. It is above all thanks to SELLIER (1969, 1975, 1982) and his high-speed photography, together with POLLAK (1982), POLLAK and ROPOHL (1991) and THALI et al. (2002) that we now have impressive evidence for the actual causes of this ring-shaped skin wound: The sudden crushing effect displaces skin surface particles and parts of the dermis in front of and adjacent to the tip of the bullet tangentially, at high speed, leaving a ring-shaped tissue defect. In addition, the radial acceleration of the tissue as the projectile penetrates causes the skin to be overstretched briefly in a manner analogous to the phenomena observed in the temporary cavity. This results in tiny radial stretching cracks in the epidermis, accompanied in part by tears visible with the naked eye, which favours the drying process. The result is a change to the skin, some 2 to 3 mm wide, caused by surface abrasion and stretching. In the case of a corpse, this ring is easily recognisable from the brownish-red desiccation that has occurred.

4. **Margin of distension**. This reddish ring is contiguous with the contusion ring. It is generally smaller than the contusion ring and forms the boundary of the skin stretched by the radial acceleration forces. External stretch lacerations are absent. In a manner similar to that observed in an extravasation zone, the capillaries of the skin are broken. Examination under a magnifying glass generally reveals an absence of homogeneous colouration, the effect being caused by large numbers of petechial haemorrhages.

Depending on the situation, there may be considerable variation in the above-mentioned characteristics of an entry wound. The form of the entry wound will depend on the design of the bullet, the distance from which it was fired, the angle of attack and any materials through which the bullet passes between the weapon and the victim.

The form and structure of the bullet will determine the sectional density and the amount of energy transferred. The experiments using glycerine soap discussed in 3.2 show how a bullet with a high energy level and low sectional density creates a large wound cavity that starts right at the entry wound, which may result in a large, open entry wound. The same effect may be caused by an unstable bullet (e.g. one that has struck a surface before hitting the victim, or one that has passed through another medium) or by an unfavourable interaction between bullet and weapon (the development of the US M16 being a case in point – see 3.2.3.3). In the case of unstable handgun bullets, the lower energy level by comparison with rifle bullets means that changes in entry wound morphology will be limited to the entry wound being elongated or irregular (in the case of ricochets and other bullets that strike the body side-on) or an absence of bullet wipe (if the bullet has passed through another medium before hitting the body). As a result, these entry wounds can look very similar to exit wounds.

Multiple hits from the same bullet constitute another special case. There have been a number of cases in which a single bullet passed through the trunk and then hit the arm or leg of the same person (or vice versa). When bullets are fired into a crowd, during mass executions or in other situations where people are located close together, a bullet may pass through the body of one person and then come to rest in the body of another. Bullet wipe is often absent where a bullet re-enters a body. Depending on the ammunition used, and the length of the wound channel through which the bullet has already passed, it may strike the body tip-first at a slight angle (producing a round or elliptical entry wound), side-on or tail first (producing a large, elongated, irregular entry wound). In the case of higher-energy rifle bullets in particular, the position of the temporary cavity plays a major role in determining the morphology of the second entry wound.

The area of the body also affects the morphology of the wound. Especially in the case of areas covered in hard skin, such as the palms of the hands and the soles of the feet, the entry wound is usually small, with no marginal abrasion (POLLAK 1980), sometimes with a bullet wipe under the raised epidermis (SIGRIST et al. 1992). Where a flat bone lies underneath the skin (especially in the case of the skull) there may be significant tearing at the perimeter of the wound, regardless of the distance from which the shot was fired.

Given the above points, an experienced investigator generally has no difficulty differentiating between entry and exit wounds. Bullet wipe and signs of a bullet fired from close range indicate an entry wound. Signs of a bullet having been fired from close range include above all traces of gunshot residues on the skin around the wound. In the case of contact shots there may be the imprint of the muzzle and powder gas cavities and in the case of close range shots there may be heat-related changes to the skin and tissue.

The lower pressure behind the bullet sucks explosive and propellant particles a certain distance into the wound. Under certain circumstances, these particles may be found in the wound channel or even at the exit wound. The concentration of particles decreases rapidly with distance, which means that it is almost always possible to determine the direction in which the bullet travelled by looking at how the quantity of gunshot residues present changes over the length of the wound channel. The same applies to textile fibres in cases where the bullet has passed through clothing. Regardless of the distance from which the shot was fired, the bullet often drags textile fibres into the wound and the vacuum created by the temporary cavity sucks them in further. For the same reason, it is not unusual to find textile fibres at the exit wound, albeit at lower concentrations, as STRASSMANN already pointed out in 1912.

At very short ranges, primary or secondary flash (see 2.3.3.2) causes short-lived thermal effects in the area of the entry wound. The high concentrations of carbon monoxide in the propellant gases transform haemoglobin (Hb, red blood cells) and myoglobin (Mb, red muscle cells) into CO–Hb (carboxy-haemoglobin) and CO–Mb (carboxy-myoglobin). This effect may be more or less marked, depending on the ammunition and weapon used (especially the length of the barrel). However, as the flame only acts on the skin for a few milliseconds, major damage is unlikely, and damage at a significant depth more so. As a result, significant concentrations of CO–Hb and CO–Mb are found only in the uppermost thin layers of tissue around the entry wound. Nonetheless, increased levels of carbon monoxide (CO) may also be observed in the vicinity of the exit wound (WOJAHN 1968). This is because not only the thermal components but also the concentration of CO in the propellant gases plays a role. In the case of a shot from close range, propellant gases will enter the wound channel. These will be sucked deep into the wound channels by the temporary cavity, bringing the inner layers of tissue at the exit wound into contact with these gases and the CO they contain. It sometimes happens that routine measurements of CO in tissue at the entry and exit wounds reveal higher CO-Hb concentrations at what is known (from other sources) to be the exit wound than at the entry wound. This means that CO levels cannot be used to determine the direction of shot.

It would be pointless to differentiate between entry and exit wound purely by comparing their respective sizes. The often-heard statement that entry wounds are smaller than exit wounds applies only to harder, more brittle materials, and even then only under certain conditions. The size of the entry and exit wounds is determined by how the amount of energy transferred by the bullet changes as it passes along the wound channel. Depending on the ammunition and the orientation of the bullet on impact (e.g. high-energy bullet with a low sectional density or high angle of incidence), a large temporary cavity may be formed on entry, leading to a large entry wound. In the case of a contact shot, the pressure of the propellant gases has an additional effect, often causing a large, torn entry wound. Conversely, if the wound channel is short the exit wound may lie in the narrow channel or, in the case of a longer wound channel, it may lie beyond the temporary cavity. In such cases the exit wound will be small. In the case of fragment wounds, the entry wound is always larger than the exit wound.

5.1.3.2 Exit wounds

Exit wounds also display a high degree of variation. The mechanism by which they are created differs from that of entry wounds. The projectile exits from the interior of the body and the skin bulges outwards before breaking. As a result, exit wounds often take the form of a slit and are irregular. Bullet wipe is usually absent.

Bullets that lack the energy density to penetrate the skin from the inside are often found just under the skin. Their position is sometimes indicated by a small bruise, and it is often possible to feel them from outside. If the bullet has just enough energy to exit the body, the skin often tears along a natural tear line, which can lead to a slit-shaped tear with smooth edges, which on first examination can easily be mistaken for a stab wound (see Fig. 5-1).

If a bullet that has undergone more deformation or fragmentation has just enough energy to leave the body, the exit wound will be correspondingly irregular and torn. Unlike the edges of entry wounds, those of exit wounds can generally be matched up, as there is generally no significant removal of matter. Matter is lost only when a bullet exits at high speed, in the case of an exit wound co-located with a large temporary cavity or where the exit area was in contact with a hard surface, e.g. when a bullet passes through the trunk of a person lying on their back on a stone floor, when the exit wound is in the back or possibly when the person is wearing heavy, close-fitting clothing. In such cases, the skin is crushed at the exit wound by being trapped briefly between the projectile and the surface behind. In such cases, it is no longer possible to match up the edges of the wound, as tissue has been removed by the crushing process. The epidermis in a ring around the hole undergoes abrasion when forced against the external surface, and this may produce an imitation of a contusion ring (SELLIER 1982, RESCHELEIT et al. 2001). However, this generally does not complicate differentiation between the entry and exit wounds, as the projectile remains in the vicinity of the exit wound in cases where the surface of the body was touching a hard surface (such as stone or concrete) at the time of the shot.

5.1.3.3 Grazing shots

A grazing shot is one in which the bullet passes along the surface of the body without penetrating. Injury to the surface of the body is generally limited to the skin itself and to the subcutaneous connective tissue. However, the subcutaneous fatty tissue may be affected, as may the upper layers of muscle, depending on the area of the body concerned and the thickness of the fatty tissue. The resulting defect often takes the form of a channel, with abraded tissue. The shorter the area of contact between projectile and skin, the smaller and more insignificant-looking such a graze is likely to be.

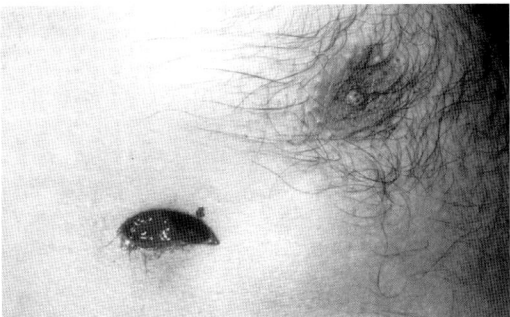

Fig. 5-1. Exit wound in the form of a slit, very similar to a stab wound (ribcage).

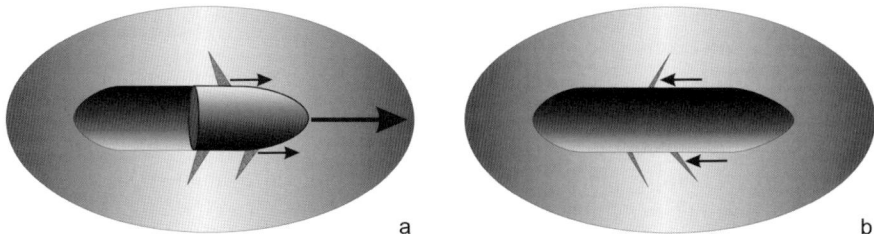

Fig. 5-2. Grazing shot: **a.** The skin is dragged forwards by the bullet, generating shear stresses in the skin near the channel. **b.** This results in small tears in the epidermis on both sides of the channel. Once the skin is no longer under tension, these tears point in the direction of shot.

Such grazes, while comparatively harmless from the victim's point of view, may be useful in determining the direction of shot. DIXON (1980, 1982, 1984a,b) has looked at this topic in detail, and pointed out the presence of small, lateral tears in the skin. As the bullet scrapes over the skin, it drags the skin along with it for a short distance (see Fig. 5-2 a). The skin in direct contact with the bullet is abraded, while the neighbouring areas are subjected to shear stresses, which may be considerable in some cases, and may tear. This results in small tears in the epidermis on both sides of the channel, pointing in the direction of travel (see Fig. 5-2 b).

5.1.3.4 Indicators of muzzle-target distance

Determining muzzle-target distance is of considerable significance in reconstructing an incident. This applies not only to cases in which another person is known to have been involved, but also to accidents and suicides. For instance, if a suspected suicide victim has suffered a bullet wound to the temple from a distance greater than the length of their arm, the police and prosecution service will initially have to assume that another person was involved.

Conventionally, muzzle-target distances are divided into three main ranges, based primarily on the density and distribution of gunshot residues on the surface of the body (skin and/or clothing): 1. Contact shot. 2a. Close range shot. 2b. Intermediate range shot. 3. Distant shot. The largest number of wound ballistics phenomena are observed in the case of a contact shot.

1. A **contact shot** is one in which the muzzle is in contact with the body. At very short distances between muzzle and skin (up to a few mm), the typical characteristics of a contact shot will also be present. Such shots are also classified as contact shots therefore.

If the muzzle is in contact with the body, large quantities of propellant gas and particles will accompany the bullet into the entry wound. The gases – which are under high pressure – can follow close behind the bullet, partly on account of the temporary cavity. In addition, the skin in the vicinity of the entry wound suffers undermining, with gunshot residues spreading into the subcutaneous fatty tissue

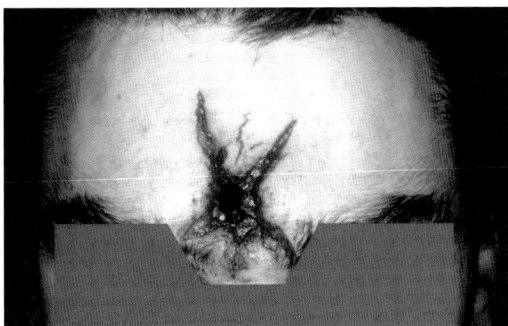

Fig. 5-3. Contact shot to a flat bone under soft tissue. The star-shaped wound is a typical indicator.

or muscle (depending on the anatomy of the area). The pressure may also generate radial tears centred on the entry wound. Primary flash (see 2.3.3.2) may also cause heat-induced changes in the tissue. The effects are particularly pronounced where flat bones lie beneath the soft tissue (e.g. skull and sternum). At such locations the bone impedes the penetration of propellant gases into the body, increasing the pressure effects on the soft tissue, which often lifts away from the bone. The periosteum around the hole in the bone is generally "blown" away radially and its underside (and the outer surface of the bone) will show black gunshot residues marking, the degree of such marking depending on the ammunition used (FALLER-MARQUARDT et al. 2004). See Fig. 5-3.

The pressure in the powder gas cavity under the skin can be so high that with the muzzle in contact (or nearly so) the skin is pressed against the muzzle with considerable force, causing abrasion. The pressure and abrasion effects cause minute losses of epidermis and micro-tears, leaving an image of the muzzle and possibly the area surrounding it. WERKGARTNER (1924, 1928) gave such an imprint the name of "Waffengesicht" (literally, "weapon face") in German and HAUSBRAND (1943) and ELBEL (1958) conducted intensive experiments on such marks. Under favourable circumstances, these imprints make it possible to determine not only the exact positioning of the weapon (e.g. from the imprint of the foresight) but also, in some cases, to identify the type of weapon used (see Fig. 5-4).

As the clarity of the traces left by a contact shot are determined above all by the quantity of propellant gas and its pressure, these indications will be more pro-

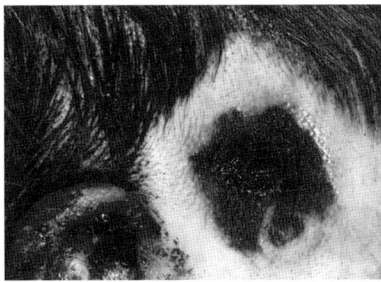

Fig. 5-4. The muzzle impression that is often visible in the case of a contact shot. The skin is pressed up against the muzzle by the pressure in the powder gas cavity. Here we can see the imprint of the foresight, the bolt spring rod and a square slide. The weapon was a Glock 9 mm Luger pistol.

nounced if the weapon is pressed firmly against the body, the volume of the barrel is smaller (i.e. the barrel is shorter) and the propellant load is greater (see Table 2-12). Conversely, no signs of a contact may be visible to the naked eye in the case, for instance, of a contact shot from a 22 revolver with a 4-inch barrel, or a 6.35 Browning pistol.

A contact shot to a clothed area of the body usually causes radial tears in the cloth. Thermal damage generally affects only a limited area. If the muzzle is only in light contact or a short distance from the body, however, limited singeing of the cloth and superficial burns to the skin may result.

2a. In the case of a **near-contact shot**, one will observe both morphological indicators of a close range shot and gunshot residues in the form of molten particles of metal and combustion residues. Increased concentrations of CO–Hb and CO-Mb will be found in tissue around the entry wound at distances of up to 5 or 15 cm (WOJAHN 1968). These are caused primarily by the high concentration of CO in propellant gas (see 2.2.2.1 and 5.1.3.1), a phenomenon noted by PALTAUF as early as 1890. At closer ranges, primary flash may also play a role. The relatively short duration of primary or secondary flash means that major direct thermal damage to the skin in the vicinity of the entry wound is not to be expected, although cloth may undergo significant damage (BERG 1959, BONTE and KIJEWSKI 1976). Aside from the effect of muzzle flash, the degree of thermal damage to clothing depends primarily on the type of propellant and the type of fabric. Natural fibres such as cotton are relatively heat-stable and do not melt, though they may continue to smoulder after burning. This often results in a wider area of damage. Black powder is likely to create impressive thermal damage due to penetration by hot powder particles. Synthetic fibres such as polyamide and polyester can melt, but by doing so they prevent smouldering, which means that the area affected by heat will be smaller. The skin may suffer secondary heat damage caused not by primary flash itself but indirectly by the fabric (POLLAK 1982, POLLAK and STELLWAG-CARION 1988, ROTHSCHILD et al. 1998).

2b. In the case of an **intermediate range shot**, propellant particles accelerated by the muzzle gases may still be visible, but in contrast to a close range shot there will be little or no trace of gunshot residues. The sensitive technology available today makes it possible to identify an intermediate range shot at distances of up to 1 to 3 m, depending on the weapon (barrel length) and ammunition (type and quantity of propellant and explosive).

3. If there are no signs of a close range shot, and if normal means of detection do not reveal any propellant residues, the shot is classed as "distant." With current technology, this applies to shots fired from approx. 3 m or more. In such cases, it is only possible to estimate muzzle-target distance under particular conditions, such as when one knows what weapon and ammunition were used and when the wound channel has been precisely recorded (including length and whether or not it passes through bone). The loss of energy (and hence of velocity) of the bullet

can be calculated per cm of wound channel, right back to the entry wound. This yields the impact velocity, and if the muzzle velocity is known, ballistics tables can be used to narrow down the range of possible muzzle-target distances (SELLIER 1982).

5.1.4 The wound channel

5.1.4.1 Wound morphology

As a bullet passes through a body, a number of dynamic processes take place, not only in the tissue in direct contact with the bullet, but also behind and in front of the bullet (see 3.2).

When the bullet strikes the body, it initially creates a shock wave, which propagates through tissue at the speed of sound. Even though the pressure amplitudes are high, the velocity means that the effect is very short-lived. As a result, no tissue can be displaced, and no significant injuries related to the shock wave have so far been identified. We shall therefore not discuss the shock wave further in this context (see 4.3.2.3 regarding the significance of the shock wave as regards tissue). Two fluid dynamics processes are of great significance, however. These occur in parallel when a bullet penetrates and passes through a body: 1. Crushing. 2. Stretching/shear.

1. Crushing results from direct contact between body tissue and the leading surface of the bullet. Tissue is deformed inelastically, torn, compressed and destroyed. This results in a permanent wound channel, which is clearly visible at dissection and is filled with blood and destroyed tissue particles.

2. Tissue is pushed away radially from the leading surface of the bullet, with major displacement effects in the surrounding tissue. This phenomenon is especially significant in the case of rifle bullets (see 3.2.2.1). Simultaneous stretch and shear movements cause a temporary cavity to form behind the bullet. The compression of tissue that results from this stretching has significant wounding potential. Pulsations in the temporary cavity (see 3.2.3.1) and the presence of boundaries between different types of tissue create additional shear forces, which cause the tissue to tear if the limit of elasticity is exceeded. Wound pockets are created, typical of a temporary cavity, pointing radially outwards from the permanent wound channel.

Even though the temporary cavity eventually collapses, most of the tissue in this area is also destroyed. The tissue in the permanent wound channel immediately adjacent to the geometrical wound channel undergoes extensive tearing and devitalization (see 3.2.3.1 and Fig. 3-19 b, c, Zone 1). In the next zone, surrounding the temporary cavity, we mainly observe histological evidence of compression effects with minor bleeding into the tissue, but no damage visible to the naked eye (extravasation zone, see Fig. 3-19 b, c, Zone 2). The tissue around this zone is stretched, but remains undamaged (Fig. 3-19 b, c, Zone 3).

From the above, it is clear that (aside from the characteristics of the ammunition and weapon) the degree of tissue damage depends above all on the tissue itself. The more elastic the tissue, the better it will be able to compensate for the stretch and shear forces generated when the temporary cavity is created, up to a certain point. Blood vessels, muscle, connective tissue, skin and such internal organs as the lung and bowel are fairly elastic. Such parenchymatous organs as the liver, spleen, kidneys and brain are inelastic.

5.1.4.2 The relationship between the wound channel and the direction of shot

If it is not immediately possible to differentiate between the entry and exit wound (e.g. if the body has been in water for an extended period, or has been burned, or has decomposed), the edges of wounds to bones often give a good indication of the direction of shot. However, if the bullet has not struck bone, or if it is not possible to draw reliable conclusions from bone damage, the only indicator is damage to soft tissue in the vicinity of the wound channel. However, the amount of information that can be derived from this is extremely limited. While entry wounds in bone are generally smaller than exit wounds, this is not the case with soft tissue. Even in the case of inelastic soft tissue such as the liver, there will rarely be a funnel-shaped exit wound such as one often finds in bone.

One should also be cautious about drawing conclusions from the positions of bone and bullet fragments along the wound channel. When the temporary cavity is created, the stretch forces – and especially the shear forces – tear pockets in the tissue, in which a vacuum exists briefly. Furthermore, when a bullet hits a hollow bone, hydraulic pressure is generated in the marrow. As a result, bone fragments are to be found in tissue both in front of and behind the bone (LORENZ 1948, KARGER et al. 1998). However, the majority of bone fragments will lie on the exit side of the bone, where the vacuum created by the temporary cavity and the direct displacement of bone fragments through contact with the bullet combine. If metal is abraded from the bullet through contact with the bone, the resulting particles may be found both in front of and behind the bone. The vacuum created by the temporary cavity may suck metal particles from the bullet right through to the exit wound, at the end of the wound channel. If the jacket material is soft, and if the bullet is travelling at over 600 m/s, the bending and compression stresses caused as the bullet yaws may crush or break it, and this may result in lead particles being squeezed out. Under certain circumstances, these may be found closer to the end of the wound channel, depending on the position of the maximum temporary cavity (see 3.2.2.1).

However, the above-mentioned forces may lead to erroneous conclusions. When performing a reconstruction of a shooting death, another factor of importance (alongside the differentiation between entry and exit wound) is the angle of shot. As it is not always possible to determine this from the morphology of the entry wound, the track of the wound channel in the body may be of considerable significance. One must exercise caution in the case of long, slender bullets, as their direction of travel can easily change by up to 30° in the vicinity of the maximum temporary cavity (see Fig. 3-8).

5.1.5 Bullet wounds to the head

5.1.5.1 Brain injuries

The brain links together a number of vital structures. Direct damage to these structures causes instant death. On the other hand, the brain also contains a number of structures that can be destroyed without threat to life. Despite this, the vast majority of bullet wounds to the head encountered in the field of forensic pathology are fatal. This is because of the particular structure of the head from a wound ballistics point of view.

1. Brain tissue is particularly inelastic and hence has little capacity to absorb the radial displacement of tissue caused by the temporary cavity. This results in tearing of the brain tissue, the extent of which corresponds approximately to the dimensions of the temporary cavity. In his neurohistopathological studies, OEHMICHEN (1985, 1992) demonstrated that the nerve cells are also destroyed and the axons crushed in the area of the temporary cavity.

2. The brain is firmly enclosed by the cranium at almost all points. If a projectile penetrates the skull and causes radial displacement of tissue, it is not possible for this displacement to be transmitted outwards by expansion or deformation. The brain undergoes brief pressure peaks and large pressure gradients, as confirmed empirically by WATKINS et al. (1988) (see 3.2.3.1, Fig. 3-20, 3-21 and Table 3-3) and DITTMANN (1989). This accentuates the compression that the brain undergoes as the temporary cavity is formed, leading to further deformation and shearing of the brain tissue. This explains why even brain structures at some distance from the tissue can sustain damage. Brain structures adjacent to bony edges and projections can easily be pressed up against them, causing secondary brain injury. It is precisely the sensitive regions of the brain that suffer the effects of this mechanism, such as the pons and the brainstem, with their vital breathing and circulatory centres. These structures are located at the foramen magnum, and exhibit clusters of small regions of perivascular bleeding, The basal parts of the brain in the frontal, temporal and occipital lobes, together with the cerebellum, are also affected by these indirect, amplified compression effects. This very often results in bleeding due to contusion of the cerebral cortex (HENN and LIEBHARDT

1969, KIRKPATRICK and DIMAIO 1978, ALLEN et al. 1983, FINNIE 1993, KARGER et al. 1998). The pressure generated by the temporary cavity also results petechiae in the conjunctiva, especially in the case of contact and close range shots.

3. Depending on the ammunition used, the relatively hard bone of the skull may cause deformation or even fragmentation of the bullet. The skull may also make the bullet more likely to yaw. All three effects will tend to reduce the sectional density of the bullet, further increasing the amount of energy transferred to the tissue through which it passes.

In most cases, the effects of bullet wounds to the cranium mentioned above are instantly fatal. The bullet may, however, cause death indirectly. Like reactive cerebral swelling, bleeding within the brain causes the brain to increase in volume within the rigid cranium, resulting in fatal compression of the brain stem at the foramen magnum.

5.1.5.2 Skull injuries

In addition to the primary bone defect in the form of a bullet hole, one very often observes secondary fracture lines radiating from the hole. These burst fracture lines are created after the bullet has penetrated the skull, and are the result of hydraulic pressure (see Fig. 5-5).

This pressure causes local outwards flexion, which can be recorded using high-speed photography. Once the maximum bending stress has been exceeded, cracks begin to propagate from the existing defect, i.e. the bullet hole.

These radial cracks propagate rapidly. If they become sufficiently long, they can easily reach the opposite side of the skull before the bullet.

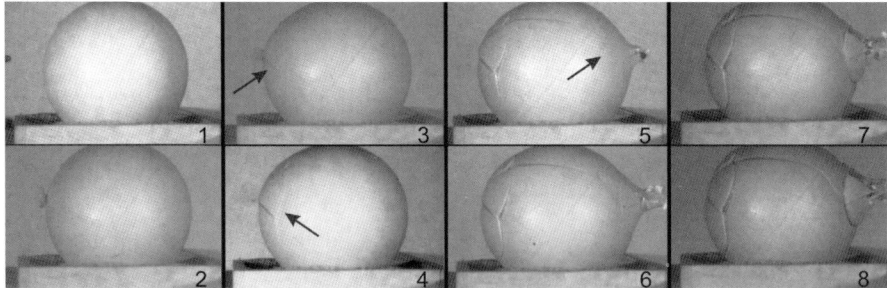

Fig. 5-5. Development of fracture pattern in the calvarium, demonstrated on a model head. Frame rate: 5 kHz. In the second picture, the bullet has already penetrated fully, but no fracture lines are visible. However, measurement of the calvarium indicates the presence of a slight bulge on the side from which the shot has come. In the third picture, this bulge has become more prominent, and a radial crack has appeared. The fourth picture (taken approx. 0.7 ms after impact) shows the tertiary, radiate fractures. In the fifth picture (taken immediately after the bullet exited the model), we see that one of the radial cracks has already reached the opposite side (9 mm Luger, full metal-jacketed bullet), (KNEUBUEHL et al. 1999, unpublished).

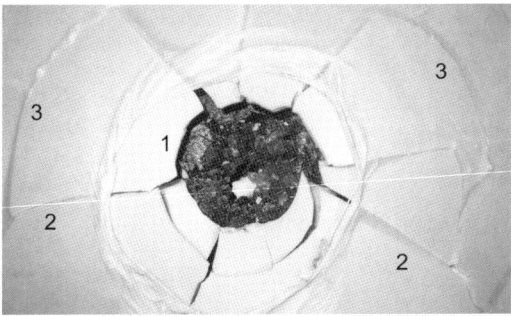

Fig. 5-6. Development of fracture pattern in the calvarium, demonstrated on a model head. Spherical lead projectile.
1 Primary bone defect, here showing an additional depressed fracture at the lower edge, because the sphere struck at a slight angle.
2 Radial fracture lines.
3 Radial bending fractures.

If the internal pressure is sufficiently high, and is maintained for a sufficient period, the segments created by the radial cracks will be pushed out far enough to create further, tertiary fractures. These bending fractures appear immediately after the secondary fractures concentrically around the bullet hole. In conjunction with the secondary, radial burst fracture lines, they often form a complex bending/burst fracture system, in the form of a globe fracture (see Fig. 5-6) The pressure generated by the passage of the bullet can lead to indirect fractures, especially in the base of the skull and, in this instance in the anterior cranial fossa (KLAUE 1949).

The wounding mechanisms of bullet wounds to the head are indicative of a hydraulic explosive effect. However, not every bullet wound to the head that causes injury to the brain results in the indirect or tertiary fractures described above. According to HARVEY et al. (1947), SELLIER (1982) and WATKINS et al. (1988), the pressure applied to the brain depends on the ballistic parameters of the bullet. The larger the area of the leading face and the higher the velocity of the bullet, the higher the pressure generated in the cranium.

This is immediately clear if we consider the basic laws of wound ballistics: Energy transfer per unit distance (i.e. effectiveness) is directly proportional to the energy of the bullet and inversely proportional to its sectional density (see 3.2.3.4).

In the case of handguns, even a 9 mm Luger bullet can pass right through the head, causing clear bursting fractures. The relatively high velocity of rifle bullets means that they will generally cause extensive fragmentation of bone when passing through the head, because of the high pressure gradients they generate in the brain. While shotgun slugs travel at lower velocities, they do have a low sectional density, which means that they also generally cause considerable damage to the head. The most extreme type of brain injury is that named after the researcher who first described it (KRÖNLEIN 1899). The transfer of energy from the bullet to the brain tissue creates such a high pressure that almost the entire brain is ejected from the cranium (see Fig. 5-7). This phenomenon is particularly common when the muzzle of a rifle is placed close to the base of the skull (lower forehead, mouth, etc.) (PANKRATZ and FISCHER 1985, RISSE and WEILER 1988, KARGER 2003).

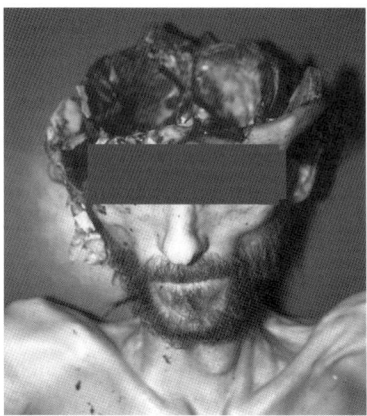

Fig. 5-7. Contact shot, shotgun pellets. Entry in lower forehead. Complete removal of the brain. This is known as a Krönlein shot.

Injuries caused by bullets that ricochet off the skull constitute another special category. In the case of an external ricochet, the projectile strikes the head at a tangent, penetrating the epicranium and rebounding from the tabula externa (KNEUBUEHL 1999c). Even though the bullet does not enter the skull, it can cause a depressed fracture of the calvaria ("skull-cap") with bleeding due to contusion of the cerebral cortex and intracranial bleeding. The situation is different in the case of an interior rebounding shot. In such cases, the bullet enters the skull, passes through the brain and rebounds from the tabula interna with a limited amount of energy. The bullet may rebound a certain distance into the wound channel or rebound at an angle into another section of the brain, coming to a halt at the end of a second wound channel. If the bullet strikes the internal surface of the skull at a very obtuse angle and with a small amount of energy, it may follow the interior surface of the skull, a phenomenon described by von HOFMANN in 1898. This may cause damage to the sinus veins and the meningeal arteries, followed by fatal bleeding.

5.1.6 Bullet wounds to the trunk

5.1.6.1 The ribcage

The ribcage contains a number of vital structures: the heart, the aorta and the main pulmonary arteries. Damage to any of these can cause rapid, absolute incapacitation (see 5.1.9). From a wound ballistics point of view, the structures in the ribcage consist primarily of elastic tissue. The elasticity of this soft tissue, and of the lungs, allows them to compensate for the stretch and shear effects that result from the creation of the temporary cavity. The extent to which they are able to do so depends on the energy deposited by the bullet. In the case of handgun bullets, for instance, the extravasation zone is often comparatively limited.

As the ribs form a cage around the organs, damage to the ribs is not unusual where a person is hit in the trunk. This can make it easier to distinguish between entry and exit wounds. It can be difficult to identify the wound channel in the case of injury to the lungs if one or both lungs collapse because of air or blood entering the pleural cavity (phenomena known as pneumothorax or haematothorax respectively). As it is relatively unusual for a bullet to come to rest in a lung, the path of the wound channel through the lung can generally be established with a good degree of certainty from the start and end points of the bullet's passage through the parietal pleura.

The volume of the chest cavities previously occupied by the now-collapsed lungs is easily sufficient to accommodate major bleeding from damaged blood vessels or the heart. It is perfectly possible for the victim to exanguinate into one or both chest cavities. No significant quantities of blood need be visible outside the body.

Elastic arteries display particular behaviour. Bullet holes in the aorta remain very small because of the elastic restoring force but have a characteristic, star shape with actinomorphic radial tears (POLLAK 1987, KARGER et al. 1997).

5.1.6.2 Abdomen

The abdomen contains a large number of closely-packed organs, each with a strong blood supply. In contrast to the organs in the ribcage, most of the organs in the abdomen (such as the liver, spleen and kidneys) are relatively inelastic. The shear and stretch forces generated by the formation of the temporary cavity are hence most visible in these organs, which often show star-shaped burst damage (METTER and SCHULZ 1983). A bullet wound to the perimeter of one of these organs may cause a broad tear in the form of a channel, and the resulting blood loss may quickly prove fatal (see Fig. 5-8).

Just as the chest cavities can hold the blood resulting from hits to the ribcage, so can large quantities of blood be almost completely accommodated by the abdominal cavity and the retroperitoneum, with little blood escaping to the outside.

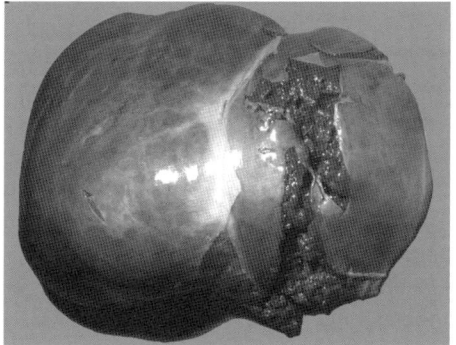

Fig. 5-8. Liver following passage of a 9 mm Luger full metal-jacketed round-nose bullet through the organ, close to the surface.

5.1.7 Bullet wounds to bones

5.1.7.1 General

From a forensic pathology perspective, bullet wounds to bones have the advantage that they often constitute a shape-stable medium for investigation, and one which generally gives a more reliable indication of direction of shot, for instance. Where the body has undergone extensive change because of decomposition or heat, whereby the soft tissues (including the skin) no longer allow reliable conclusions to be drawn, injuries to the skeleton provide a relatively sound basis for analysis.

Bone is significantly denser and harder than soft tissue, and is less elastic. Depending on the structure, velocity and sectional density of the bullet at the time of impact, changes may occur in the bullet itself and in its movement. Contact with bone may cause a projectile to tumble, deform, or even fragment, resulting in a decrease in sectional density and hence in an increase in energy transferred.

As with entry wounds in skin (see 5.1.3.1), one should be very cautious about trying to determine the calibre of the bullet from the diameter of an entry wound in bone. The diameter of the bullet hole in bone depends on a large number of parameters, such as the thickness and hydroxyapatite content of the bone, and the calibre, velocity, structure, material and angle of attack of the bullet. There appears to be a tendency for full metal-jacketed bullets from handguns to produce holes smaller than their calibre, while semi-jacketed and solid lead bullets tend to produce holes larger than their calibre (BERG 1955, KLAGES and WEITHOENER 1973, KUHL and JANSSEN 1977, KIJEWSKI 1979, ROSE et al. 1988, POLLAK and RITT 1992). The implication of this is that if one is asked to state the calibre on the basis of bullet holes in bone, one should give no more than a rough estimate for guidance (see also 2.3.7).

It is worth mentioning two common fallacies related to the effect of a bullet on bone: Fallacy No. 1: Handgun bullets, which travel more slowly, cause smaller defects in bone than do high-velocity rifle bullets, which cause extensive bone damage (see 3.2.5 and Fig. 3-36). Fallacy No. 2: The bone fragments created when a bullet strikes bone act as secondary projectiles and are capable of causing significant damage (see Fig. 3-39).

If a bullet strikes a flat bone (such as the calvarium or shoulder blade) approximately at right angles, one can normally expect to see a simple hole, with no significant cracking. The amount of energy that the bullet loses is virtually independent of its impact energy and is very low (60 to 70 J). The energy transferred is used primarily to create the hole in the bone; acceleration of bone fragments is a secondary effect. If we assume, for instance, that half the energy transferred from the bullet as it passes through the bone is used to break through the bone and that the other half is used to accelerate the fragments, and that approximately 10 fragments are created, then the mean energy imparted to each fragment will be approx. 3.5 J. Furthermore, the bone fragments generally have such low masses, that they are unlikely to possess any significant destructive potential, despite the energy transferred to them.

The amount of energy transferred to a hollow bone is significantly greater (up to approx. 300 J in the case of a rifle bullet passing through a femur). If a hollow bone of this sort is hit approximately dead centre by a bullet of which the calibre is less than the diameter of the bone, fracture is caused primarily by hydraulic pressure in the bone marrow. The bone splinters will be accelerated both in the direction of the shot and in the opposite direction. But even in such a case, the energy transferred to the bone fragments will be insufficient to accelerate them to the point at which they cause wound channels (see argumentation above).

The amount of energy transferred to the bone depends on both the total energy available to the bullet and the time for which the bullet remains near the bone. This time is inversely proportional to the velocity of the bullet, i.e. the slower the bullet, the more time it has to transfer its energy to the bone. It is therefore possible for a relatively slow-moving full metal-jacketed bullet from a handgun to cause more damage to a bone than a similar rifle bullet travelling at a higher velocity.

However, if one looks at real-life cases where bullets have caused bone damage, we often do find bone fragments embedded in soft tissue around the wound channel. As we mentioned above, the bullet breaks fragments out of the bone, and these fragments follow the bullet into the temporary cavity before coming to rest, often at some distance from the bone. When the temporary wound cavity collapses, the bone fragments are enclosed in the wound pockets as they form. As a result, they appear to lie at some distance from the actual permanent wound channel, each at the end of its own "wound channel." The above is confirmed by high-speed images, which show that the bone fragments follow the bullet as it passes through, rather than preceding it. They in fact follow the bullet along the vacuum left by the temporary cavity as it develops.

A somewhat different mechanism may come into play in the case of larger bone fragments. Here, contact with the bullet may cause a piece of bone to separate from the bone itself as if attached by a hinge. Such fragments may then be pressed into the soft tissue immediately adjacent to the bone. Even here, however, the term "secondary projectile" would not appear to be appropriate.

5.1.7.2 *Flat bones*

Flat bones have a solid structure, because of their dense, compact layers. The compact layers are separated by cancellous bone, a dense, interconnected system of fibres forming the diploës. Flat bones are to be found primarily in the skull, the shoulder blades, the sternum and the iliac fossae.

A bullet passing through a flat bone generally punches a hole in its outer surface. The edges of the hole are clearly defined, and the entry hole will be circular or elliptical, depending on the angle at which the bullet strikes. On the exit side of the bone, one will observe a funnel-shaped defect, wider in the direction of the shot, which is created by displacement of matter.

Exceptions do arise in very rare cases, e.g. funnel-shaped widening of the defect may be observed on what is known to be the entry side of the bone. In the case of one autopsy performed by the author, the outer surface of the bone displayed funnel-shaped widening on the side from which the shot had come, an effect for which the author not able to account with any degree of certainty. It is possible that the structure of the bullet (possibly the jacket) was primarily responsible (FRITZ 1933, COE 1982, PETERSEN 1991, BHOOPAT 1995). If, however, the entry hole in the outer surface of the bone displays a funnel-shaped opening on one edge only, this may be because the projectile has struck the bone at an angle. If a bullet strikes the skull at an obtuse angle, pieces of bone will be chipped out of the edge of the entry wound furthest from the weapon, resulting in funnel-shaped widening of the entry hole on that edge. The edge of the entry hole nearest to the weapon will look as if it has been punched out. If we observe the bone defect orthogonally, it will appear to take the form of a keyhole (MEIXNER 1923, MEIXNER and WERKGARTNER 1928, WEIMANN 1931, MAYER 1932).

The calvarium ("skull-cap") consists of a relatively large, contiguous surface made up of flat bones. If the skull has suffered multiple hits, under favourable conditions it is possible to work out the order in which the shots were fired. PUPPE's rule (named after G. PUPPE) is obvious from a physical point of view. This rule states that the bursting fracture lines emanating from a later bullet hole cannot cross those emanating from an earlier hole (see Fig. 5-9). This may enable one to distinguish between entry and exit wounds, as the speed at which the bursting fracture lines propagate is greater than that at which the bullet is moving (MADEA et al. 1987, KÖNIG and SCHMIDT 1989).

5.1.7.3 Long hollow bones

Hollow bones consist of a tube of bone, with the wall being made up of quite solid cortical bone. This tube contains cancellous bone. This in turn contains the bone

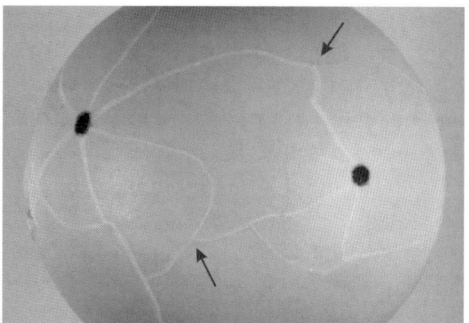

Fig. 5-9. PUPPE's rule, demonstrated on a model head. The radial fracture lines from the second shot end at those from the first.

marrow, especially in the area of the shaft. From a ballistics point of view, bone marrow behaves almost like a fluid. The long hollow bones are found primarily in the limbs.

Bullets generally cause drill hole or butterfly fractures, but any other type of fracture is also possible, including comminuted fractures (HUELKE and DARLING 1964, ROSE et al. 1988).

As far as identification of the direction of shot is concerned, the compact components of the bone (the "tube") behave like flat bones: the entry hole looks as if it has been punched out, while the exit hole widens in the form of a funnel. However, this is not often observed, as a bullet passing through the bone somewhere near the centreline creates hydraulic pressure in the bone marrow, and it is this pressure that actually causes the bone to fracture. The entry and exit holes leave a number of fragments, which make it difficult or impossible to decide which is which.

5.1.7.4 Vertebrae

From a wound ballistics point of view, the reaction of vertebrae to the passage of a bullet (and likewise of the calcaneus and parts of the pelvis) lies between that of flat bones and of long hollow bones. It is often possible to determine the track of the wound channel from the funnel-shaped expansion of the defect in the direction of the bullet.

5.1.8 Peculiarities of shotgun wounds

5.1.8.1 General

The discussion concerning wound ballistics above also applies to shotgun blasts. However, the wounds caused may differ significantly from those caused by bullets. The volume of a shotgun barrel is comparatively large, and the ammunition may consist either of individual pellets with a filler or of a slug. After leaving the barrel (which is smooth and may include a choke), the shot sheaf spreads out longitudinally and radially. The rate at which it does so depends on the filler used, e.g. plastic cup or plastic crosspiece. After approximately 6 m, the sheaf has fully separated. Being spherical, the individual pellets lose energy very rapidly, so the penetration depth in the human body of a pellet from a conventional hunting cartridge decreases quickly with distance. At distances of approx. 15 to 20 metres, relatively small pellets penetrate no further than the boundary between subcutaneous fatty tissue and muscle if the victim is wearing light to medium clothing. If the pellets are larger (i.e. 4 mm or more in diameter), they will penetrate significantly further, and may reach the internal organs. If buckshot is used (8 to 16 pellets per cartridge), significant and even lethal wounding may occur even at greater distances. As the cohesion of the shot sheaf plays a decisive role in the wound ballistics of shotgun ammunition, wounding potential depends to a large degree on the type of ammunition used and the design of the barrel (length, whether it has been sawn off and whether it is choked). See Fig. 4-38 for details of effects.

5.1.8.2 Morphology of entry wounds

The morphology of the entry wound is very different from that caused by a bullet. On the basis of their studies and systematic observations, BREITENECKER (1967, 1969, 1973), SELLIER (1982) and JAUHARI et al. (1983) have all described the particular features associated with the penetration of a large number of pellets, either as a sheaf or as individual but closely-spaced pellets. The latter is of particular significance in determining the distance from which a shot was fired, as it is possible to conduct test shots and measure the spread of the pellets at a given distance (see Fig. 4-38).

Qualitatively speaking, the results of a contact shot with a shotgun generally differ little from those of a bullet. However, the effects are generally more intense and/or more extensive. Filler (felt wads, plastic cup, plastic crosspiece, etc.) will be found in the wound.

At distances of up to approx. 1 m, the wound will be up to approx. 5 cm in diameter, and from approximately 30 to 40 cm onwards, the wound will clearly have a jagged edge, indicating that "fliers" have begun to separate from the shot sheaf. The wound may still contain filler, depending on the ammunition used and the distance. At distances of approximately 1 to 3 m, the fliers become more distinct. The components of the filler will only very rarely enter the wound at this distance, but do still have enough energy to graze or tear the skin in the vicinity of the wound. Occasionally, a piece of filler will perforate the skin and penetrate slightly.

At distances of approx. 3 m and above, the morphology of the entry wound depends to a very large degree on the characteristics of the weapon and the ammunition (choke/half-choke/no choke, pellet size and filler). The degree to which the pellets scatter and penetrate various considerably. At distances of about 5 to 10 m and above, one generally observes an impact area, within which it is possible to distinguish the individual pellets (see Fig. 5-10). Being un-aerodynamic, the filler materials generally do not reach the victim at these ranges.

With the exception of buckshot, the pellets have a low mass and hence a low sectional density. Penetration is therefore limited. As a result, conventional shotgun ammunition does not usually cause an exit wound.

5.1.8.3 Internal morphology of shotgun wounds

At close range, where the pellets still constitute a cohesive sheaf when they penetrate the skin, a mechanical effect occurs that is known as the "billiard ball effect."

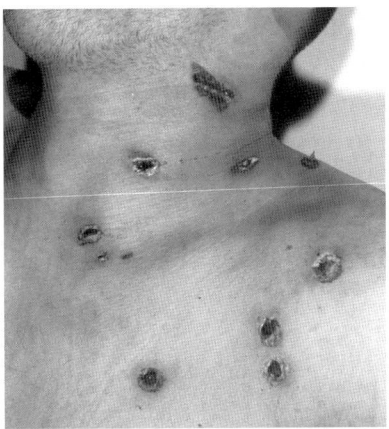

Fig. 5-10. Shotgun wound (buckshot). Distance: approx. 12 m. See Fig. 4-37 for the effects of a shotgun at close range.

When the leading pellets are slowed by contact with a dense medium such as skin, the pellets behind catch up. The pellets collide, and deflect each other. This causes the pellets to scatter much more pronouncedly inside the body than outside, in a conical pattern. If we ignore slugs (POLLAK 1990) and larger calibres of buckshot (e.g. 12/76), it is this effect that determines the wounding potential of a shotgun at close range. The sheaf and the billiard ball effect cause a huge wound cavity, with extensive damage to blood vessels. If organs are affected, they will exhibit tearing and shredding.

In the case of contact shots, the relatively large volume of the barrel, the quantity of propellant and the comparatively large entry wound mean that large quantities of expanding propellant gases will be projected into the wound, in addition to the pellets or slug. This may generate additional pressure effects, especially in the case of the head.

5.1.9 Causes of death and incapacitation

5.1.9.1 Causes of death

In principle, the fatal effects of a gunshot wound are no different from those of other forms of mechanical violence in which an object penetrates the human body. Depending on the area of the body affected, there are two types of lethal effect: *1.* Direct, mechanical destruction of important regions of the brain. *2.* Failure of the brain to function because blood loss has deprived the brain of oxygen.

1. If the parts of the central nervous system responsible for controlling breathing and circulation are destroyed, central regulatory failure (brain death) is to be expected. However, it is also possible for sensitive parts of the brain to undergo compression in the foramen magnum because of gunshot wounds to other parts of the brain causing swelling or releasing blood into the cranium.

2. If the gunshot wound damages blood vessels and/or organs that contain large quantities of blood, then significant blood loss – exanguination (haemorrhagic shock) – may lead to internal suffocation and hence to brain death (cerebral ischaemia).

An air embolism is another, albeit rarer phenomenon. When a person is standing, there is a negative pressure in the veins above the heart. If a peripheral wound to the head or neck of a person who is standing, sitting upright or squatting opens up larger blood vessels, such as the jugular vein or the venous cranial blood vessels, air from outside may be sucked into those vessels. Air then collects in the right half of the heart and the heart muscles contract around this air, without sufficient blood being pumped towards the lungs. This results in failure of the right side of the heart. According to the literature, 70 to 130 ml of air is sufficient to cause a fatal embolism.

Inhalation of blood is a much-feared complication associated with gunshot wounds to the neck and face. If the victim remains in the prone position, incapacitated, either because he is unconscious or has suffered other injuries, or for any other reason, blood from injuries to the visceral cranium may flow into the pharynx, or blood from damaged blood vessels in the neck may flow into the damaged trachea. This may result in death by suffocation – the victim "drowns" in his or her own blood.

Fatal blood poisoning (septic shock) originating from serious infection to a wound is a further complication that may arise after the victim has initially survived the shooting. As we saw in 3.2.6, a bullet is not sterilized by the process of being fired, and indeed will accumulate additional germs as it passes through clothing and skin. These germs are then introduced into the wound. As the permanent wound channel is surrounded by an extravasation zone containing torn blood vessels (see Fig. 3-19), the germs encounter an ideal environment: a warm, moist matrix filled with devitalized, destroyed tissue, where circulation is poor, and where antibiotics penetrate only slowly, on account of damage to the capillaries. If it is not possible to control the local infection, the rest of the circulatory system rapidly becomes affected (septicaemia), followed by fatal septic multiple organ failure.

Laypersons frequently raise the question of "death by shock," i.e. the idea that a bullet to a particular part of the body can cause instant death. In medical terminology, "shock" is in fact circulatory shock, defined in PSCHYREMBEL 2007 as acute to sub-acute progressive and general circulatory failure. This can be caused by a reduction in blood volume in the circulatory system (hypovolaemic shock), heart failure (cardiogenic shock), blood poisoning (septic shock), an allergy (anaphylactic shock) or a neurogenic disturbance to the regulation of the blood vessels (neurogenic shock).

From the point of view of medicine, and especially forensic medicine, there are grounds for considerable scepticism regarding the concept of "immediate death by shock following gunshot wound that does not damage directly vital structures," particularly as there have so far been no properly documented cases in which such a cause of death was established on a solid scientific basis. If one looks at stories of persons dying of shock – most of which are battlefield accounts from soldiers – one quickly starts to doubt the testimony of witnesses, and reliable autopsy results (including histology) are usually missing as well. On the other hand, one cannot ignore the results of the experiments regarding the stimulation of nerves by shock waves (see 4.3.2.3).

To date, true "death by shock" in the above sense has been proven only in the case of animals up to the size of a deer (the most common animal concerned being the rabbit) and even then, all cases involved the use of shot from a shotgun. If, for instance, a rabbit is hit by 3 mm shot at a range of approx. 40 m, it will tumble and, at least, will lie motionless, incapable of moving. If the rabbit is dissected, or opened up by the hunter, one finds that all the pellets are just under the skin and that none has entered the body cavities or touched the organs. There is generally no lethal injury (brain injury or exanguination, for instance). A similar reaction by a human being to pellet injuries of this sort has not been observed to date.

5.1.9.2 Incapacitation

Whether a person was incapacitated after being hit by one or more bullets depends on the factors regarding the effect and effectiveness of bullets discussed in 4.1. Only when it is clear from a post-mortem examination which body structures were damaged, and in what way, and when sufficient information is available regarding traces of blood, the position of the body and any bullet damage, will it be possible to say whether the person was incapacitated.

There is no standard definition of incapacitation. Given the different viewpoints, this is hardly surprising (WALCHER 1929, PETERSOHN 1967, KARGER 1995a). For a marksman, it is important to know which parts of a hostage-taker's body he must hit, for instance, in order to be sure that he will not be able to stab his victim or pull the trigger. In investigating a murder, on the other hand, the point of interest is whether, for instance, a person was capable of moving a certain distance, hiding an object or shouting to someone after being hit.

How an injury affects a person's consciousness and reactions depends on several factors. The main psychological factors are the degree to which the victim expects to be hit, their degree of excitation and any psychoactive substances or stimulants they may have taken. Clearly, it will often be difficult or impossible to take account of these factors when determining whether a person was incapacitated, or when predicting the likelihood of incapacitation in a future situation. By contrast, it is somewhat easier to determine whether the victim was incapacitated from physical evidence. If we are interested in knowing whether a person was capable of specific, individual acts, the morphology and position of injuries will generally answer the question (wounds to the limbs, the spinal cord, etc.). If we wish to assess general incapacitation, i.e. for how long a person remained capable of any kind of coordinated movement, or the point at which one could assume he or she became completely incapacitated, we need to assess the degree to which the central nervous system was capable of functioning. Absolute incapacitation can only be caused by two mechanisms: stopping the central nervous system from functioning directly, or stopping it functioning indirectly, via blood loss.

The time factor is of great significance in relation to the deactivation of the central nervous system. The delay between the shot and total incapacitation depends from the part of the body affected. KARGER and BRINKMANN (1997) divide incapacitation into three time categories, and relate these to the area that has suffered the injury: 1. Immediate incapacitation. 2. Rapid incapacitation. 3. Delayed incapacitation.

One can only assume immediate, absolute incapacitation in the case of significant injury to the brain or medulla oblongata where the head meets the neck. Immediate incapacitation requires the destruction of large parts of the brain or vital parts of the central nervous system, such as stem ganglions, midbrain, interbrain, cerebellum, brain stem, pons or medulla oblongata.

If the above regions escape injury, the victim may survive a gunshot wound to the head, at least initially. There have even been cases in which the victim remained capable of voluntary action despite a gunshot wound to the brain. However, these rare cases only involved injury to the frontal lobe or a temporal lobe, and the relatively low energy and energy release of the bullet meant that it was not capable of causing damage to areas outside the permanent wound channel via shear or stretching forces (e.g. SMITH 1943, KRAULAND 1952, FRYC and KROMPECHER 1979, BRATZKE et al. 1985, ROTHSCHILD 2000). In an analysis of 38 cases where victims of bullet wounds to the cranium initially remained capable of voluntary action, KARGER (1995b) established that 34 cases involved the use of pocket pistols, small-calibre bullets or 6.35 mm and 7.65 mm full metal-jacketed bullets. An additional factor is that in many cases the ammunition used had been obtained illegally and was out of date, which meant that the propellant had lost some of its power. However, it is possible for the victim to survive a rifle bullet wound, if the entire wound channel lies within the narrow channel. This may be the case where the shot was fired from a great distance (see 6.3.2, Case C) or from just a few centimetres (from cases studied by Berne University's Institute of Forensic Medicine, see 5.3.2.3 on suicide and attempted suicide).

In the latter case, the bullet enters the head before the gases emerging from the barrel can initiate precession and nutation. This results in a very long narrow channel. The gap between muzzle and head allows muzzle pressure to dissipate to the sides, without entering the skull.

Rapid absolute incapacitation in the case of indirect deactivation of the central nervous system is to be expected in the case of injuries to the heart or other major, central blood vessels such as the main branch of the pulmonary artery or the aorta.

Here, the rapid loss of blood leads to a shortage of blood – and hence of oxygen – in the brain, resulting in unconsciousness followed by brain death. Leaving aside other elements (psychological factors, drugs, alcohol, etc.), even complete destruction of the heart or severing of the aorta leaves a few second's oxygen reserve in the brain, which means that the victim will not necessarily become incapacitated or lose consciousness immediately. The literature reports numerous cases in which fatal gunshot wounds opened up both sides of the heart, or opened the left side radically, and yet left the victim capable of walking a number of metres, reloading a weapon or even firing a number of shots (e.g. SCHRADER 1942, HUDSON 1981, MARSH et al. 1989, MISSLIWETZ 1990, KARGER and BRINKMANN 1997). In a case known to the author, a 9 mm Luger full metal-jacketed bullet passed through the left side of the heart. The victim was a muscular young man who worked as a nightclub bouncer. Despite his injury, he was capable of calling out to his friends and walking some 20 m at a brisk pace before collapsing.

Delayed incapacitation is most often associated with injuries to organs and to vital arteries (those of the organs, together with arteries of the limbs located close to the trunk). It may take several minutes for the victim to lose consciousness, even if organs have been completely pierced, the delay depending on the number of wounds and the structures affected. Here again, the mechanism is that of indirect deactivation of the brain resulting from blood loss.

5.1.10 Particular projectiles

5.1.10.1 Gas-powered weapons

Gas-powered weapons fire projectiles weighing approximately 0.5 g (diabolo, pointed or round pellets) using pressurized gas – air, carbon dioxide or nitrogen. The pressure is generated either by mechanical work (i.e. compression, or cocking the weapon) or taken from a reservoir. Pistols deliver muzzle velocities of 50 to 150 m/s, rifles 170 to 300 m/s or more.

Under German legislation, the energy of the projectile may not exceed 7.5 J. This restricts muzzle velocity to about 175 m/s.

A pellet striking clothed skin is unlikely to cause any significant injury. Local bruising may be observed, however, occasionally accompanied by superficial removal of the epithelium resulting from abrasion via the clothing. Airgun pellets are capable of penetrating clothing at close range, however. A round or diabolo pellet striking the unprotected skin at 90° will cause reddening and superficial grazing because of the crushing effect of the projectile on the epithelium. According to MISSLIWETZ (1987), the minimum velocity at which a gas-powered projectile will penetrate the skin is approximately 100 m/s (see 4.2.1.2 and Tables 4-11, 4-12 and 4-13). This means that if a pointed projectile strikes the skin at 90°, or a diabolo pellet strikes the skin with the sharp edge leading, they can completely pass through the skin and penetrate deep into the body, on account of their relatively high sectional density. The degree of penetration will depend on the area of skin concerned. Serious and even fatal injuries have been described (e.g. MARKERT and RÖMER 1973, HARRIS et al. 1983, WASCHER and GWINN 1995, BOND et al. 1996, NAUDE and BONGART 1996).

5.1.10.2 Alarm pistols and weapons firing irritants

For many years, alarm pistols were seen as "harmless toys." Such weapons are particularly common in Germany. They are intended purely for self-defence, but generally look exactly like real firearms. Their mode of operation is based on that of the weapons they imitate. The blank or irritant cartridges used in such weapons do not contain a projectile. As we saw in 4.2.3, however, it is not true to say that nothing comes out of the muzzle when an alarm pistol is discharged. The gases leave the muzzle at a high velocity. They hence have a high energy flow density and deliver significant kinetic energy on striking the body.

If the muzzle is in contact with the skin, and at distances of up to approximately 1 cm, the jet is capable of penetrating the skin and, depending on the part of the body, may reach the soft tissue underneath. Wounding potential depends not only on the part of the body concerned but also on the length of the barrel and on the ammunition used.

The visible effects of a contact shot with an alarm pistol are the same as those of a contact shot from a real weapon. The central hole may be round or irregular, the black "bullet wipe" is caused by the outer boundary of the gas jet as it penetrates, the "contusion ring" is caused by the edges of the gas jet being pushed outwards, which displaces the epithelium, and the "margin of distension" are often imitated by subsequent bleeding beneath the skin.

Contact shots from an alarm pistol to soft skin tissue supported by flat bone often leave an imprint similar to that of a bullet wound. This effect will be especially pronounced if the gas jet does not break through the underlying bone, and no gas enters the skull. All the gas generated by the propellant undermines the soft tissue, separating it from the outer surface of the bone. Even the periosteum, which is firmly bonded to the bone, is often separated and displaced. This results in a clear gas cavity, often accompanied by an impressive imprint. Unlike bullet wounds, wounds from blank ammunition often leave imprints even in areas where the skin is not supported by flat bone (see Fig. 5-11). Here again, the reason is that there is no wound channel into which the gases can escape, leaving them to expand in the tissue under the opening in the skin that they have caused. As a result, the tissue is briefly "inflated" and the skin pressed up against the muzzle, causing radial abrasion.

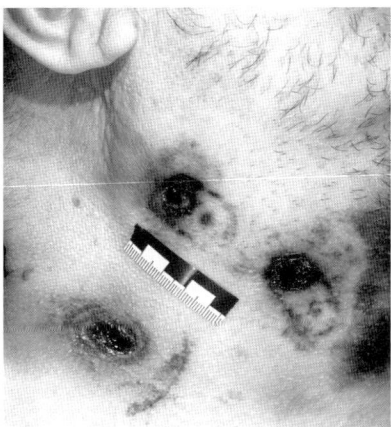

Fig. 5-11. Contact shots with blank ammunition fired from different alarm pistols, showing clear imprints on the neck. The muzzle impression is also visible on skin overlying soft tissue.

If it penetrates the body, a gas jet has sufficient energy to injure soft tissue such as fat and muscle, and if the barrel is short and the cartridge contains enough propellant, it can cause damage to organs and break through thin bones. ROTHSCHILD (1999) contains an overview and a collection of cases. The wound cavity consists of tissue that has been destroyed by pressure and heat, torn and devitalized. Other than in the case of shots to the skull, when penetration of the brain by the gas jet can cause brain death, the most common cause of death is peripheral circulatory collapse resulting from exanguination.

At distances of a few centimetres, the gas jet is usually no longer capable of penetrating the skin. The main effects seen here are those of propellant particles embedded in the skin and heat effects from primary flash (see 2.3.3.2). The latter are caused by propellant that is still burning. By contrast, secondary flash (the ignition of propellant gases on combining with atmospheric oxygen) plays no significant role in the case of alarm pistols. At short range, foreign bodies can also cause a "tattoo" effect. Depending on the distance between muzzle and skin, the length of the barrel and the quantity of propellant, the gas jet may have a velocity of over 3,000 m/s (see Figs 4-22 and 4-23), which is sufficient to force propellant particles into the dermis. Even at several tens of centimetres, there is a risk of propellant particles causing irreversible damage to the unprotected eye.

5.1.10.3 Arrow wounds

Descriptions of injuries and fatalities due to arrows/quarrels from longbows and crossbows are plentiful in the literature. KARGER (2003) contains an overview. Because of the actions needed to use a longbow (and, to some extent, a crossbow) the vast majority of cases involve accidents and deliberate killings, with self-inflicted wounds being the exception.

The penetration mechanism is equivalent to stabbing. If the arrowhead also has lateral blades (e.g. the three-bladed broad-head), there will also be a cutting effect.

In contrast to the effects of projectiles from firearms, cutting and piercing play the primary role, with that of blunt force being secondary. Skin and tissue undergo splitting rather than crushing. The morphology of the wound is therefore similar to that of a stab or slash wound.

Their length and form give arrows very good ballistic characteristics. The mass of the shaft can give the tip a high sectional density. As a result, arrows can penetrate deep into the body, and can easily pass through organs, major blood vessels, cartilage and flat bones. Bone slows down an arrow considerably, and arrows generally come to rest in more compact bone, such as vertebrae. Although the kinetic energy of an arrow is considerably less than that of a handgun bullet, an arrow can cause a wound channel of comparable length if it only passes through soft tissue. However, the wound channel will be limited to the area in direct contact with the arrowhead or shaft. No significant temporary cavity is to be expected, and there is no significant shock wave effect.

The form of the entry wound will vary according to the shape of the arrowhead. Field tips are simply sharpened at the front, in various ways, with no parts projecting beyond the perimeter of the shaft. Such arrows can easily create something resembling a contusion ring by abrasion between the shaft and the skin (MISSLIWETZ and WIESER 1985, HAIN 1989). If the arrowhead also has lateral blades (e.g. in the case of a broad-head), it will also make lateral cuts, which can at first sight look very similar to stab wounds (KARGER et al. 2004).

As with other cases in which an object penetrates the body, the cause of death is usually peripheral circulatory collapse resulting from loss of blood. There may be a delay before this occurs if the arrow remains in the body and hence plugs the wound channel. If the arrow passes through the heart, it may physically prevent the heart muscle from operating and hence preventing the heart from pumping blood. Central regulatory failure is to be expected if an arrow hits the medulla. As for gunshot wounds, infection may cause death at a later stage.

5.1.10.4 Captive bolt pistols and bolt-firing tools

A **captive bolt pistol** is a device used to slaughter animals in an abattoir. A bolt is fired 8 to 10 cm by a blank cartridge, before being pulled back into the device by a return spring. The pistol is supposed to be placed in contact with the skull of the animal before firing. The bolt penetrates deep into the skull. Literature in German contains numerous examples of suicide, fatal accidents and murder (e.g. ISFORT 1961, JANSSEN and STIEGER 1964, POLLAK 1977, WIRTH et al. 1983, KOOPS et al. 1987, LIGNITZ et al. 1988, FALLER-MARQUARDT 1991, NADJEM and POLLAK 1993, BETZ et al. 1993).

In virtually all cases mentioned, the device was in contact with the victim when fired. One widely-used model, the KERNER, has a bolt either 10.5 or 12 mm in diameter. The bolt is concave on the underside and has sharp edges. A gas vent is located on each side of the bolt aperture. As a result, the entry wound has a central hole, which looks as if material has been punched out. On either side of this hole, one observes a smoke mark (CZURSIEDEL 1937). Models with different gas vents produce similar marks (LIEBEGOTT 1948/49, NADJEM and POLLAK 1999). The bolt leaves a combination of a "bullet wipe" and a true marginal abrasion around the central hole. The hole in the skull also appears to have been punched out, with a diameter similar to that of the bolt. If the device was perpendicular to the skull at the time of firing, the hole will be circular. The greater the angle between the device and the perpendicular, the more elliptical the hole. The hole will widen in a funnel shape towards the inner surface of the skull, as for a bullet wound. The bolt almost always pushes the bone fragment it has removed further into the brain, and this fragment remains at the end of the wound channel when the bolt is withdrawn.

The velocity and energy of the bolt are not sufficient to cause a significant temporary wound cavity or shock wave.

Bolt-firing tools and nailers are used on building sites to insert nails, pins, bolts, etc. into timber, metal or concrete. There are two different mechanisms: Cartridge tools insert the connector using an explosive cartridge, while bolt insertion tools and nailers press the connector into the material via a piston. This piston may be powered by an explosive cartridge or compressed air.

To prevent accidental firing, or deliberate misuse, virtually all models are so designed that they can be fired only with the tip of the tool pressed firmly against a solid surface. Nevertheless, the literature contains numerous cases of fatal accidents and suicide (e.g. MAURER 1961, CRAGG 1967, DIMAIO and SPITZ 1972, SCHMIDT and GÖB 1981, WEEDN and MITTLEMAN 1984, PANKRATZ et al. 1986, KOOPS et al. 1987, JACOB et al. 1990, BOCK et al. 2002). In cases of suicide, the need to press the tool firmly against the body in order to fire it means that one often finds a broad imprint, caused by abrasion of the skin against the tip of the tool (KARGER and TEIGE 1995). It is also possible to render these tools capable of firing at targets further away by modifying the safety device. Connectors fired in this fashion will have an initial velocity of approximately 150 m/s, but are extremely unstable in flight. However, a nail or pin that strikes the body at 90° is perfectly capable of passing through bone, penetrating deep into the body and causing fatal injuries.

5.2 Modern graphical methods
M. J. THALI

5.2.1 Surface documentation

In forensics, the standard way of documenting wounds is to present a verbal description, sketches and photographs showing a scale. Two-dimensional photography has the disadvantage that 3D data is projected onto a single plane. The Virtopsy project (www.virtopsy.com) has developed optical 3D documentation techniques with which it is possible to show objects and bodies in three dimensions, to scale and in colour. Using photogrammetric 3D surface scanners or laser scanners, it is possible to document three-dimensional objects without touching them, to scale and in colour. The resolution of the 3D scanner can be adapted to the object under study. It is already possible to achieve precision of less than a millimetre. Using these methods, it is possible to record the surface of a shooting victim (living or dead), of the instruments that caused the injury (such as weapons) or of the entire scene (see Fig. 5-12). The main advantage of this method is that all the evidence can be recorded with a high degree of spatial accuracy, without touching anything, to scale and in colour, and that it is possible to analyse the 3D datasets for the injured areas of the body surface and for the suspected instrument at any time (see Figs 5-13 and 5-14). (BRUESCHWEILER et al. 2003, THALI et al. 2003a,b,c).

5.2.2 Radiological documentation

Radiology has been used for forensic purposes ever since X-rays were discovered. The first forensics case involved a shooting victim. The associated history is covered in BROGDON's "Forensic Radiology" and further publications (BROGDON 1998, DI MAIO 1993, THALI and DIRNHOFER 2004).

Just like conventional photography, however, normal X-ray techniques reduce

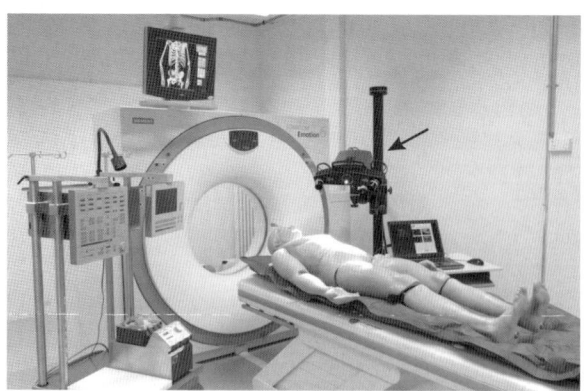

Fig. 5-12. Forensics laboratory with 3D surface scanner (arrowed) and CT scanner.

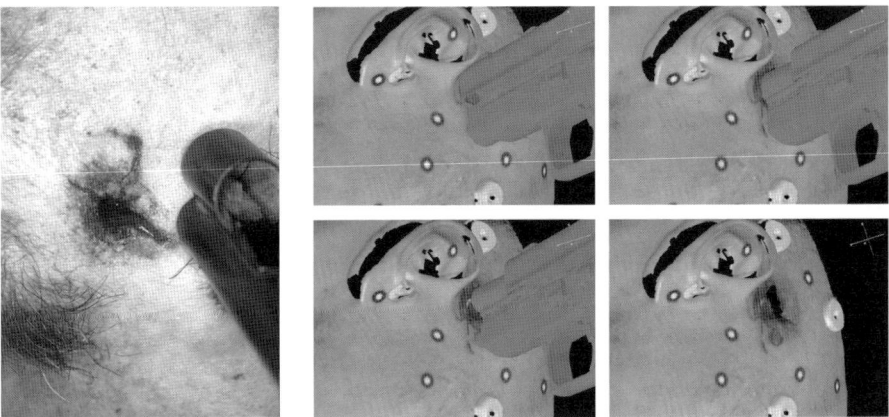

Fig. 5-13. (*left*) Conventional documentation of a bullet wound, using photography (note the muzzle impression). Three-dimensional evidence is reduced to a single plane.

Fig. 5-14. (*right*) Three-dimensional documentation of a bullet wound and weapon using 3D surface scanning.

the internal 3D evidence in the body to a 2D image on an X-ray plate (see Figs 5-15 and 5-16). This makes it more difficult to locate the projectile, for instance. The Virtopsy project introduces 3D techniques capable of generating a cutaway image – computer tomography (CT) and magnetic resonance imaging (MRI) – into the field of forensics.

Computer tomography has been available for medical purposes since about 1970, MRI since about 1980.

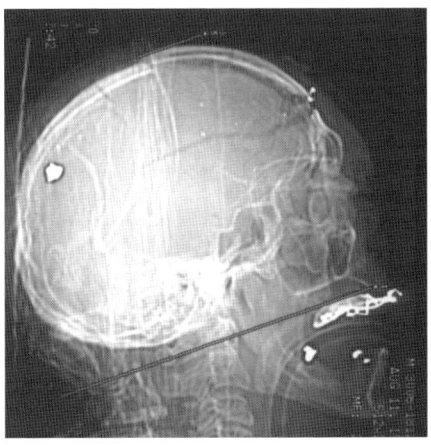 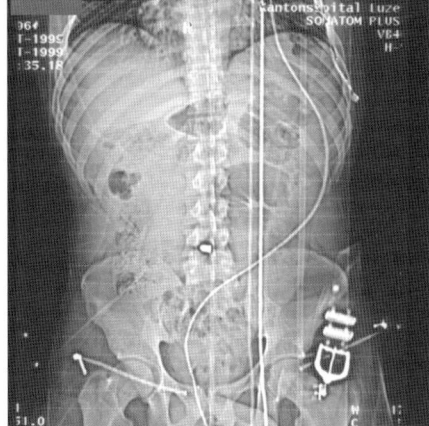

Fig. 5-15. (*left*) Lateral X-ray image of a head, showing entry wound in the temporal region and the final position of the projectile in front of the rear skull bone.

Fig. 5-16. (*right*) X-ray image of the trunk. The bullet is projected onto the spine. From the two-dimensional image, it is not possible to tell exactly where it is located.

Computer tomography uses X-rays. A rotating source generates X-rays that pass through the body in the x and y planes. The x-axis is created by the forward movement of the table on which the body is lying. By applying methods from the fields of mathematics and physics, it is possible to convert variations in radiological density at different points into cutaway images and 3D information. A body – living or dead – can be divided into thousands of virtual slices, each less than a millimetre in thickness, which can be stored as a 3D dataset. Multislice or multidetector computer tomography allows us to examine a body rapidly and scroll through it in all planes and directions using software, and hence to find and display evidence related to injury. Computer technology also makes it possible to "remove" the skin and soft tissue and to "slice open" the calvarium (see Fig. 5-17). CT is also a suitable method of producing 2D and 3D images of bone lesions and accumulations of air, and a basic overview of organ injuries.

Magnetic resonance imaging (MRI) was introduced in the early 1980s. In this technique, temporary changes in magnetic fields are used to generate 3D images.

Unlike CT, MRI also provides high-resolution imaging of soft tissue. Organs, muscles and other tissue are displayed in even more detail than when using CT.

The higher the intensity of the magnetic field generated by the device, the higher the resolution of the image. It is perfectly possible to achieve resolution similar to that of a microscope.

MRI is therefore even more suitable than CT when it comes to visualizing soft tissues and organ damage.

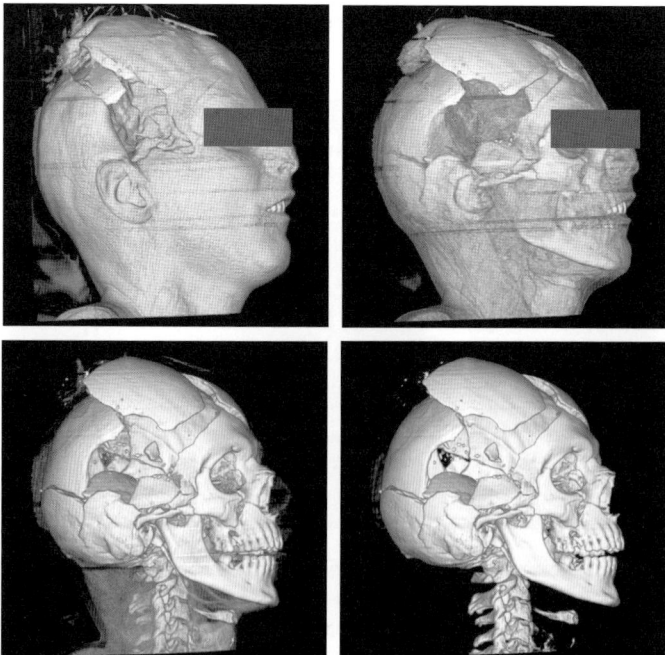

Fig. 5-17. A combination of CT and software makes it possible to "remove" the skin and soft tissue from the 3D model, exposing the fracture caused by the bullet. It is also possible to "open" the calvarium.

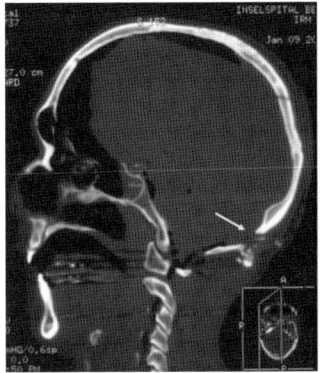 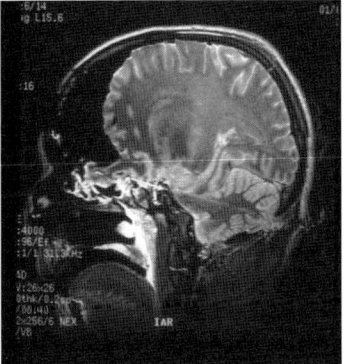

Fig. 5-18. 3D representations of a bullet wound to the head. *Left*: CT. *Right*: MRI. The CT image shows the pathological accumulation of gas in the frontal region, plus the funnel-shaped widening of the wound channel in the bony posterior region of the head (arrowed). MRI produces a higher-resolution image of the soft tissue, showing the wound channel in the cerebellum more clearly (arrowed).

CT and MRI make it possible to visualize the inside of a body (living or dead) following bullet or other injuries (see Fig. 5-18) (WULLENWEBER et al. 1977, SCHUMACHER et al. 1985, THALI et al. 2003e, THALI et al. 2007, BOLLIGER et al. 2007, DIRNHOFER et al. 2006, THALI et al. 2003f, THALI et al. 2002a, THALI et al. 2009, www.virtopsy.com).

5.2.3 Combining surface and radiological documentation

The findings obtained by 3D surface scanning, CT and MRI can be combined into one dataset. This is achieved using radiological "landmarks," which link the external and internal body datasets obtained from CT and MRI. Combining the two datasets results in a scale representation of the inside and outside of the body in 3D. In the case of a living person the wounds will eventually heal, while in the case of a corpse the evidence will undergo the normal process of decay. In both cases, however, the information can be stored to scale, in 3D, and used at a later stage for the purposes of reconstruction (see Figs 5-19 and 5-20).

5.2.4 Documenting crime scenes using modern graphics techniques

As well as providing non-invasive documentation of the outside and inside of the body, modern surface scanning and laser scanning techniques make it possible to document the scene of a crime (e.g. a shooting). If necessary (e.g. in connection with bloodstain patterns) the images can be produced in colour. The 3D data model hence forms the basis for virtual reconstructions, in which analysis of geometrical data from the scene make it possible to say whether a given action or

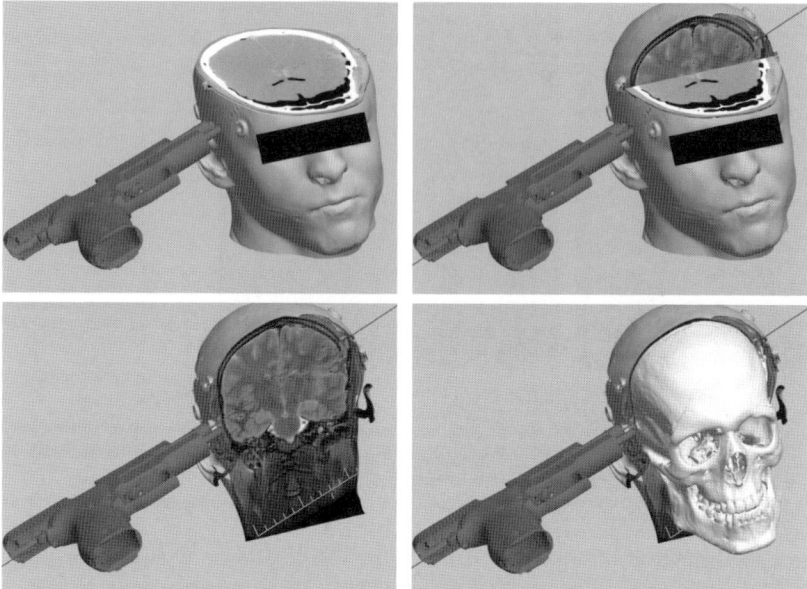

Fig. 5-19. 3D representation of weapon and head injuries generated using 3D surface and radiological scanning.

series of actions would have been possible. Combining 3D data of the scene and the persons involved (alive or dead) in the form of surface or radiological data makes it possible to run through scenarios on the computer without the need to keep the crime scene under seal for weeks or even months. (Fig. 5-21).

This is known as the "real 3D data-based approach." Rather than using computer models of the body taken from the Internet or software such as Poser, one is using real-life data (THALI et al. 2005, BUCK et al. 2007, www.virtopsy.com).

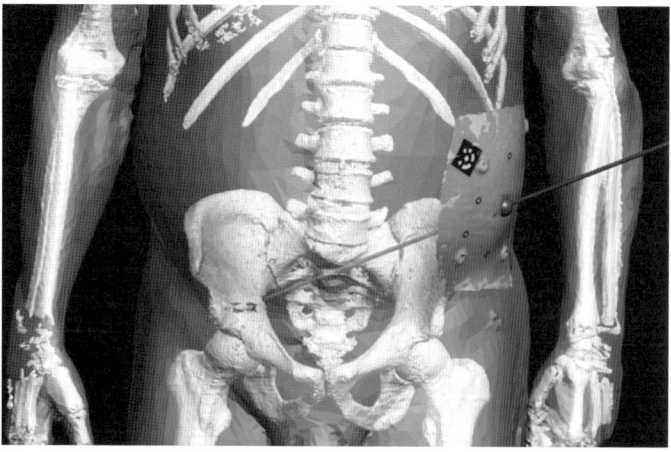

Fig. 5-20. 3D representation of injuries to the trunk generated using 3D surface and radiological scanning.

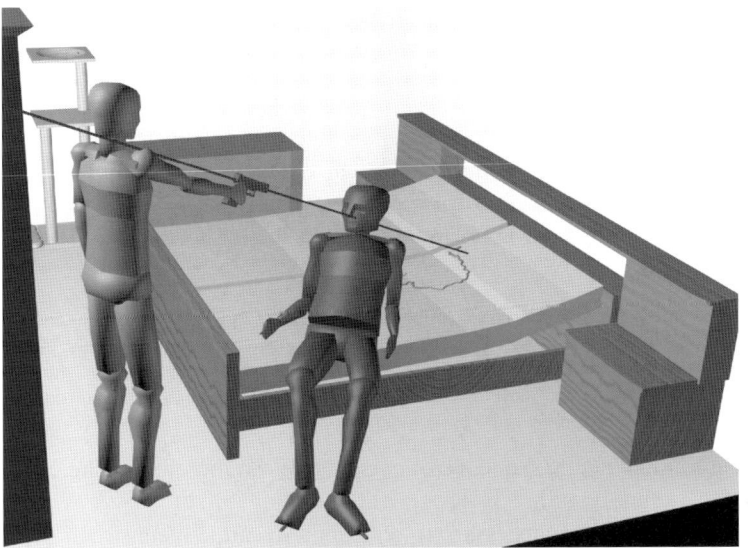

Fig. 5-21. Victim and perpetrator were documented using surface scanning, the victim subjected to CT and MRI scanning and the crime scene was scanned. This 3D data can be used to present sequences of events on the computer and check them for plausibility.

5.3 Experimental reconstruction
B. P. KNEUBUEHL, M. J. THALI

5.3.1 Introduction

For obvious reasons, crime and accident reconstructions at the scene or on the computer are carried out without injuring the victim. However, the dynamics of the damage done to the victims body often remain unclear, leaving questions unanswered regarding the manner in which the injuries were caused.

In such cases, a reconstruction of the actual physical action (shot, blow, stabbing, etc.) may yield essential information as to what happened. In order to achieve this, one requires a representation of the body that can accurately simulate the interaction between the object (bullet, blunt instrument, etc.) and the structures of the human body. For bullet wounds, one uses the simulants described in 3.3, which can be used to simulate bone and most soft tissue.

Extensive experimentation has shown the materials used for studying gunshot wounds in human tissue to be suitable – under certain conditions – for the simulation of impact and stabbing injuries. In particular, the fracture characteristics of plastic bones subjected to impact and stabbing are very close to those of real bones.

The damage to simulants can be studied using the techniques described in 5.2 (CT, MRI), enabling direct comparisons between simulated and real-life injuries.

5.3.2 Reconstructing shooting incidents

5.3.2.1 Preliminary remarks

A gunshot is an extraordinarily rapid process. The interaction between bullet and medium as the bullet penetrates and passes through the medium lasts only a few milliseconds. As a result, normal physical processes such as heat transfer simply to not take place – because there is no time – or are subject to totally different rules if they do take place.

The physical characteristics of a material under highly dynamic loads (such as strength and elasticity) are often very different from those of the same material under quasistatic loading. These characteristics can only be determined with considerable effort – if at all. For instance, the dynamic tensile strength of copper rods measured using explosive techniques is more than double that of the same rods under conventional, quasistatic loading.

When attempting to imagine the sequence of events, one can therefore make completely erroneous assumptions.

In some cases, it is therefore necessary to conduct a dynamic reconstruction of the incident. This involves repeating the shots in question using the same weapon and ammunition to simulate the incident as it is suspected to have occurred, and to repeat this process until the simulated wound corresponds to that which actually occurred. Once one has achieved this, one has at least identified a situation which, taking all the dynamics of the shot into account, produces very similar injuries.

5.3.2.2 Points to bear in mind

When conducting reconstructions of shooting incidents, it is important to use the same ammunition (i.e. same manufacturer, type and lot) as was used in the original incident. Except in the case of contact or close range shots, the weapon is generally less significant, as long as the muzzle velocity and bullet stability (spin) remain the same. Where shot is fired from a shotgun, it is also important to use the same choke.

It is perfectly permissible – and indeed advisable – to replace human shapes by geometric shapes (head → sphere, hollow bone → hollow cylinder, local area of body surface → flat surface). This will generally have little or no effect on the interaction between the bullet and the structure and will make it possible to reproduce the same experimental conditions as many times as required, which is very helpful when one is trying to identify decisive parameters.

If one suspects that the geometric form (e.g. the shape of the parts of the skeleton involved) could have a significant effect on the reconstruction, it may well be appropriate and useful to construct a target structure that reproduces the original.

If other objects are to be included in the reconstruction, the simulated versions must be as identical to the original as possible in all respects that have a significant effect on the trajectory of the bullet or the marks created.

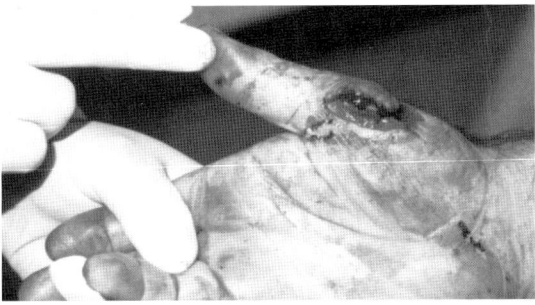 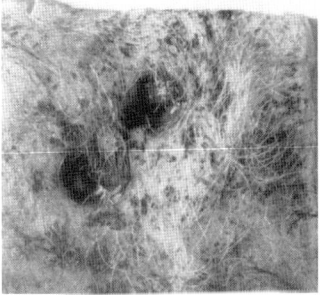

Fig. 5-22. (*left*) Unusual injury to the ball of the right thumb, for which there was initially no explanation.

Fig. 5-23. (*right*) Two adjacent entry wounds to the chest.

If a bullet passes through the side window of a car, for instance, the thickness and type of glass are important, as is the extent to which the window was open or closed. On the other hand, it is quite irrelevant whether the window was on the driver's or passenger's side.

5.3.2.3 Examples

We shall now examine three cases in which experimental reconstruction made a substantial contribution.

An unexplained hand injury. A man had apparently been shot from the inside of a car, through a window, while standing next to the door on the driver's side. In addition to three entry wounds, superficial bleeding and perforations of the skin on the left cheek and the inside of the left upper arm, the man had an unusual injury to the right thumb (see Fig. 5-22) which could not readily be explained by what was known about the incident.

The man had two entry wounds close together, in the chest, with no exit wounds (see Fig. 5-23) and a third in the right outer wall of the ribcage, with a corresponding exit wound at the rear.

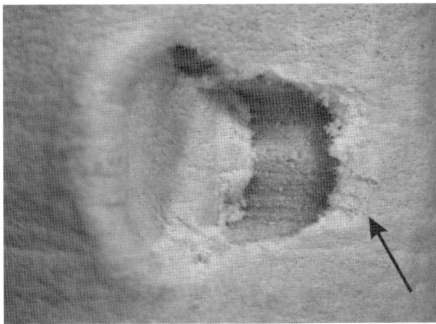 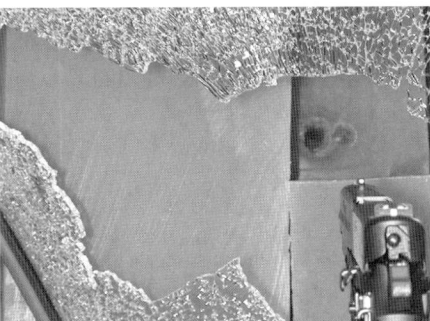

Fig. 5-24. (*left*) The simulated thumb injury (Fig. 5-22) corresponds to the real injury, even at the level of the finer structures (arrowed).

Fig. 5-25. (*right*) Separation of a bullet of the type used in the incident on passing through car glass.

It was thought that the victim might have tried to prevent the shots being fired by reaching through the partly open driver's door with his right hand. The injury to his hand could then have been caused by the slide of the pistol as the weapon recocked.

Experimental reconstruction using a number of variants made it possible to simulate an exceptionally similar wound, caused not by the slide itself but by the blade of the foresight mounted on the slide (see Fig. 5-24). It was also found, by firing bullets of the type used by the suspect into soap, that the lead core and the jacket always separate on passing through car glass, which meant that a single bullet of this type could cause two adjacent entry wounds (Fig. 5-25). The two entry wounds in Fig. 5-23 could therefore have been caused by a single shot. The simulation also showed that the superficial skin damage was caused by glass fragments. The experiments therefore resulted in a possible explanation for the incident that was not in contradiction with any of the wounds caused and left no questions open.

Who fired the shots? A police patrol stopped a suspect car. Before the two officers could ask the man for his papers, he opened fire on them with a handgun. In the ensuing firefight, one of the officers got between his colleague and the suspect. A bullet passed through each femur, just above the knee (see Fig. 5-26). The question was, from which weapon the shots had come.

The suspect had been using a 45 pistol, firing 45 Auto full metal-jacketed bullets. The police officers were using 9 mm Luger full metal-jacketed bullets, with the rear of the bullet covered by a brass plate to reduce lead emissions. It was not possible to tell from the police officer's clothing or from the entry or exit wounds which type of bullet had hit him. However, X-rays showed radio-opaque particles in both wound channels.

By firing both types of bullets through plastic bones embedded in gelatine, it was possible to demonstrate that the type of bullet used by the suspect left lead fragments on passing through a bone, whereas the bullets used by the police did not (see Fig. 5-27). This evidence therefore showed that the second police officer had not shot his colleague. In this instance, experimental simulation was the only means of deciding which weapon had caused the injuries.

Suicide and attempted suicide. A suicide and an attempted suicide, in two unrelated cases, both involving the same part of the body (the temple), the same type of weapon (a 7.5 mm Stgw 57 Swiss assault rifle with a muzzle energy of around 3,200 J) and the same ammunition, resulted in two completely different wounds. In one case, the bullet passed cleanly through the head, causing little damage, and the victim was able to leave hospital after just a few days. In the other case the skull burst, with fatal results.

By conducting experiments to simulate the two situations in a reproducible fashion, it was possible to understand the reasons for the different injuries (see

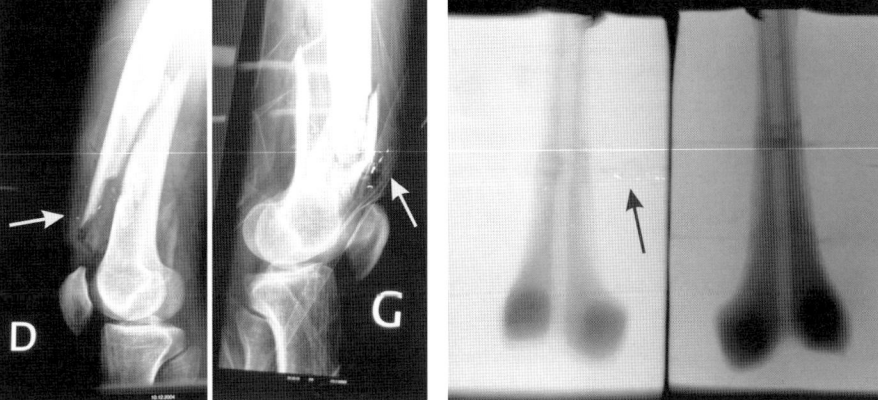

Fig. 5-26. (*left*) X-rays showing the knee joints of the injured police officer (D = Droite = Right, G = Gauche = Left). Radio-opaque particles are visible in the wound channels in both legs.

Fig. 5-27. (*right*) Simulation using anatomically-correct plastic bones embedded in gelatine. Bullets of the type used by the suspect (left) deposited radio-opaque particles in the wound channel, in all cases. Bullets of the type fired by the other police officer (right) did not.

Figs 5-28 and 5-29). The first case was a near-contact shot, as the victim had moved his head a few centimetres away from the muzzle just before pulling the trigger. Most of the gas pressure at the muzzle could escape laterally, and the bullet had to time to begin its precession and nutation movement. The result was a vanishingly small angle of incidence and hence a very long narrow channel. The second case was a contact shot, with the muzzle gases entering the head under pressure.

The skull will also burst when hit from longer range, when the bullet has started its precession and nutation movements. In such cases, however, the phenomenon is caused by the large temporary

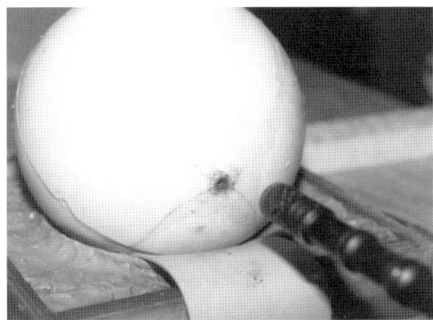
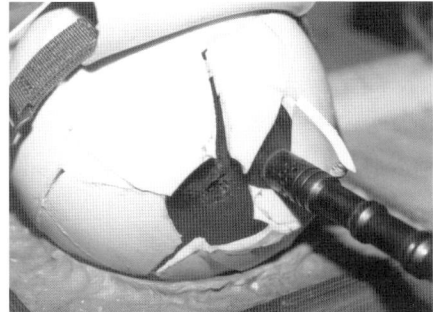

Fig. 5-28. (*left*) Near-contact shot into head model, fired from a 7.5 mm Stgw 57 assault rifle. Distance between muzzle and target: approx. 5 cm. The bullet has passed through cleanly, depositing little energy on the way. The fracture pattern corresponds to that observed in the real-life case.

Fig. 5-29. (*right*) Contact shot with the same weapon. The head model has burst as a result of the gas pressure at the muzzle.

Table 5-1. Velocities obtainable using vertical tubes and towers

Initial height [m]	Impact velocity [m/s]	Impact velocity [km/h]
5	9.9	35.7
10	14.0	50.4
15	17.2	61.7
20	19.8	71.3
25	22.1	79.7
30	24.3	87.3

cavity that the bullet creates. Only at much longer ranges (approx. 100 m and above, depending on calibre and type of bullet) does it once again become possible for the bullet to pass cleanly through the skull (see 6.3.2, Case C).

5.3.3 Blunt force

5.3.3.1 Equipment used and experimental options

Blunt force is generally applied at a velocity much lower than that of a bullet. In order to simulate blunt force, one therefore requires special equipment to accelerate the object. It is possible to use vertical tubes and towers, and compressed air cannons with a calibre of 10 cm or more are also available. Table 5-1 shows the velocities that can be achieved using vertical tubes and towers, as a function of initial height.

Fig. 5-30. Falling body with part of a hammer. It is possible to obtain any desired impact angle by positioning the head model appropriately.

In the case of a tube, the falling body is guided. It can therefore be fitted with the relevant object (hammer, pistol grip, etc.), which can allowed to fall onto the body model under precisely determined impact conditions (see Fig. 5-30). Towers are only suitable for objects that remain stable in free-fall, such as spheres, fin-stabilized bodies and those stabilized along the principle of an arrow.

Compressed-air cannons make it possible to obtain velocities and energies that would be virtually impossible to achieve using gravity equipment, as they are capable of accelerating objects of a few kg to velocities of up 120 m/s.

There are limits to the simulation of blunt force, as neither gelatine nor soap simulate the reaction of soft tissue to this type of effect correctly. Gelatine can be used for a rough simulation of the dynamic penetration depth of a non-penetrating body in soft tissue. Soap is suitable for determining the energy of blunt force, as there is a closely-correlated linear relationship between the volume of soap displaced and the energy applied.

Synthetic bone models, however, do allow accurate simulation of bone injuries due to blunt force. These models include the skin/skull/brain model (THALI et al. 2002a), plus hollow bones and ribcage/spine combinations suitably covered and embedded in gelatine.

5.3.3.2 Examples

Attacks involving hammers and similar tools. Where an assault involves the use of a hammer or similar against the head, there is often a requirement to determine the energy used. Experiments using the head model simulate the fracture pattern very accurately (see Fig. 5-31). The relevant part of the tool is fixed to the falling body, which is then dropped onto the head model from various heights until the simulated injury corresponds to the injury observed in real life. From the height and the mass of the falling body, it is then possible to calculate the approximate energy required to produce the injury.

Falls. It is also possible to determine the impact velocity (and hence the height) required to cause a given skull fracture pattern. Here, the head model is dropped from a tower onto the relevant object (post, pavement slab, etc), as this is the only way to ensure that the impact occurs with the model oriented in a particular fashion (being spherical, the head model remains stable in free-fall). It is also possible to equip the model with accelerometers and record the accelerations to which it is subjected.

Projection of debris. When defining danger areas for such sites as ammunition dumps and explosives factories, it is necessary to calculate the danger posed by flying debris. Here, one needs to determine percentage lethality as a function of

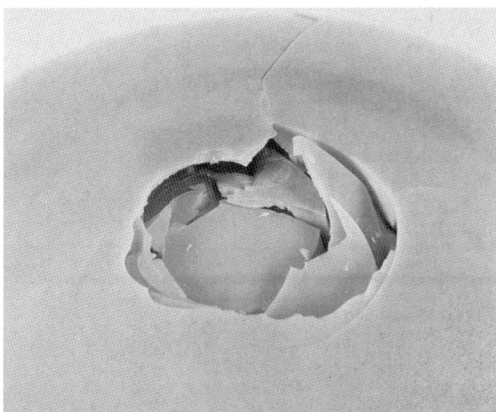

Fig. 5-31. (*left*) Depressed fracture on a head model, caused by a hammer blow simulated using a vertical tube.

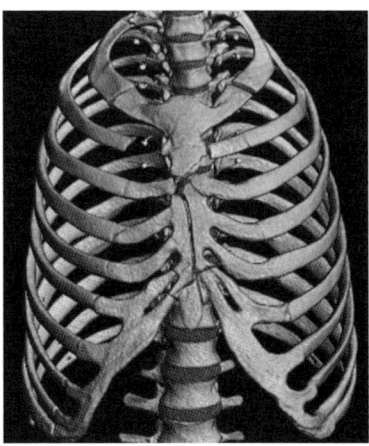

Fig. 5-32. CT image of a ribcage embedded in gelatine after being struck by flying debris, simulated using a compressed-air cannon. Mass of object: 1 kg).

impact energy. As there is virtually no real-life data on this, simulation is the obvious approach.

Another question, which is analogous from a forensics point of view, is that of the danger posed by an object that is thrown, such as a cobblestone.

Fig. 5-32 shows an example taken from an extensive series of experiments involving models of heads, torsos and limbs. This type of experiment is perfectly suitable for the purposes mentioned above, and the results make it possible to draw relevant conclusions.

5.3.4 Using Virtopsy in practice

5.3.4.1 Documentation and visualization

Modern graphics methods can also be used for non-invasive examination of soft-tissue simulants such as gelatine and soap, plus synthetic models of body parts. It is possible to visualize the effects of bullets, blunt force and stabbing using radiological techniques (KORAC et al. 2002, THALI et al. 2008, www.virtopsy.com).

Reconstructions can be performed using either anatomically-correct models of body parts or, as mentioned above, simple geometric forms (see Fig. 5-33).

Figures 5-34 to 5-43 show examples corresponding to images taken from clinical practice.

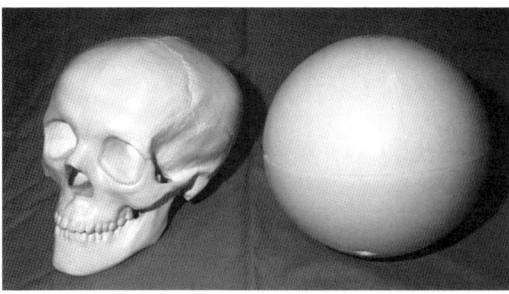

Fig. 5-33. Anatomical and geometric skull models.

5.3 Experimental reconstruction 299

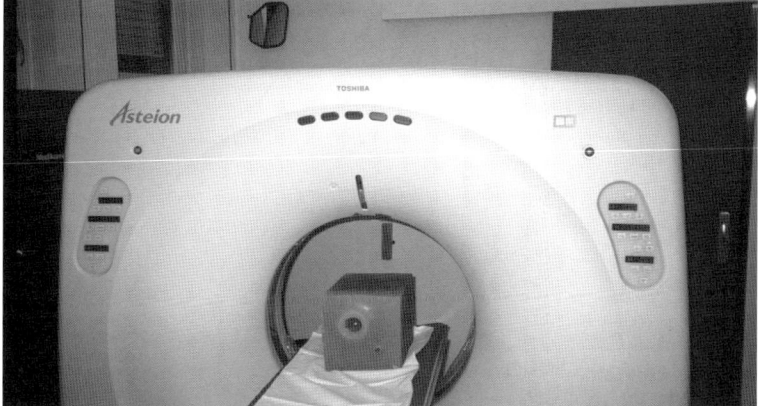

Fig. 5-34. Soap and gelatine blocks can be studied using CT, as can models of body parts (see also Fig. 5-32).

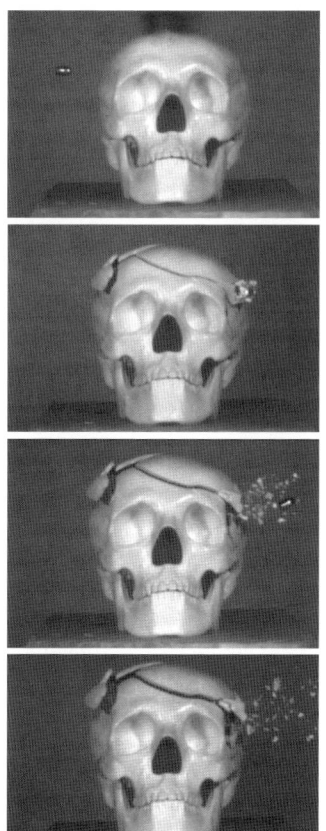

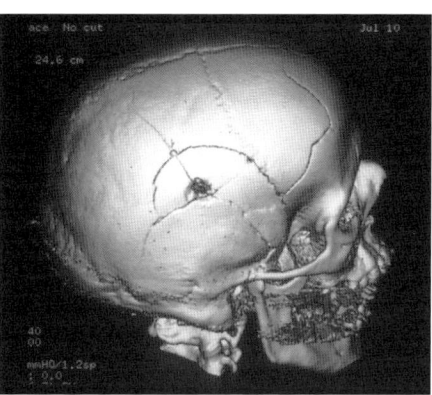

Fig. 5-35. Real-life case: A 9 mm Luger full metal-jacketed bullet has passed through this man's skull.

Fig. 5-36. High-speed images of a simulation of the case shown in Fig. 5-37 using the skull/brain model.

300   5 Wound ballistics and forensic medicine

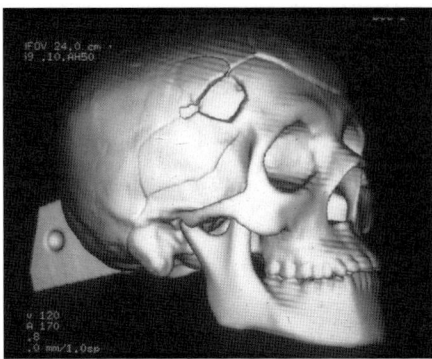 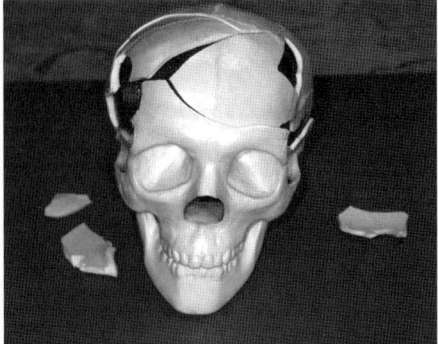

Fig. 5-37.   (*left*) CT image of a bullet wound analogous to that shown in Fig. 5-37, simulated using a skull/brain model.

Fig. 5-38.   (*right*) Same model as in Fig. 5-38, showing fracture pattern caused by bullet.

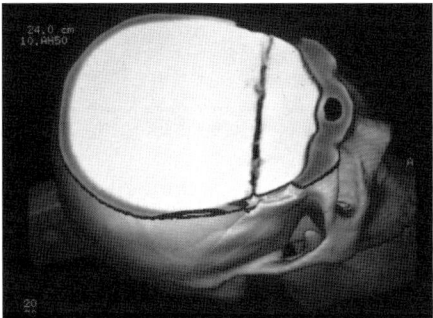 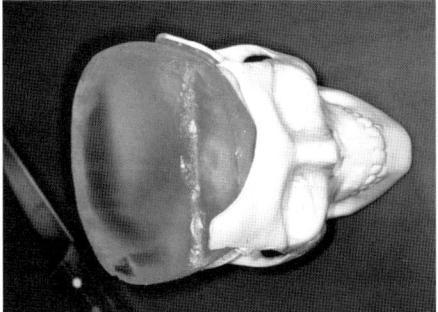

Fig. 5-39.   (*left*) Visualization of the wound channel in a synthetic, anatomical head model, "sliced through" using software.

Fig. 5-40.   (*right*) Same wound channel in the model itself, with the wound channel exposed in conventional fashion, using a knife and/or scalpel.

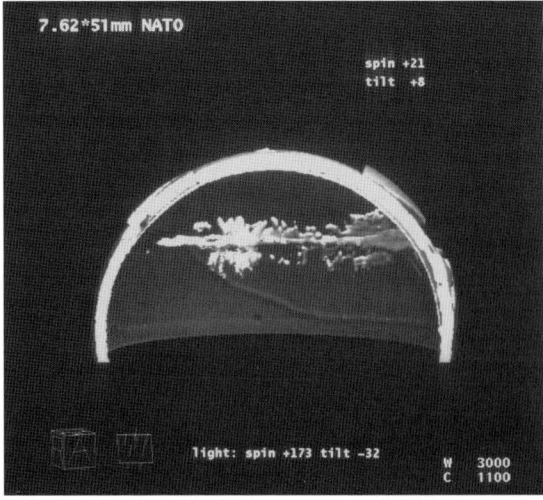

Fig. 5-41.   3D radiological visualization of a wound channel in a synthetic, geometrical head model (7.62 mm NATO full metal-jacketed rifle bullet).

5.3 Experimental reconstruction 301

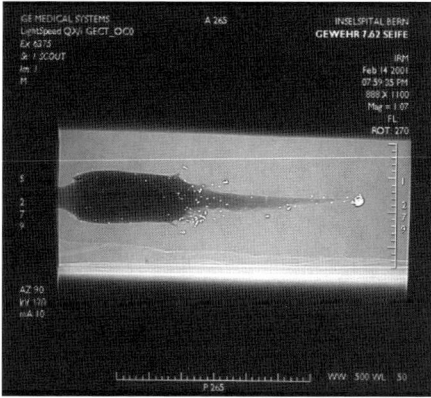 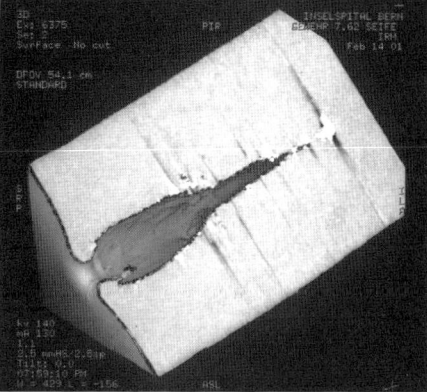

Fig. 5-42. (*left*) CT image of a soap block following penetration by a hunting bullet. The position of the remains of the bullet and of the fragments in the wound channel are clearly visible, as is the temporary cavity.

Fig. 5-43. (*right*) 3D CT image of the soap block from Fig. 5-35. It is possible to calculate the volume of the temporary wound cavity using suitable software.

5.3.4.2 Example

A man sitting at a bar was hit several times by full metal-jacketed bullets fired from a 5.6 mm GP 90 assault rifle (Stgw 90). The post-mortem revealed that three bullets had passed through his body and one had grazed his head (see Figs 5-44 and 5-45). One bullet entered via the right shoulder blade, passed through it and

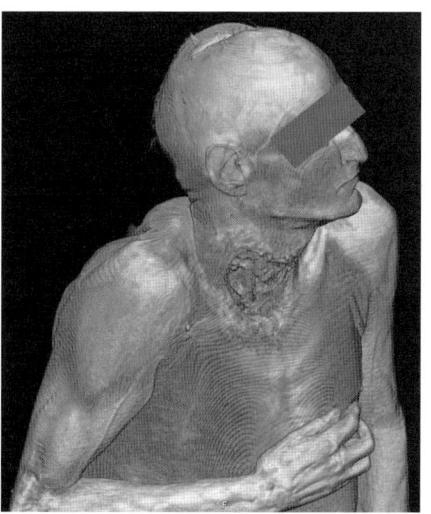 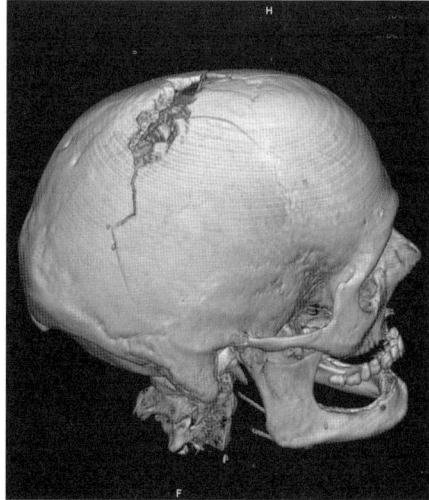

Fig. 5-44. (*left*) Semi-transparent volume rendering view of the three-dimensional CT dataset, showing a grazing shot to the head and an exit wound in the throat.

Fig. 5-45. (*right*) Bone damage associated with the grazing shot to the head.

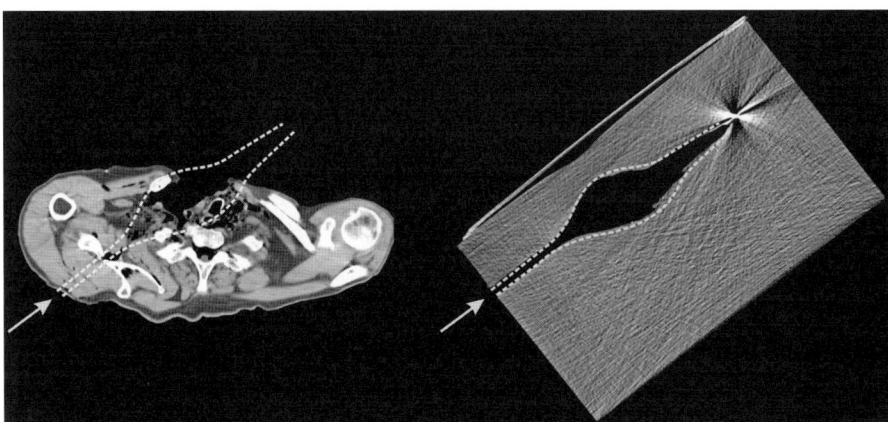

Fig. 5-46. *Left:* Wound channel associated with the exit wound shown in Fig. 5-44, displayed using multiplanar CT reconstruction. There is little damage to soft tissue in the narrow channel, but the scan clearly shows the damage done in the region of the temporary cavity, which formed after the bullet had passed through the shoulder blade. *Right:* The channel created by the same type of bullet in soap.

exited via the front of the throat, creating a gaping exit wound. The projectile destroyed the blood vessels in the neck, causing external bleeding. In addition, air entered the circulatory system via the damaged blood vessels, causing an air embolism. Another bullet entered via the left buttock and exited via the left hand side of the left thigh, creating a gaping exit wound. This bullet shattered the femur. A bullet also passed through the right thigh, causing a superficial wound. In addition, a bullet grazed the skull, opening up the calvarium. The causes of death were the loss of blood due to damage to the blood vessels, together with the air embolism.

The first step in conducting the post mortem was to take a number of photographs. These served to record the wounds caused by the bullets.

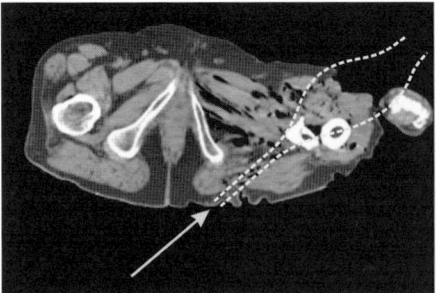 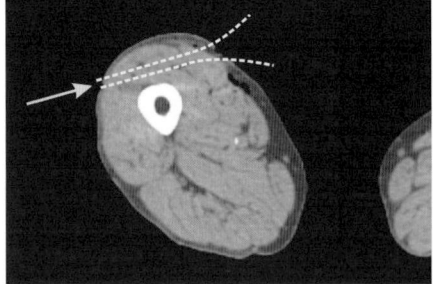

Fig. 5-47. Shot through buttocks and thigh: little damage to soft tissue in the region of the entry wound, more damage around the temporary cavity.

Fig. 5-48. Shot through right thigh: Little damage, as the wound channel lies almost entirely within the narrow channel.

Shots were fired into a soap block, using the same type of bullet. The block was then subjected to CT scanning. This made it possible to document the channel of the bullet in the block non-invasively, in three dimensions.

It was then possible to compare the radiological data from the soap block with the finding from the body (see Figs 5-46 to 5-48 and accompanying captions). One can see that the energy transfer profile of the bullet in the simulant corresponds to the damage in the body resulting from the temporary cavity. The size is visible primarily in the form of air and deformation of tissue, rather than as a permanent cavity, as biological tissue collapses following passage of the bullet.

This example clearly shows that the information regarding the energy transfer (effectiveness) of a bullet obtained by wound ballistics simulations is of value in investigating and assessing the wound caused by that bullet. This is true not only for radiological visualization of the wound but also in the case of a conventional post-mortem examination.

6 Wound ballistics and surgery
R. M. COUPLAND

6.1 The historical connection between wound ballistics and surgery

There is a long and close association between surgery and wound ballistics studies (KOCHER 1875a,b, SPENCER 1908, OWEN-SMITH 1981, FACKLER et al. 1985, COOPER 1990, BELLAMY et al. 1991, PROKOSCH 1995, BOWYER 1997, RYAN et al. 1997). The reason for this tradition is obvious. Surgeons – whether military or civilian – who are working in any area where there might be acts of armed violence must prepare themselves to treat wounded people.

Thousands of wound ballistics studies have been undertaken since the nineteenth century. A variety of experimental targets have been shot including corpses, live animals, wet clay, wet telephone directories, soap and gelatine. The current understanding of wound ballistics therefore has two sources: wound ballistics experiments, and the results of ballistic trauma witnessed by health professionals in surgical hospitals and mortuaries. In the past, many assumptions have been made when applying findings from one domain to the other. Although extensive knowledge has been gained about the interaction between projectiles and tissue, it is not obvious whether or how these studies have contributed to improving the surgical care of wounded people. This chapter aims to highlight those studies that have important clinical implications. It also identifies some aspects of surgical care that have not, as yet, been helped by wound ballistics studies. This may indicate where new models are needed and possible areas in which to focus research.

Many wound ballistics studies relate more to weapon design than to the treatment of wounded people. Such studies might pertain to, for example, the 1899 Hague Declaration prohibiting bullets that expand or flatten easily in the human body (see Chapter 7). Others may generate information pertaining to weapon development (COUPLAND 1999b, OGSTON 1899, BLACK et al. 1941). There is therefore an ethical dimension to surgeons' involvement in wound ballistics studies (WORLD MEDICAL ASSOCIATION 1997).

On a broader scale, surgeons treating people wounded in conflict are also in a position to document the nature of wounds, which may well reflect the kinds of weapons or ammunition used, the number of people injured, who the wounded people are and the circumstances of injury; this in turn pertains to how the weapons in question were used. Thus the surgeon is an important witness to the "human cost" of the conflict. The role of the "surgeon as witness" clearly has implications for preventive policies and laws (COUPLAND et al. 1997, MEDDINGS 1997 und 1999). It can also, in some circumstances, have implications for the personal safety of the surgeon. This chapter also discusses possible roles for health professionals who witness ballistic trauma.

6.2 Wound ballistics and ballistic trauma – what's the difference?

Wound ballistics is a scientific discipline. It investigates the interaction between a projectile or blast and tissues. As mentioned above, two sources of data feed this discipline; the laboratory, and observations of the effects on people, either in hospital or in a mortuary. Ballistic trauma is both a field of study and a clinical specialty. It consists of observing both the effects of a projectile or blast on human tissue and the pathophysiological sequelae (the biological response to that trauma).

Wound ballistics pertains only to the physics of the interaction between a projectile or blast and simulated or real tissues. Ballistic trauma includes the pathophysiological outcome of the physical process, such as cell death through crushing and laceration of tissue leading to tissue necrosis; resulting blood loss; fragmentation of bone and loss of corresponding periosteum. Later pathophysiological effects of ballistic trauma include inflammation of the wound, wound infection and any disability.

The premise upon which this chapter is written is that effective treatment of people suffering ballistic trauma involves understanding wound ballistics, i.e. the physical interaction between projectiles and tissues. Understanding wound ballistics does not automatically generate treatment regimes for wounded people. The knowledge that a bullet might deposit a certain number of joules of energy at a certain point along its track in a soft tissue simulant does not automatically translate into a policy for treating such a wound in any part of the body. Before surgeons can draw conclusions about wound management from wound ballistics studies, it is necessary to ensure that laboratory findings in simulated wounds constitute a valid comparison for wounds that are observed in people coming to hospitals.

6.3 Comparing simulated wounds and real wounds

6.3.1 Preliminary remarks

When comparing simulated and real wounds, one makes a very important assumption: that the deposit of energy at a given point along a simulated wound track is in some way representative of the degree of tissue damage at the same point in an equivalent, real wound. In other words, one assumes that the point at which the simulated track is widest is the point at which there is most tissue damage in a real wound. This assumption can be supported by the clinical cases presented here from ICRC hospitals.

The appearance of the wound, combined in some cases with the x-ray of the wounded part, give a clear indication of the wound track. These can then be compared with equivalent simulated wounds.

6.3.2 Case studies

Case A: Bullet wound, left buttock, perineum and right thigh. The photograph (Fig. 6-1 a) shows a long bullet track. The bullet entered the patient high on the left buttock, traversed the pelvic soft tissues without causing any damage to viscera or major vessels, exited the perineum and then re-entered the right thigh where it caused a wider channel. It then made a smaller exit. Together, these wounds represent the simulated wound in Fig. 6-1 b. [ICRC wound score (see 6.6.2): pelvic wound: E1, X1, C0, F0, V0, M0, right thigh E21, X5, C1, F0, V0, M0.]

Case B: Bullet wound, left thigh, with femoral fracture. This patient was shot in the left thigh at close range (less than one metre) with an assault rifle. There is a small entry on the medial side of the thigh and a large exit on the lateral side (Fig. 6-2 a). The x-ray (Fig. 6-2 c) shows a femoral fracture and multiple metallic fragments of the bullet, indicating that it broke up early in the track. The simulated equivalent is presented in Fig. 6-2 b. [Wound score: E1, X8, C1, F2, V0, M2.]

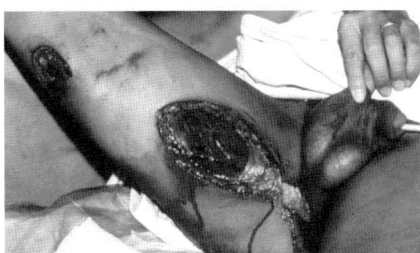

Fig. 6-1 a. Case A. Real wound. Exit wound top left.

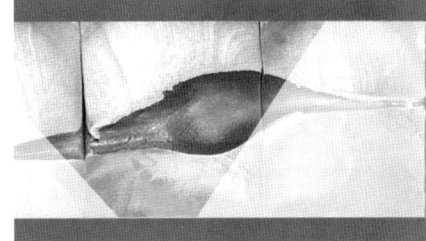

Fig. 6-1 b. Case A. Simulation in glycerine soap. Bullet from right. The relevant part of the bullet track is highlighted.

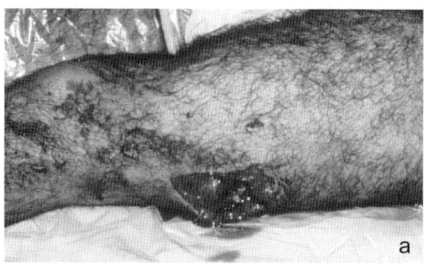
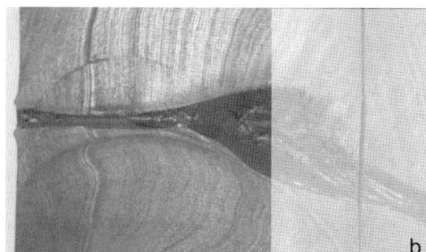

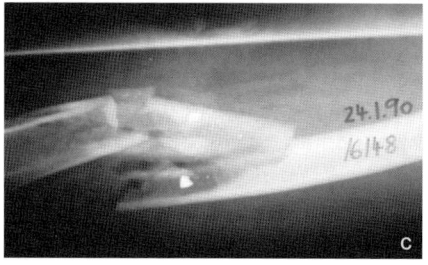

Fig. 6-2. Case B. Passage of bullet through thigh:
a. Real wound. The bullet entered from the top.
b. Simulation in glycerine soap (highlighted). Bullet from left.

Case C: Cranio-cerebral bullet wound. Fig. 6-3 a shows a patient with a bullet wound of the head being prepared for surgery. He had a similar sized wound over the occiput. X-ray (Fig. 6-3 b) shows the cranial entry and exit wounds and a linear cranial fracture. The patient survived with a left hemiplegia. It is only possible to survive such a wound if the energy deposited along the track is minimal. This wound represents the equivalent of the long narrow channel as seen in Fig. 6-3 c. [Wound score: E1, X1, C0, F1, V1, M0]

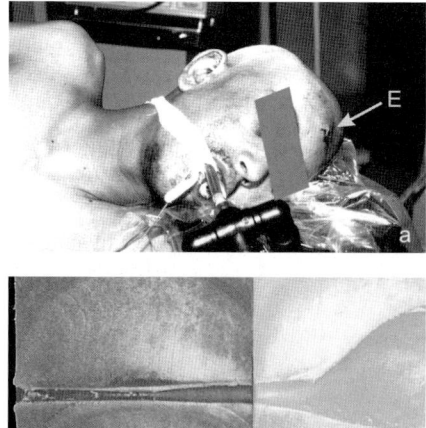
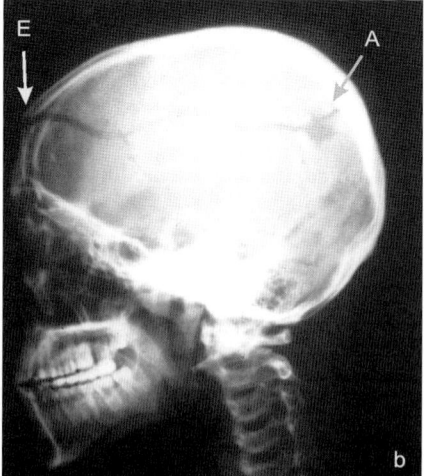

Fig. 6-3. Case C, Head: **a.** Entry wound in forehead (E). **b.** X-ray showing entry wound (E) and exit wound (A). **c.** Corresponding simulation in glycerine soap (highlighted).

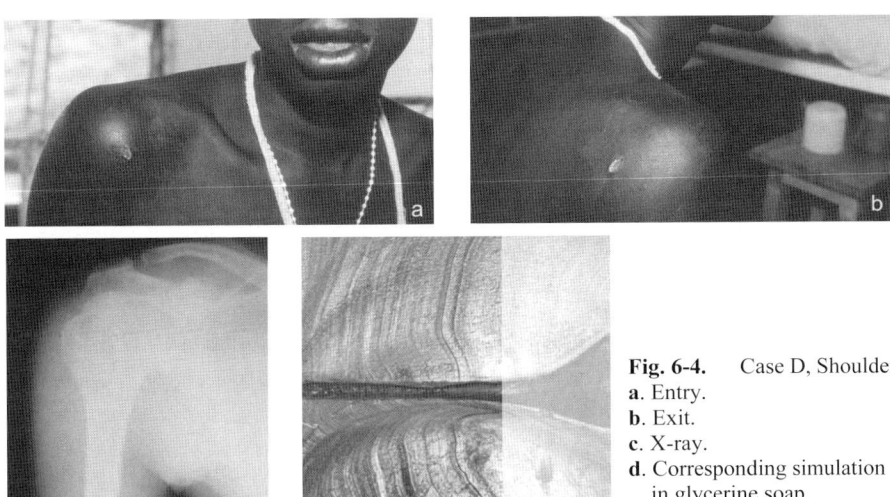

Fig. 6-4. Case D, Shoulder:
a. Entry.
b. Exit.
c. X-ray.
d. Corresponding simulation in glycerine soap (highlighted).

Case D. Bullet wound, right shoulder. Entry of a rifle bullet on the front of the right shoulder (Fig. 6-4 a). Exit on the back of the right shoulder (Fig. 6-4 b). X-ray of the shoulder. No fracture is visible (Fig. 6-4 c). The patient required no surgical intervention. He recovered good shoulder movement with physiotherapy. This is the equivalent of the simulated wound in Fig. 6-4 d. [Wound score: E1, X1, C0, F1, V0, M0. (The fracture – F1 – is presumed, but is not visible on x-ray.)]

Case E: Bullet wound, right chest and abdomen. This patient (Fig. 6-5 a) said he had been shot. The wound of the lower right chest was the entry wound. There was no exit wound. The large wound represents the entry wound of a bullet with

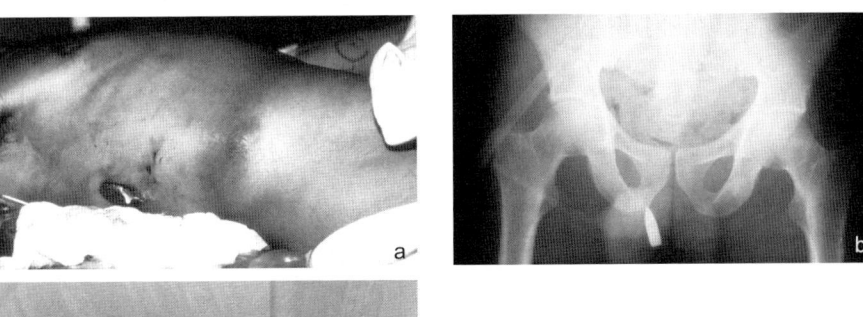

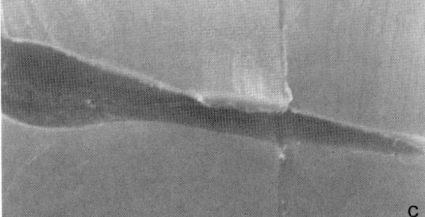

Fig. 6-5. Case E. Ricochet:
a. Entry wound.
b. X-ray, showing bullet in pelvis.
c. Corresponding bullet track in glycerine soap.

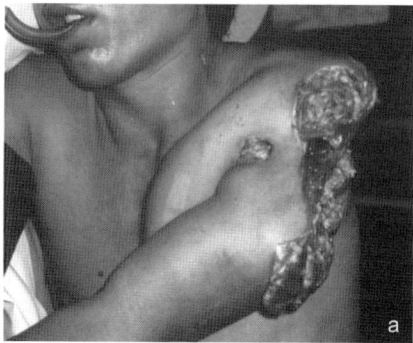
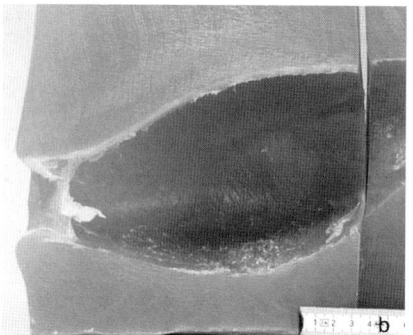

Fig. 6-6. Case F. Wound caused by a ricochet passing through shoulder: **a**. Real wound, **b**. Simulation.

low stability, possibly a ricochet (Fig. 6-5 b). X-ray (Fig. 6-5 c) shows the bullet in the perineum having traversed the ribs, right lung, liver and small bowel. [Wound classification E4, X0, C1, F1, V1, M1.]

Case F: Bullet wound, left shoulder. The bullet entered on the front of the shoulder and caused complete destruction of the shoulder joint and all surrounding soft tissues (Fig. 6-6 a). (No x-ray was taken.) The tip of the full metal jacket was found in the wound. This was almost certainly a ricochet from a shot fired at close range. The patient survived despite developing clinical gas gangrene and undergoing a forequarter amputation. (For the corresponding simulation, see Fig. 6-6 b). [Wound classification E3, X30, C1, F2, V0, M1]

Case G: Multiple fragment wounds, right thigh and leg. (Fig. 6-7 a) This woman suffered multiple small wounds of her right thigh and leg from a hand grenade. The x-ray shows multiple small metallic fragments. She did not undergo an operation and the wounds healed without complication. The "funnel" shape of a simulated small fragment wound can be seen in Fig. 6-7 b. [Wound classification for each wound: E1, X0, C0, V0, F0, M1.]

Case H Cranio-cerebral fragment wound. (Fig. 6-8 a) Photograph of a patient with a cranio-cerebral fragment wound being prepared for surgery. (Fig. 6-8 b) X-ray showing the cranial entry wound, the cranial exit wound, linear cranial fractures and the fragment itself. The wound track is about 12 cm long. The fragment

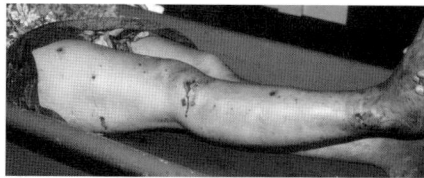
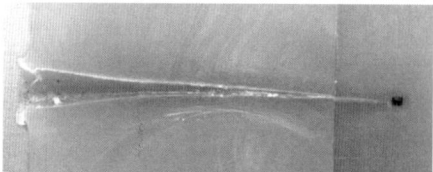

Fig. 6-7 a. Case G: Fragment wounds caused by a hand-grenade.

Fig. 6-7 b. Case G: Track of a small fragment in glycerine soap.

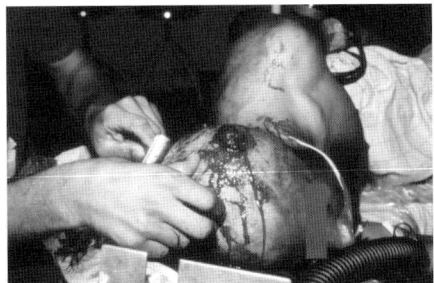

Fig. 6-8 a. Case H:. Cranio-cerebral injury caused by a fragment. Entry wound.

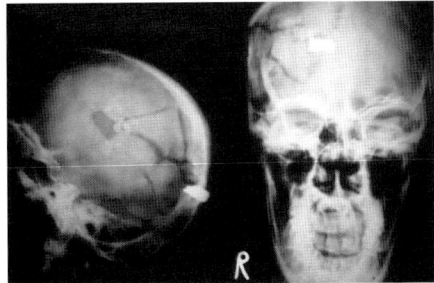

Fig 6-8 b. Case H: X-ray. The entry hole and the final position of the fragment under the eipcranium are clearly visible.

was removed via a separate occipital incision as it was palpable beneath the skin, having exited the cranium. The "funnel" shape of a simulated small fragment wound can be seen in Fig. 6-7 b. [Wound score E3, X0, C0, F2, V1, M1.]

6.3.3 Conclusions

The above comparisons go further than showing tracks in real wounds that are similar to wounds simulated in a laboratory. They confirm two laboratory phenomena that are counter-intuitive from a clinical perspective. The first is that when a projectile passes through tissue at high speed, not only does the bullet do work on the tissue but the tissue also does work on the bullet. This is the reason why the bullet deforms or fragments. This has important legal implications (for example, the degree of deformation may indicate the range from which the bullet was fired). The second is that the length and breadth of the wound track created when a bullet passes through bone may not differ significantly from those observed where there was no contact between bullet and bone. This second phenomenon has important clinical implications (see below.)

We can draw four main clinical conclusions from the above comparisons. The first is that projectiles can cause a variety of wounds. The second is that there is not always significant tissue damage along the wound track. The third is that the tissue damage in a wound is not uniformly distributed along the wound track. The fourth is that real and simulated wound tracks would appear to be similar. These conclusions support the assumption above, namely that the point of maximum energy transfer in a simulated track corresponds to the point of maximum tissue damage in a real, equivalent wound.

Whilst wound ballistics studies may tell us where in a wound track (made by a certain projectile travelling at a certain velocity) we will most likely find the greatest volume of dead or damaged tissue, (see 6.5.1) the same studies do not tell us how much tissue is dead. These studies also do not tell us how much tissue is damaged but reparable or which tissue, if not removed, could act as a culture medium for dangerous pathogens (COUPLAND 1993b). These are matters of judgement for the surgeon and represent difficult clinical decisions. Furthermore, these factors can change over time with changes in the pathophysiological reaction to the physical wounding.

6.4 Clinical features of real wounds

Hospital staff working with people wounded by weapons of war see a perplexing variety of wounds of all parts of the body. The above comparison of real wounds with simulated wounds permits us to summarize clinical features of the bullet and fragment wounds that may be seen in a hospital treating wounded people.

Features of individual bullet wounds:
- There is usually a small entry wound if the bullet is stable in flight and is not designed to expand on impact;
- There is rarely a large volume of tissue damage at a small entry wound;
- Extensive tissue damage at the entry most likely results from a ricochet or impact of an expanding bullet;
- There may be no exit wound; if there is an exit wound, the size is variable;
- The length of the track is variable; this is influenced by both energy of the bullet and its construction;
- The amount of tissue damage in a wound is variable;
- Whatever the bullets' construction, an x-ray of the wounded part commonly shows small metal fragment (COUPLAND 1999).

In armed conflict, bullet wounds are usually single, because the wounds tend to be sustained at longer range. Where there are multiple bullet wounds, the person has usually been shot at close range. This means that the wound severity is compounded not only by the multiple wounds but also by the higher energy deposit from the bullets having been fired at closer range.

Features of individual fragment wounds:
- The wound track is always widest at the entry and therefore the most tissue damage is near the entry.
- The length of each track and the degree of tissue damage are determined by the mass and velocity of the fragment.
- The entry is more or less circular whatever the shape of the fragment.

When people suffer wounds from fragmentation weapons, the wounds are usually multiple. The number of wounds is determined by the kind of weapon and how close the person is to the source of the fragments. There may also be accompanying blast, burn or crush injuries.

Some anatomical features of wounds can be "scored" using the Red Cross wound classification. (See 6.6)

6.5 The contribution of wound ballistics to the care of wounded people

Certain observations regarding simulated wound tracks have important implications for the care of wounded people, as do the comparisons in 6.3.

6.5.1 The "wound profile"

The concept of the wound profile is an important conceptual tool when considering wounds (FACKLER et al. 1985). Its origin lies in the observation that the deposit of energy along a wound track is not uniform. The concept allows people with minimal experience of ballistic trauma to understand where there may be the most energy deposit along a track and therefore where most tissue damage is likely to be located. It also makes a clinically important distinction between bullet wounds and fragment wounds.

The most clinically relevant element of this is that both bullets and fragments can penetrate the body and yet cause little tissue damage.

Bullets of whatever velocity, if stable in flight (see 2.3.5.1), may make a wound track that begins with a narrow channel. If the total length of the wound track is short (e.g. across a limb) then there may be little tissue damage along the track. Likewise, the tissue damage in a small fragment wound may be confined to the first centimetre of its track, with little tissue damage from there on. Such wounds, if they only affect soft tissue, may not require an operation to remove dead or damaged tissue because the volume of such tissue is minimal. This has important implications for treating people with multiple small fragment wounds (COUPLAND 1993a, BOWYER 1997). In turn, this has important implications for surgical workload, triage and the management of multiple admissions to hospital. People with multiple small fragment wounds often arrive in groups; a policy of initially not operating in the case of small soft tissue wounds has important implications for running a hospital in a conflict area.

6.5.2 What causes tissue damage?

Wound ballistics studies, such as those using high speed photography of projectiles passing through gelatine, show that as the moving projectile does work on the tissue, the tissue accelerates away from the presenting point of the projectile.

This acceleration is the result of the work done on the medium by the bullet (see also 3.2, especially 3.2.1.1).

The two possible mechanisms of tissue damage therefore are crush or laceration. A projectile that causes minimal tissue damage at a given point in its track is causing less crush by comparison with laceration. This is compatible with the notion that narrow wound tracks contain little tissue damage (see 6.5.1).

Laceration of tissue may also lacerate blood vessels. This may also result in necrosis near to the track, but from ischaemia as opposed to crush.

The question is often raised of whether a shock wave (see 4.3.2) is clinically significant. Shock waves generally do not cause tissue damage; they do not actually move tissue and therefore carry little energy and therefore cannot do physical work. However, clinical cases are occasionally seen in which a projectile passes close to tissue that conducts electrical impulses, without obviously damaging the tissue, and yet the electrical function of the tissues suffers disruption.

One example is paraplegia resulting from a wound track that passes close to but not through the vertebral column. Recovery from such paraplegia may be possible. However, this remains largely a clinical observation.

How the electrical conductivity of certain tissues is disrupted by the shock wave has been investigated *in vitro* (see 4.3.2.3) but has not been extensively investigated *in vivo* (see 6.5.7).

6.5.3   Gas in tissues on a clinical x-ray

Gas is frequently seen on x-rays of a wounded part. Wound ballistics studies show how the passage of a projectile causes a vacuum which, on collapsing, sucks in air. This is also the mechanism by which a wound becomes contaminated.

One of the many misperceptions in the overlapping domains of surgery and wound ballistics is that all "high velocity" wounds cause massive tissue destruction. This myth, combined with the surgical history of World War I, has left the notion in the collective surgical mind that gas gangrene in penetrating trauma is common, if not inevitable. This is not the case. The belief has led to unnecessarily aggressive surgical regimes. All wounds are contaminated; they are not all infected (MELLOR 1997). They are certainly not all infected with gas-producing clostridial bacteria.

6.5.4   The "hot bullet" theory

Another surgical myth is that the heat of a bullet sterilizes tissue as it passes through. Whilst it is true that a bullet recovered soon after being shot is hot to the touch, wound ballistics studies show that the time of contact between the bullet

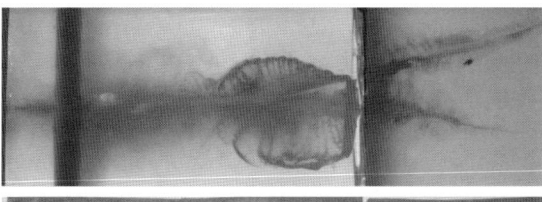
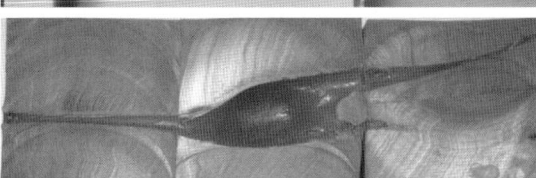

Fig. 6-9. Comparison between a wound profile where the bullet has passed through bone (top) and one where it has not (bottom). The length of the narrow channel and the point of maximum energy transfer are virtually identical in the two cases.

and tissue is so short that there is simply not enough time for the bullet to transfer its heat to the tissue (see also 3.2.6.4).

6.5.5 Long bone fractures

Wound ballistics studies have significantly enhanced our understanding of the mechanism of fractures (see 3.2.5). Simulated wounds show that the wound profile changes little if the projectile encounters bone in its passage through the body.

The laboratory demonstration that best illustrates this comprises shooting into a gelatine block which incorporates a polyurethane "bone" covered with a latex "periosteum" (see 3.2.5 and 3.3.5).

The effects of low energy transfer from a full-metal jacket with high energy transfer from a semi-jacket bullet of the same weight and velocity are compared in COUPLAND et al. (2000). The fact that the wound profile is little changed by contact with bone (see Fig. 6-9), shows that comminuted fractures seen on clinical x-rays can be associated with low-energy wounds. In such cases, the surrounding structures, including the periosteum, are likely to remain intact. By contrast, a high-energy wound not only causes a comminuted fracture but also destroys the surrounding structure. In the low-energy wound, most bone fragments will still be attached to soft tissue, maintaining their blood supply. The high-energy wound will contain many loose bone fragments, which should be removed during wound excision. Furthermore, the two wounds have quite different bone-healing capacities. This has implications for wound healing, the method of fracture stabilization, physiotherapy, duration of hospital stay, risk of amputation and residual disability.

A clinical x-ray is not by itself a reliable indicator of the extent of tissue damage around the fracture. When describing low-energy fractures to surgeons unfamiliar with wound ballistics, it is sometimes useful to use the analogy of a wine bottle hanging in a sock. This parallel describes tapping a wine bottle (the bone) which is hanging in a sock (the surrounding soft tissues) with a hammer. The bottle will fracture into many pieces but the sock remains intact. Whilst this is a crude analogy, it serves the purpose of illustrating how a comminuted fracture can be produced with little energy transfer and so little energy transfer to surrounding tissues.

The same wound ballistics experiments show that bone fragments do not act as secondary projectiles; they do not inherit enough kinetic energy from the projectile to penetrate tissue beyond the temporary cavity. This means that the detached bone fragments that the surgeon should remove during wound excision lie in the wound cavity and not in their own tracks.

6.5.6 Cranio-cerebral wounds

The head model shown in 3.3.5.3 is important for surgeons as well as for forensic specialists. The model shows the sort of cranio-cerebral wounds that a person might survive. The head model shows how bone fragments that appear to have been driven into the brain are in fact embedded in lacerated tissue (see Fig. 3-40). The head wound model also shows how deeply these bone fragments may be embedded. Removing bone fragments to avoid later brain abscess or epilepsy need not involve traversing undamaged brain tissue.

6.5.7 Unresolved issues

The above shows how wound ballistics can be used to draw up science-based treatment guidelines and operative guidelines for the surgeon. There are a number of issues of critical importance to the surgical care of wounded people, to which wound ballistics has yet to make a major contribution. This stems from the fact that these issues pertain more to understanding the pathophysiology resulting from the physical process rather than to understanding the physical process itself. Developing valid models to simulate such processes will always be extremely difficult.

One example has already been given in 6.5.2: the question of how the normal electrical function of certain cells might be disrupted by a shock wave. Two other such issues are addressed here.

How much tissue needs to be removed during wound excision is still a matter of surgical judgement. Whilst tissue is clearly damaged along a wound track and the degree of damage is not uniform, there is little evidence which can be brought to bear on the questions of which tissue is dead, which is damaged but viable and which tissue must be removed. As we saw in 6.5.1, it may be possible to manage wounds with minimal tissue damage without operating. It is far from clear what volume of potential "culture medium" in a wound can be accepted; this will certainly change according to the part of the body injured, whether muscles were contracted at the time of impact and context (which might determine for example the nature and extent of wound contamination). In the same vein, it is a matter of surgical judgment as to which bone fragments in a wound should be removed. Clearly, bone fragments that are completely detached and lying free in the wound cavity must be removed and those still firmly attached to soft tissue must be left in place. However, one frequently finds a large piece of bone with minimal attachment to, and therefore potential blood supply from, surrounding soft tissues. Whether to remove such partially attached bone fragments can be a difficult decision.

Blood vessel injury is common in ballistic trauma. The projectile can lacerate a major vessel. The surgical decision as to whether to repair a lacerated vessel is relatively easy. A more difficult clinical decision relates to a vessel that has not been lacerated but is near to the wound track and has been crushed. There is a presumed risk of thrombus formation following intimal damage. Should the affected part of the vessel be excised or repaired? This is a very difficult decision in a field hospital, which may not have the facilities for performing an arteriogram. There is no adequate model of this wound at present.

In brief, wound ballistics studies tell us why we have damaged tissue in the track of a projectile and where we will find damaged tissue; such studies have not as yet told us what to do with this damaged tissue.

6.6 Documenting ballistic trauma

6.6.1 Overview

Surgeons see (and can document) a variety of wounds. The parameters that vary between wounds are as follows:

The mass of the projectile.

The velocity of the projectile.

The design of the projectile (in the case of bullets).

The width of the body between the entry and exit (if there is an exit.)

The tissue in the track of the projectile.

As this section has shown, wound ballistics studies provide the basis for understanding this heterogeneity. The surgical task presented by any wound depends on the its severity, i.e. the degree of tissue damage and the structure(s) injured. Recognising the heterogeneity of wounds is an important step in their surgical management. This demands a clinical classification of wounds based on the features of the wound, rather than on the weaponry or the presumed velocity of the projectile. This is the basis of the Red Cross wound classification (COUPLAND 1991).

6.6.2 Scoring wounds in the field

The Red Cross wound classification is a system whereby certain features of a wound are scored: the size of the skin wound(s) (E and X); whether there is a cavity (C), a fracture (F), or a vital structure injured (V); and the presence or absence of metallic foreign bodies (M). A numerical value is given to each feature. This scoring system is intended for quick and easy use in the field (BOWYER et al. 1993, BOWYER 1995).

See 6.3 for the wound scores for cases A – H.

From the score, the wound can be graded according to size by the E, X and C scores (grade 1 – small penetrating wounds; grade 2 – wounds with a cavity but skin wounds measuring less than 10 cm; grade 3 – wounds with a cavity and skin wounds measuring 10 cm or more) and typed according to structures injured (type ST – soft tissue only: (F = 0 and V = 0); type F – fracture; type V – vital injury, type VF – vital injury and fracture. The three grades and the four types then combine to give twelve possible categories. Any wound, from any weapon, of any part of the body, can be placed in one of these categories. All wounds in a given category carry comparable clinical significance.

By scoring wounds, a surgeon is assessing wounds in clinical terms. Surgical communication about wounds is facilitated. The classification permits comparison between treatments of identical wounds and prognoses for similar wounds. In the case of a "gunshot wound of the thigh," for instance, the treatment and prognosis differs according to the amount of tissue damage, the degree of bone comminution and whether the femoral vessels are injured. Before one can determine which kind

Table 6-1. Red Cross wound classification: Points to be scored

Type		Score	Description
E	Entry	in cm	Estimate the maximum diameter of the **E**ntry wound (in centimetres)
X	Exit	in cm	Estimate the diameter of the e**X**it wound (in centimetres) (X = 0 if no exit)
C	Cavity	C 0 C 1	Can the **C**avity of the wound take two fingers before surgery? No: C = 0; Yes: C = 1; this may be obvious before operation or established only after skin incision; for chest or abdomen wounds it refers to the wound of the chest or abdominal wall.
F	Fracture	F 0 F 1 F 2	No **F**racture: F = 0; simple fracture, hole or insignificant comminution: F = 1; clinically significant comminution: F = 2.
V	Vital structure	V 0 V 1	Are **V**ital structures injured – brain, viscera (breach of dura, pleura or peritoneum) or major vessels? No: V = 0; Yes: V = 1.
M	Metallic body	M 0 M 1 M 2	Bullet or metal fragments visible on radiographs. None: M = 0; one **M**etallic body: M = 1; multiple metallic bodies: M = 2.

of fracture fixation is best for a femur fracture, the patient's wounds must be categorized according to grade and type.

Wound scoring has implications beyond its application to individual patients in a hospital. This procedure may make it possible to audit surgical care or a surgical hospital. It also provides a means of transferring wound data from the field. e. g. for promoting the application of international humanitarian law.

6.6.3 The role of surgeons and the application of international humanitarian law

One of the most common accusations of IHL violations is that the adversary is using bullets prohibited by the 1899 Hague Declaration. These claims are frequently initiated by surgeons.

Any healthcare professional making, supporting or investigating such claims must be familiar with wound ballistics. The kind of information that is required to substantiate such a claim includes: documentation of large wounds including clinical photographs, x-ray evidence or removal at operation of the typical expanding bullet (see Fig. 3-11, for instance). The presence of metal fragments on an x-ray of a bullet wound is not an indication that the bullet was of a prohibited type (COUPLAND 1999a). This is an example of how wound ballistics can form the critical interface between the care of wounded people, the ethical responsibilities of healthcare professionals and international law.

6.6.4 Documenting ballistic trauma – a wider responsibility for health professionals?

One has only to watch the news to see that how injury and suffering result from the use of weapons that inflict injury by ballistic trauma; artillery, mortars, bombs, grenades, assault rifles and handguns. It is not only the technical features of these weapons such as rate of fire, bullet velocity or area covered by explosive force that are responsible for these effects but also their widespread availability and, above all, how they are used. Civilians suffer, as well as combatants. Ballistic trauma is a major factor in how international affairs are played out. This is a massive global health issue, which demands prevention. That prevention lies largely in the hands of politicians. Sections 6.5.2 and 6.5.3 raise the following question: given that surgeons frequently witness the effects of armed violence – ballistic trauma – do they bear a wider responsibility in terms of prevention and the pursuit of justice?

The surgeon's responsibility extends beyond the treatment of wounded people; he or she also has a responsibility to document in objective terms the ballistic trauma they observe. Apart from putting surgeons in a strong position to advocate immediate preventive policies, the collection, analysis and communication of such data underpin the *raison d'être* of the international legal instruments that constitute the broader mechanisms of prevention. These mechanisms include international humanitarian law, the Charter of the United Nations, the international laws of arms control and disarmament, and human rights law. These bodies of law represent management of ballistic trauma in a much wider sense.

In considering this wider view of surgeons' responsibilities, it is worth bearing in mind recent developments in international law. The International Criminal Court has been set up, with the aim of limiting the absolute worst excesses of armed violence. The court has jurisdiction over the crime of aggression, the crime of genocide, crimes against humanity and war crimes. The crimes have been precisely defined in legal terms and, not surprisingly, virtually every crime listed represents the result of ballistic trauma or the threat of it (INTERNATIONAL CRIMINAL COURT 1998). A surgeon treating ballistic trauma cannot do so in isolation from the context in which that trauma was suffered.

Documenting ballistic trauma is the surest way to bring science to the political concept of human security (TABACK and COUPLAND 2007).

7 Wound ballistics and international agreements
B. P. KNEUBUEHL

7.1 Introduction

All the international agreements related to the effects of military firearms came into being during the second half of the 19th Century, which is exactly when these weapons and the associated ammunition were undergoing their period of most rapid development. This is no coincidence; the agreements had to be drawn up in response to the developments.

We shall therefore start this chapter with an overview of the history of such weapons and their ammunition. The emphasis will be on the consequences of the technological developments and related design in terms of ballistics and wound ballistics, rather than on development and design *per se*.

7.2 History of firearms and ammunition

7.2.1 General

The history of firearms is inextricably intertwined with that of gunpowder. The oldest known European formulation for a usable propellant – a mixture of charcoal, sulphur and saltpetre – is said to have been recorded by the English monk Roger Bacon in 1250. German records credit Franciscan monk Berthold Schwarz with the invention of "black powder." He experimented with the same three ingredients at the beginning of the 14th Century, firing stones and iron balls into the air from containers. The same mixture had been described in the Middle and Far East centuries before, however, where uses had included rockets.

The history of firearms becomes interesting in terms of physics and ballistics from around 1800. A number of major inventions appeared during the 19th Century, with the result that, by the end of that century, the effectiveness of small arms had almost reached today's levels.

7.2.2 The development of ammunition

7.2.2.1 The situation in 1800

At the beginning of the 19th Century, most European armies were using smooth-bored muskets with calibres of around 18 mm. These were flintlock muzzleloaders, firing lead balls weighing 30 to 35 g. Their ammunition therefore consists of separate components: lead balls and black powder. To simplify loading, the diameter of the ball was approximately 1 mm less than that of the barrel. The resulting gap was closed off using a "patch" of greased cotton. However, the patch had a negative effect on accuracy, as it caused considerable variation in the direction the ball took on leaving the muzzle. Sectional density (mass per unit area), an important factor in assessing the ballistics of a projectile, was approximately 0.135 g/mm^2.

However, weapons with rifled barrels were already in use, known as rifles. At 14 to 17 mm, their calibre was generally smaller than that of a musket. However, these rifles were more difficult to load, as the ball had to be hammered into the barrel to seat it in the rifling. This meant that rifles were not in widespread service, and were mainly used for hunting.

Rifles also fired lead balls, albeit much more accurately than did muskets, as the projectile was guided along the barrel. This state of affairs lasted until around the middle of the 19th Century. This was the point at which paper cartridges came into use, their aim being to speed up the loading process. The powder was pre-measured and was rolled up in a piece of paper, along with the ball.

7.2.2.2 The elongated bullet

Considerable thought was given to the question of how best to load a rifle. As early as 1828, French Captain DELVIGNE had invented a solution consisting of a lead ball with a cylindrical shaft, the diameter of which was slightly smaller than the calibre. This projectile could easily be pushed down the barrel to a point just in front of the powder chamber. Here, it was pressed onto a spike that extended into the powder chamber with a few blows. This pressed the lead into the rifling, ensuring that the projectile was guided as it passed along the barrel on firing. However, the first idea to be adopted was that of another Frenchman. In 1848, Captain MINIÉ also designed a projectile with a cylindrical shaft, but his model included a hollow in the base. The pressure developed by the expanding gases pressed the walls of the projectile into the rifling, causing the projectile to rotate. The fact that the rear of the bullet expanded, led to its becoming known as the "expanding bullet." The name referred only to the behaviour of the bullet in the barrel, not to any expansion on impact.

This innovation was intended to facilitate loading, but in fact became one of the most ballistically significant in the history of firearms. The ball with the cylindrical shaft was, in effect, the first bullet, a projectile with a significantly higher sectional density for the same calibre. There was, however, an interior ballistics problem: An elongated bullet with the same diameter as the musket balls in use at the time would have been so heavy that the typical powder load of 8 to 12 g would have been incapable of imparting sufficient initial velocity.

Increasing the load was not an option, as this would have led to excessively high pressure in the powder chamber. The solution was to reduce the calibre. The new rifles had calibres of some 10.5 to 13.9 mm, and the bullets they fired weighed approximately 18 to 20 g (in the case of a 10.5 mm bullet). This resulted in a sectional density of 0.20 to 0.23 g/mm^2, which was still an increase over the sectional density of a musket ball, despite the reduced calibre.

7.2.2.3 The primer

The means of igniting the powder had also undergone some change in the meantime. In 1800, most muskets, rifles and pistols were equipped were flintlocks or wheel-locks, making the ignition of the powder the task of the weapon. The discovery of substances that could be detonated by impact, such as potassium chlorate or mercury fulminate, led in 1805 to the invention of percussion ignition by Scots clergyman Alexander FORSYTH. His mechanism used potassium chlorate tablets mounted on paper strips. However, percussion ignition only entered widespread use after an Englishman by the name of EGG invented the percussion cap in 1818. This was a copper cup containing mercury fulminate that was pushed onto the flash hole. When the weapon was fired, the hammer caused the mercury fulminate to ignite and the resulting flame ignited the powder via the flash hole. This meant that the weapon was no longer responsible for igniting the powder – it had only to perform mechanical work. The igniter system finally became part of the ammunition with the invention of the Dreyse needle gun in 1841. This was one of the first breech loaders and the very first weapon to use a paper cartridge containing the bullet, propellant and means of ignition. The bullet was pressed into a cardboard sabot, the base of which contained the percussion cap. The black powder charge was behind this. When the weapon was fired, a needle was rammed through the charge into the percussion cap. In the French Chassepot system, the percussion cap was located behind the powder, as in a modern cartridge.

7.2.2.4 The metal cartridge

Around the middle of the 19th century, the French rifle gunsmith FLOBERT presented the first cartridges made of metal. These were in effect elongated primer caps reinforced with a rim at the base, with a lead sphere pressed into the opening. The primer incorporated into the rim was ignited by striking, hence the name "rimfire cartridge," although "Flobert cartridge" is also used. This type of cartridge contains no powder, the bullet being propelled purely by the primer cap.

The rimfire principle invented by FLOBERT is still in use, the 22 long rifle (22 L.R.) being the best-known example.

The first large-calibre metal cartridges were developed in the United States, and already saw service in the US Civil War (1861-1865). These cartridges used FLOBERT's rimfire system. Shortly afterwards, metal cartridges with primer caps were introduced. The primer cap was incorporated into the base of the cartridge and fired the propellant via a flash hole.

The metal cartridge also solved the obturation problem that had hitherto confronted the designers of breech loaders. The pressure generated as the powder burned pressed the walls of the cartridge firmly against the walls of the chamber, thereby preventing the escape of gases. As a result, metal cartridges were often described as "self-sealing." This innovation meant that the cartridge case had taken over a further function previously performed by the weapon.

7.2.2.5 Smokeless powder

Percussion ignition and elongated bullets formed the basis for a period of intensive development in the field of ammunition during the second half of the 19th century. At the beginning of the 1880s, there was little room for further improvement to cartridges using the technology of the time. The calibre had been reduced to between 10.5 and 13.6 mm. Bullets weighed between 20 and 30 g. A black powder load of approximately 5 g made it possible to attain initial velocities of approximately 450 m/s. However, the trajectories of these bullets were still very sharply curved. To achieve flatter trajectories, it would have been necessary to increase the velocities dramatically. This was clear from the (then) young science of ballistics. There were, however, a number of obstacles to increasing the initial velocity: Increasing the load would have required a heavier weapon, and would also have led to increased recoil, as recoil is proportional to muzzle momentum. Tests were conducted with smaller calibres, as the correspondingly reduced sectional densities were more favourable in terms of interior ballistics.

It was during precisely this period that nitrocellulose-based smokeless powder was invented, nitrocellulose having been invented by German chemist SCHÖNBEIN in 1846. This new powder (also known as nitrocellulose powder) had a number of major advantages: When black powder is ignited, only about half of it is converted into gas, with the remainder being turned into soot and smoke. By contrast, nitro powder leaves less than 1% solid residue. This meant that more powder was converted into energy and significantly reduced fouling of the weapon. Furthermore, it was easier to control the combustion of the powder by modifying the shape of the grains, thereby obtaining the best possible pressure/time curves.

The introduction of smokeless powder allowed for a number of major improvements. The higher energy yield and the more advantageous pressure/time curve gave bullet designers more freedom: bullets were made longer and thinner.

7.2 History of firearms and ammunition   325

Despite the reduction in calibre, sectional densities rose to approximately 0.3 g/mm² and the new propellant made it possible to achieve muzzle velocities of 600 m/s. The new weapons had calibres of 7 to 8 mm and fired bullets weighing 10 to 15 g. The limiting factor on further increases in performance was now the bullet.

Fig. 7-1 shows the development of small arms ammunition from powder horn to metal cartridge and from lead ball to jacketed bullet in the case of Swiss army ammunition.

7.2.2.6 Bullets

A requirement for more effective bullets emerged at the beginning of the second half of the 19th century. Munitions and other military assets could easily be protected against the relatively low-velocity bullets used at the time. Various types of canister and incendiary bullet were developed between 1850 and 1860. Canister rounds consisted of four to five smaller projectiles that separated on leaving the muzzle. Incendiary bullets were hollow projectiles filled with a primer or fine powder, and incorporated some means of igniting the charge they carried. Such projectiles had horrific effects if they happened to hit persons rather than inanimate targets. As a result, such projectiles were (and still are) prohibited under the St Petersburg Declaration of 1868.

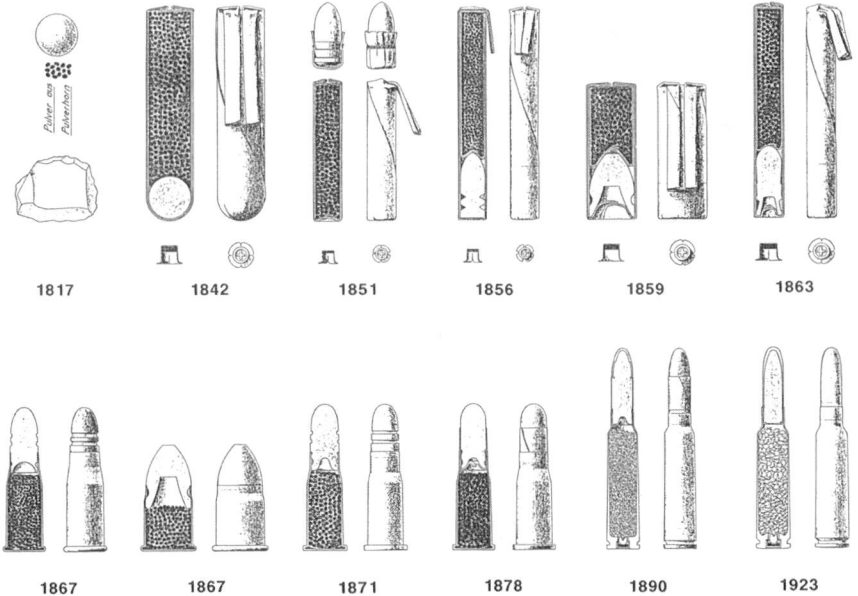

Fig. 7-1. Development of small arms ammunition since the begin of the 19th century in the case of Swiss army ammunition. In other countries the development proceeded very similar.

A technical problem arose following the introduction of smokeless powder. Conventional bullets were unable to withstand the forces generated by the new propellant. As a result, it was not possible to achieve muzzle velocities of more than 600 m/s, even with hardened lead alloys. The acceleration to which the bullet was subjected was such that it was pushed straight down the barrel without engaging in the rifling. If the precision of spin-stabilized bullets were to be maintained, a new design was required that would ensure that the bullet started to rotate in the barrel, even when subjected to high levels of acceleration.

The solution was to encase the bullet in a tougher material that could both resist the shear forces and allow the bullet to slide easily. The most suitable materials were copper and copper alloys. The new jacketed bullet did not deform under pressure, in contrast to its plain lead predecessor. To ensure that the rifling could impart spin to the bullet, its diameter was increased by comparison with that of the barrel. This is why bullets almost always have a diameter slightly greater than the nominal calibre, which corresponds approximately to the diameter of the bore, measured land to land. These jacketed bullets made it possible to increase muzzle velocity still further.

In 1895, however, the British Indian Army reported that, contrary to expectations, these jacketed bullets were virtually incapable of stopping enemy soldiers, despite their relatively high muzzle velocity. A munitions factory in Dum Dum, near Calcutta, was commissioned to design and manufacture more effective ammunition, leading to the development of the dum-dum bullet.

7.2.2.7 "Dum-dum" bullets

Rarely can a bullet have attracted so much attention – well beyond the realms of ballistics specialists – as the dum-dum bullet. The term is well known to the general public. This being the case, it is worth looking at the history of this bullet in more detail. Comprehensive sources on the subject include SPIERS (1975) and GREENWOOD (1980).

In 1871, the British Army introduced the Martini-Henry rifle. This was initially 577 (14.7 mm) single-shot weapon, but was later converted to 577/450 (11.4 mm). The high recoil produced by this cartridge because of the heavy bullet (31.1 g), high velocity (410 m/s) and resulting high energy level (2614 J), together with the introduction of smaller-calibre magazine-fed rifles in other countries rendered the Martini-Henry old-fashioned and it was replaced by the 303 Lee-Metford in 1888. The round-tipped bullet with a full cupro-nickel jacket (the Mark I) weighed 13.9 g and with a black powder charge had an initial velocity of approximately 565 m/s, delivering 2218 J. Black powder was replaced by nitro powder in 1891. The bullet and cartridge case remained unchanged. Muzzle velocity rose to about 600 m/s. Woolwich Arsenal produced the Mark II bullet, which was used with an improved cartridge case by comparison with the Mark I.

In 1895, the British Indian Army conducted its Chitral campaign (close to the Afghan border in what is now northern Pakistan) using the Lee-Metford rifle and Mark II bullet. This bullet proved surprisingly ineffective.

The story is told, for instance, of the enemy who sustained six hits, walked 14 km to the nearest dressing station and was soon back in action. Surgeons described the wounds as being small, clean and with no significant inflammation. The entry and exit wounds were of similar size, and the bullet drilled through bones rather than breaking them. The military view was that these bullets were incapable of causing shock or immediate death.

The north-west frontier was the scene of frequent rebellions. The British forces (Indian soldiers with British officers) were heavily outnumbered and the enemy was highly motivated. An effective bullet was therefore a matter of life and death.

In view of this dangerous situation, the Indian arsenal was ordered to make the Mark II bullet more effective. A suitable bullet was developed by Captain (later Lieutenant-Colonel) Neville BERTIE-CLAY at the munitions factory near Dum Dum, a small town a few kilometres north-east of Calcutta. The only modification was that 1 mm of the lead core was exposed at the tip of the bullet. This was the "dum-dum" bullet, and was only used in India.

To understand the history of this bullet, one needs to be aware that the armed forces of the British Empire were not entirely centralized. The Indian Civil Service had a certain degree of autonomy, while the Indian Army (as it was known, even under British rule) had its own budget and its own supply chain. Development of the dum-dum bullet was therefore undertaken by the Indian Ordnance Department.

The new bullet first saw service in the Tirah campaign of 1897 to 1898, which took place west of Peshawar in what is now Pakistan, close to the Afghan border. The effect was exactly as hoped for.

The enemy – an Afghan tribe known as the Afridi – were so impressed by the effects of this weapon and ammunition that they were prepared to pay 70 sixpences for a rifle and 1 sixpence per dum-dum cartridge (£12 and 2.5 pence respectively in modern British currency).

Meanwhile, efforts were also underway in Britain to produce a more effective bullet, as the British Army itself was encountering problems with the Mark II full metal-jacketed bullet. As a result, a hollow-point bullet was produced at Woolwich Arsenal. The first version, the Mark III, was introduced in 1897, but was only produced in small numbers. An improved version (the Mark IV) appeared in October 1897. Like its predecessor, this bullet weighted 13.9 g and consisted of a solid lead core with a cupro-nickel jacket. The tip had a cylindrical hole bored into it, approximately 9 mm deep and 2.5 mm in diameter. Large quantities of this ammunition were used, in a number of different theatres.

The first major use of this ammunition came on 2 September 1898 near Omdurman in Sudan, when Lord Kitchener's British-Sudanese army fought the Mahdi's dervish army. The Egyptian and Sudanese soldiers were still using the old Martini-Henry rifle. Some British battalions were firing the new Mark IV bullet from their Lee-Metfords, while the remainder used the Mark II. Here again, the Mark IV proved substantially more effective.

However, the new bullet had a serious defect. In a certain number of cases, the jacket and the core separated in the barrel. When the next round was fired, the remains of the jacket caused damage to the weapon and serious injury to the firer.

This defect was remedied by using antimony lead and crimping the jacket tighter at the tail of the bullet. This Mark V bullet also had a slightly differently-shaped tip compared to its predecessor. The many millions of Mark IV bullets already produced were exported or sold for hunting and target shooting, as this type of user had time to check that the barrel was clear after each shot.

The British and Indian Armies were satisfied with the effects of the dum-dum, the Mark IV and the Mark V, as observed in their colonial wars. The rest of the world was not so impressed.

The name "dum-dum" was initially used only for the bullets actually produced at the munitions factory in Dum Dum. The Mark III, IV and V bullets are therefore not actually "dum-dum" bullets. Only later did the name come to be used for any bullet with the same characteristics, at least colloquially.

The accusation was levelled at this type of bullet that it was inhumane. The British responded by pointing to the provisions of the 1868 St Petersburg Declaration, which they had signed, under which projectiles weighing less than 400 g were only prohibited if they were "either explosive or charged with fulminating or inflammable substances." The British government therefore maintained that these bullets were in conformity with both the letter and the spirit of the declaration, and were hence legitimate.

The British were once again attacked at the May 1899 Hague Peace Conference, primarily on the basis of experiments conducted by VON BRUNS (1898a,b), who was the chief surgeon of the army of Württemberg.

In the absence of original dum-dum ammunition, VON BRUNS used Mauser hunting ammunition (8 mm, known as S calibre). The semi-jacketed bullet was heavier than the dum-dum, 5 mm of lead was exposed at the tip and the initial velocity was higher (715 m/s). Experiments on animal cadavers led VON BRUNS to conclude that a limb hit by a dum-dum bullet (by which he meant those produced by the Dum Dum factory in India) would have to be amputated. The British criticized these experiments in the strongest terms. They pointed out that the Mauser bullet had a much higher velocity (715 m/s instead of 600 m/s), that it was heavier than the dum-dum bullet and that 5 mm of the lead core was exposed, as against 1 mm in the case of the dum-dum. They also maintained that their combat experience with the dum-dum round contradicted the conclusions of VON BRUNS' experiments. And in any case, they were only using these bullets in colonial wars, and then only because one could not defend oneself against the barbarians otherwise.

The introduction of the Mark IV hollow-tip bullet prompted VON BRUNS to undertake further experiments (VON BRUNS 1899). This time, he compared the Mark II full metal-jacketed bullet not with the Mauser bullet but with the Mark IV. He fired these bullets into wood, clay and living horses. Publication of these results led VON ESMARCH (1899), who was also a leading surgeon, to address a publication to the Peace Conference. He wrote:

"The employment of such missiles is, perhaps, excusable in a war with fanatical barbarians, who, ignorant of the rules of international law, give and take no quarter [...] but it would be a matter for the deepest regret were such barbarous engines of destruction ever to come into use in European wars."

This brought about a reaction from the French. On 18 January 1899, "La Semaine Médicale" published an article in which the author condemned both the British point of view and that of VON ESMARCH:

"Two principles of humanity, two weights and two measures, one for the civilized peoples and one for the barbarians (by which is meant any other race) and far-off lands."

The Hague Conference finally produced three conventions, together with three declarations that do not form part of any convention. Convention II included a clause to the effect that bullets which cause unnecessary suffering were prohibited. Declaration III was aimed specifically at the British ammunition, and provided that ammunition that expanded or flattened easily was not to be used (see 7.3.2.4 for precise wording). Despite the protests of the other European nations, the governments of the United Kingdom, the United States and Portugal all refused to sign this part of Declaration III.

The Boer War broke out on 11 October 1899. Huge quantities of Mark IV ammunition were shipped off to South Africa, followed by quantities of Mark V. However, this war was being fought not against "barbarians" but against "civilized" troops. In view of the Hague Conference, which was still under way at the time, all soft-tipped ammunition (i.e. Mark IV and V) was withdrawn and replaced with the "ineffective" Mark II.

7.2.3 The development of firearms in the 19th century

7.2.3.1 Muzzle loaders and their problems

The basic principle of firearms had been the same for centuries: Powder was poured into a tube closed off at one end, a lead ball was placed on top and the powder was ignited via a flash hole. It was discovered at an early stage (16th century) that accuracy could be improved significantly if the ball was made a tight fit in the barrel and spin imparted to it thereby. Throughout the history of firearms, there had been rifles – so called on account of their rifled barrels – and smooth-barrelled weapons – shotguns and muskets.

Both had their advantages and disadvantages. While rifles were far more accurate than muskets, with four or five times less scatter, their more complex loading process gave them a lower rate of fire – muskets could be fired two to three times faster.

These two parameters – accuracy and rate of fire – determined the further development of shoulder weapons. What was needed was a rifle with a high rate of fire.

7.2.3.2 Breech-loaders

The need to load the weapon via the muzzle was seen as the main factor limiting the rate of fire. Accordingly, numerous attempts were made to load the weapon from the rear. However, designers repeatedly came up against the same problem: that of effectively obturating the joint against the propellant gases. Part of the gas escaped via the joint in the chamber. This both caused problems for the user and considerably reduced the range of the weapon, as less gas was available to accelerate the projectile. This problem proved to be the downfall of the Crespi breech-loading flintlock introduced into the Austrian army in 1770. In 1811, an American by the name of HALL patented a similar system. HALL's weapons were manufactured to higher standards; the components were produced using templates and were hence interchangeable, and the system was more successful. The Hall was the first functioning military breech-loader.

However, the breech-loading principle only took off in 1841 when the Dreyse needle gun was produced, with its percussion-ignition cartridge. The effects of the increased rate of fire are often illustrated by reference to the Battle of Königgrätz (1866), when Prussian soldiers with needle guns fought against the Austrians with their muzzle loaders. The Prussians suffered 9,150 dead and wounded, the Austrians 24,400.

Towards the end of the 1860s, France introduced the Chassepot rifle, with an improved cartridge. Its smaller calibre and improved obturation allowed this weapon to achieve higher initial velocities, which meant better ballistic performance. The rifle suddenly took on huge military significance as a result of this development, and considerable effort was devoted to further improvements.

7.2.3.3 Repeaters

Like a number of other developments, the next step in the process of increasing the rate of fire came from the United States. During the Civil War, both sides had used rifles that no longer required each round to be loaded separately; the cartridges were chambered from a magazine by means of a simple cocking motion. The cartridges were stored in a tubular magazine, which was either built into the butt, as in the case of the Spencer (1861) or fitted under the barrel, as with the Henry (also 1861). The Henry system is still in use, and is found on the famous Winchester lever-action rifles. The theoretical rate of fire rose to over 60 rounds per minute. European gunmakers began producing their own repeaters as soon as news of these weapons reached the continent. The Swiss Army was the first to use a repeating rifle, theirs being based on the Henry-Winchester system. The Vetterli rifle, named after its designer, had a tubular magazine under the barrel and an innovative mechanism with a lug-locking system.

Another American – LEE – invented the box magazine, which was easier to charge than the tubular magazines used hitherto. This system is so much safer and easier to use, that almost all repeaters produced since then (with the exception of Winchesters) use box magazines

7.2.3.4 Handguns

Until the start of the 19th century, handguns developed along virtually the same path as that of shoulder weapons. Like rifles and muskets, handguns were muzzle-loaders, with smooth or rifled bores. Initially, they were fired by wheel-locks or flintlocks, with percussion ignition being introduced later on. As a single-shot weapon is of limited use in an emergency, a number of designs were developed whereby multiple barrels could be fired, simultaneously or singly. These weapons never saw widespread use, however.

The handgun took a significant step forward in 1836 when Samuel COLT invented the revolver. The idea of arranging a number of chambers in a rotating cylinder behind the barrel had already been mooted in connection with multiple-fire weapons. The innovation in COLT's design was that it linked together the movement of the hammer and that of the cylinder, in a principle known as "single action." Considerable effort was immediately devoted to producing a "double action" revolver, in which the trigger operated both the hammer and the cylinder.

The breech-loading revolver appeared immediately after the invention of the metal cartridge, and the basic principle of the revolver remains the same today. Revolvers do have one major disadvantage: The gap between the cylinder and the barrel allows part of the gas to escape at the sides, reducing the efficiency of the weapon. This has to be compensated for by a larger powder load. A number of gas-tight revolvers were produced, in which barrel and cylinder were only separated when the cylinder rotated and the cartridge case was used to seal off the gap between the two.

The only successful revolver of this type was the Belgian Nagant, which was used by the Russian army. Despite the problem outlined above, the revolver became the most widely-used type of handgun in the second half of the 19th century. A number of repeating pistols were produced, along the same principles as repeating rifles, but they were considerably less reliable than revolvers. The first self-loading pistols with rates of fire similar to those of revolvers appeared at the end of the 19th and the beginning of the 20th century. However, it was to be decades before pistols were seen as reliable.

7.2.4 The 20th century

7.2.4.1 Ammunition

There have been no innovations in handgun and rifle ammunition with significant effects on ballistics since the beginning of the 20th century. Calibre, bullet mass and muzzle velocity have varied at times, but have remained within the regular physical laws of ballistics. The history of Swiss army weapons illustrates this quite clearly.

The 7.5 mm 1890 cartridge with a bullet mass of 13.8 g formed the basis for the GP 11. This cartridge was of the same calibre, but used a full metal-jacketed bullet weighing 11.3 g, i.e. 2.5 g less. This reduced the sectional density from 0.312 g/mm^2 to 0.256 g/mm^2. Further technical enhancements (regarding the guiding of the bullet and the gas seal) made it possible to increase muzzle velocity

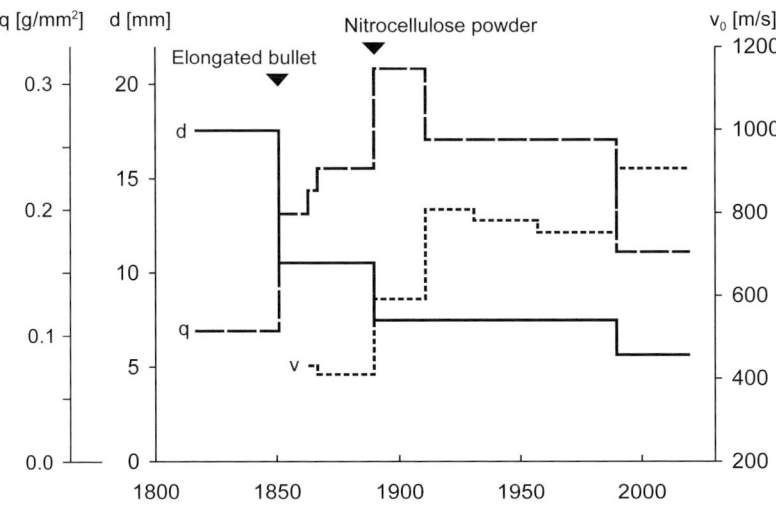

Fig. 7-2. Changes in calibre (d), sectional density (q) and muzzle velocity (v_0) since the beginning of the 19th century. There were only two departures from the general rule, whereby reduced calibre = reduced sectional density and reduced sectional density = increased muzzle velocity: introduction of the elongated bullet and of smokeless powder.

from approximately 590 m/s to 805 m/s in the case of the Modell 11. Later rifles were given progressively shorter barrels, which led to lower muzzle velocities (785 m/s for the Karabiner 31 and 750 m/s for the Sturmgewehr 57). The only way to achieve an even flatter trajectory was to increase the initial velocity. This meant reducing the sectional density and hence the calibre. The Sturmgewehr 90 achieves an initial velocity of 905 m/s with the Swiss army Gewehrpatrone 90 cartridge.

Fig. 7-2 shows the history of Swiss army muskets/rifles and ammunition since the beginning of the 19th century. The effects of the two most significant inventions in the field of ammunition – elongated bullets and nitro powder – are very apparent. One cannot expect that sectional density will increase as calibre decreases (as was the case when elongated bullets were introduced) and that muzzle velocity will increase as sectional density increases (as occurred when black powder was replaced by nitro).

These two inventions brought about a fundamental change in the physics and ballistics of firearms.

Developments in other countries were completely analogous. Germany went from an 7.9 mm bullet to the 7.62 NATO round, while France started with the 8 mm Lebel and ended with the 5.56 × 45 (223 Rem.). NATO also switched from the 7.62 × 51 to the 5.56 × 45. The former Warsaw Pact countries followed the same trend, changing over from the 7.62 × 39 (Kalashnikov) to the 5.45 × 39 (Kalashnikov).

The latest developments in ammunition, such as caseless systems, remain subject to the physical framework that emerged at the end of the 19th century. Changes that bring advantages in one respect always involve disadvantages in some other respect.

7.2.4.2 Weapons

Unlike ammunition, weapons underwent a number of fundamental changes in the course of the 20th century. As it was virtually impossible to reduce spread, development efforts focused on increasing rate of fire and reducing the weight of the weapon. One obvious idea was to fully automate the loading process.

The first self-loading principles were developed at the end of the 19th century, with two principles coming into widespread use. In the case of a blowback operated weapon, part of the propellant gas is allowed to push the cartridge back against the moving parts, forcing them to the rear, ejecting the empty case and chambering the next round. On a gas operated weapon, some of the combustion gas is diverted onto the head of a piston via an aperture, the gas port. The piston operates a mechanism that unlocks the breech block and forces it back. Reliable weapons of both types were developed. Initially, weapons based on these systems were heavy, and hence were only suitable for use as support weapons. The Maxim was the first usable machine gun and operated on the blowback principle. Following its invention in 1883, this weapon remained in service with many armies until after the First World War. Its nominal rate of fire was 600 rounds per minute, and the barrel was water cooled. This was followed by the Hotchkiss machine gun, a gas operated weapon. It had an air-cooled barrel, which had to be changed every 500 to 600 rounds to prevent overheating.

In 1898, the German company Mauser patented a light self-loading rifle, and in 1907 Winchester brought out a self-loading carbine. It was to take several decades, however, before self-loading rifles became standard issue. Throughout the two World Wars, such weapons were issued only to special forces. Self-loading rifles did not enter widespread use until after the Second World War. There was nothing to be gained by raising rates of fire still further. On the contrary, the current trend is to limit burst length in order to conserve ammunition. Most new assault rifles have a setting that allows the user to fire three-round bursts. Current development efforts focus primarily on reduced weight and increased reliability.

The first self-loading ("semi-automatic") pistols also appeared at the end of the 19th century (the first being the Bergmann in 1896), all of them using the blowback principle. Their greater complexity meant that for many years pistols were heavier and less reliable than revolvers. The German Luger Parabellum was a milestone in the history of the pistol. Luger 9 mm ammunition (9×19 mm) was developed in parallel with this pistol and is still probably the most widely-used pistol and submachine gun cartridge in the world.

Pistols became steadily more reliable during the course of the 20th century. Incorporating the double-action principle meant that the user could fire the first shot from a pistol as quickly as from a revolver and increased magazine capacity brought further advantages. As a result, the double-action pistol is currently the most popular type of handgun.

7.3 International treaties

7.3.1 Basic principles

Two principles that began to emerge in the 18th century recur in all subsequent agreements regarding inter-State conflict. Both are derived from the "natural law." The first principle is that of *proportionality of the means*. According to this principle, no warring party has unlimited freedom in the choice of the means used to damage the adversary.

One example of the general acceptance and application of this rule at that time is the use of pairs of identical duelling pistols.

The second principle is that of *proportionality in the effect* applied to the adversary. This principle requires that unnecessary suffering be avoided. Clearly, "proportionality" depends on one's understanding of ethics and morality, and is hence open to divergences of interpretation. One of the main aims of wound ballistics research is therefore to identify physically measurable parameters defining the effect of a weapon, in order to be able to incorporate significantly more precise wording into future conventions.

In the next sub-section, we shall discuss some of the most important instruments, together with those passages from those instruments that are of fundamental importance or contain clauses regarding the effects of ammunition.

7.3.2 The instruments

7.3.2.1 The original Geneva Convention (1864)

After the June 1859 Battle of Solferino (during the war between Sardinia and France on the one side and Austria on the other) over 40,000 wounded soldiers were left behind on the battlefield. Henri DUNANT from Geneva arrived on the scene just after the battle. After helping to organize spontaneous assistance for the wounded, DUNANT formulated proposals the creation of societies for the protection of wounded soldiers. A five-member committee of the Public Welfare Committee of Geneva was set up, which eventually became the International Committee of the Red Cross.

At DUNANT's instigation, an international conference was held in Geneva in 1863. That conference called for humanity towards the wounded, even in time of war, and for the convening of a general conference of States. In 1864, the Swiss government invited States to participate in a diplomatic congress in Geneva, which resulted in the *Convention for the Amelioration of the Condition of the Wounded and Sick in Armies in the Field.* This first convention consisted of just ten articles, setting out certain basic humanitarian norms. These were expressed in such general terms that they were able to form the basis for all subsequent treaties.

7.3.2.2 The St Petersburg Declaration (1868)

In the years following 1850, many States developed rifle ammunition designed to have a significantly enhanced effect on protected targets. This mainly involved the production of explosive and incendiary bullets, in which the projectile was hollowed out, filled with gunpowder or an incendiary compound, and fitted with an ignition device. Clearly, it was difficult to ensure that such projectiles only hit war material. The injuries these bullets caused to enemy soldiers were totally at odds with the principles mentioned above. It is therefore not surprising that demands for such bullets to be prohibited were made very soon after they were introduced, and that these demands were successful. The initiative to impose restrictions on weapons emanated from the Russian Empire, which convened a conference of European States in St Petersburg, in 1868. The St Petersburg Declaration signed at this conference on 11 December 1868 regulated the principles of proportionality. One of its provisions stipulated that "the employment of arms which uselessly aggravate the sufferings of disabled men, or render their death inevitable" would exceed the object of war. The Contracting Parties also undertook not to use "any projectile of a weight below 400 grammes, which is either explosive or charged with fulminating or inflammable substances."

7.3.2.3 The Brussels Conference (1874)

Another conference took place at the instigation of the Russian government six years later in Brussels. The conference protocol was signed in August 1874. A draft convention was attached. While this convention was never ratified, it was of significance, as it formed the basis for later instruments. It is interesting to note that the two basic principles had now been stated explicitly. Article 12 reads as follows:

> "The laws of war do not recognize in belligerents an unlimited power in the adoption of means of injuring the enemy."

> In the French original: *"Les lois de la guerre ne reconnaissent pas aux belligérants un pouvoir illimité quant aux choix des moyens de nuire à l'ennemi."*

Prohibitions set out in Article 13 included:

> "The employment of arms, projectiles or material calculated to cause unnecessary suffering …"
>
> French original: *D'après ce principe sont notamment interdits:*
>
> …
>
> e: *"L'emploi d'armes, de projectiles ou de matières propres à causer des maux superflus, ainsi que l'usage des projectiles prohibés par la Déclaration de Saint-Petersbourg de 1868."*

This was thus an explicit reference to the projectiles mentioned in the St Petersburg Declaration. The conference also decided to prohibit the use of poison or poisoned weapons.

7.3.2.4 The Hague Convention (1899)

Russia once again called a peace conference, this time in 1898. One of the main topics of this conference, which took place in The Hague in 1899, was that of arms limitations. The conference discussed proposals not to develop the then existing artillery and rifles, or at least to suspend further development for five years. Differences in quality between small arms would then be ascertained by specifying a type of rifle in terms of the mass of the weapon, its calibre, the mass of the bullet, the muzzle velocity and the rate of fire. These proposals were not adopted, but because the Brussels Convention had not been ratified, the topic of limiting the effect of weapons was once more placed on the agenda for the conference. Article 12 of the Brussels Declaration formed Article 22 of the new convention, almost word-for-word. The article concerning restrictions on certain weapons was a point of discussion, however. The dum-dum bullet had been developed in the meantime, and there were differences of opinion as to whether its effects came constituted the explosive force mentioned in the St Petersburg declaration. The section on small arms and their ammunition was eventually amended to read:

> "The Contracting Parties agree to abstain from the use of bullets which expand or flatten easily in the human body, such as bullets with a hard envelope which does not entirely cover the core or is pierced with incisions."
>
> French original: *"Les Puissances contractantes s'interdisent l'emploi de balles qui s'épanouissent ou s'aplatissent facilement dans le corps humain, telles que les balles à enveloppes dures dont l'enveloppe ne couvrirait pas entièrement le noyau ou serait pourvue d'incisions."*

This was a direct reference to the dum-dum bullet.

7.3.2.5 The Regulations concerning the Laws and Customs of War on Land (The Hague, 1907)

After an extended period of preparation, the second Hague Peace Conference took place from 15 June to 18 October 1907. Eight years previously, only 26 States had taken part, whereas now 44 States signed the instruments that were drawn up. The "Convention respecting the Laws and Customs of War on Land" consists of a convention and an annex, the "Regulations concerning the Laws and Customs of War on Land." The content of the regulations is based largely on the 1899 convention. Article 22, for instance, specifies that:

> "The right of belligerents to adopt means of injuring the enemy is not unlimited."

> French original: *"Les belligérants n'ont pas un droit illimité quant au choix des moyens de nuire à l'ennemi."*

This is almost exactly the wording used in Brussels in 1874. The article limiting the use of weapons was amended, however, reverting to the 1874 wording but without mentioning the projectiles prohibited in St Petersburg in 1868. Article 23 (based on Article 13 from 1874) now read:

> "In addition to the prohibitions provided by special Conventions, it is especially forbidden
>
> ....
>
> e: To employ arms, projectiles, or material calculated to cause unnecessary suffering;"

> French original: *"Outre les prohibitions établies par des conventions spéciales, il est notamment interdit:*
>
> ...
>
> *e: d'employer des armes, des projectiles ou des matières propres à causer des maux superflus."*

This meant that neither dum-dum nor exploding bullets were expressly forbidden. Proportionality of effect was therefore only described very vaguely.

7.3.2.6 The Geneva Conventions of 1949

At the invitation of the Swiss Federal Council, a diplomatic conference was held in 1949, bringing together 59 States for the purpose of revising the existing international treaties.

In the articles common to all four conventions, it is specified that they apply to all cases of armed conflict between two or more of the contracting States. In the case of civil war or internal disturbance, the main principles of humanity were to apply as a minimum.

The four conventions are as follows:

Convention (I) for the Amelioration of the Condition of the Wounded and Sick in Armed Forces in the Field.

Convention (II) for the Amelioration of the Condition of Wounded, Sick and Shipwrecked Members of Armed Forces at Sea.

Convention (III) relative to the Treatment of Prisoners of War.

Convention (IV) relative to the Protection of Civilian Persons in Time of War.

7.3.2.7 The 1977 protocols additional to the Geneva Conventions

Between 1974 and 1977, the "Diplomatic Conference on the Reaffirmation and Development of International Humanitarian Law applicable in Armed Conflicts" drew up two protocols additional to the Geneva Conventions of 1949. Protocol I covers the protection of victims of international armed conflicts, while Protocol II addresses the protection of victims of non-international armed conflicts.

Protocol I, Part III, Section I ("Methods and Means of Warfare") repeats the principles of the 1907 "Convention Respecting the Laws and Customs of War on Land."

Article 35 reads:

"Basic rules

1. In any armed conflict, the right of the Parties to the conflict to choose methods or means of warfare is not unlimited.
2. It is prohibited to employ weapons, projectiles and material and methods of warfare of a nature to cause superfluous injury or unnecessary suffering.
3. It is prohibited to employ methods or means of warfare which are intended, or may be expected, to cause widespread, long-term and severe damage to the natural environment."

Article 36, entitled "New weapons." sets out the following obligation:

"In the study, development, acquisition or adoption of a new weapon, means or method of warfare, a High Contracting Party is under an obligation to determine whether its employment would, in some or all circumstances, be prohibited by this Protocol or by any other rule of international law applicable to the High Contracting Party."

In other words, Parties must verify that any new weapon is in conformity with the conventions.

7.3.2.8 The United Nations Conference (Geneva, 1980)

During the 1960s, the 223 Rem. (5.56 × 45) cartridge was in widespread use, especially in conjunction with the American M 16 assault rifle. As far as its design was concerned, the bullet used appeared to comply with the conventions in force. However, it displayed a tendency to fragment in soft tissue. This once again raised the question of proportionality with regard to the effects of small arms. Many thought (wrongly) that this ammunition was having a greater effect than one would expect from a 5.56 round because of its high muzzle velocity, or even because of the calibre itself. In the run-up to the UN conference, there were proposals to limit the muzzle velocity of small arms to 800 m/s. There were even demands to prohibit 5.56 mm rounds. However, a large number of experiments were conducted in the 1970s, in many countries which resulted to a relatively clear qualitative picture of the behaviour of a bullet in a living being. On the basis of this knowledge, a number of countries submitted proposals to the 1980 UN conference regarding the projectiles used with small arms:

– limits on the energy transferred to a soft target;
– no fragmentation of the bullet in the target;
– verification of these characteristics using a simulant (glycerine soap);
– prohibition of exploding bullets.

The tests were to have been conducted using a soap block 14 cm thick, at a range of 100 m. The conference did not accept these proposals, and did not adopt a protocol regarding small arms. However, the convention that was adopted leaves open the possibility of producing additional protocols to prohibit or restrict certain weapons at any time in the future. When the convention was drawn up, provision was made for its being revised at a future date, to ensure that the law of war could keep up with developments in weapons technology. The conference did result in one protocol of direct relevance to wound ballistics. This was the Protocol on Non-Detectable Fragments, i.e. munitions deliberately designed in such a way that the fragments they produce cannot be detected in the human body. One result of this protocol is that plastic bullets have to incorporate supplementary materials to render them radio-opaque.

7.3.2.9 The relevance of international instruments to wound ballistics

The wound ballistics implications of those international instruments with a direct bearing on bullets can be summarized as follows: Both the 1868 St Petersburg Declaration and the 1899 Hague Convention prohibit projectiles which exhibit a high degree of effectiveness immediately after striking the body (see Fig. 7-3). If such a bullet hits a limb, it is probable that the limb will have to be amputated, and this can certainly be classified as "unnecessary suffering."

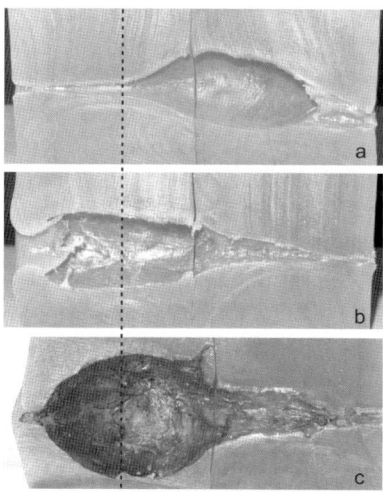

Fig. 7-3. Bullet tracks in ballistic soap:
a. Legal full metal-jacketed bullet.
b. Semi-jacketed bullet, illegal under The Hague Convention of 1899.
c. Explosive bullet, illegal under the St Petersburg Declaration of 1868.
The dotted line indicates the approximate mean diameter of one extremity.

According to the comprehensive statistics that have been collected regarding wounds sustained in conflict, over half of all bullet and fragment wounds affect the limbs.

7.3.3 A basis for formulating future instruments of international humanitarian law

7.3.3.1 The disadvantages of the wording of existing conventions

In addition to a comprehensive description of the principles of wound ballistics, this book contains (in 7.3.2) an account of the multiple attempts that States have undertaken to include humanitarian considerations in the waging of war. These efforts have resulted in a number of instruments. However, the statements and definitions in those instruments are often difficult to verify, and do not reflect recent developments in munitions.

One disadvantage of the current conventions is that they refer to types of bullet that were in use when they were formulated. By definition, such formulations cannot cover future, unknown developments.

For instance, at the beginning of the 1970s the bullet designated 5.56 × 45 or 223 Rem. (or M193) was accused with no scientific justification of violating the applicable conventions, despite being full metal-jacketed.

It would therefore be appropriate to formulate humanitarian aspects of any future convention using a parameter that is as general, design-independent, and measurable as possible.

The second disadvantage of existing conventions is the term – first used in the annex to the 1974 Brussels protocol – of "superfluous injuries." A verbal criterion such as this does not lend itself to objective assessment, and will always be a source of fruitless discussion, if it cannot be associated with some measurable parameter.

7.3.3.2 Projectile-independent assessment processes

We saw in 4.1.1 that if we leave aside the mental state of the victim, the direct effect of a bullet depends on certain physical characteristics of the bullet, on the point of impact, on the path of the wound channel in the body and on the physical condition of the victim. Of these factors, only the physical characteristics can be influenced by design and measured. The following parameters could be used to set limits on the effect of a bullet:

1. The energy of the bullet on impact.
2. The total energy transferred E_{ab} over a specified distance s.
3. Maximum potential effect E'_{ab}.
4. Maximum total potential effect (E'_{ab} as a function of distance travelled).

1. The impact energy only tells us about maximum possible energy transfer, but not about the effectiveness of the bullet or its behaviour in the body. It therefore cannot be used for this purpose.

2. The option of limiting energy transfer was suggested by a number of countries (including Sweden) at the 1980 UN conference in Geneva. See 7.3.2.8. This would require specifying the total energy transferred and the distance "s" over which the transfer occurred. From a medical point of view, this definition would not be appropriate. We saw in Chapter 3 that the potential effect of a bullet E'_{ab} at a given point s along its track is of considerable significance in determining the severity of the injury at that point. Limiting the severity of injuries therefore means limiting the maximum value of E'_{ab} over the total track of the bullet.

If one considers two different bullets, with the same value of E_{ab} over the same distance, and if one bullet transfers its energy evenly over the entire distance while the other concentrates energy transfer at one point (as does a semi-jacketed bullet, for instance), the latter bullet will be more problematic from a medical point of view.

3. Specifying a maximum value for E'_{ab} avoids disproportionately severe local injuries and hence takes account of the arguments set out under 2. above. However, the maximum value of E'_{ab} can occur at very different points along the wound channel, depending on the design of the bullet and on how stable it is. In particular, certain bullets would only attain this maximum value after creating a long wound channel, which only rarely occurs in the human body. Bullets with such characteristics would then be unfairly excluded, despite having an acceptable E'_{ab} over normal channel lengths.

4. Limiting the potential effect (i.e. limiting the effectiveness function E'_{ab} as a function of distance s) would avoid the possibility of excluding bullets that only

transfer a small percentage of their energy over normal wound channel lengths. Furthermore, using this "profile" makes it possible to measure and assess the overall behaviour of a bullet.

Given the range of possible effects that a bullet may have under different impact conditions and anatomical conditions in the wound channel, any discussion regarding limits on potential effect is pointless. It is therefore only sensible to compare the effectiveness functions of bullets with one another, i.e. to conduct relative measurements. Measurements will only be reproducible, however, if homogenous target materials are used. As we explained at some length in 3.3, the materials that could be used are gelatine and ballistic soap, with the latter being particularly well-suited because the effectiveness function $E'_{ab}(s)$ is directly visible. A further advantage of this medium is that the function $E'_{ab}(s)$ can be measured quantitatively, on account of the relationship that exists between channel volume and energy. Furthermore, comparison with war surgery cases shows that the effectiveness function gives a very faithful picture of the true nature of the injury (COUPLAND et al. 2000). This therefore constitutes a simple means of verifying the criteria specified.

7.3.3.3 Formulation of standards

The only remaining question is the choice of values to use in drawing up a curve of maximum permissible effectiveness. One could, for instance, look for a bullet known to be relatively benign in practice.

This would raise the question as to which bullet should be labelled "benign," as every State may be of the opinion that they have the "right" bullet. However, it would also be possible to summarize the results for a number of bullets with similar effectiveness functions and produce a mean, or envelope profile.

The proposals submitted to the 1980 UN conference (see 7.3.2.8) included the suggestion that the 7.62 mm NATO round be used as a standard. However, as there are various versions of this bullet (e.g. with steel or tombac jackets), which behave differently in soft matter, this would have to be specified in more detail. The parameters required would include such ballistic data as v_0, twist length, stability number and range.

Adopting such a limit on effectiveness would overcome the advantages mentioned above. The new definition would make no mention of bullet design or behaviour, and it would be possible to make quantitative comparisons between different bullets. There would also be a clear result: a given bullet would either meet the standard or it would not.

A further advantage of this method is that it could be extended to yield measurable effectiveness limits for any projectiles (e.g. fragments or flechettes). The permissible effectiveness functions for such projectiles could differ from those for conventional bullets. Regulations could also be drawn up for special-purpose ammunition (for the police or for anti-terror operations), with the advantage of including a defined, measurable risk of injury that was not to be exceeded.

A Tables

A.1 List of tables in the main text

Chap.	Title	Table	Page
2	Typical velocities	2-1	5
	Typical acceleration values	2-2	5
	Typical values for angular velocity ω, rotational speed v and circumferential velocity v	2-3	6
	Typical forces acting on a bullet	2-4	8
	Typical kinetic energy levels encountered in ballistics	2-5	9
	Comparison between linear and rotational motion	2-6	11
	Temperature reference points	2-7	16
	Examples of Reynolds numbers for bullets	2-8	25
	Powder data	2-9	39
	Burning rates	2-10	67
	Energy balance. 7.62 mm NATO (308 Win.)	2-11	69
	Typical gas pressures at the muzzle	2-12	71
	Extract from ballistics table, showing stability and tractability numbers.	2-13	80
	Gurney constant G_K for a number of common explosives	2-14	82
	Constant C_G for the Gurney formula	2-15	82
3	Weapons used by LAGARDE for his experiments	3-1	93
	Possible types of wound channel	3-2	107
	Pressure in a skull as a bullet passes through	3-3	115
	Length of the narrow channel as a function of the angle of incidence ψ_0 at the point of impact	3-4	120
	Temporary cavity as a function of E_a. (After CHARTERS and CHARTERS 1976)	3-5	124
	Results of experiments with 6 mm steel spheres (from FACKLER et al. 1986)	3-6	125
	Advantages and disadvantages of each simulant	3-7	150
4	Throwback velocity (v_K) and rotational velocity (v) ($m_K = 80$ kg)	4-1	167
	Form factors for StP (stopping power) and RSP (relative stopping power)	4-2	167
	Effectiveness of handgun cartridges, measured in accordance with four criteria (W_H, W_{TH}, StP and RSP)	4-3	170

Chap.	Title	Table	Page
4	Tables for PIR	4-4	172
	Comparison between the various effectiveness criteria	4-5	175
	Corresponding rankings	4-6	175
	Energy thresholds laid down by States	4-7	181
	Hit distribution for US infantrymen in the Second World War	4-8	183
	Summary of the US effectiveness criteria	4-9	186
	50% energy density thresholds for various parts of the body, in J/mm^2 (after BIR, STEWART et al. 2005)	4-10	188
	Threshold energy density and velocity (measured values)	4-11	190
	Results of firing various projectiles from an airgun into the thigh of an adult (after MISSLIWETZ 1987)	4-12	190
	Determining v_{gr}: Results of measurements conducted by TAUSCH et al. (1978)	4-13	191
	Values for C_D from the literature	4-14	194
	Threshold velocities for eyes (rabbit). With one exception, a steel sphere was used	4-15	200
	Threshold velocities for eyes (rabbit). Steel cubes (STEWART 1961)	4-16	201
	Energy densities of a number of alarm pistols as a function of distance from the muzzle	4-17	211
	Percentages of wounds caused by bullets and fragments in conventional warfare (from GANZONI 1975)	4-18	233
	Values for λ_2, γ and δ in formulas (4.4:18-21)	4-19	239
	Ballistic data for a number of "non-lethal" projectiles of low sectional density	4-20	242
	Ballistic data for various types of rubber shot	4-21	245
	Wounding potential of various projectiles, balls, pucks and similar	4-22	249
5	Velocities obtainable using vertical tubes and towers	5-1	296
6	ICRC wound classification: Points to be scored	6-1	318

A.2 Characteristics of materials

A.2.1 Fluids and materials that behave like fluids

Density (ρ), compressibility (κ), dynamic viscosity (η) kinematic viscosity (ν), speed of sound (c)

Material	T_C °C	ρ [kg/m³]	κ [1/Pa]	η [Pa·s]	ν [m²/s]	c [m/s]
Water	20	998	$4.6 \cdot 10^{-10}$	$1.0 \cdot 10^{-3}$	$1.00 \cdot 10^{-6}$	1483
	0	1000	$5.1 \cdot 10^{-10}$	1.8	$1.79 \cdot 10^{-3}$	1403
Glycerine	20	1260	$2.2 \cdot 10^{-10}$	1.48	$1.17 \cdot 10^{-3}$	1923
	0	–	$1.2 \cdot 10^{-9}$	–	–	–
Ethanol	20	789	$9.3 \cdot 10^{-10}$	$1.20 \cdot 10^{-3}$	$1.52 \cdot 10^{-6}$	1170
	0	806	$10.2 \cdot 10^{-10}$	--	–	1100
Soap	20	1080	$3.4 \cdot 10^{-10}$	$\approx 5.0 \cdot 10^{-3}$ [a]	$\approx 5.0 \cdot 10^{-6}$	1660
Gelatine 10%	20	1030	$4.2 \cdot 10^{-10}$	≈ 40.0 [a]	≈ 0.04	1520
	4	–	–	–	–	1486
Gelatine 20%	20	1060	$3.8 \cdot 10^{-10}$	$\approx 1.0 \cdot 10^{2}$ [a]	≈ 0.1	1567
	0	–	–	–	–	1541
Air	0	1.23	$7.4 \cdot 10^{-6}$	$1.72 \cdot 10^{-5}$	$1.33 \cdot 10^{-5}$	331

[a] At 30°C. Measurements not possible at lower temperatures. Viscosity increases rapidly as temperature decreases (SAHLI 1990).

A.2.2 Solid materials

The table below lists the typical density, tensile strength and Young's modulus of materials used in the manufacture of bullets. These are approximate values (for non-ferrous metals, e.g. cold-worked, 50%) which may change significantly depending on how the material is used.

Material		Density [kg/m³]	Tensile strength [N/mm²]	Young's modulus [N/mm²]	Composition [%]
Steel (C 45)	Fe	7,850	660–780	210,000	
Reinforced steel		7,850	1,200	210,000	
Copper	Cu	8,930	220	124,000	
Nickel	Ni	8,900	440	210,000	
Cupro-nickel		8,880	630	25,000	20 Ni, 1 Fe, 2 Mn
Brass	CuZn	8,600	600	110,000	20-40 Zn
Tombac	CuZn	8,800	470	120,000	5-20 Zn
Antimony lead	PbSb	10,950	45	16,000	2-5 Sb
Aluminium	Al	2,700	50–130	71,000	
Duraluminium		2,990	250–400	73,000	4 Cu, 1 Mg, 1 Mn
Tungsten	W	19,300	1,400	400,000	

A.3 Calibre designations (metric system)

Values and nomenclature from the data sheets of the CIP (Permanent International Commission for the Proof of Small Arms). English names based on the standard work "Cartridges of the World" by F. C. BARNES (7th Edition).

A.3.1 Handguns

Designation		Calibre	Bullet ∅	Case length
C.I.P.	English	[mm]	[mm]	[mm]
5.45 × 18	5.45 × 18 Soviet	5.45	5.62	18.00
6.35 Browning	25 Auto Pistol	6.17	6.38	15.55
7 × 49 GJW		6.96	7.25	49.00
7.5 Ord. Swiss	7.5 mm Swiss Army Rev.	7.65	8.00	22.80
7.62 Nagant	7.62 mm Russian Nagant Rev.	7.62	7.82	38.80
7.62 × 25 Tokarev	7.62 × 25 Russian Tokarev	7.62	7.90	25.00
7.63 Mauser	30 Mauser	7.62	7.86	25.15
7.65 Browning	32 Auto	7.63	7.85	17.20
7.65 Parabellum	30 Luger	7.62	7.85	21.59
9 mm Browning short	38 Auto	8.84	9.04	17.33
9 × 18	9 mm Ultra	8.82	9.02	18.00
9 mm Makarov	9 mm Russian Makarov	9.00	9.27	18.10
9 mm Luger[a]		8.82	9.03	19.15
9 mm Browning long	9 mm Browning long	8.92	9.09	20.20
9 × 21		8.79	9.03	21.15
10 mm Auto		9.91	10.16	25.20
32 S & W	32 Smith & Wesson	7.70	8.00	15.37
32 Long Colt		7.75	7.97	23.27
357 SIG		8.79	9.03	21.97
357 Magnum		8.79	9.12	32.77
38 S & W	38 Smith & Wesson	8.89	9.17	19.69
38 Special (38 Spl.)	38 Smith & Wesson Special	8.79	9.12	29.34
38 Super Auto		8.79	9.04	22.86
40 S & W	40 Smith & Wesson Auto	9.91	10.17	21.59
41 Rem. Mag.	41 Smith & Wesson Magnum	10.13	10.41	32.77
44 S & W Special	44 Smith & Wesson Special	10.59	10.98	29.46
44 Rem. Mag.	44 Smith & Wesson Magnum	10.59	10.97	32.64
45 Auto		11.23	11.48	22.81

[a] Formerly known as "9 mm Parabellum"

A.3.2 Military rifles

Designation		Calibre [mm]	Bullet ⌀ [mm]	Case length [mm]
Military	C.I.P.			
5.45 × 39 (Kalashnikov)	5.45 × 39	5.45	5.60	39.82
5.56 mm NATO	223 Rem. (5.56 × 45)	5.56	5.70	44.70
7.5 mm GP 11	7.5 × 55 Suisse	7.54	7.73	55.60
7.62 × 39 (Kalashnikov)	7.62 × 39	7.62	7.92	38.70
30 M 1 Carbine	30 Carbine (7.62 × 33)	7.62	7.85	32.77
7.62 mm NATO	308 Win. (7.62 × 51)	7.62	7.85	51.18
7.62 mm Mosin-Nagant	7.62 × 54 R	7.62	7.92	53.72
30-06	30-06 Spring. (7.62 × 63)	7.62	7.85	63.35
303 British	303 British (7.7 × 56 R)	7.70	7.92	56.44

A.3.3 Hunting and sporting rifles

Designation C.I.P.	Calibre [mm]	Bullet ⌀ [mm]	Case length [mm]
5.6 × 57	5.54	5.70	56.70
6 mm Rem.	6.00	6.18	56.70
6 mm BR Norma	6.02	6.18	39.60
6.5 × 57	6.45	6.70	56.70
6.5 × 68	6.45	6.70	67.50
7 × 49 GJW	6.96	7.25	49.00
7 × 57	6.98	7.25	57.00
7 × 64	6.98	7.25	64.00
7 mm Rem. Mag.	7.04	7.23	63.50
7 mm Weath. Mag.	7.02	7.22	64.75
8 × 57 J	7.80	8.09	57.00
8 × 57 JS	7.89	8.22	57.00
8 × 60 S	7.89	8.22	60.00
8 × 64 S	7.89	8.22	64.00
8 × 68 S	7.89	8.22	67.50
9.3 × 53 R	8.90	9.25	53.00
9.3 × 62	9.00	9.30	62.00
9.3 × 74 R	9.00	9.30	74.70
10.75 × 68	10.45	10.78	68.00
222 Rem.	5.56	5.70	43.20
243 Win.	6.02	6.18	51.95
270 Win.	6.86	7.06	64.50

Designation C.I.P.	Calibre [mm]	Bullet ⌀ [mm]	Case length [mm]
280 Rem.	7.04	7.20	64.50
30-30 Win.	7.62	7.85	51.80
300 H&H Mag.	7.62	7.85	72.40
300 Weath. Mag.	7.62	7.84	71.75
300 Win. Mag.	7.62	7.84	66.55
300 Savage	7.62	7.85	47.50
303 British	7.69	7.94	56.45
358 Win.	8.89	9.11	51.20
375 H&H Mag.	9.30	9.55	72.40
458 Win. Mag.	11.43	11.66	63.50
460 Weath. Mag.	11.43	11.64	74.00

Abbreviations: Mag. Magnum Win. Winchester
Rem. Remington Weath. Weatherby
Spring. Springfield

A.4 Ballistic data for cartridges (metric system)

A.4.1 Handgun cartridges

Designation	Calibre [mm]	Mass [g]	q [g/mm²]	v_0 [m/s]	I [N·s]	E [J]	ED [J/mm²]
5.45 × 18	5.45	2.45	0.1050	315	0.82	122	5.53
6.35 Browning	6.35	3.20	0.1010	230	0.74	85	3.16
7.62 mm Nagant	7.62	7.00	0.1535	285	2.00	284	6.23
7.62 × 25 Tokarev	7.62	5.60	0.1228	455	2.55	580	12.71
7.63 Mauser	7.62	5.50	0.1206	430	2.37	508	11.15
7.65 Browning	7.65	4.70	0.1023	305	1.43	219	4.76
7.65 Parabellum	7.65	6.00	0.1305	360	2.16	389	8.46
9 mm Browning short	9.00	6.10	0.0959	275	1.68	231	3.63
9 mm Browning long	9.00	7.20	0.1132	335	2.41	404	6.35
9 mm Makarov	9.00	6.10	0.0959	340	2.07	353	5.54
9 × 18	9.00	6.10	0.0959	330	2.01	332	5.22
9 mm Luger	9.00	8.00	0.1258	350	2.80	490	7.70
9 mm Luger (submachine gun)	9.00	8.00	0.1258	410	3.28	672	10.57
10 mm Auto	10.00	10.70	0.1362	365	3.91	713	9.08
		13.00	0.1655	320	4.16	666	8.47

Designation	Calibre [mm]	Mass [g]	q [g/mm^2]	v_0 [m/s]	I [N·s]	E [J]	ED [J/mm^2]
22 short	5.60	1.80	0.0731	280	0.50	71	2.86
22 long	5.60	1.80	0.0731	365	0.66	120	4.87
22 L.R. (22 long rifle)	5.60	2.55	0.1035	330	0.84	139	5.64
22 Win. Mag.	5.60	2.60	0.1056	615	1.60	492	19.96
32 S & W	7.65	5.50	0.1197	210	1.16	121	2.64
32 S & W long	7.65	6.35	0.1382	240	1.52	183	3.98
357 Magnum	9.00	10.20	0.1603	430	4.39	943	14.82
357 SIG	9.00	8.10	0.1273	410	3.32	681	10.70
38 S & W	9.00	9.40	0.1478	220	2.07	227	3.58
38 S & W Spl.	9.00	10.20	0.1603	265	2.70	358	5.63
40 S & W	10.00	10.00	0.1273	350	3.50	613	7.80
		11.70	0.1490	290	3.39	492	6.26
41 Mag.	10.40	13.60	0.1601	455	6.19	1408	16.57
44 Rem. Mag.	11.20	15.60	0.1583	440	6.86	1510	15.33
44 S & W Special	11.20	15.90	0.1614	230	3.66	421	4.27
45 Auto	11.25	14.90	0.1499	260	3.87	504	5.07
45 Colt	11.25	16.20	0.1630	265	4.29	569	5.72

A.4.2 Military ammunition

Designation	Calibre [mm]	Mass [g]	q [g/mm^2]	v_0 [m/s]	I [N·s]	E [J]	ED [J/mm^2]
4.6 × 30 (MP 7)	4.60	2.60	0.1564	600	1.56	468	28.16
5.45 × 39 (Kalashnikov).	3.45	3.45	0.1479	900	3.11	1397	59.90
5.56 × 45 (223 Rem.)	5.56	3.56	0.1466	965	3.44	1658	68.27
5.56 mm NATO	5.56	4.00	0.1647	920	3.68	1693	69.72
5.6 mm GP 90 (Swiss)	5.56	4.10	0.1689	905	3.71	1679	69.15
7.5 × 55 Suisse	7.50	11.30	0.2558	750	8.48	3178	71.94
7.62 × 39 (Kalashnikov)	7.62	8.00	0.1754	700	5.60	1960	42.98
7.62 mm NATO (7.62 × 51)	7.62	9.50	0.2083	830	7.89	3272	71.75
7.62 × 54 R	7.62	12.00	0.2631	800	9.60	3840	84.20
7.92 mm Mauser	7.92	11.50	0.2334	835	9.60	4009	81.38
30 M1 Carbine	7.62	7.10	0.1557	600	4.26	1278	28.02
300 Win. Mag.	7.62	9.70	0.2127	1035	10.04	5195	113.93
30-06 (7.62 × 63).	7.62	9.75	0.2138	835	8.14	3399	74.53
303 British	7.70	11.30	0.2427	745	8.42	3136	67.34

A.4.3 Hunting and sporting ammunition

Designation	Calibre [mm]	Mass [g]	q [g/mm^2]	v_0 [m/s]	I [N·s]	E [J]	ED [J/mm^2]
5.6 × 57	5.60	4.80	0.1991	1040	4.99	2596	107.69
5.6 × 61 SE v. H.	5.60	5.00	0.2105	1130	5.65	3192	134.36
6 mm Rem.	6.00	5.20	0.1839	1080	5.62	3033	107.26
		6.50	0.2299	972	6.32	3071	108.60
6 mm BR Norma	6.00	6.80	0.2405	850	5.78	2457	86.88
6.5 × 54 M. Sch.	6.50	10.10	0.3044	750	7.58	2841	85.60
6.5 × 55	6.50	9.00	0.2712	850	7.65	3251	97.98
6.5 × 57	6.50	6.00	0.1808	1010	6.06	3060	92.22
		10.00	0.3014	815	8.15	3321	100.08
6.5 × 68	6.50	8.50	0.2562	985	8.37	4123	124.26
7 × 49 GJW	7.00	10.90	0.2832	650	7.09	2303	59.83
7 × 57	7.00	7.50	0.1949	890	6.68	2970	77.18
		9.00	0.2339	780	7.02	2738	71.14
		11.20	0.2910	760	8.51	3235	84.05
7 × 64	7.00	11.20	0.2910	850	9.52	4046	105.13
7 × 75 R SE v. Hofe	7.00	11.00	0.2858	935	10.29	4808	124.94
7 mm Rem. Mag.	7.00	9.70	0.2520	995	9.65	4802	124.77
		11.30	0.2936	935	10.57	4939	128.35
7 mm Weath. Mag.	7.00	11.30	0.2936	890	10.06	4475	116.29
8 × 57 JS	8.00	10.30	0.2049	830	8.55	3548	70.58
		12.70	0.2527	800	10.16	4064	80.85
8 × 60	8.00	12.70	0.2527	792	10.06	3983	79.24
8 × 64 S	8.00	12.70	0.2527	803	10.20	4095	81.46
8 × 68 S	8.00	12.10	0.2407	970	11.74	5692	113.25
9 × 57	9.00	15.90	0.2499	656	10.43	3421	53.78
9.3 × 53 R	9.30	16.60	0.2444	700	11.62	4067	59.87
9.3 × 62	9.30	16.00	0.2355	800	12.80	5120	75.37
		18.50	0.2723	720	13.32	4795	70.59
9.3 × 74 R	9.30	18.50	0.2723	710	13.14	4663	68.64
10.75 × 68	10.75	22.50	0.2479	680	15.30	5202	57.31
222 Rem.	5.56	3.20	0.1318	975	3.12	1521	62.65
243 Win.	6.00	5.20	0.1839	1060	5.51	2921	103.32
		6.50	0.2299	935	6.08	2841	100.49
244 Rem.	6.00	5.80	0.2051	975	5.66	2757	97.50
250 Savage	6.35	6.50	0.2052	860	5.59	2404	75.90
270 Win.	6.85	8.40	0.2279	960	8.06	3871	105.03

Designation	Calibre [mm]	Mass [g]	q [g/mm²]	v_0 [m/s]	I [N·s]	E [J]	ED [J/mm²]
270 Win.	6.85	9.70	0.2632	850	8.25	3504	95.08
280 Rem.	7.00	10.70	0.2780	860	9.20	3957	102.82
30-30 Win.	7.62	11.00	0.2412	675	7.43	2506	54.95
30-06 Spring.	7.62	9.70	0.2127	905	8.78	3972	87.10
	7.62	11.70	0.2566	825	9.65	3982	87.31
300 H & H Mag.	7.62	11.70	0.2566	890	10.41	4634	101.61
	7.62	14.30	0.3136	800	11.44	4576	100.34
300 Weath. Mag.	7.62	9.70	0.2127	1060	10.28	5449	119.50
300 Win. Mag.	7.62	9.70	0.2127	1035	10.04	5195	113.93
	7.62	11.70	0.2566	935	10.94	5114	112.14
300 Savage	7.70	11.70	0.2513	720	8.42	3033	65.13
303 British	7.70	11.20	0.2405	725	8.12	2944	63.21
	7.70	14.00	0.3006	665	9.31	3096	66.48
358 Win.	8.90	13.00	0.2090	771	10.02	3864	62.11
375 H & H Mag.	9.30	17.50	0.2576	835	14.61	6101	89.81
	9.30	19.40	0.2856	790	15.33	6054	89.12
458 Win. Mag.	11.43	32.40	0.3158	650	21.03	6823	66.50
460 Weath. Mag.	11.43	32.40	0.3158	770	24.95	9605	93.61

A.4.4 Pre-1900 weapons and ammunition

Designation	Calibre [mm]	Mass [g]	q [g/mm²]	v_0 [m/s]	I [N·s]	E [J]	ED [J/mm²]
Flintlock rifle[a]	16.6	26.7	0.124	455	12.2	2767	12.8
Flintlock rifle[b]	17.5	30.9	0.129	495	15.3	3786	15.7
Flintlock pistol[c]	13.5	14.5	0.101	385	5.6	1071	7.5
Percussion rifle[d]	11.5	33.0	0.318	410	13.5	2774	26.7
Dreyse (D)	13.6	31.0	0.213	295	9.1	1349	9.3
Chassepot (F)	11.8	25.0	0.229	420	10.5	2200	20.2
Vetterli (CH)	10.4	19.4	0.228	435	8.4	1835	21.6
45-70 Springfield	11.4	25.9	0.254	395	10.2	2021	19.7
Gewehr 88 (D)	7.9	15.0	0.306	640	9.6	3072	62.7

Values for first three weapons from MISSLIWETZ and WIESER.

[a] Hunting rifle, second half of 18th century. Weight 3.88 kg, barrel length 680 mm.
[b] Flintlock rifle with combined match and smooth bore. Montecuccolli, built in 1686, weight 4.2 kg. barrel length 1050 mm.
[c] Smooth-bore pistol. Built around 1700. Weight 1.16 kg, barrel length 332 mm.
[d] Percussion rifle. Mid-19th-century.

A.4.5 Ballistic performance of certain bows and crossbows

A.4.5.1 Technical data[a]

Weapon	Weight [kg]	Pull force [N]	Draw length [cm]	Mass [g]	Length [cm]	Arrow/bolt diameter [mm]
Crossbow, 180 lbs	2.56	800	27	24.5	38.2	6
Crossbow, 220 lbs	3.60	987	29	24.0	37.5	8.5
Pistol crossbow	1.68	200	9	9.2	15.5	6
Delta V longbow	3.58	785	74	39.0	76.8	8.6

[a] from MISSLIWETZ and WIESER (1985a)

A.4.5.2 Ballistic data[a]

Weapon	v [m/s]	E [J]	Max. range [m]	Penetration depth Wood [cm]	Penetration depth Gelatine [cm]
Crossbow, 180 lbs	58	40.4	268	2.8	24.2
Crossbow, 220 lbs	59	41.4	274	2.4	>30
Pistol crossbow	45	9.3	165	0.8	7.5
Delta V longbow	71	98.2	337	4.6	>30

[a] from MISSLIWETZ and WIESER (1985a)

A.4.6 Ballistic data for various types of projectiles used in sport[a]

Projectile	m [g]	d [mm]	v [m/s]	E [J]	ED [J/mm^2]	Range[b] [m]
Golfball	46	41	75	130	0.230	260
Tennis ball	59	65	60	106	0.037	140
Squash ball	28	40	55	42	0.076	130
Nouss (used in hornussen[c])	78	60 × 30	80	250	0.200[d]	300
Baseball	145	74	60	180	0.090	180
Ice-hockey puck	163	76 × 25	50	205	0.210	130
Football	425	221	40	340	0.025	70
Handball	450	188	35	275	0.026	70
Shot used in shot-putting	7257	121	15	815	0.071	23

[a] Approximate values encountered at top levels in the sports concerned Source: Internet and various reference works.
[b] Ballistic calculations, based on data from tables.
[c] Swiss team sport (see KNEUBUEHL 2001).
[d] Energy density when the projectile is in the vicinity of the opposing team.

A.5 Calibre designations (British/U.S. system)

Values and nomenclature from the data sheets of the CIP (Permanent International Commission for the Proof of Small Arms). English names based on the standard work "Cartridges of the World" by F. C. BARNES (7th Edition).

A.5.1 Handguns

Designation		Calibre	Bullet ⌀	Case length
C.I.P.	English	[in]	[in]	[in]
5.45 × 18	5.45 × 18 Soviet	0.215	0.221	0.709
6.35 Browning	25 Auto Pistol	0.243	0.251	0.612
7 × 49 GJW		0.274	0.285	1.929
7.5 Ord. Swiss	7.5 mm Swiss Army Rev.	0.301	0.315	0.898
7.62 Nagant	7.62 mm Russian Nagant Rev.	0.300	0.308	1.528
7.62 × 25 Tokarev	7.62 × 25 Russian Tokarev	0.300	0.311	0.984
7.63 Mauser	30 Mauser	0.300	0.309	0.990
7.65 Browning	32 Auto	0.300	0.309	0.677
7.65 Parabellum	30 Luger	0.300	0.309	0.850
9 mm Browning short	38 Auto	0.348	0.356	0.682
9 × 18	9 mm Ultra	0.347	0.355	0.709
9 mm Makarov	9 mm Russian Makarov	0.354	0.365	0.713
9 mm Luger[a]		0.347	0.356	0.754
9 mm Browning long	9 mm Browning long	0.351	0.358	0.795
9 × 21		0.346	0.356	0.833
10 mm Auto		0.390	0.400	0.992
32 S & W	32 Smith & Wesson	0.303	0.315	0.605
32 Long Colt		0.305	0.314	0.916
357 SIG		0.346	0.356	0.865
357 Magnum		0.346	0.359	1.290
38 S & W	38 Smith & Wesson	0.350	0.361	0.775
38 Special (38 Spl.)	38 Smith & Wesson Special	0.346	0.359	1.155
38 Super Auto		0.346	0.356	0.900
40 S & W	40 Smith & Wesson Auto	0.390	0.400	0.850
41 Rem. Mag.	41 Smith & Wesson Magnum	0.399	0.410	1.290
44 S & W Special	44 Smith & Wesson Special	0.417	0.432	1.160
44 Rem. Mag.	44 Smith & Wesson Magnum	0.417	0.432	1.285
45 Auto		0.442	0.452	0.898

[a] Formerly known as "9 mm Parabellum"

A.5.2 Military rifles

Designation Military	Designation C.I.P.	Calibre [in]	Bullet ⌀ [in]	Case length [in]
5.45 × 39 (Kalashnikov)	5.45 × 39	0.215	0.220	1.568
5.56 mm NATO	223 Rem. (5.56 × 45)	0.219	0.224	1.760
7.5 mm GP 11	7.5 × 55 Suisse	0.297	0.304	2.189
7.62 × 39 (Kalashnikov)	7.62 × 39	0.300	0.312	1.524
30 M 1 Carbine	30 Carbine (7.62 × 33)	0.300	0.309	1.290
7.62 mm NATO	308 Win. (7.62 × 51)	0.300	0.309	2.015
7.62 mm Mosin-Nagant	7.62 × 54 R	0.300	0.312	2.115
30-06	30-06 Spring. (7.62 × 63)	0.300	0.309	2.494
303 British	303 British (7.7 × 56 R)	0.303	0.312	2.222

A.5.3 Hunting and sporting rifles

Designation C.I.P.	Calibre [in]	Bullet ⌀ [in]	Case length [in]
5.6 × 57	0.218	0.224	2.232
6 mm Rem.	0.236	0.243	2.232
6 mm BR Norma	0.237	0.243	1.559
6.5 × 57	0.254	0.264	2.232
6.5 × 68	0.254	0.264	2.657
7 × 49 GJW	0.274	0.285	1.929
7 × 57	0.275	0.285	2.244
7 × 64	0.275	0.285	2.520
7 mm Rem. Mag.	0.277	0.285	2.500
7 mm Weath. Mag.	0.276	0.284	2.549
8 × 57 J	0.307	0.319	2.244
8 × 57 JS	0.311	0.324	2.244
8 × 60 S	0.311	0.324	2.362
8 × 64 S	0.311	0.324	2.520
8 × 68 S	0.311	0.324	2.657
9.3 × 53 R	0.350	0.364	2.087
9.3 × 62	0.354	0.366	2.441
9.3 × 74 R	0.354	0.366	2.941
10.75 × 68	0.411	0.424	2.677
222 Rem.	0.219	0.224	1.701
243 Win.	0.237	0.243	2.045
270 Win.	0.270	0.278	2.539

Designation C.I.P.	Calibre [in]	Bullet ⌀ [in]	Case length [in]
280 Rem.	0.277	0.283	2.539
30-30 Win.	0.300	0.309	2.039
300 H&H Mag.	0.300	0.309	2.850
300 Weath. Mag.	0.300	0.309	2.825
300 Win. Mag.	0.300	0.309	2.620
300 Savage	0.300	0.309	1.870
303 British	0.303	0.313	2.222
358 Win.	0.350	0.359	2.016
375 H&H Mag.	0.366	0.376	2.850
458 Win. Mag.	0.450	0.459	2.500
460 Weath. Mag.	0.450	0.458	2.913

Abbreviations: Mag. Magnum Win. Winchester
Rem. Remington Weath. Weatherby
Spring. Springfield

A.6 Ballistic data for cartridges (British/U.S. units)

A.6.1 Handgun cartridges

Designation	Calibre [in]	Mass [gr]	q [lb/in²]	v_0 [ft/s]	I [lbf·s]	E [ft·lbf]	ED [a]
5.45 × 18	0.21	38	0.139	1035	0.184	90	2631
6.35 Browning	0.25	50	0.133	755	0.166	62	1504
7.62 mm Nagant	0.30	108	0.203	935	0.450	210	2965
7.62 × 25 Tokarev	0.30	86	0.162	1495	0.573	428	6048
7.63 Mauser	0.30	85	0.159	1410	0.533	375	5306
7.65 Browning	0.30	73	0.135	1000	0.321	161	2265
7.65 Parabellum	0.30	93	0.172	1180	0.486	287	4026
9 mm Browning short	0.35	94	0.127	900	0.378	170	1727
9 mm Browning long	0.35	111	0.149	1100	0.542	298	3022
9 mm Makarov	0.35	95	0.127	1115	0.465	260	2636
9 × 18	0.35	95	0.127	1085	0.452	245	2484
9 mm Luger	0.35	125	0.166	1150	0.629	361	3664
9 mm Luger (submachine gun)	0.35	125	0.166	1345	0.737	496	5030
10 mm Auto	0.39	165	0.180	1200	0.879	526	4321
	0.00	200	0.218	1050	0.935	491	4030

[a] [ft·lbf/in²]

Designation	Calibre [in]	Mass [gr]	q [lb/in²]	v_0 [ft/s]	I [lbf·s]	E [ft·lbf]	ED [a]
22 short	0.22	28	0.097	920	0.112	52	1361
22 long	0.22	28	0.097	1200	0.148	88	2317
22 L.R. (22 long rifle)	0.22	40	0.137	1085	0.189	102	2684
22 Win. Mag.	0.22	40	0.139	2020	0.360	363	9498
32 S & W	0.30	85	0.158	690	0.261	89	1256
32 S & W long	0.30	98	0.182	785	0.342	135	1894
357 Magnum	0.35	158	0.212	1410	0.987	696	7052
357 SIG	0.35	125	0.168	1345	0.746	502	5092
38 S & W	0.35	145	0.195	720	0.465	168	1704
38 S & W Spl.	0.35	158	0.212	870	0.607	264	2679
40 S & W	0.39	154	0.168	1150	0.787	452	3712
	0.00	180	0.197	950	0.762	363	2979
41 Mag.	0.41	210	0.211	1495	1.392	1038	7885
44 Rem. Mag.	0.44	240	0.209	1445	1.542	1114	7295
44 S & W Special	0.44	245	0.213	755	0.823	310	2032
45 Auto	0.45	230	0.198	855	0.870	371	2413
45 Colt	0.45	250	0.215	870	0.964	420	2722

A.6.2 Military ammunition (U.S.)

Designation	Calibre [in]	Mass [gr]	q [lb/in²]	v_0 [ft/s]	I [lbf·s]	E [ft·lbf]	ED [a]
4.6 × 30 (MP 7)	0.18	40	0.206	1970	0.351	345	13400
5.45 × 39 (Kalashnikov).	0.21	53	0.195	2955	0.699	1030	28503
5.56 × 45 (223 Rem.)	0.22	55	0.194	3165	0.773	1223	32486
5.56 mm NATO	0.22	62	0.217	3020	0.827	1249	33176
5.6 mm GP 90 (Swiss)	0.22	63	0.223	2970	0.834	1238	32905
7.5 × 55 Suisse	0.30	174	0.338	2460	1.906	2344	34232
7.62 × 39 (Kalashnikov)	0.30	123	0.232	2295	1.259	1446	20452
7.62 mm NATO (308 Win.)	0.30	147	0.275	2725	1.774	2413	34142
7.62 × 54 R	0.30	185	0.347	2625	2.158	2832	40066
7.92 mm Mauser	0.31	177	0.308	2740	2.158	2957	38724
30 M1 Carbine	0.30	110	0.206	1970	0.958	943	13333
300 Win. Mag.	0.30	150	0.281	3395	2.257	3832	54213
30-06 (7.62 × 63).	0.30	150	0.282	2740	1.830	2507	35465
303 British	0.30	174	0.320	2445	1.893	2313	32043

[a] [ft·lbf/in²]

A.6.3 Hunting and sporting ammunition

Designation	Calibre [in]	Mass [gr]	q [lb/in²]	v_0 [ft/s]	I [lbf·s]	E [ft·lbf]	ED [ª]
5.6 × 57	0.22	74	0.263	3410	1.122	1915	51244
5.6 × 61 SE v. H.	0.22	77	0.278	3705	1.270	2354	63935
6 mm Rem.	0.24	80	0.243	3545	1.263	2237	51039
		100	0.304	3190	1.421	2265	51677
6 mm BR Norma	0.24	105	0.317	2790	1.299	1812	41341
6.5 × 54 M. Sch.	0.26	156	0.402	2460	1.704	2095	40732
6.5 × 55	0.26	139	0.358	2790	1.720	2398	46623
6.5 × 57	0.26	93	0.239	3315	1.362	2257	43883
		154	0.398	2675	1.832	2449	47623
6.5 × 68	0.26	131	0.338	3230	1.882	3041	59129
7 × 49 GJW	0.28	168	0.374	2135	1.594	1699	28470
7 × 57	0.28	116	0.257	2920	1.502	2191	36726
		139	0.309	2560	1.578	2019	33852
		173	0.384	2495	1.913	2386	39995
7 × 64	0.28	173	0.384	2790	2.140	2984	50026
7 × 75 R SE v. Hofe	0.28	170	0.377	3070	2.313	3546	59452
7 mm Rem. Mag.	0.28	150	0.333	3265	2.169	3542	59371
		174	0.388	3070	2.376	3643	61075
7 mm Weath. Mag.	0.28	174	0.388	2920	2.262	3301	55336
8 × 57 JS	0.31	159	0.270	2725	1.922	2617	33585
		196	0.334	2625	2.284	2997	38472
8 × 60	0.31	196	0.334	2600	2.262	2938	37706
8 × 64 S	0.31	196	0.334	2635	2.293	3020	38762
8 × 68 S	0.31	187	0.318	3180	2.639	4198	53890
9 × 57	0.35	245	0.330	2150	2.345	2523	25591
9.3 × 53 R	0.37	256	0.323	2295	2.612	3000	28489
9.3 × 62	0.37	247	0.311	2625	2.878	3776	35864
		285	0.359	2360	2.994	3537	33590
9.3 × 74 R	0.37	285	0.359	2330	2.954	3439	32662
10.75 × 68	0.42	347	0.327	2230	3.440	3837	27271
222 Rem.	0.22	49	0.174	3200	0.701	1122	29812
243 Win.	0.24	80	0.243	3480	1.239	2154	49164
		100	0.304	3070	1.367	2095	47818
244 Rem.	0.24	90	0.271	3200	1.272	2033	46395
250 Savage	0.25	100	0.271	2820	1.257	1773	36117
270 Win.	0.27	130	0.301	3150	1.812	2855	49978

360 Appendix A: Tables

Designation	Calibre [in]	Mass [gr]	q [lb/in^2]	v_0 [ft/s]	I [lbf·s]	E [ft·lbf]	ED [a]
270 Win.	0.27	150	0.347	2790	1.855	2584	45243
280 Rem.	0.28	165	0.367	2820	2.068	2919	48926
30-30 Win.	0.30	170	0.318	2215	1.670	1848	26148
30-06 Spring.	0.30	150	0.281	2970	1.974	2930	41446
		181	0.339	2705	2.169	2937	41546
300 H & H Mag.	0.30	181	0.339	2920	2.340	3418	48351
		221	0.414	2625	2.572	3375	47746
300 Weath. Mag.	0.30	150	0.281	3480	2.311	4019	56864
300 Win. Mag.	0.30	150	0.281	3395	2.257	3832	54213
	0.00	181	0.339	3070	2.459	3772	53361
300 Savage	0.30	181	0.332	2360	1.893	2237	30992
303 British	0.30	173	0.317	2380	1.825	2171	30078
	0.00	216	0.397	2180	2.093	2283	31634
358 Win.	0.35	201	0.276	2530	2.253	2850	29555
375 H & H Mag.	0.37	270	0.340	2740	3.284	4500	42736
		299	0.377	2590	3.446	4465	42407
458 Win. Mag.	0.45	500	0.417	2135	4.728	5032	31644
460 Weath. Mag.	0.45	500	0.417	2525	5.609	7084	44544

[a] [ft·lbf/in^2]

A.6.4 Pre-1900 weapons and ammunition

Designation	Calibre [in]	Mass [gr]	q [lb/in^2]	v_0 [ft/s]	I [lbf·s]	E [ft·lbf]	ED [e]
Flintlock rifle[a]	0.65	412	0.164	1495	2.743	2041	6091
Flintlock rifle[b]	0.69	477	0.170	1625	3.440	2792	7471
Flintlock pistol[c]	0.53	224	0.133	1265	1.259	790	3569
Percussion rifle[d]	0.45	509	0.420	1345	3.035	2046	12705
Dreyse (D)	0.54	478	0.281	970	2.046	995	4425
Chassepot (F)	0.46	386	0.302	1380	2.361	1623	9612
Vetterli (CH)	0.41	299	0.301	1425	1.888	1353	10278
45-70 Springfield	0.45	400	0.335	1295	2.293	1491	9374
Gewehr 88 (D)	0.31	231	0.404	2100	2.158	2266	29836

Values for first three weapons from MISSLIWETZ and WIESER.

[a] Hunting rifle, second half of 18th century. Weight 3.88 kg, barrel length 680 mm.
[b] Flintlock rifle with combined match and smooth bore. Montecuccolli, built in 1686, weight 4.2 kg. barrel length 1050 mm.
[c] Smooth-bore pistol. Built around 1700. Weight 1.16 kg, barrel length 332 mm.
[d] Percussion rifle. Mid-19th-century.
[e] [ft·lbf/in^2]

A.6.5 Ballistic performance of certain bows and crossbows

A.6.5.1 Technical data[a]

Weapon	Weight [lb]	Pull force [lbf]	Draw length [in]	Mass [gr]	Arrow Length [in]	Arrow/bolt diameter [in]
Crossbow, 180 lbs	5.64	180	10.6	380	15.0	0.24
Crossbow, 220 lbs	7.94	222	11.4	370	14.8	0.33
Pistol crossbow	3.70	45	3.5	142	6.1	0.24
Delta V longbow	7.89	176	29.1	600	30.2	0.34

[a] from MISSLIWETZ and WIESER (1985a)

A.6.5.2 Ballistic data[a]

Weapon	v [ft/s]	E [ft·lbf]	Max. range [yd]	Penetration depth Wood [in]	Penetration depth Gelatine [in]
Crossbow, 180 lbs	190	29.8	293	1.1	9.5
Crossbow, 220 lbs	195	30.5	300	0.9	> 11.8
Pistol crossbow	150	6.9	180	0.3	3.0
Delta V longbow	235	72.4	369	1.8	> 11.8

[a] from MISSLIWETZ and WIESER (1985a)

A.6.6 Ballistic data for various types of projectiles used in sport[a]

Projectile	m [lb]	d [in]	v [ft/s]	E [ft·lbf]	ED [c]	Range[b] [yd]
Golfball	0.10	1.61	245	96	109.4	284
Tennis ball	0.13	2.56	195	78	17.6	153
Squash ball	0.06	1.57	180	31	36.2	142
Nouss (used in hornussen[d])	0.17	2.4 x 1.2	260	184	95.2[e]	328
Baseball	0.32	2.91	195	133	42.8	197
Ice-hockey puck	0.36	3 x 1	165	151	99.9	142
Football	0.94	8.70	130	251	11.9	77
Handball	0.99	7.40	115	203	12.4	77
Shot used in shot-putting	16.00	4.76	50	601	33.8	25

[a] Approximate values encountered at top levels in the sports concerned Source: Internet and various reference works.
[b] Ballistic calculations, based on data from the table.
[c] [ft·lbf/in²]
[d] Swiss team sport (see KNEUBUEHL 2001).
[e] Energy density when the projectile is in the vicinity of the opposing team.

A.7 Bullet designations

A.7.1 Bullet form

English	Abbreviation	German	Abbreviation
round ball		Kugel	
round nose	RN	Rundkopf	R
flat point	FP	Flachkopf	F
spire point	SP	Spitzkopf	S
wadcutter	WC	Scharfrand	
semi-wadcutter	SWC		
truncated cone	TC	Kegelstumpf	KS
boattailed	BT	mit Heckkonus	

A.7.2 Bullet material

English	Abbreviation	German	Abbreviation
lead	Pb	Blei	Pb
copper	Cu	Kupfer	Cu
brass		Messing	Ms
sintered iron		Sintereisen	
steel		Stahl	Fe
aluminium	Al	Aluminium	Al
tungsten	W	Wolfram	W
plastic		Kunststoff	P
wood		Holz	

A.7.3 Bullet structure

English	Abbreviation	German	Abbreviation
full metal-jacketed	FMJ	Vollmantel	VM
jacketed	J	Mantel	M
copper-jacketed	CJ	Kupfermantel	CuVM
semi-jacketed	SJ	Teilmantel (kurz)	TM
soft-point	SP	Teilmantel (lang)	TM
hollow-point	HP	Hohlspitze	HSp
lead hollow-point	LHP	Vollblei-	PbHSp
		Vollkupfer-	CuHSp
semi-jacketed hollow-point	SJHP	Teilmantel-	TMHSp
metal-point	MP	Metallkappe	MK

A.8 Geometric data for selected bullets

The following values were obtained, by measurement, by the weapon systems and munitions department of armasuisse, W+T, Thun Switzerland.

Abbreviations:
m : mass of bullet
J_a : axial moment of inertia
J_q : radial moment of inertia
ℓ_g : length of bullet
x_{SP} : distance between centre of gravity and tip

A.8.1 Military bullets

Designation	Bullet	m [g]	$J_a \cdot 10^8$ [kg·m^2]	$J_q \cdot 10^7$ [kg·m^2]	ℓ_g [mm]	x_{SP} [mm]
5.45 × 39 (Kalashnikov)	FMJ SP	3.45	1.22	1.22	25.3	15.2
5.56 × 45 SS 109	FMJ SP	4.0	1.096	1.123	23.1	14.5
5.56 × 45 SS 92	FMJ SP	3.56	1.165	0.763	19.0	11.3
5.6 mm GP 90 (CH)	FMJ SP	4.1	1.28	1.15	22.0	13.3
7.5 × 55 (Suisse)	FMJ SP	11.3	7.91	7.32	35.2	14.7
7.62 × 39 (Kalashnikov)	FMJ SP	8.0	5.74	2.49	22.6	13.7
7.62 mm NATO	FMJ SP	9.5	5.96	4.10	28.4	17.4

A.8.2 Other bullets

Designation	Bullet	m [g]	$J_a \cdot 10^8$ [kg·m^2]	$J_q \cdot 10^7$ [kg·m^2]	ℓ_g [mm]	x_{SP} [mm]
9 mm Luger	FMJ RN	8.0	6.5	1.5	15.7	9.1
357 Mag.	FMJ TC	10.3	8.5	3.1	19.5	11.9
38 Spl.	LRN	10.3	12.0	4.7	21.2	11.0
38 Spl.	FMJ FP	9.5	9.3	2.4	16.0	8.7
38 Spl.	Finishing shot	7.5	10.6	1.1	13.0	6.7
44 Rem. Mag.	TM FP	15.5	18.3	4.4	17.5	9.7
7 × 64	Partitioned	11.2	7.4	15.2	34.0	19.7
7 × 64	TIG	10.5	7.9	11.8	30.0	15.0
7 × 64	TC	8.0	7.3	5.3	24.6	14.6
7 mm Rem. Mag.	TC	10.5	6.3	12.0	30.5	17.5
243 Win.	FMJ SP	5.8	2.9	4.2	24.0	14.5
308 Win.	RWS Match	12.3	7.2	7.8	34.7	20.4
308 Win.	Sierra MK	10.9	6.6	5.7	31.2	19.0
308 Win.	Soft-point	7.1	1.7	3.4	19.0	11.7
460 Weath. Mag.	FMJ RN	32.4	290.0	31.0	35.3	19.4

A.9 Twist length, angle of twist and rotation

A.9.1 Handguns

Designation	Calibre [mm]	v_0 [m/s]	Twist length [mm]	Twist length [Kal]	Twist length [in]	Angle [°]	Rotation [1/s]
22 L.R.	5.60	330	350	63	13.75	2.90	943
			381	68	15	2.65	866
7.65 Browning	7.65	305	250	33	10	5.50	1220
7.65 Parabellum	7.65	360	250	33	10	5.50	1440
9 mm Browning, short	9.00	275	250	28	10	6.45	1100
9 mm Browning, long	9.00	335	250	28	10	6.45	1340
9 × 18	9.00	330	250	28	10	6.45	1320
9 mm Luger	9.00	350	250	28	10	6.45	1400
9 mm Luger (subm. gun)		420	305	34	12	5.30	1377
32 S & W	7.65	210	476	62	18.75	2.90	441
38 Spl.	9.00	265	476	53	18.75	3.40	557
357 Magnum	9.00	430	476	53	18.75	3.40	903
41 Mag.	10.40	455	476	46	18.75	3.95	956
44 Rem. Mag.	11.20	440	508	45	20	3.95	866
45 Auto	11.25	260	406	36	16	5.00	640

A.9.2 Rifles

A.9.2.1 Military rifles

Designation	Calibre [mm]	v_0 [m/s]	Twist length [mm]	Twist length [Cal]	Twist length [in]	Angle [°]	Rotation [1/s]
4.73 × 33	4.73	930	155	33	6.1	5.50	6000
5.45 × 39 (AKS-74)	5.45	900	196	36	7.75	5.00	4592
5.56 × 45 (223 Rem.)	5.56	965	305	55	12	3.30	3164
5.56 mm NATO	5.56	920	178	32	7	5.60	5169
5.6 mm GP 90 (Swiss)	5.56	905	250	45	10	4.00	3620
7.5 × 55 Suisse	7.50	750	270	36	10.6	5.00	2778
7.62 × 39 (AK-47)	7.62	700	235	31	9.25	5.80	2979
7.62 mm NATO	7.62	830	305	40	12	4.50	2721
	7.62	830	270	35	10.6	5.05	3074
7.62 × 54 R	7.62	800	250	33	10	5.45	3200
30 M1 Carbine	7.62	600	406	33	16	3.35	1478
300 Win. Mag.	7.62	1035	254	33	10	5.40	4075
30-06 (7.62 × 63)	7.62	835	254	33	10	5.40	3287

A.9.2.2 Hunting and sporting rifles

Designation	Calibre	v_0	Twist length			Angle of twist	Rotation
	[mm]	[m/s]	[mm]	[Cal]	[in]	[°]	[1/s]
5.6 × 57	5.54	1040	250	45	10	4.00	4160
6.5 × 54 M. Sch.	6.50	750	200	31	8	5.75	3695
6.5 × 57	6.50	1010	200	31	8	5.85	5050
6.5 × 68	6.50	985	280	43	11	4.15	3518
7 × 57	7.00	760	220	31	8.6	5.70	3455
7 × 64	7.00	850	220	31	8.6	5.70	3864
7 mm Rem. Mag.	7.00	935	229	33	9	5.50	4083
7 mm Weath. Mag.	7.00	890	305	44	12	4.10	2918
8 × 57 JS	8.00	800	235	29	9.25	6.10	3404
8 × 60 S	8.00	792	240	30	9.5	6.00	3300
8 × 68 S	8.00	970	280	35	11	5.15	3464
9.3 × 62	9.30	720	360	39	14.2	4.65	2000
9.3 × 74 R	9.30	710	360	39	14.2	4.65	1972
10.75 × 68	10.75	680	420	39	16.5	4.60	1619
222 Rem.	5.56	975	356	64	14	2.80	2739
243 Win.	6.00	935	254	42	10	4.25	3681
270 Win.	6.85	960	254	37	10	4.85	3780
280 Rem.	7.00	860	254	36	10	4.95	3386
30-30 Win.	7.62	675	305	40	12	4.50	2213
30-06 (7.62 × 63)	7.62	825	254	33	10	5.40	3248
300 H & H Mag.	7.62	890	254	33	10	5.40	3504
300 Weath. Mag.	7.62	1060	254	33	10	5.40	4173
300 Win. Mag.	7.62	935	254	33	10	5.40	3681
300 Savage	7.70	720	305	40	12	4.55	2361
303 British	7.70	665	254	33	10	5.45	2618
358 Win.	8.90	771	305	34	12	5.25	2528
375 H & H Mag.	9.30	835	305	33	12	5.45	2738
458 Win. Mag.	11.43	649	356	31	14	5.75	1823
460 Weath. Mag.	11.43	770	406	36	16	5.05	1897

A.10 Ballistics tables (metric system)

A.10.1 Notes

These tables use mils for the angle of both the angle of shot and the angle of descent. $360° \Leftrightarrow 6400$ mils

At the small angles encountered with flat-trajectory weapons, 1 mil corresponds to 1 mm change in altitude per metre of flight.

The ballistics tables are valid at sea level, and were generated using k-ballistics 4 software (www.kneubuehl.com, S. ROTHE).

Abbreviations:

x	:	Range	t	:	Time of flight
as	:	Angle of shot	Cw	:	Deviation with a cross-wind of 10 m/s
v	:	Velocity	y_s	:	Vertex height
E	:	Energy	x_s	:	Vertex distance
ED	:	Energy density	ad	:	Angle of descent

A.10.2 Handguns

6.35 Browning, m = 3.2 g

x [m]	as [Mil]	v [m/s]	E [J]	ED [J/mm^2]	t [s]	Cw [cm]	y_s [cm]	x_s [m]	ad [Mil]
0	0.00	245	96	3.03	0.00	0	0	0	0.0
10	0.84	241	93	2.94	0.04	0	0	5	0.9
20	1.70	238	90	2.86	0.08	1	1	10	1.7
30	2.57	234	88	2.77	0.13	3	2	15	2.7
40	3.47	231	85	2.69	0.17	5	4	20	3.6
50	4.38	227	83	2.61	0.21	8	6	25	4.6
60	5.31	224	80	2.53	0.26	11	8	30	5.6
70	6.25	221	78	2.46	0.30	16	11	36	6.7
80	7.22	218	76	2.39	0.35	20	15	41	7.8
90	8.21	214	74	2.32	0.39	26	19	46	9.0
100	9.22	211	71	2.25	0.44	32	24	51	10.2
110	10.24	208	69	2.19	0.49	39	29	57	11.4
120	11.29	205	67	2.12	0.54	47	35	62	12.7
130	12.36	202	65	2.06	0.59	55	42	67	14.1
140	13.45	199	64	2.00	0.64	64	50	72	15.4
150	14.57	196	62	1.95	0.69	74	58	78	16.9

7.62 × 25 Tokarev, m = 5.6 g

x [m]	as [Mil]	v [m/s]	E [J]	ED [J/mm²]	t [s]	Cw [cm]	y_s [cm]	x_s [m]	ad [Mil]
0	0.00	455	580	12.71	0.00	0	0	0	0.0
10	0.25	441	544	11.93	0.02	0	0	5	0.3
20	0.50	427	510	11.18	0.05	1	0	10	0.5
30	0.77	413	478	10.48	0.07	3	1	15	0.8
40	1.05	400	448	9.82	0.09	6	1	20	1.2
50	1.35	387	420	9.21	0.12	9	2	26	1.5
60	1.65	375	394	8.64	0.15	14	3	31	1.9
70	1.97	364	371	8.13	0.17	19	4	36	2.3
80	2.31	353	350	7.67	0.20	25	5	42	2.7
90	2.66	344	331	7.26	0.23	31	6	47	3.2
100	3.02	335	315	6.91	0.26	39	8	53	3.7
110	3.40	329	302	6.62	0.29	47	10	58	4.2
120	3.79	323	292	6.39	0.32	56	13	64	4.8
130	4.20	318	283	6.19	0.35	65	15	69	5.4
140	4.62	313	274	6.02	0.38	75	18	75	6.0
150	5.05	309	267	5.85	0.41	85	21	80	6.6

7.65 Browning, m = 4.6 g

x [m]	as [Mil]	v [m/s]	E [J]	ED [J/mm²]	t [s]	Cw [cm]	y_s [cm]	x_s [m]	ad [Mil]
0	0.00	305	219	4.76	0.00	0	0	0	0.0
10	0.54	300	212	4.61	0.03	0	0	5	0.6
20	1.10	296	205	4.47	0.07	1	1	10	1.1
30	1.66	291	199	4.33	0.10	2	1	15	1.7
40	2.24	287	194	4.21	0.14	4	2	20	2.3
50	2.83	283	188	4.09	0.17	7	4	25	3.0
60	3.43	279	183	3.98	0.21	9	5	30	3.6
70	4.04	275	178	3.87	0.24	13	7	36	4.3
80	4.67	272	173	3.77	0.28	16	10	41	5.0
90	5.30	268	169	3.67	0.32	21	12	46	5.8
100	5.95	265	164	3.58	0.35	25	15	51	6.5
110	6.61	261	160	3.48	0.39	31	19	56	7.3
120	7.28	258	156	3.39	0.43	37	23	62	8.1
130	7.97	254	152	3.31	0.47	43	27	67	9.0
140	8.66	251	148	3.22	0.51	50	32	72	9.8
150	9.37	248	144	3.14	0.55	57	37	78	10.7

9 mm Makarov, m = 6.1 g

x [m]	as [Mil]	v [m/s]	E [J]	ED [J/mm²]	t [s]	Cw [cm]	y_s [cm]	x_s [m]	ad [Mil]
0	0.00	340	353	5.54	0.00	0	0	0	0.0
10	0.44	329	331	5.20	0.03	0	0	5	0.5
20	0.90	320	313	4.91	0.06	2	1	10	0.9
30	1.38	312	297	4.66	0.09	4	1	15	1.5
40	1.88	304	283	4.44	0.12	7	2	20	2.0
50	2.39	298	270	4.24	0.16	11	3	26	2.6
60	2.92	291	259	4.06	0.19	16	5	31	3.2
70	3.47	285	248	3.90	0.23	21	6	36	3.9
80	4.03	279	238	3.74	0.26	27	8	41	4.6
90	4.61	274	229	3.60	0.30	34	11	47	5.3
100	5.21	269	220	3.46	0.34	41	14	52	6.1
110	5.82	264	212	3.33	0.37	49	17	57	6.9
120	6.46	259	204	3.21	0.41	58	21	63	7.7
130	7.11	254	197	3.10	0.45	68	25	68	8.6
140	7.77	250	190	2.99	0.49	78	30	74	9.5
150	8.46	245	183	2.88	0.53	89	35	79	10.4

9 mm Luger FMJ RN, m = 8.0 g

x [m]	as [Mil]	v [m/s]	E [J]	ED [J/mm²]	t [s]	Cw [cm]	y_s [cm]	x_s [m]	ad [Mil]
0	0.00	350	490	7.70	0.00	0	0	0	0.0
10	0.42	341	465	7.31	0.03	0	0	5	0.4
20	0.84	333	444	6.98	0.06	1	0	10	0.9
30	1.29	327	427	6.72	0.09	3	1	15	1.3
40	1.74	321	413	6.50	0.12	6	2	20	1.8
50	2.21	317	401	6.30	0.15	8	3	25	2.4
60	2.69	312	390	6.13	0.18	12	4	31	2.9
70	3.17	308	380	5.97	0.22	15	6	36	3.4
80	3.67	304	370	5.82	0.25	19	8	41	4.0
90	4.18	301	362	5.69	0.28	24	10	46	4.6
100	4.69	297	353	5.55	0.31	29	12	51	5.2
110	5.22	294	345	5.43	0.35	34	15	57	5.8
120	5.75	291	338	5.31	0.38	40	18	62	6.5
130	6.29	287	331	5.20	0.42	46	21	67	7.1
140	6.84	284	323	5.08	0.45	52	25	72	7.8
150	7.41	281	317	4.98	0.49	59	29	77	8.5

38 Spl. LRN, m = 10.2 g

x [m]	as [Mil]	v [m/s]	E [J]	ED [J/mm²]	t [s]	Cw [cm]	y_s [cm]	x_s [m]	ad [Mil]
0	0.00	265	360	5.65	0.00	0	0	0	0.0
10	0.72	262	351	5.51	0.04	0	0	5	0.7
20	1.45	259	343	5.38	0.08	1	1	10	1.5
30	2.19	256	334	5.26	0.12	2	2	15	2.2
40	2.94	253	327	5.14	0.15	4	3	20	3.0
50	3.70	250	319	5.02	0.19	6	5	25	3.9
60	4.48	247	312	4.90	0.23	8	7	30	4.7
70	5.27	244	305	4.79	0.28	11	9	35	5.6
80	6.07	241	298	4.68	0.32	15	12	41	6.5
90	6.89	239	291	4.58	0.36	19	16	46	7.4
100	7.71	236	285	4.48	0.40	23	20	51	8.3
110	8.55	233	278	4.38	0.44	28	24	56	9.3
120	9.41	231	272	4.28	0.49	34	29	61	10.3
130	10.27	228	266	4.18	0.53	39	34	67	11.3
140	11.15	225	260	4.09	0.57	46	40	72	12.4
150	12.04	223	254	4.00	0.62	53	47	77	13.5

357 Magnum SJHP, m = 8 g

x [m]	as [Mil]	v [m/s]	E [J]	ED [J/mm²]	t [s]	Cw [cm]	y_s [cm]	x_s [m]	ad [Mil]
0	0.00	425	723	11.36	0.00	0	0	0	0.0
10	0.28	412	677	10.64	0.02	0	0	5	0.3
20	0.58	399	636	10.00	0.05	2	0	10	0.6
30	0.89	387	600	9.43	0.07	3	1	15	0.9
40	1.21	377	568	8.93	0.10	6	1	20	1.3
50	1.54	367	540	8.48	0.13	10	2	26	1.7
60	1.88	359	515	8.09	0.15	14	3	31	2.1
70	2.24	351	492	7.73	0.18	18	4	36	2.5
80	2.61	343	472	7.41	0.21	24	6	41	3.0
90	2.99	337	453	7.13	0.24	29	7	47	3.5
100	3.39	331	437	6.87	0.27	36	9	52	4.0
110	3.79	325	422	6.63	0.30	43	11	58	4.5
120	4.21	320	408	6.42	0.33	50	14	63	5.1
130	4.64	315	396	6.22	0.36	58	16	68	5.6
140	5.07	310	384	6.03	0.40	67	19	74	6.2
150	5.52	305	373	5.86	0.43	76	23	79	6.8

40 S & W SJHP, m = 10.0 g

x [m]	as [Mil]	v [m/s]	E [J]	ED [J/mm^2]	t [s]	Cw [cm]	y_s [cm]	x_s [m]	ad [Mil]
0	0.00	350	613	7.57	0.00	0	0	0	0.0
10	0.41	342	584	7.22	0.03	0	0	5	0.4
20	0.84	334	558	6.90	0.06	1	0	10	0.9
30	1.28	327	536	6.62	0.09	3	1	15	1.3
40	1.74	321	515	6.37	0.12	5	2	20	1.8
50	2.20	315	497	6.14	0.15	8	3	25	2.4
60	2.68	310	480	5.93	0.18	12	4	31	2.9
70	3.17	305	464	5.74	0.22	16	6	36	3.5
80	3.67	300	450	5.56	0.25	20	8	41	4.1
90	4.19	296	437	5.40	0.28	25	10	46	4.7
100	4.71	291	424	5.24	0.32	31	12	52	5.3
110	5.25	287	412	5.09	0.35	37	15	57	6.0
120	5.80	283	401	4.95	0.39	43	18	62	6.7
130	6.36	279	390	4.82	0.42	50	22	67	7.4
140	6.93	276	380	4.69	0.46	58	26	73	8.1
150	7.51	272	370	4.57	0.49	66	30	78	8.8

44 Rem. Mag., SJHP, m = 15.6 g

x [m]	as [Mil]	v [m/s]	E [J]	ED [J/mm^2]	t [s]	Cw [cm]	y_s [cm]	x_s [m]	ad [Mil]
0	0.00	450	1574	15.98	0.00	0	0	0	0.0
10	0.25	439	1500	15.23	0.02	0	0	5	0.3
20	0.51	429	1430	14.51	0.05	1	0	10	0.5
30	0.78	419	1363	13.83	0.07	2	1	15	0.8
40	1.05	409	1300	13.19	0.09	4	1	20	1.1
50	1.34	399	1240	12.59	0.12	7	2	26	1.4
60	1.63	390	1184	12.02	0.14	10	3	31	1.8
70	1.94	381	1131	11.48	0.17	14	4	36	2.2
80	2.25	373	1082	10.98	0.20	18	5	41	2.6
90	2.58	365	1036	10.51	0.22	23	6	47	3.0
100	2.91	358	994	10.09	0.25	28	8	52	3.4
110	3.25	351	955	9.70	0.28	34	10	57	3.8
120	3.61	344	920	9.34	0.31	41	12	63	4.3
130	3.97	338	887	9.01	0.34	48	14	68	4.8
140	4.35	332	858	8.71	0.37	56	17	74	5.3
150	4.74	327	831	8.43	0.40	64	19	79	5.9

45 Auto FMJ RN, m = 14.9 g

x [m]	as [Mil]	v [m/s]	E [J]	ED [J/mm^2]	t [s]	Cw [cm]	y_s [cm]	x_s [m]	ad [Mil]
0	0.00	245	447	4.34	0.00	0	0	0	0.0
10	0.84	243	440	4.27	0.04	0	0	5	0.8
20	1.68	241	433	4.21	0.08	1	1	10	1.7
30	2.54	239	427	4.14	0.12	1	2	15	2.6
40	3.40	237	420	4.08	0.17	3	3	20	3.5
50	4.27	236	413	4.01	0.21	4	5	25	4.4
60	5.15	234	407	3.95	0.25	6	8	30	5.3
70	6.05	232	401	3.89	0.29	8	11	35	6.3
80	6.95	230	394	3.83	0.34	11	14	40	7.2
90	7.86	228	388	3.77	0.38	13	18	46	8.2
100	8.78	227	382	3.71	0.42	17	22	51	9.2
110	9.70	225	377	3.66	0.47	20	27	56	10.3
120	10.64	223	371	3.60	0.51	24	32	61	11.3
130	11.59	221	365	3.55	0.56	28	38	66	12.4
140	12.55	220	359	3.49	0.60	33	45	71	13.5
150	13.52	218	354	3.44	0.65	37	52	76	14.6

Flintlock pistol, 13.5 mm, m = 14.5 g

x [m]	as [Mil]	v [m/s]	E [J]	ED [J/mm^2]	t [s]	Cw [cm]	y_s [cm]	x_s [m]	ad [Mil]
0	0.00	385	1075	7.51	0.00	0	0	0	0.0
10	0.35	364	962	6.72	0.03	1	0	5	0.4
20	0.73	346	866	6.05	0.05	3	0	10	0.8
30	1.13	329	785	5.48	0.08	7	1	15	1.3
40	1.56	314	716	5.00	0.12	12	2	21	1.8
50	2.03	301	656	4.58	0.15	18	3	26	2.4
60	2.52	289	605	4.23	0.18	26	4	31	3.0
70	3.04	278	560	3.91	0.22	36	6	37	3.8
80	3.60	268	521	3.64	0.25	46	8	43	4.6
90	4.18	259	486	3.39	0.29	58	11	48	5.4
100	4.80	250	455	3.18	0.33	72	14	54	6.3
110	5.45	243	427	2.98	0.37	86	17	59	7.3
120	6.13	235	402	2.81	0.41	102	21	65	8.4
130	6.85	229	379	2.65	0.46	119	26	71	9.5
140	7.60	222	358	2.50	0.50	138	31	77	10.8
150	8.38	216	338	2.36	0.55	157	37	82	12.1

A.10.3 Rifles

5.45 × 39 AK-74, m = 3.45 g

x [m]	as [Mil]	v [m/s]	E [J]	ED [J/mm^2]	t [s]	Cw [cm]	y_s [cm]	x_s [m]	ad [Mil]
0	0.00	910	1429	61.2	0.00	0	0	0	0.0
20	0.12	889	1363	58.4	0.02	0	0	10	0.1
40	0.25	868	1300	55.7	0.05	1	0	20	0.3
60	0.38	848	1239	53.1	0.07	2	1	30	0.4
80	0.51	828	1181	50.6	0.09	4	1	41	0.5
100	0.65	808	1126	48.3	0.12	7	2	51	0.7
120	0.80	788	1072	46.0	0.14	10	3	61	0.9
140	0.95	769	1021	43.8	0.17	14	3	72	1.1
160	1.10	751	972	41.7	0.19	18	5	83	1.3
180	1.26	732	925	39.7	0.22	23	6	93	1.5
200	1.42	714	880	37.7	0.25	29	8	104	1.7
220	1.59	697	837	35.9	0.28	35	9	115	1.9
240	1.77	679	796	34.1	0.31	42	12	126	2.1
260	1.95	662	756	32.4	0.34	50	14	137	2.4
280	2.14	645	718	30.8	0.37	59	17	148	2.7
300	2.34	628	681	29.2	0.40	68	20	160	3.0

5.56 × 45 M 193 (SS 92), m = 3.56 g

x [m]	as [Mil]	v [m/s]	E [J]	ED [J/mm^2]	t [s]	Cw [cm]	y_s [cm]	x_s [m]	ad [Mil]
0	0.00	960	1640	67.6	0.00	0	0	0	0.0
20	0.11	933	1550	63.9	0.02	0	0	10	0.1
40	0.23	907	1464	60.3	0.04	1	0	20	0.2
60	0.34	881	1383	57.0	0.07	3	1	30	0.4
80	0.47	856	1305	53.7	0.09	5	1	41	0.5
100	0.60	832	1231	50.7	0.11	8	2	51	0.7
120	0.73	807	1160	47.8	0.14	11	2	62	0.8
140	0.87	784	1093	45.0	0.16	16	3	72	1.0
160	1.02	760	1029	42.4	0.19	21	4	83	1.2
180	1.17	738	969	39.9	0.21	27	6	94	1.4
200	1.33	715	911	37.5	0.24	33	7	105	1.6
220	1.49	693	856	35.3	0.27	41	9	116	1.9
240	1.66	672	804	33.1	0.30	49	11	128	2.1
260	1.84	651	754	31.1	0.33	59	13	139	2.4
280	2.03	630	707	29.1	0.36	69	16	151	2.7
300	2.23	610	661	27.2	0.39	81	19	162	3.0

A.10 Ballistics tables (metric system)

5.56 × 45 SS 109 (M 855), m = 4.0 g

x [m]	as [Mil]	v [m/s]	E [J]	ED [J/mm²]	t [s]	Cw [cm]	y_s [cm]	x_s [m]	ad [Mil]
0	0.00	920	1693	69.7	0.00	0	0	0	0.0
20	0.12	900	1619	66.7	0.02	0	0	10	0.1
40	0.24	880	1548	63.8	0.04	1	0	20	0.3
60	0.37	860	1480	60.9	0.07	2	1	30	0.4
80	0.50	841	1414	58.2	0.09	4	1	41	0.5
100	0.64	822	1350	55.6	0.12	6	2	51	0.7
120	0.78	803	1289	53.1	0.14	9	2	61	0.8
140	0.92	784	1230	50.7	0.16	13	3	72	1.0
160	1.07	766	1173	48.3	0.19	17	5	83	1.2
180	1.22	748	1118	46.1	0.22	22	6	93	1.4
200	1.38	730	1065	43.9	0.24	27	7	104	1.6
220	1.54	712	1014	41.8	0.27	33	9	115	1.8
240	1.71	694	964	39.7	0.30	40	11	126	2.1
260	1.89	677	917	37.8	0.33	47	13	137	2.3
280	2.07	660	871	35.9	0.36	55	16	148	2.6
300	2.26	643	826	34.0	0.39	64	19	159	2.9

6 mm BR Norma, m = 6.9 g

x [m]	as [Mil]	v [m/s]	E [J]	ED [J/mm²]	t [s]	Cw [cm]	y_s [cm]	x_s [m]	ad [Mil]
0	0.00	860	2552	90.3	0.00	0	0	0	0.0
20	0.14	849	2489	88.0	0.02	0	0	10	0.1
40	0.28	839	2427	85.8	0.05	1	0	20	0.3
60	0.42	828	2366	83.7	0.07	1	1	30	0.4
80	0.56	818	2307	81.6	0.10	2	1	40	0.6
100	0.70	807	2249	79.5	0.12	4	2	51	0.7
120	0.85	797	2192	77.5	0.14	5	3	61	0.9
140	1.00	787	2137	75.6	0.17	7	4	71	1.1
160	1.16	777	2082	73.6	0.20	10	5	81	1.2
180	1.31	767	2029	71.8	0.22	12	6	92	1.4
200	1.47	757	1976	69.9	0.25	15	8	102	1.6
220	1.63	747	1925	68.1	0.27	19	9	113	1.8
240	1.80	737	1875	66.3	0.30	22	11	123	2.0
260	1.97	728	1826	64.6	0.33	27	13	134	2.2
280	2.14	718	1778	62.9	0.36	31	16	144	2.4
300	2.31	708	1731	61.2	0.38	36	18	155	2.6

7 × 64 partition, m = 11.2 g

x [m]	as [Mil]	v [m/s]	E [J]	ED [J/mm²]	t [s]	Cw [cm]	y_s [cm]	x_s [m]	ad [Mil]
0	0.00	810	3674	95.5	0.00	0	0	0	0.0
20	0.15	794	3526	91.6	0.02	0	0	10	0.2
40	0.31	777	3384	87.9	0.05	1	0	20	0.3
60	0.48	761	3245	84.3	0.08	2	1	30	0.5
80	0.64	746	3112	80.9	0.10	4	1	41	0.7
100	0.82	730	2983	77.5	0.13	7	2	51	0.9
120	0.99	714	2858	74.3	0.16	10	3	61	1.1
140	1.18	699	2738	71.1	0.19	13	4	72	1.3
160	1.37	684	2621	68.1	0.22	18	6	82	1.5
180	1.56	669	2509	65.2	0.24	22	7	93	1.8
200	1.76	655	2401	62.4	0.27	28	9	104	2.0
220	1.96	640	2296	59.7	0.31	34	12	114	2.3
240	2.18	626	2195	57.0	0.34	41	14	125	2.6
260	2.40	612	2098	54.5	0.37	49	17	136	2.9
280	2.62	598	2004	52.1	0.40	57	20	147	3.2
300	2.85	585	1914	49.7	0.44	66	23	159	3.5

7 mm Rem. Mag. spire point, m = 9.0 g

x [m]	as [Mil]	v [m/s]	E [J]	ED [J/mm²]	t [s]	Cw [cm]	y_s [cm]	x_s [m]	ad [Mil]
0	0.00	945	4019	104.4	0.00	0	0	0	0.0
20	0.11	928	3876	100.7	0.02	0	0	10	0.1
40	0.23	911	3738	97.1	0.04	1	0	20	0.2
60	0.35	895	3604	93.7	0.07	2	1	30	0.4
80	0.47	879	3474	90.3	0.09	3	1	40	0.5
100	0.60	863	3348	87.0	0.11	5	2	51	0.6
120	0.72	847	3225	83.8	0.13	7	2	61	0.8
140	0.85	831	3106	80.7	0.16	10	3	72	0.9
160	0.99	815	2990	77.7	0.18	13	4	82	1.1
180	1.13	800	2879	74.8	0.21	17	5	93	1.3
200	1.27	785	2770	72.0	0.23	21	7	103	1.4
220	1.41	770	2665	69.3	0.26	25	8	114	1.6
240	1.56	755	2563	66.6	0.28	30	10	125	1.8
260	1.72	740	2465	64.0	0.31	36	12	136	2.0
280	1.87	726	2369	61.6	0.34	42	14	146	2.2
300	2.04	711	2276	59.2	0.37	49	17	157	2.5

7.5 × 55 Swiss (GP11), m = 11.3 g

x [m]	as [Mil]	v [m/s]	E [J]	ED [J/mm²]	t [s]	Cw [cm]	y_s [cm]	x_s [m]	ad [Mil]
0	0.00	750	3178	71.9	0.00	0	0	0	0.0
20	0.18	739	3088	69.9	0.03	0	0	10	0.2
40	0.36	729	3001	67.9	0.05	1	0	20	0.4
60	0.55	718	2916	66.0	0.08	2	1	30	0.6
80	0.74	708	2833	64.1	0.11	3	2	40	0.8
100	0.93	698	2751	62.3	0.14	5	2	51	1.0
120	1.13	688	2672	60.5	0.17	7	3	61	1.2
140	1.33	678	2595	58.7	0.20	10	5	71	1.4
160	1.54	668	2519	57.0	0.23	13	6	82	1.7
180	1.75	658	2445	55.3	0.26	16	8	92	1.9
200	1.96	648	2373	53.7	0.29	20	10	102	2.2
220	2.18	638	2302	52.1	0.32	25	12	113	2.4
240	2.40	629	2233	50.5	0.35	30	15	124	2.7
260	2.63	619	2165	49.0	0.38	35	18	134	3.0
280	2.86	610	2099	47.5	0.41	41	21	145	3.3
300	3.10	600	2035	46.1	0.45	47	25	156	3.6

7.62 × 39 AK-47 (Kalashnikov), m = 8.0 g

x [m]	as [Mil]	v [m/s]	E [J]	ED [J/mm²]	t [s]	Cw [cm]	y_s [cm]	x_s [m]	ad [Mil]
0	0.00	710	2016	44.2	0.00	0	0	0	0.0
20	0.20	688	1893	41.5	0.03	0	0	10	0.2
40	0.41	666	1774	38.9	0.06	2	0	20	0.4
60	0.63	645	1662	36.5	0.09	4	1	30	0.7
80	0.87	624	1555	34.1	0.12	8	2	41	0.9
100	1.11	603	1454	31.9	0.15	12	3	51	1.2
120	1.36	583	1358	29.8	0.19	18	4	62	1.5
140	1.62	563	1266	27.8	0.22	24	6	73	1.9
160	1.90	543	1180	25.9	0.26	32	8	84	2.3
180	2.19	524	1099	24.1	0.30	42	11	95	2.7
200	2.50	505	1022	22.4	0.33	52	14	106	3.1
220	2.82	487	949	20.8	0.37	65	17	117	3.6
240	3.16	469	881	19.3	0.42	78	21	129	4.2
260	3.51	452	817	17.9	0.46	94	26	141	4.7
280	3.89	435	756	16.6	0.50	111	31	152	5.4
300	4.28	418	700	15.3	0.55	129	38	164	6.1

7.62 mm NATO (308 Win.), m = 9.5 g

x [m]	as [Mil]	v [m/s]	E [J]	ED [J/mm²]	t [s]	Cw [cm]	y_s [cm]	x_s [m]	ad [Mil]
0	0.00	830	3272	71.8	0.00	0	0	0	0.0
20	0.15	815	3153	69.1	0.02	0	0	10	0.1
40	0.30	800	3037	66.6	0.05	1	0	20	0.3
60	0.45	785	2925	64.2	0.07	2	1	30	0.5
80	0.61	770	2817	61.8	0.10	4	1	41	0.6
100	0.77	756	2711	59.5	0.13	6	2	51	0.8
120	0.94	741	2609	57.2	0.15	8	3	61	1.0
140	1.11	727	2509	55.0	0.18	12	4	72	1.2
160	1.29	713	2412	52.9	0.21	15	5	82	1.4
180	1.47	699	2319	50.9	0.24	20	7	93	1.6
200	1.65	685	2228	48.9	0.27	24	9	103	1.9
220	1.84	671	2140	46.9	0.29	30	11	114	2.1
240	2.04	658	2055	45.1	0.33	36	13	125	2.4
260	2.24	644	1972	43.2	0.36	42	16	136	2.6
280	2.44	631	1891	41.5	0.39	50	18	147	2.9
300	2.66	618	1813	39.8	0.42	58	22	158	3.2

300 Win. Mag. Sierra MK, m = 12.3 g

x [m]	as [Mil]	v [m/s]	E [J]	ED [J/mm²]	t [s]	Cw [cm]	y_s [cm]	x_s [m]	ad [Mil]
0	0.00	880	4763	104.4	0.00	0	0	0	0.0
20	0.13	869	4643	101.8	0.02	0	0	10	0.1
40	0.26	858	4525	99.2	0.05	1	0	20	0.3
60	0.40	847	4409	96.7	0.07	1	1	30	0.4
80	0.53	836	4296	94.2	0.09	2	1	40	0.6
100	0.67	825	4185	91.8	0.12	4	2	51	0.7
120	0.82	814	4077	89.4	0.14	5	3	61	0.9
140	0.96	804	3970	87.1	0.17	7	3	71	1.0
160	1.11	793	3866	84.8	0.19	10	5	81	1.2
180	1.26	782	3763	82.5	0.22	12	6	92	1.4
200	1.41	772	3663	80.3	0.24	15	7	102	1.5
220	1.56	761	3565	78.2	0.27	19	9	113	1.7
240	1.72	751	3469	76.1	0.30	23	11	123	1.9
260	1.88	741	3375	74.0	0.32	27	13	134	2.1
280	2.05	731	3282	72.0	0.35	31	15	144	2.3
300	2.21	721	3192	70.0	0.38	36	17	155	2.5

8 × 68 S FMJ RN, m = 14.25 g

x [m]	as [Mil]	v [m/s]	E [J]	ED [J/mm²]	t [s]	Cw [cm]	y_s [cm]	x_s [m]	ad [Mil]
0	0.00	870	5393	107.3	0.00	0	0	0	0.0
20	0.13	856	5217	103.8	0.02	0	0	10	0.1
40	0.27	842	5047	100.4	0.05	1	0	20	0.3
60	0.41	828	4881	97.1	0.07	2	1	30	0.4
80	0.55	814	4719	93.9	0.10	3	1	40	0.6
100	0.70	800	4561	90.7	0.12	5	2	51	0.7
120	0.85	787	4408	87.7	0.15	7	3	61	0.9
140	1.00	773	4259	84.7	0.17	10	4	71	1.1
160	1.16	760	4114	81.8	0.20	13	5	82	1.3
180	1.32	747	3973	79.0	0.22	17	6	92	1.5
200	1.48	734	3836	76.3	0.25	21	8	103	1.7
220	1.65	721	3702	73.7	0.28	25	10	114	1.9
240	1.82	708	3573	71.1	0.31	30	12	124	2.1
260	2.00	696	3447	68.6	0.33	36	14	135	2.3
280	2.18	683	3325	66.2	0.36	42	16	146	2.6
300	2.36	671	3207	63.8	0.39	48	19	157	2.8

9.3 × 62 TC, m = 16.0 g

x [m]	as [Mil]	v [m/s]	E [J]	ED [J/mm²]	t [s]	Cw [cm]	y_s [cm]	x_s [m]	ad [Mil]
0	0.00	800	5120	75.37	0.00	0	0	0	0.0
20	0.16	781	4879	71.86	0.03	0	0	10	0.2
40	0.32	763	4648	68.49	0.05	1	0	20	0.3
60	0.49	744	4426	65.24	0.08	3	1	30	0.5
80	0.67	726	4214	62.12	0.10	5	1	41	0.7
100	0.85	709	4010	59.12	0.13	8	2	51	0.9
120	1.03	691	3815	56.23	0.16	11	3	62	1.1
140	1.23	674	3628	53.46	0.19	16	5	72	1.4
160	1.43	657	3449	50.81	0.22	21	6	83	1.6
180	1.63	640	3278	48.26	0.25	27	8	93	1.9
200	1.85	624	3113	45.82	0.28	33	10	104	2.2
220	2.07	608	2956	43.47	0.32	41	12	115	2.5
240	2.30	592	2804	41.23	0.35	49	15	126	2.8
260	2.54	576	2660	39.08	0.38	59	18	138	3.2
280	2.79	561	2521	37.02	0.42	69	22	149	3.5
300	3.04	546	2388	35.05	0.45	80	26	160	3.9

338 Win. Mag. jacketed spitzer, m = 16.2 g

x [m]	as [Mil]	v [m/s]	E [J]	ED [J/mm^2]	t [s]	Cw [cm]	y$_s$ [cm]	x$_s$ [m]	ad [Mil]
0	0.00	810	5314	95.9	0.00	0	0	0	0.0
20	0.15	800	5179	93.5	0.02	0	0	10	0.2
40	0.31	789	5046	91.1	0.05	1	0	20	0.3
60	0.47	779	4916	88.7	0.08	1	1	30	0.5
80	0.63	769	4789	86.4	0.10	3	1	40	0.7
100	0.80	759	4665	84.2	0.13	4	2	51	0.8
120	0.96	749	4543	82.0	0.15	6	3	61	1.0
140	1.13	739	4424	79.8	0.18	8	4	71	1.2
160	1.31	729	4307	77.7	0.21	11	5	81	1.4
180	1.48	719	4192	75.6	0.24	14	7	92	1.6
200	1.66	710	4081	73.6	0.26	17	9	102	1.8
220	1.85	700	3971	71.6	0.29	21	11	113	2.0
240	2.03	691	3864	69.7	0.32	25	13	123	2.3
260	2.22	681	3759	67.8	0.35	29	15	134	2.5
280	2.42	672	3657	66.0	0.38	34	18	144	2.7
300	2.62	663	3557	64.2	0.41	39	21	155	3.0

12.7 × 99 jacketed spitzer, m = 42 g

x [m]	as [Mil]	v [m/s]	E [J]	ED [J/mm^2]	t [s]	Cw [cm]	y$_s$ [cm]	x$_s$ [m]	ad [Mil]
0	0.00	890	16634	131.3	0.00	0	0	0	0.0
50	0.32	866	15731	124.2	0.06	1	0	25	0.3
100	0.66	842	14872	117.4	0.12	3	2	50	0.7
150	1.00	818	14053	110.9	0.18	7	4	76	1.1
200	1.36	795	13271	104.8	0.24	13	7	102	1.5
250	1.74	772	12525	98.9	0.30	21	11	128	1.9
300	2.12	750	11813	93.3	0.37	30	17	154	2.4
350	2.53	728	11133	87.9	0.44	42	23	181	2.9
400	2.95	707	10483	82.8	0.50	55	31	208	3.4
450	3.39	685	9862	77.9	0.58	71	41	235	4.0
500	3.85	664	9264	73.1	0.65	89	52	263	4.7
550	4.32	643	8690	68.6	0.73	109	65	291	5.4
600	4.82	623	8139	64.3	0.81	132	80	319	6.1
650	5.34	602	7610	60.1	0.89	158	97	348	6.9
700	5.89	582	7102	56.1	0.97	186	117	377	7.8
750	6.46	561	6615	52.2	1.06	217	139	406	8.8
800	7.06	541	6149	48.5	1.15	252	163	436	9.8

A.10.4 Old rifles

Flintlock rifle, 18th century, 16.6 mm, m = 26.7 g

x [m]	as [Mil]	v [m/s]	E [J]	ED [J/mm²]	t [s]	Cw [cm]	y_s [cm]	x_s [m]	ad [Mil]
0	0.00	455	2764	12.8	0.00	0	0	0	0.0
20	0.52	411	2258	10.4	0.05	2	0	10	0.6
40	1.11	374	1865	8.6	0.10	9	1	21	1.3
60	1.78	343	1567	7.2	0.15	21	3	31	2.2
80	2.55	317	1340	6.2	0.21	38	6	43	3.2
100	3.40	295	1164	5.4	0.28	60	10	54	4.5
120	4.36	277	1024	4.7	0.35	86	15	65	6.0
140	5.41	261	911	4.2	0.42	116	22	77	7.7
160	6.56	247	817	3.8	0.50	151	31	89	9.7
180	7.83	235	739	3.4	0.59	190	43	100	11.8
200	9.20	224	671	3.1	0.67	233	56	112	14.3
220	10.68	214	613	2.8	0.76	280	73	124	16.9
240	12.28	205	561	2.6	0.86	332	92	136	19.9
260	13.99	197	516	2.4	0.96	388	115	149	23.1
280	15.84	189	475	2.2	1.06	448	141	161	26.7
300	17.81	181	438	2.0	1.17	512	172	173	30.6

Percussion rifle, mid-19th-century, 11.5 mm, m = 33 g

x [m]	as [Mil]	v [m/s]	E [J]	ED [J/mm²]	t [s]	Cw [cm]	y_s [cm]	x_s [m]	ad [Mil]
0	0.00	410	2774	26.7	0.00	0	0	0	0.0
20	0.61	399	2629	25.3	0.05	1	0	10	0.6
40	1.23	389	2495	24.0	0.10	3	1	20	1.3
60	1.88	379	2372	22.8	0.15	6	3	30	2.0
80	2.55	370	2260	21.8	0.21	11	5	41	2.7
100	3.25	362	2157	20.8	0.26	17	8	51	3.5
120	3.97	354	2063	19.9	0.32	24	12	62	4.4
140	4.71	346	1978	19.0	0.37	32	17	72	5.3
160	5.47	339	1900	18.3	0.43	42	23	83	6.2
180	6.26	333	1830	17.6	0.49	52	30	93	7.2
200	7.07	327	1767	17.0	0.55	64	37	104	8.2
220	7.90	322	1709	16.5	0.61	77	46	115	9.3
240	8.75	317	1656	15.9	0.68	91	56	125	10.4
260	9.63	312	1607	15.5	0.74	106	67	136	11.5
280	10.53	308	1561	15.0	0.80	122	80	147	12.7
300	11.45	303	1519	14.6	0.87	138	93	158	13.9

45-70 U.S. Government, m = 26.2 g

x [m]	as [Mil]	v [m/s]	E [J]	ED [J/mm²]	t [s]	Cw [cm]	y_s [cm]	x_s [m]	ad [Mil]
0	0.00	390	1993	19.42	0.00	0	0	0	0.0
20	0.67	380	1891	18.43	0.05	1	0	10	0.7
40	1.36	371	1798	17.52	0.11	3	1	20	1.4
60	2.08	362	1713	16.69	0.16	6	3	30	2.2
80	2.82	353	1636	15.94	0.22	11	6	41	3.0
100	3.58	346	1567	15.27	0.27	17	9	51	3.9
120	4.37	339	1503	14.65	0.33	24	14	61	4.8
140	5.18	332	1446	14.10	0.39	32	19	72	5.8
160	6.01	326	1395	13.60	0.45	42	25	82	6.8
180	6.87	321	1349	13.14	0.51	52	32	93	7.8
200	7.75	316	1306	12.72	0.58	64	41	104	8.9
220	8.65	311	1266	12.34	0.64	76	50	114	10.1
240	9.58	306	1229	11.98	0.71	90	61	125	11.2
260	10.52	302	1195	11.64	0.77	104	73	136	12.4
280	11.49	298	1163	11.33	0.84	120	86	146	13.7
300	12.48	294	1133	11.04	0.91	136	101	157	15.0

10.4 mm Vetterli rifle, m = 19.4 g

x [m]	as [Mil]	v [m/s]	E [J]	ED [J/mm²]	t [s]	Cw [cm]	y_s [cm]	x_s [m]	ad [Mil]
0	0.00	435	1835	21.6	0.00	0	0	0	0.0
20	0.54	423	1732	20.4	0.05	1	0	10	0.6
40	1.10	411	1637	19.3	0.10	3	1	20	1.1
60	1.68	400	1548	18.2	0.14	6	3	30	1.8
80	2.28	389	1467	17.3	0.20	11	5	41	2.5
100	2.91	379	1393	16.4	0.25	17	7	51	3.2
120	3.56	370	1324	15.6	0.30	25	11	62	4.0
140	4.23	361	1262	14.9	0.36	33	15	72	4.8
160	4.93	353	1206	14.2	0.41	43	21	83	5.7
180	5.65	345	1155	13.6	0.47	55	27	94	6.6
200	6.39	338	1109	13.1	0.53	67	34	104	7.6
220	7.16	332	1067	12.6	0.59	81	42	115	8.6
240	7.95	326	1029	12.1	0.65	96	52	126	9.6
260	8.76	320	995	11.7	0.71	112	62	137	10.7
280	9.60	315	964	11.3	0.77	129	73	148	11.9
300	10.46	310	935	11.0	0.84	147	86	159	13.1

A.10.5 Various

4.5 mm air pistol, Diabolo, m = 0.53 g

x [m]	as [Mil]	v [m/s]	E [J]	ED [J/mm^2]	t [s]	Cw [cm]	y$_s$ [cm]	x$_s$ [m]	ad [Mil]
0	0.00	90	2.1	0.14	0.00	0	0	0	0
5	3.15	87	2.0	0.13	0.06	1	0	3	3
10	6.44	84	1.9	0.12	0.12	4	2	5	7
15	9.87	82	1.8	0.11	0.18	8	4	8	11
20	13.45	79	1.7	0.11	0.24	15	7	10	15
25	17.19	77	1.6	0.10	0.30	23	11	13	19
30	21.11	74	1.5	0.09	0.37	34	17	16	24
35	25.20	72	1.4	0.09	0.44	47	23	18	29
40	29.48	70	1.3	0.08	0.51	62	32	21	35
45	33.95	67	1.2	0.08	0.58	79	41	24	41
50	38.64	65	1.1	0.07	0.66	99	53	26	48
55	43.54	63	1.1	0.07	0.73	122	66	29	55
60	48.68	61	1.0	0.06	0.81	147	82	32	63
65	54.07	59	0.9	0.06	0.90	174	99	35	71
70	59.72	57	0.9	0.06	0.98	205	119	38	80
75	65.79	56	0.8	0.05	1.08	238	143	41	90

4.5 mm air rifle, Diabolo, m = 0.53 g

x [m]	as [Mil]	v [m/s]	E [J]	ED [J/mm^2]	t [s]	Cw [cm]	y$_s$ [cm]	x$_s$ [m]	ad [Mil]
0	0.00	175	8.1	0.51	0.00	0	0	0	0
5	0.83	170	7.6	0.48	0.03	0	0	3	1
10	1.70	165	7.2	0.45	0.06	2	0	5	2
15	2.60	160	6.8	0.43	0.09	4	1	8	3
20	3.54	155	6.4	0.40	0.12	7	2	10	4
25	4.52	151	6.0	0.38	0.15	11	3	13	5
30	5.54	146	5.7	0.36	0.19	16	4	15	6
35	6.60	142	5.3	0.33	0.22	23	6	18	8
40	7.71	137	5.0	0.31	0.26	30	8	21	9
45	8.86	133	4.7	0.30	0.30	38	11	24	11
50	10.07	129	4.4	0.28	0.33	48	14	26	12
55	11.33	125	4.2	0.26	0.37	59	17	29	14
60	12.64	122	3.9	0.25	0.41	70	21	32	16
65	14.01	118	3.7	0.23	0.46	84	26	35	18
70	15.45	114	3.5	0.22	0.50	98	31	38	21
75	16.95	111	3.2	0.20	0.54	114	36	41	23

4.5 mm air rifle, lead sphere, m = 0.53 g

x [m]	as [Mil]	v [m/s]	E [J]	ED [J/mm²]	t [s]	Cw [cm]	y$_s$ [cm]	x$_s$ [m]	ad [Mil]
0	0.00	175	8.1	0.51	0.00	0	0	0	0
5	0.84	169	7.6	0.47	0.03	1	0	3	1
10	1.71	163	7.0	0.44	0.06	2	0	5	2
15	2.63	157	6.5	0.41	0.09	5	1	8	3
20	3.60	152	6.1	0.38	0.12	9	2	10	4
25	4.61	147	5.7	0.36	0.16	14	3	13	5
30	5.67	142	5.3	0.34	0.19	20	5	16	7
35	6.78	137	5.0	0.31	0.23	27	6	18	8
40	7.95	133	4.7	0.29	0.26	35	9	21	10
45	9.17	129	4.4	0.28	0.30	45	11	24	11
50	10.45	125	4.1	0.26	0.34	56	14	26	13
55	11.78	121	3.9	0.24	0.38	68	18	29	15
60	13.18	117	3.6	0.23	0.42	82	22	32	17
65	14.65	113	3.4	0.21	0.47	96	27	35	20
70	16.19	110	3.2	0.20	0.51	113	32	38	22
75	17.79	106	3.0	0.19	0.56	131	39	41	25

8 mm arrow (shot from a bow), m = 40 g

x [m]	as [Mil]	v [m/s]	E [J]	ED [J/mm²]	t [s]	Cw [cm]	y$_s$ [cm]	x$_s$ [m]	ad [Mil]
0	0.00	70.0	98.0	1.95	0.00	0.0	0	0	0.0
5	5.10	69.9	97.6	1.94	0.07	0.2	1	3	5.1
10	10.22	69.7	97.2	1.93	0.14	0.5	3	5	10.2
15	15.35	69.6	97.0	1.93	0.22	0.9	6	8	15.4
20	20.49	69.5	96.5	1.92	0.29	1.3	10	10	20.6
25	25.65	69.4	96.2	1.91	0.36	1.8	16	13	25.8
30	30.82	69.2	95.9	1.91	0.43	2.4	23	15	31.1
35	36.01	69.1	95.5	1.90	0.50	3.2	31	18	36.3
40	41.22	69.0	95.2	1.89	0.58	4.2	41	20	41.6
45	46.44	68.9	94.8	1.89	0.65	5.4	52	23	47.0
50	51.69	68.7	94.5	1.88	0.72	6.8	64	25	52.3
55	56.95	68.6	94.1	1.87	0.80	8.3	78	28	57.7
60	62.22	68.5	93.8	1.87	0.87	9.9	92	30	63.1
65	67.52	68.4	93.4	1.86	0.94	11.5	109	33	68.6
70	72.84	68.2	93.1	1.85	1.02	13.2	126	35	74.1
75	78.18	68.1	92.7	1.85	1.09	15.0	146	38	79.6

Cuboid steel fragment, m = 0.1 g

x [m]	as [Mil]	v [m/s]	E [J]	ED [J/mm²]	t [s]	Cw [cm]	y$_s$ [cm]	x$_s$ [m]	ad [Mil]
0	0.00	1000	50.0	6.10	0.00	0	0	0	0
5	0.03	754	28.4	3.47	0.01	1	0	3	0
10	0.08	559	15.6	1.91	0.01	3	0	6	0
15	0.15	409	8.4	1.02	0.02	9	0	9	0
20	0.26	308	4.7	0.58	0.04	18	0	12	1
25	0.43	244	3.0	0.36	0.06	32	0	16	1
30	0.69	197	1.9	0.24	0.08	49	1	20	2
35	1.06	160	1.3	0.16	0.11	73	2	24	3
40	1.58	131	0.9	0.11	0.14	102	3	28	5
45	2.32	108	0.6	0.07	0.18	139	5	32	8
50	3.34	89	0.4	0.05	0.24	186	8	36	12
55	4.76	73	0.3	0.03	0.30	243	13	41	18
56	5.09	70	0.2	0.03	0.31	255	14	42	20
60	6.74	60	0.2	0.02	0.37	314	20	45	28
65	9.49	50	0.1	0.02	0.47	400	32	50	42
70	13.31	41	0.1	0.01	0.58	506	50	56	63
75	18.62	34	0.1	0.01	0.71	636	77	61	94

Cuboid steel fragment, m = 1 g

x [m]	as [Mil]	v [m/s]	E [J]	ED [J/mm²]	t [s]	Cw [cm]	y$_s$ [cm]	x$_s$ [m]	ad [Mil]
0	0.00	1000	500.0	13.18	0.00	0	0	0	0
5	0.03	879	386.5	10.19	0.01	0	0	3	0
10	0.06	770	296.8	7.82	0.01	1	0	5	0
15	0.10	673	226.2	5.96	0.02	3	0	8	0
20	0.15	585	171.3	4.51	0.03	6	0	11	0
25	0.20	508	128.8	3.39	0.04	11	0	14	0
30	0.28	439	96.2	2.54	0.05	16	0	17	0
35	0.36	380	72.2	1.90	0.06	23	0	21	1
40	0.47	333	55.4	1.46	0.07	32	1	24	1
45	0.61	296	43.7	1.15	0.09	43	1	28	1
50	0.77	265	35.1	0.92	0.11	56	2	32	2
55	0.97	239	28.5	0.75	0.13	71	2	35	2
60	1.21	216	23.4	0.62	0.15	88	3	39	3
65	1.50	196	19.2	0.51	0.17	108	4	43	4
70	1.84	178	15.9	0.42	0.20	129	5	47	5
75	2.24	162	13.2	0.35	0.23	154	7	51	6

A.11 Ballistics tables (British/U.S. system)

A.11.1 Notes

These tables use mils for the angle of both the angle of shot and the angle of descent. $360° \Leftrightarrow 6400$ mils

At the small angles encountered with flat-trajectory weapons, 1 mil corresponds to 1 mm change in altitude per metre of flight.

The ballistics tables are valid at sea level, and were generated using k-ballistics 4 software (www.kneubuehl.com, S. ROTHE).

Abbreviations:

x	:	Range	t	:	Time of flight
as	:	Angle of shot	Cw	:	Deviation with a cross-wind of 10 m/s
v	:	Velocity	y_s	:	Vertex height
E	:	Energy	x_s	:	Vertex distance
ED	:	Energy density	ad	:	Angle of descent

A.11.2 Handguns

25 Auto Pistol, m = 50 gr

x [yd]	as [mils]	v [ft/s]	E [ft lbf]	ED [a]	t [s]	Cw [in]	y_s [in]	x_s [yd]	ad [mils]
0	0.00	804	71	1442	0.00	0.0	0.0	0	0.0
10	0.77	793	69	1404	0.04	0.0	0.1	5	0.8
20	1.55	782	67	1365	0.08	0.4	0.3	10	1.6
30	2.35	771	65	1328	0.11	0.8	0.6	15	2.4
40	3.16	761	64	1294	0.15	1.6	1.1	20	3.3
50	3.98	751	62	1259	0.19	2.8	1.8	25	4.2
60	4.83	741	60	1224	0.23	3.9	2.6	30	5.1
70	5.68	731	59	1192	0.27	5.1	3.6	36	6.1
80	6.56	721	57	1161	0.32	6.7	4.8	41	7.1
90	7.45	711	55	1129	0.36	8.7	6.2	46	8.1
100	8.35	702	54	1100	0.40	10.6	7.7	51	9.2
110	9.28	692	53	1070	0.44	12.6	9.5	56	10.3
120	10.22	683	51	1043	0.49	15.4	11.4	62	11.4
130	11.17	674	50	1015	0.53	18.1	13.6	67	12.6
140	12.15	665	49	988	0.58	20.9	16.0	72	13.8
150	13.14	656	47	962	0.62	24.0	18.6	78	15.0

[a] [ft·lbf/in²]

7.62 × 25 Russian Tokarev, m = 85 gr

x [yd]	as [mils]	v [ft/s]	E [ft lbf]	ED [ª]	t [s]	Cw [in]	y_s [in]	x_s [yd]	ad [mils]
0	0.00	1493	428	6048	0.00	0.0	0.0	0	0.0
10	0.23	1450	403	5706	0.02	0.0	0.0	5	0.2
20	0.46	1408	380	5380	0.04	0.4	0.1	10	0.5
30	0.70	1367	359	5072	0.06	1.2	0.2	15	0.7
40	0.96	1327	338	4779	0.09	2.0	0.4	20	1.0
50	1.22	1288	318	4503	0.11	3.1	0.6	26	1.4
60	1.49	1251	300	4247	0.13	4.3	0.8	31	1.7
70	1.78	1216	284	4011	0.16	6.3	1.2	36	2.0
80	2.08	1183	268	3796	0.18	8.3	1.6	42	2.4
90	2.39	1152	255	3600	0.21	10.2	2.1	47	2.8
100	2.71	1124	242	3426	0.23	12.6	2.6	53	3.3
110	3.04	1099	232	3279	0.26	15.4	3.3	58	3.7
120	3.39	1078	223	3156	0.29	18.5	4.0	64	4.2
130	3.75	1061	216	3054	0.32	21.7	4.8	69	4.7
140	4.12	1045	210	2967	0.34	24.8	5.7	75	5.3
150	4.50	1031	204	2886	0.37	28.3	6.8	80	5.8

32 Auto, m = 71 gr

x [yd]	as [mils]	v [ft/s]	E [ft lbf]	ED [ª]	t [s]	Cw [in]	y_s [in]	x_s [yd]	ad [mils]
0	0.00	1001	161	2263	0.00	0.0	0.0	0	0.0
10	0.50	986	157	2198	0.03	0.0	0.0	5	0.5
20	1.00	972	152	2136	0.06	0.4	0.2	10	1.0
30	1.52	959	148	2079	0.09	0.8	0.4	15	1.6
40	2.04	946	144	2024	0.12	1.6	0.7	20	2.1
50	2.58	934	141	1972	0.16	2.0	1.2	25	2.7
60	3.12	922	137	1922	0.19	3.1	1.7	30	3.3
70	3.67	911	134	1874	0.22	4.3	2.3	36	3.9
80	4.24	899	130	1828	0.25	5.5	3.1	41	4.6
90	4.81	888	127	1784	0.29	6.7	4.0	46	5.2
100	5.40	878	124	1741	0.32	8.3	5.0	51	5.9
110	5.99	867	121	1700	0.36	10.2	6.1	56	6.6
120	6.59	857	118	1659	0.39	12.2	7.4	62	7.3
130	7.21	847	115	1620	0.43	14.2	8.7	67	8.0
140	7.83	837	113	1582	0.46	16.1	10.3	72	8.8
150	8.46	827	110	1545	0.50	18.9	11.9	77	9.6

[a] [ft·lbf/in^2]

9 mm Russian Makarov, m = 95 gr

x [yd]	as [mils]	v [ft/s]	E [ft lbf]	ED [a]	t [s]	Cw [in]	y_s [in]	x_s [yd]	ad [mils]
0	0.00	1116	260	2637	0.00	0.0	0.0	0	0.0
10	0.40	1084	245	2489	0.03	0.0	0.0	5	0.4
20	0.82	1055	233	2360	0.06	0.8	0.1	10	0.9
30	1.26	1030	222	2247	0.08	1.2	0.3	15	1.3
40	1.70	1007	212	2148	0.11	2.4	0.6	20	1.8
50	2.17	986	203	2059	0.14	3.5	1.0	26	2.4
60	2.65	966	195	1977	0.17	5.1	1.5	31	2.9
70	3.14	947	187	1900	0.21	7.1	2.0	36	3.5
80	3.64	929	180	1829	0.24	9.1	2.7	41	4.1
90	4.16	912	174	1764	0.27	11.0	3.5	47	4.7
100	4.70	896	168	1703	0.30	13.8	4.5	52	5.4
110	5.25	881	162	1644	0.34	16.5	5.5	57	6.1
120	5.81	866	157	1588	0.37	19.3	6.7	63	6.8
130	6.38	851	151	1535	0.41	22.4	8.0	68	7.6
140	6.98	837	146	1485	0.44	26.0	9.5	73	8.4
150	7.58	823	142	1436	0.48	29.5	11.1	79	9.2

9 mm Luger FMJ RN, m = 124 gr

x [yd]	as [mils]	v [ft/s]	E [ft lbf]	ED [a]	t [s]	Cw [in]	y_s [in]	x_s [yd]	ad [mils]
0	0.00	1148	361	3665	0.00	0.0	0.0	0	0.0
10	0.38	1121	345	3494	0.03	0.0	0.0	5	0.4
20	0.77	1098	330	3348	0.05	0.4	0.1	10	0.8
30	1.17	1077	318	3227	0.08	1.2	0.3	15	1.2
40	1.59	1060	308	3126	0.11	2.0	0.6	20	1.7
50	2.01	1045	300	3037	0.14	2.8	0.9	25	2.1
60	2.44	1031	292	2957	0.17	3.9	1.3	31	2.6
70	2.88	1019	284	2884	0.20	5.1	1.9	36	3.1
80	3.33	1007	278	2817	0.23	6.7	2.5	41	3.6
90	3.79	996	272	2755	0.26	7.9	3.2	46	4.2
100	4.25	985	266	2696	0.29	9.8	3.9	51	4.7
110	4.72	974	260	2639	0.32	11.4	4.8	56	5.2
120	5.20	964	255	2585	0.35	13.4	5.8	62	5.8
130	5.69	955	250	2532	0.38	15.4	6.9	67	6.4
140	6.18	945	245	2483	0.41	17.3	8.1	72	7.0
150	6.69	936	240	2434	0.44	19.7	9.4	77	7.6

[a] [ft·lbf/in^2]

38 S & W Spl. LRN, m = 158 gr

x [yd]	as [mils]	v [ft/s]	E [ft lbf]	ED [a]	t [s]	Cw [in]	y_s [in]	x_s [yd]	ad [mils]
0	0.00	869	265	2689	0.00	0.0	0.0	0	0.0
10	0.66	860	259	2630	0.03	0.0	0.1	5	0.7
20	1.32	850	254	2572	0.07	0.4	0.2	10	1.3
30	2.00	841	248	2517	0.11	0.8	0.5	15	2.0
40	2.68	832	243	2463	0.14	1.2	1.0	20	2.8
50	3.38	823	238	2412	0.18	2.0	1.5	25	3.5
60	4.08	815	233	2361	0.21	2.8	2.2	30	4.3
70	4.80	806	228	2312	0.25	3.9	3.0	35	5.0
80	5.52	798	223	2265	0.29	4.7	4.0	41	5.9
90	6.26	790	219	2218	0.33	6.3	5.1	46	6.7
100	7.00	781	214	2172	0.36	7.5	6.4	51	7.5
110	7.76	773	210	2128	0.40	9.4	7.8	56	8.4
120	8.53	765	206	2084	0.44	11.0	9.4	61	9.3
130	9.31	757	201	2040	0.48	13.0	11.2	67	10.2
140	10.10	750	197	1999	0.52	15.0	13.1	72	11.1
150	10.90	742	193	1958	0.56	17.3	15.2	77	12.1

357 Magnum SJHP, m = 125 gr

x [yd]	as [mils]	v [ft/s]	E [ft lbf]	ED [a]	t [s]	Cw [in]	y_s [in]	x_s [yd]	ad [mils]
0	0.00	1394	533	5404	0.00	0.0	0.0	0	0.0
10	0.26	1354	502	5093	0.02	0.0	0.0	5	0.3
20	0.53	1315	474	4808	0.04	0.4	0.1	10	0.6
30	0.81	1280	449	4551	0.07	1.2	0.2	15	0.9
40	1.10	1247	427	4325	0.09	2.0	0.4	20	1.2
50	1.40	1218	407	4123	0.12	3.1	0.6	26	1.5
60	1.71	1191	389	3943	0.14	4.3	1.0	31	1.9
70	2.03	1166	373	3780	0.17	5.9	1.3	36	2.3
80	2.36	1143	358	3631	0.19	7.9	1.8	41	2.7
90	2.70	1121	345	3495	0.22	9.8	2.3	47	3.1
100	3.05	1101	333	3372	0.25	11.8	2.9	52	3.6
110	3.41	1083	322	3260	0.27	14.2	3.6	57	4.0
120	3.78	1066	312	3159	0.30	16.9	4.4	63	4.5
130	4.16	1050	302	3065	0.33	19.7	5.2	68	5.0
140	4.55	1035	294	2977	0.36	22.4	6.2	74	5.5
150	4.95	1021	286	2895	0.39	25.2	7.3	79	6.1

[a] [ft·lbf/in^2]

40 S & W Auto SJHP, m = 155 gr

x [yd]	as [mils]	v [ft/s]	E [ft lbf]	ED [a]	t [s]	Cw [in]	y$_s$ [in]	x$_s$ [yd]	ad [mils]
0	0.00	1148	452	3602	0.00	0.0	0.0	0	0.0
10	0.38	1123	432	3448	0.03	0.0	0.0	5	0.4
20	0.77	1100	415	3307	0.05	0.4	0.1	10	0.8
30	1.17	1080	399	3183	0.08	1.2	0.3	15	1.2
40	1.58	1060	385	3071	0.11	2.0	0.6	20	1.7
50	2.00	1042	372	2967	0.14	2.8	0.9	25	2.1
60	2.43	1025	360	2872	0.17	3.9	1.3	31	2.6
70	2.88	1010	349	2784	0.20	5.1	1.9	36	3.1
80	3.33	995	339	2702	0.23	6.7	2.5	41	3.7
90	3.79	981	330	2627	0.26	8.3	3.2	46	4.2
100	4.26	967	321	2556	0.29	10.2	4.0	51	4.8
110	4.74	955	312	2489	0.32	12.2	4.9	57	5.4
120	5.23	942	304	2426	0.35	14.6	5.9	62	6.0
130	5.74	930	297	2365	0.38	16.5	7.1	67	6.6
140	6.25	919	289	2307	0.41	19.3	8.3	73	7.2
150	6.77	908	282	2251	0.45	22.0	9.7	78	7.9

44 S & W Magnum SJHP, m = 240 gr

x [yd]	as [mils]	v [ft/s]	E [ft lbf]	ED [a]	t [s]	Cw [in]	y$_s$ [in]	x$_s$ [yd]	ad [mils]
0	0.00	1476	1161	7604	0.00	0.0	0.0	0	0.0
10	0.23	1444	1111	7275	0.02	0.0	0.0	5	0.2
20	0.47	1413	1063	6962	0.04	0.4	0.1	10	0.5
30	0.71	1382	1017	6662	0.06	0.8	0.2	15	0.7
40	0.96	1352	974	6379	0.08	1.6	0.3	20	1.0
50	1.22	1323	933	6110	0.11	2.4	0.6	25	1.3
60	1.48	1296	894	5855	0.13	3.1	0.8	31	1.6
70	1.75	1269	857	5614	0.15	4.7	1.1	36	1.9
80	2.04	1243	823	5387	0.18	5.9	1.5	41	2.3
90	2.33	1218	790	5173	0.20	7.5	2.0	46	2.6
100	2.62	1194	759	4973	0.23	9.4	2.5	52	3.0
110	2.93	1172	731	4789	0.25	11.4	3.1	57	3.4
120	3.24	1151	705	4619	0.28	13.4	3.7	63	3.8
130	3.57	1131	681	4461	0.30	15.7	4.5	68	4.3
140	3.90	1112	659	4316	0.33	18.5	5.3	73	4.7
150	4.24	1095	639	4183	0.36	21.3	6.2	79	5.2

[a] [ft·lbf/in^2]

45 Auto FMJ RN, m = 230 gr

x [yd]	as [mils]	v [ft/s]	E [ft lbf]	ED [^{a}]	t [s]	Cw [in]	y_s [in]	x_s [yd]	ad [mils]
0	0.00	804	330	2066	0.00	0.0	0.0	0	0.0
10	0.77	798	325	2037	0.04	0.0	0.1	5	0.8
20	1.54	792	320	2008	0.08	0.4	0.3	10	1.6
30	2.32	787	316	1979	0.11	0.4	0.6	15	2.4
40	3.10	781	311	1950	0.15	0.8	1.1	20	3.2
50	3.90	775	307	1923	0.19	1.2	1.7	25	4.0
60	4.70	770	303	1896	0.23	2.0	2.5	30	4.8
70	5.51	764	298	1869	0.27	2.8	3.5	35	5.7
80	6.33	759	294	1842	0.31	3.5	4.6	40	6.6
90	7.15	754	290	1816	0.35	4.3	5.8	45	7.5
100	7.99	748	286	1791	0.39	5.5	7.2	51	8.4
110	8.83	743	282	1766	0.43	6.7	8.8	56	9.3
120	9.68	738	278	1741	0.47	7.9	10.6	61	10.3
130	10.54	733	274	1716	0.51	9.1	12.5	66	11.2
140	11.40	727	270	1692	0.55	10.6	14.6	71	12.2
150	12.28	722	266	1669	0.59	12.2	16.9	76	13.2

Flintlock pistol, 13.5 mm, m = 224 gr

x [yd]	as [mils]	v [ft/s]	E [ft lbf]	ED [^{a}]	t [s]	Cw [in]	y_s [in]	x_s [yd]	ad [mils]
0	0.00	1263	793	3572	0.00	0.0	0.0	0	0.0
10	0.32	1200	716	3226	0.02	0.4	0.0	5	0.3
20	0.66	1144	650	2929	0.05	0.8	0.1	10	0.7
30	1.03	1093	593	2674	0.08	2.4	0.3	15	1.1
40	1.41	1047	544	2453	0.10	3.9	0.5	21	1.6
50	1.83	1005	502	2262	0.13	5.9	0.9	26	2.1
60	2.26	967	465	2095	0.16	8.7	1.3	31	2.7
70	2.72	933	432	1948	0.20	11.8	1.9	37	3.3
80	3.21	901	404	1819	0.23	15.4	2.5	42	4.0
90	3.73	872	378	1703	0.26	19.3	3.3	48	4.7
100	4.27	845	355	1600	0.30	23.6	4.3	53	5.5
110	4.83	820	334	1506	0.33	28.7	5.4	59	6.4
120	5.43	797	315	1421	0.37	33.9	6.7	65	7.3
130	6.05	775	298	1344	0.41	39.4	8.1	70	8.3
140	6.70	754	283	1273	0.45	45.7	9.8	76	9.3
150	7.38	735	268	1208	0.49	52.0	11.6	82	10.4

^{a} [ft·lbf/in^2]

A.11.3 Rifles

5.45 × 39 AK-74, m = 53 gr

x [yd]	as [mils]	v [ft/s]	E [ft lbf]	ED [^{a}]	t [s]	Cw [in]	y_s [in]	x_s [yd]	ad [mils]
0	0.00	2986	1054	29138	0.00	0.0	0.0	0	0.0
20	0.11	2922	1009	27910	0.02	0.1	0.0	10	0.1
40	0.23	2860	967	26729	0.04	0.4	0.1	20	0.2
60	0.35	2798	926	25595	0.06	0.8	0.2	30	0.4
80	0.47	2738	886	24500	0.08	1.4	0.3	41	0.5
100	0.59	2678	848	23449	0.11	2.2	0.5	51	0.6
120	0.72	2620	811	22434	0.13	3.2	0.8	61	0.8
140	0.86	2562	776	21458	0.15	4.4	1.1	72	1.0
160	0.99	2505	742	20518	0.18	5.8	1.5	82	1.1
180	1.14	2449	709	19613	0.20	7.5	1.9	93	1.3
200	1.28	2394	678	18739	0.22	9.3	2.4	104	1.5
220	1.43	2340	647	17902	0.25	11.4	3.0	115	1.7
240	1.59	2287	618	17091	0.28	13.7	3.7	126	1.9
260	1.75	2234	590	16314	0.30	16.2	4.4	137	2.1
280	1.92	2182	563	15562	0.33	19.0	5.3	148	2.4
300	2.09	2131	537	14843	0.36	22.0	6.2	159	2.6

223 Rem. M 193 (SS 92), m = 55 gr

x [yd]	as [mils]	v [ft/s]	E [ft lbf]	ED [^{a}]	t [s]	Cw [in]	y_s [in]	x_s [yd]	ad [mils]
0	0.00	3150	1210	32150	0.00	0.0	0.0	0	0.0
20	0.10	3069	1149	30531	0.02	0.1	0.0	10	0.1
40	0.21	2990	1091	28982	0.04	0.4	0.1	20	0.2
60	0.31	2913	1035	27502	0.06	0.9	0.2	30	0.3
80	0.43	2837	982	26086	0.08	1.6	0.3	41	0.5
100	0.54	2763	931	24733	0.10	2.5	0.5	51	0.6
120	0.66	2689	882	23439	0.12	3.7	0.7	62	0.7
140	0.79	2617	836	22204	0.15	5.1	1.0	72	0.9
160	0.92	2547	791	21021	0.17	6.8	1.4	83	1.1
180	1.05	2478	749	19895	0.19	8.7	1.8	94	1.2
200	1.19	2410	708	18816	0.22	10.8	2.3	105	1.4
220	1.34	2343	669	17787	0.24	13.3	2.9	116	1.6
240	1.49	2277	632	16804	0.27	16.0	3.5	127	1.8
260	1.64	2212	597	15864	0.30	19.0	4.2	138	2.1
280	1.81	2149	563	14965	0.32	22.4	5.1	149	2.3
300	1.98	2086	531	14107	0.35	26.0	6.0	161	2.6

5.56 mm NATO SS 109 (M 855), m = 62 gr

x [yd]	as [mils]	v [ft/s]	E [ft lb.wt]	ED [a]	t [s]	Cw [in]	y_s [in]	x_s [yd]	ad [mils]
0	0.00	3018	1249	33175	0.00	0.0	0.0	0	0.0
20	0.11	2958	1199	31852	0.02	0.1	0.0	10	0.1
40	0.22	2898	1151	30574	0.04	0.3	0.1	20	0.2
60	0.34	2838	1104	29338	0.06	0.7	0.2	30	0.4
80	0.46	2780	1059	28145	0.08	1.3	0.3	41	0.5
100	0.58	2723	1016	26992	0.10	2.1	0.5	51	0.6
120	0.70	2666	974	25876	0.13	3.0	0.8	61	0.8
140	0.83	2609	933	24795	0.15	4.1	1.1	72	0.9
160	0.97	2554	894	23750	0.17	5.5	1.4	82	1.1
180	1.10	2499	856	22738	0.20	7.0	1.9	93	1.3
200	1.25	2445	819	21760	0.22	8.7	2.4	104	1.4
220	1.39	2391	783	20811	0.25	10.7	2.9	114	1.6
240	1.54	2337	749	19895	0.27	12.8	3.6	125	1.8
260	1.69	2285	715	19007	0.30	15.2	4.3	136	2.0
280	1.85	2232	683	18149	0.32	17.9	5.1	147	2.3
300	2.02	2181	652	17317	0.35	20.7	6.0	159	2.5

6 mm BR Norma, m = 107 gr

x [yd]	as [mils]	v [ft/s]	E [ft lbf]	ED [a]	t [s]	Cw [in]	y_s [in]	x_s [yd]	ad [mils]
0	0.00	2822	1882	42943	0.00	0.0	0.0	0	0.0
20	0.12	2790	1839	41971	0.02	0.0	0.0	10	0.1
40	0.25	2758	1798	41020	0.04	0.2	0.1	20	0.3
60	0.38	2726	1757	40084	0.06	0.4	0.2	30	0.4
80	0.51	2695	1717	39167	0.09	0.8	0.4	40	0.5
100	0.64	2664	1677	38266	0.11	1.2	0.6	50	0.7
120	0.78	2633	1638	37383	0.13	1.8	0.8	61	0.8
140	0.91	2602	1600	36515	0.16	2.4	1.2	71	1.0
160	1.05	2571	1563	35664	0.18	3.2	1.5	81	1.1
180	1.19	2541	1527	34831	0.20	4.0	2.0	92	1.3
200	1.34	2511	1491	34012	0.23	5.0	2.5	102	1.4
220	1.48	2481	1456	33211	0.25	6.1	3.0	112	1.6
240	1.63	2452	1421	32424	0.27	7.3	3.6	123	1.8
260	1.78	2422	1387	31653	0.30	8.6	4.3	133	2.0
280	1.93	2393	1354	30895	0.32	10.1	5.0	144	2.2
300	2.09	2364	1322	30154	0.35	11.7	5.9	155	2.4

[a] [ft·lbf/in^2]

7 × 64 partition, m = 173 gr

x [yd]	as [mils]	v [ft/s]	E [ft lbf]	ED [ª]	t [s]	Cw [in]	y_s [in]	x_s [yd]	ad [mils]
0	0.00	2658	2710	45429	0.00	0.0	0.0	0	0.0
20	0.14	2608	2610	43756	0.02	0.1	0.0	10	0.1
40	0.29	2559	2513	42133	0.05	0.3	0.1	20	0.3
60	0.43	2511	2420	40561	0.07	0.8	0.2	30	0.5
80	0.59	2463	2329	39037	0.09	1.4	0.4	41	0.6
100	0.74	2416	2241	37560	0.12	2.2	0.7	51	0.8
120	0.90	2370	2155	36127	0.14	3.1	1.0	61	1.0
140	1.07	2324	2072	34737	0.17	4.3	1.4	72	1.2
160	1.24	2278	1992	33389	0.20	5.7	1.8	82	1.4
180	1.41	2233	1914	32087	0.22	7.3	2.4	93	1.6
200	1.59	2189	1839	30824	0.25	9.1	3.0	103	1.8
220	1.77	2145	1766	29604	0.28	11.1	3.7	114	2.0
240	1.96	2102	1696	28425	0.30	13.3	4.5	125	2.3
260	2.15	2059	1627	27282	0.33	15.8	5.4	136	2.6
280	2.35	2017	1562	26177	0.36	18.5	6.4	147	2.8
300	2.56	1976	1498	25108	0.39	21.4	7.5	158	3.1

7 mm Rem. Mag. spire point, m = 140 gr

x [yd]	as [mils]	v [ft/s]	E [ft lbf]	ED [ª]	t [s]	Cw [in]	y_s [in]	x_s [yd]	ad [mils]
0	0.00	3100	2964	49689	0.00	0.0	0.0	0	0.0
20	0.10	3050	2868	48078	0.02	0.1	0.0	10	0.1
40	0.21	3000	2774	46510	0.04	0.3	0.1	20	0.2
60	0.32	2950	2683	44985	0.06	0.6	0.2	30	0.3
80	0.43	2901	2595	43500	0.08	1.0	0.3	40	0.5
100	0.54	2852	2509	42055	0.10	1.6	0.5	51	0.6
120	0.66	2804	2425	40648	0.12	2.4	0.7	61	0.7
140	0.78	2757	2343	39280	0.14	3.2	1.0	71	0.8
160	0.90	2710	2264	37949	0.17	4.3	1.3	82	1.0
180	1.02	2663	2187	36655	0.19	5.4	1.7	92	1.1
200	1.15	2617	2111	35396	0.21	6.8	2.1	103	1.3
220	1.28	2571	2039	34174	0.23	8.2	2.6	114	1.5
240	1.41	2526	1968	32987	0.26	9.9	3.2	124	1.6
260	1.55	2482	1899	31833	0.28	11.7	3.8	135	1.8
280	1.69	2438	1832	30712	0.31	13.7	4.5	146	2.0
300	1.83	2394	1767	29624	0.33	15.9	5.3	157	2.2

[a] [ft·lbf/in²]

A.11 Ballistics tables (English/U.S. system)

7.5 × 55 Swiss (GP11), m = 175 gr

x [yd]	as [mils]	v [ft/s]	E [ft lbf]	ED [ᵃ]	t [s]	Cw [in]	y_s [in]	x_s [yd]	ad [mils]
0	0.00	2461	2344	34232	0.00	0.0	0.0	0	0.0
20	0.16	2429	2284	33347	0.02	0.1	0.0	10	0.2
40	0.33	2397	2224	32482	0.05	0.3	0.1	20	0.3
60	0.50	2366	2167	31638	0.07	0.6	0.3	30	0.5
80	0.67	2335	2110	30813	0.10	1.0	0.5	40	0.7
100	0.85	2304	2055	30006	0.13	1.6	0.8	51	0.9
120	1.03	2273	2001	29216	0.15	2.3	1.1	61	1.1
140	1.21	2243	1948	28443	0.18	3.2	1.5	71	1.3
160	1.40	2213	1896	27687	0.21	4.2	2.0	81	1.5
180	1.58	2183	1845	26946	0.23	5.3	2.6	92	1.7
200	1.78	2154	1796	26220	0.26	6.6	3.3	102	1.9
220	1.97	2124	1747	25511	0.29	8.1	4.0	113	2.2
240	2.17	2095	1699	24814	0.32	9.7	4.9	123	2.4
260	2.38	2066	1653	24134	0.35	11.4	5.8	134	2.7
280	2.58	2037	1607	23466	0.38	13.3	6.8	145	2.9
300	2.79	2009	1562	22812	0.41	15.4	7.9	155	3.2

7.62 × 39 AK-47 (Kalashnikov), m = 124 gr

x [yd]	as [mils]	v [ft/s]	E [ft lbf]	ED [ᵃ]	t [s]	Cw [in]	y_s [in]	x_s [yd]	ad [mils]
0	0.00	2329	1487	21040	0.00	0.0	0.0	0	0.0
20	0.19	2263	1404	19855	0.03	0.2	0.0	10	0.2
40	0.38	2197	1323	18721	0.05	0.6	0.1	20	0.4
60	0.58	2133	1247	17639	0.08	1.4	0.3	30	0.6
80	0.79	2069	1174	16603	0.11	2.5	0.6	41	0.9
100	1.00	2007	1104	15616	0.14	3.9	0.9	51	1.1
120	1.23	1945	1037	14675	0.17	5.7	1.4	62	1.4
140	1.46	1885	974	13778	0.20	7.9	1.9	73	1.7
160	1.71	1826	914	12925	0.23	10.5	2.6	83	2.0
180	1.97	1768	856	12114	0.27	13.5	3.4	94	2.4
200	2.23	1711	802	11345	0.30	17.0	4.4	105	2.7
220	2.52	1655	750	10615	0.34	20.9	5.5	117	3.2
240	2.81	1600	701	9923	0.37	25.3	6.8	128	3.6
260	3.12	1546	655	9269	0.41	30.1	8.2	140	4.1
280	3.44	1494	611	8650	0.45	35.5	9.9	151	4.6
300	3.78	1442	570	8065	0.49	41.5	11.7	163	5.2

ᵃ [ft·lbf/in²]

7.62 mm NATO (308 Win.), m = 147 gr

x [yd]	as [mils]	v [ft/s]	E [ft lbf]	ED [a]	t [s]	Cw [in]	y_s [in]	x_s [yd]	ad [mils]
0	0.00	2723	2414	34144	0.00	0.0	0.0	0	0.0
20	0.13	2677	2333	33004	0.02	0.1	0.0	10	0.1
40	0.27	2632	2255	31897	0.04	0.3	0.1	20	0.3
60	0.41	2587	2179	30821	0.07	0.7	0.2	30	0.4
80	0.56	2543	2105	29774	0.09	1.2	0.4	40	0.6
100	0.70	2499	2033	28755	0.12	1.9	0.6	51	0.7
120	0.85	2456	1963	27764	0.14	2.8	0.9	61	0.9
140	1.01	2412	1894	26797	0.16	3.8	1.3	71	1.1
160	1.17	2370	1828	25858	0.19	5.0	1.7	82	1.3
180	1.33	2328	1763	24946	0.21	6.4	2.2	92	1.5
200	1.49	2286	1701	24057	0.24	7.9	2.8	103	1.7
220	1.66	2244	1640	23194	0.27	9.7	3.4	114	1.9
240	1.84	2203	1580	22355	0.29	11.6	4.2	124	2.1
260	2.01	2163	1522	21538	0.32	13.8	5.0	135	2.4
280	2.20	2122	1466	20742	0.35	16.2	5.9	146	2.6
300	2.38	2082	1412	19969	0.38	18.7	6.9	157	2.9

300 Win. Mag. Sierra MK, m = 190 gr

x [yd]	as [mils]	v [ft/s]	E [ft lbf]	ED [a]	t [s]	Cw [in]	y_s [in]	x_s [yd]	ad [mils]
0	0.00	2887	3513	49695	0.00	0.0	0.0	0	0.0
20	0.12	2854	3432	48549	0.02	0.0	0.0	10	0.1
40	0.24	2820	3352	47423	0.04	0.2	0.1	20	0.2
60	0.36	2787	3274	46316	0.06	0.4	0.2	30	0.4
80	0.49	2754	3197	45230	0.09	0.8	0.3	40	0.5
100	0.61	2722	3122	44163	0.11	1.2	0.6	50	0.6
120	0.74	2689	3048	43116	0.13	1.8	0.8	61	0.8
140	0.87	2657	2975	42088	0.15	2.4	1.1	71	0.9
160	1.01	2625	2904	41079	0.17	3.2	1.5	81	1.1
180	1.14	2593	2834	40087	0.20	4.1	1.9	92	1.2
200	1.28	2561	2765	39115	0.22	5.0	2.4	102	1.4
220	1.42	2530	2697	38160	0.24	6.1	2.9	112	1.6
240	1.56	2499	2631	37224	0.27	7.4	3.5	123	1.7
260	1.70	2468	2566	36304	0.29	8.7	4.1	133	1.9
280	1.85	2437	2503	35403	0.32	10.2	4.8	144	2.1
300	2.00	2406	2440	34520	0.34	11.6	5.6	155	2.3

[a] $[ft \cdot lbf/in^2]$

8 × 68 S FMJ RN, m = 220 gr

x [yd]	as [mils]	v [ft/s]	E [ft lbf]	ED [a]	t [s]	Cw [in]	y_s [in]	x_s [yd]	ad [mils]
0	0.00	2854	3978	51053	0.00	0.0	0.0	0	0.0
20	0.12	2812	3859	49532	0.02	0.1	0.0	10	0.1
40	0.25	2769	3744	48048	0.04	0.3	0.1	20	0.3
60	0.37	2727	3631	46602	0.06	0.6	0.2	30	0.4
80	0.50	2686	3521	45191	0.09	1.1	0.4	40	0.5
100	0.64	2644	3414	43814	0.11	1.6	0.6	51	0.7
120	0.77	2603	3309	42470	0.13	2.4	0.8	61	0.8
140	0.91	2563	3207	41158	0.16	3.3	1.2	71	1.0
160	1.05	2523	3107	39880	0.18	4.3	1.5	82	1.1
180	1.19	2483	3010	38634	0.20	5.5	2.0	92	1.3
200	1.34	2444	2915	37419	0.23	6.8	2.5	103	1.5
220	1.49	2405	2823	36236	0.25	8.3	3.1	113	1.7
240	1.64	2366	2733	35083	0.28	9.9	3.7	124	1.9
260	1.80	2328	2646	33960	0.30	11.8	4.4	135	2.1
280	1.96	2290	2561	32867	0.33	13.8	5.2	145	2.3
300	2.13	2253	2478	31804	0.36	15.9	6.1	156	2.5

9.3 × 62 TC, m = 247 gr

x [yd]	as [mils]	v [ft/s]	E [ft lbf]	ED [a]	t [s]	Cw [in]	y_s [in]	x_s [yd]	ad [mils]
0	0.00	2625	3776	35866	0.00	0.0	0.0	0	0.0
20	0.15	2568	3614	34321	0.02	0.1	0.0	10	0.2
40	0.29	2511	3457	32833	0.05	0.4	0.1	20	0.3
60	0.45	2456	3306	31400	0.07	0.9	0.2	30	0.5
80	0.61	2401	3161	30021	0.10	1.6	0.4	41	0.6
100	0.77	2348	3021	28694	0.12	2.6	0.7	51	0.8
120	0.94	2295	2887	27418	0.15	3.8	1.0	61	1.0
140	1.11	2243	2758	26190	0.17	5.2	1.4	72	1.2
160	1.29	2192	2634	25012	0.20	6.8	1.9	82	1.5
180	1.48	2142	2514	23880	0.23	8.7	2.5	93	1.7
200	1.67	2092	2400	22791	0.26	10.9	3.2	104	1.9
220	1.86	2044	2289	21744	0.29	13.3	3.9	115	2.2
240	2.07	1996	2183	20734	0.32	16.0	4.8	126	2.5
260	2.28	1948	2081	19763	0.35	19.0	5.8	137	2.8
280	2.49	1902	1982	18828	0.38	22.2	6.9	148	3.1
300	2.72	1856	1888	17929	0.41	25.8	8.1	159	3.4

[a] $[\text{ft·lbf/in}^2]$

338 Win. Mag. jacketed spitzer, m = 250 gr

x [yd]	as [mils]	v [ft/s]	E [ft lbf]	ED [ᵃ]	t [s]	Cw [in]	y_s [in]	x_s [yd]	ad [mils]
0	0.00	2658	3920	45632	0.00	0.0	0.0	0	0.0
20	0.14	2626	3828	44566	0.02	0.1	0.0	10	0.1
40	0.28	2595	3738	43520	0.05	0.2	0.1	20	0.3
60	0.43	2565	3650	42495	0.07	0.5	0.2	30	0.4
80	0.58	2534	3564	41489	0.09	0.9	0.4	40	0.6
100	0.72	2504	3479	40503	0.12	1.3	0.7	51	0.8
120	0.88	2474	3396	39537	0.14	1.9	1.0	61	0.9
140	1.03	2444	3315	38589	0.16	2.6	1.3	71	1.1
160	1.19	2414	3235	37659	0.19	3.5	1.7	81	1.3
180	1.35	2385	3157	36747	0.21	4.5	2.2	92	1.5
200	1.51	2356	3080	35853	0.24	5.5	2.8	102	1.6
220	1.68	2327	3004	34975	0.27	6.7	3.4	112	1.8
240	1.84	2298	2930	34115	0.29	8.1	4.1	123	2.0
260	2.01	2269	2858	33272	0.32	9.5	4.9	134	2.2
280	2.19	2241	2787	32447	0.34	11.1	5.7	144	2.5
300	2.36	2213	2718	31639	0.37	12.8	6.7	155	2.7

50 Browning jacketed spitzer, m = 648 gr

x [yd]	as [mils]	v [ft/s]	E [ft lbf]	ED [ᵃ]	t [s]	Cw [in]	y_s [in]	x_s [yd]	ad [mils]
0	0.00	2920	12269	62484	0.00	0.0	0.0	0	0.0
50	0.29	2846	11658	59374	0.05	0.4	0.1	25	0.3
100	0.60	2774	11075	56406	0.11	1.2	0.5	50	0.6
150	0.91	2704	10517	53564	0.16	2.4	1.2	76	1.0
200	1.24	2634	9982	50839	0.22	4.3	2.3	102	1.3
250	1.57	2565	9470	48230	0.27	6.7	3.6	128	1.7
300	1.92	2498	8979	45732	0.33	9.8	5.4	154	2.1
350	2.28	2432	8509	43335	0.39	13.8	7.5	180	2.6
400	2.66	2366	8058	41037	0.46	18.1	10.1	207	3.1
450	3.05	2302	7625	38834	0.52	23.2	13.1	234	3.6
500	3.45	2238	7209	36715	0.59	29.1	16.7	262	4.1
550	3.87	2175	6807	34670	0.66	35.4	20.8	289	4.7
600	4.31	2112	6421	32699	0.73	42.9	25.4	317	5.3
650	4.76	2050	6048	30801	0.80	50.8	30.8	345	6.0
700	5.24	1988	5689	28972	0.87	59.8	36.8	374	6.8
750	5.73	1927	5343	27211	0.95	70.1	43.6	403	7.6
800	6.24	1866	5010	25514	1.03	80.7	51.2	432	8.4

ᵃ [ft·lbf/in^2]

A.11.4 Old rifles

Flintlock rifle, 18th century, .65 in, m = 410 gr

x [yd]	as [mils]	v [ft/s]	E [ft lbf]	ED [a]	t [s]	Cw [in]	y_s [in]	x_s [yd]	ad [mils]
0	0.00	1493	2039	6077	0.00	0	0.0	0	0.0
20	0.47	1361	1694	5049	0.04	1	0.1	10	0.5
40	1.00	1246	1420	4233	0.09	3	0.4	21	1.1
60	1.60	1149	1207	3599	0.14	7	0.9	31	1.9
80	2.28	1067	1042	3105	0.19	13	1.8	42	2.8
100	3.02	998	911	2715	0.25	20	3.1	54	3.9
120	3.85	939	806	2402	0.31	28	4.8	65	5.2
140	4.76	887	720	2147	0.38	38	7.0	76	6.7
160	5.76	842	649	1934	0.45	50	9.8	88	8.3
180	6.84	802	589	1755	0.52	63	13.2	100	10.1
200	8.01	766	537	1602	0.60	77	17.5	112	12.2
220	9.28	734	492	1468	0.68	93	22.5	124	14.4
240	10.63	704	453	1350	0.76	110	28.4	136	16.9
260	12.09	676	418	1246	0.85	128	35.3	148	19.5
280	13.64	650	387	1153	0.94	148	43.3	160	22.5
300	15.30	626	359	1069	1.03	169	52.5	172	25.6

Percussion rifle, mid-19th-century, .45 in, m = 510 gr

x [yd]	as [mils]	v [ft/s]	E [ft lbf]	ED [a]	t [s]	Cw [in]	y_s [in]	x_s [yd]	ad [mils]
0	0.00	1345	2046	12706	0.00	0.0	0.0	0	0.0
20	0.55	1312	1947	12094	0.05	0.4	0.1	10	0.6
40	1.12	1281	1855	11524	0.09	0.8	0.4	20	1.2
60	1.71	1252	1771	11000	0.14	2.0	0.9	30	1.8
80	2.32	1224	1693	10515	0.19	3.5	1.7	41	2.5
100	2.95	1197	1621	10069	0.24	5.5	2.7	51	3.2
120	3.60	1173	1555	9660	0.29	7.9	4.0	61	3.9
140	4.26	1150	1495	9283	0.34	10.6	5.6	72	4.7
160	4.95	1128	1439	8939	0.39	13.8	7.4	82	5.6
180	5.65	1108	1389	8624	0.45	17.3	9.6	93	6.4
200	6.38	1090	1342	8337	0.50	21.3	12.1	104	7.3
220	7.12	1072	1300	8073	0.56	25.6	14.9	114	8.3
240	7.88	1056	1261	7829	0.61	30.3	18.1	125	9.3
260	8.66	1041	1224	7605	0.67	35.4	21.7	136	10.3
280	9.46	1026	1191	7398	0.73	40.6	25.6	147	11.3
300	10.28	1013	1160	7206	0.79	46.1	29.9	157	12.4

[a] [ft·lbf/in^2]

45-70 U.S. Government, m = 405 gr

x [yd]	as [mils]	v [ft/s]	E [ft lbf]	ED [a]	t [s]	Cw [in]	y_s [in]	x_s [yd]	ad [mils]
0	0.00	1280	1470	9240	0.00	0.0	0.0	0	0.0
20	0.61	1249	1401	8808	0.05	0.4	0.1	10	0.6
40	1.24	1221	1337	8408	0.10	0.8	0.4	20	1.3
60	1.89	1194	1279	8041	0.15	2.0	1.0	30	2.0
80	2.56	1168	1226	7705	0.20	3.5	1.9	41	2.7
100	3.25	1145	1177	7399	0.25	5.5	3.0	51	3.5
120	3.96	1123	1132	7118	0.30	7.9	4.4	61	4.3
140	4.69	1103	1091	6862	0.36	10.6	6.1	72	5.2
160	5.44	1084	1055	6630	0.41	13.8	8.1	82	6.1
180	6.21	1067	1021	6420	0.47	17.3	10.5	93	7.0
200	7.00	1050	990	6224	0.52	21.3	13.2	103	8.0
220	7.80	1035	961	6043	0.58	25.6	16.3	114	9.0
240	8.63	1020	934	5875	0.64	29.9	19.7	125	10.0
260	9.47	1007	909	5717	0.70	34.6	23.5	135	11.1
280	10.33	994	886	5570	0.76	39.8	27.8	146	12.2
300	11.21	981	864	5433	0.82	45.3	32.4	157	13.3

10.4 mm Vetterli rifle, m = 300 gr

x [yd]	as [mils]	v [ft/s]	E [ft lbf]	ED [a]	t [s]	Cw [in]	y_s [in]	x_s [yd]	ad [mils]
0	0.00	1427	1354	10282	0.00	0.0	0.0	0	0.0
20	0.49	1390	1285	9757	0.04	0.4	0.1	10	0.5
40	1.00	1355	1220	9265	0.09	0.8	0.4	20	1.0
60	1.53	1321	1159	8803	0.13	2.0	0.8	30	1.6
80	2.07	1288	1102	8372	0.18	3.5	1.5	41	2.2
100	2.64	1257	1050	7977	0.22	5.5	2.4	51	2.9
120	3.22	1228	1002	7612	0.27	7.9	3.6	62	3.6
140	3.82	1201	958	7278	0.32	11.0	5.0	72	4.3
160	4.45	1175	918	6972	0.37	14.2	6.7	83	5.1
180	5.09	1151	881	6690	0.42	18.1	8.7	93	5.9
200	5.75	1129	847	6434	0.48	22.0	11.0	104	6.7
220	6.43	1108	817	6201	0.53	26.8	13.6	115	7.6
240	7.14	1089	788	5988	0.59	31.9	16.6	126	8.5
260	7.86	1071	763	5792	0.64	37.0	19.9	136	9.5
280	8.60	1055	739	5613	0.70	42.9	23.5	147	10.5
300	9.36	1039	717	5448	0.75	48.8	27.5	158	11.5

[a] [ft·lbf/in^2]

A.11.5 Various
.177 in air pistol, Diabolo, m = 8.2 gr

x [yd]	as [mils]	v [ft/s]	E [ft lbf]	ED [ᵃ]	t [s]	Cw [in]	y_s [in]	x_s [yd]	ad [mils]
0	0.00	295	1.58	64	0.00	0.0	0.0	0	0.0
5	2.88	287	1.49	61	0.05	0.2	0.2	3	3.0
10	5.86	279	1.41	57	0.10	0.5	0.6	5	6.1
15	8.97	271	1.33	54	0.16	1.1	1.3	8	9.5
20	12.21	263	1.25	51	0.22	1.9	2.3	10	13.2
25	15.57	255	1.18	48	0.27	3.0	3.6	13	17.2
30	19.08	248	1.11	45	0.33	4.4	5.4	15	21.5
35	22.72	241	1.05	43	0.39	6.1	7.6	18	26.1
40	26.52	234	0.99	40	0.46	8.0	10.2	21	31.0
45	30.48	227	0.93	38	0.52	10.2	13.3	24	36.4
50	34.61	220	0.88	36	0.59	12.7	16.9	26	42.1
55	38.92	214	0.83	34	0.66	15.6	21.1	29	48.3
60	43.41	208	0.78	32	0.73	18.8	25.9	32	54.9
65	48.09	202	0.74	30	0.80	22.3	31.4	35	62.0
70	52.98	196	0.70	28	0.88	26.2	37.6	37	69.6
75	58.07	190	0.66	27	0.96	30.4	44.7	40	77.9

.177 in air rifle, Diabolo, m = 8.2 gr

x [yd]	as [mils]	v [ft/s]	E [ft lbf]	ED [ᵃ]	t [s]	Cw [in]	y_s [in]	x_s [yd]	ad [mils]
0	0.00	574	5.99	243	0.00	0.0	0.0	0	0.0
5	0.76	559	5.66	230	0.03	0.0	0.0	3	0.8
10	1.55	544	5.36	218	0.05	0.4	0.1	5	1.6
15	2.36	529	5.08	206	0.08	1.2	0.3	8	2.5
20	3.21	514	4.80	195	0.11	2.4	0.6	10	3.5
25	4.09	500	4.55	185	0.14	3.5	0.9	13	4.5
30	5.01	487	4.30	175	0.17	5.5	1.4	15	5.6
35	5.96	473	4.07	165	0.20	7.5	2.0	18	6.8
40	6.94	460	3.85	156	0.23	9.8	2.6	21	8.0
45	7.97	448	3.64	148	0.27	12.6	3.4	23	9.4
50	9.03	435	3.44	140	0.30	15.7	4.4	26	10.9
55	10.14	423	3.25	132	0.34	18.9	5.5	29	12.4
60	11.29	412	3.07	125	0.37	22.8	6.7	32	14.1
65	12.49	400	2.91	118	0.41	27.2	8.1	35	15.9
70	13.74	389	2.75	111	0.45	31.9	9.7	37	17.8
75	15.03	378	2.59	105	0.49	37.0	11.4	40	19.9

ᵃ [ft·lbf/in²]

.177in air rifle, lead sphere, m = 8.2 gr

x [yd]	as [mils]	v [ft/s]	E [ft lbf]	ED [a]	t [s]	Cw [in]	y_s [in]	x_s [yd]	ad [mils]
0	0.00	574	5.99	243	0.00	0.0	0.0	0	0.0
5	0.76	555	5.60	227	0.03	0.4	0.0	3	0.8
10	1.56	538	5.24	213	0.05	0.7	0.1	5	1.6
15	2.39	521	4.92	199	0.08	1.6	0.3	8	2.6
20	3.26	504	4.62	187	0.11	3.0	0.6	10	3.6
25	4.17	489	4.34	176	0.14	4.6	1.0	13	4.6
30	5.12	474	4.08	165	0.17	6.6	1.4	15	5.8
35	6.11	460	3.84	156	0.21	8.9	2.0	18	7.1
40	7.15	446	3.61	147	0.24	11.6	2.7	21	8.5
45	8.22	433	3.41	138	0.27	14.6	3.6	24	9.9
50	9.35	420	3.21	130	0.31	18.3	4.6	26	11.5
55	10.52	408	3.02	123	0.34	22.3	5.7	29	13.2
60	11.75	396	2.85	116	0.38	26.6	7.0	32	15.0
65	13.02	385	2.69	109	0.42	31.7	8.5	35	17.0
70	14.35	374	2.54	103	0.46	36.6	10.2	38	19.0
75	15.74	363	2.39	97	0.50	42.5	12.1	41	21.3

.31 in arrow (shot from a bow), m = 617 gr

x [yd]	as [mils]	v [ft/s]	E [ft lbf]	ED [a]	t [s]	Cw [in]	y_s [in]	x_s [yd]	ad [mils]
0	0.00	229.7	72.3	928	0.00	0.0	0.0	0	0.0
5	4.67	229.2	72.0	924	0.07	0.1	0.2	3	4.7
10	9.34	228.8	71.7	921	0.13	0.2	0.8	5	9.4
15	14.03	228.4	71.6	919	0.20	0.3	1.9	8	14.1
20	18.73	228.0	71.3	915	0.26	0.5	3.3	10	18.8
25	23.44	227.7	71.0	912	0.33	0.6	5.2	12	23.6
30	28.16	227.4	70.8	909	0.39	0.8	7.5	15	28.4
35	32.90	227.0	70.6	906	0.46	1.1	10.2	17	33.2
40	37.65	226.6	70.4	903	0.53	1.4	13.4	20	38.0
45	42.42	226.2	70.1	900	0.59	1.8	17.0	23	42.8
50	47.20	225.8	69.9	897	0.66	2.2	21.0	25	47.7
55	51.99	225.4	69.7	894	0.73	2.7	25.5	28	52.6
60	56.80	225.1	69.4	891	0.79	3.3	30.4	30	57.6
65	61.63	224.7	69.2	888	0.86	3.8	35.7	33	62.5
70	66.47	224.3	69.0	885	0.93	4.4	41.5	35	67.5
75	71.33	223.9	68.7	882	0.99	5.0	47.7	38	72.5

[a] $[\text{ft} \cdot \text{lbf}/\text{in}^2]$

Cuboid steel fragment, m = 1.54 gr

x [yd]	as [mils]	v [ft/s]	E [ft lbf]	ED [a]	t [s]	Cw [in]	y_s [in]	x_s [yd]	ad [mils]
0	0.00	3281	36.88	2904	0.00	0.0	0.0	0	0.0
5	0.03	2536	22.11	1741	0.01	0.4	0.0	3	0.0
10	0.07	1934	12.77	1006	0.01	1.2	0.0	5	0.1
15	0.13	1455	7.26	572	0.02	2.8	0.0	9	0.2
20	0.21	1106	4.19	330	0.03	5.9	0.1	12	0.4
25	0.35	881	2.65	209	0.05	9.8	0.1	16	0.8
30	0.54	720	1.78	140	0.07	15.7	0.2	20	1.4
35	0.82	595	1.21	95	0.09	22.8	0.4	23	2.3
40	1.21	494	0.84	66	0.12	31.9	0.7	27	3.6
45	1.73	411	0.58	46	0.15	43.3	1.2	31	5.5
50	2.44	344	0.40	32	0.19	57.1	2.0	36	8.3
55	3.41	287	0.28	22	0.24	74.4	3.1	40	12.3
60	4.72	241	0.20	16	0.30	94.9	4.9	44	18.2
65	6.48	202	0.14	11	0.36	120.1	7.5	49	26.6
70	8.87	169	0.10	8	0.45	150.4	11.5	54	38.7
75	12.09	142	0.07	5	0.54	186.6	17.2	59	56.0

Cuboid steel fragment, m = 15.4 gr

x [yd]	as [mils]	v [ft/s]	E [ft lbf]	ED [a]	t [s]	Cw [in]	y_s [in]	x_s [yd]	ad [mils]
0	0.00	3281	368.8	6272	0.00	0.0	0.0	0	0.0
5	0.03	2917	291.5	4957	0.00	0.0	0.0	3	0.0
10	0.05	2586	229.1	3896	0.01	0.4	0.0	5	0.1
15	0.09	2286	179.0	3044	0.02	1.2	0.0	8	0.1
20	0.13	2015	139.1	2366	0.02	2.0	0.0	11	0.2
25	0.18	1771	107.4	1826	0.03	3.5	0.0	14	0.3
30	0.24	1552	82.5	1403	0.04	5.1	0.1	17	0.4
35	0.31	1358	63.2	1075	0.05	7.5	0.1	20	0.6
40	0.40	1194	48.9	832	0.06	10.2	0.2	24	0.8
45	0.50	1062	38.6	656	0.08	13.8	0.3	28	1.1
50	0.63	954	31.2	531	0.09	17.7	0.4	31	1.4
55	0.78	864	25.6	435	0.11	22.4	0.6	35	1.9
60	0.97	786	21.2	361	0.13	28.0	0.8	39	2.4
65	1.19	717	17.6	299	0.15	33.9	1.1	42	3.1
70	1.44	656	14.7	250	0.17	40.9	1.5	46	3.9
75	1.74	601	12.4	211	0.19	48.4	1.9	50	4.8

[a] [ft·lbf/in^2]

A.12 Shotguns and shot

A.12.1 Calibres of shotgun barrels

Designation ("bore")	10	12	14	16	20	24	28	32	36[a]
Diameter[b] [mm]	19.3	18.2	17.2	16.8	15.7	14.7	13.8	12.7	10.2
Diameter[b] [in]	0.760	0.717	0.677	0.661	0.618	0.579	0.543	0.500	0.402

[a] Also known as 410 (= .41 in.).
[b] Minimum bore diameter.

A.12.2 Ballistic data for shot pellets

Pellet Ø [mm]	Mass [g]	Mass [gr]	q [g/mm^2]	q [lb/in^2]	E_0[a] [J]	E_0[a] [ft lbf]	ED[a] [J/mm^2]	ED[a] [ft lbf/in^2]
4½	0.534	8.2	0.0336	0.0444	32.7	24.1	2.06	980
4	0.375	5.8	0.0299	0.0395	23.0	17.0	1.83	871
3½	0.251	3.9	0.0261	0.0345	15.4	11.4	1.60	761
3	0.158	2.4	0.0224	0.0296	9.7	7.2	1.37	652
2½	0.092	1.4	0.0187	0.0247	5.6	4.1	1.14	542
2.41	0.082	1.3	0.0180	0.0238	5.0	3.7	1.10	523
2¼	0.067	1.0	0.0168	0.0222	4.1	3.0	1.03	490
2	0.047	0.7	0.0149	0.0197	2.9	2.1	0.92	438

[a] The above values are valid where $v_0 = 350$ m/s.

A.12.3 Designations for buckshot pellets

Pellet Ø [in]	Mass [g]	US	GB	CAN	B	NL	F	D
0.36	4.44	000	LG				Posten II	
0.33	3.42	00	SG	SSG	B8	B8	Posten III	
0.32	3.12	0						
0.30	2.57	1 Buck	Spec. LG	SG	B6	B6		
0.27	1.87	2 Buck			B5	B5	Posten IV	
0.25	1.49	3 Buck	SSG	AAAA				
0.24	1.32	4 Buck						
0.23	1.16	FF						
0.22	1.01	F						
0.21	0.88	TT						
0.20	0.76	T	AAA	AAA	OV9	OV9	5/0	5/0

A.12.4 Designations for normal shotgun pellets: British/U.S. system

Pellet Ø [in]	Mass [g]	US	GB	CAN	B	NL	F	D
0.19	0.65	BBB						
0.18	0.56	BB						
0.175	0.51	B [a]	BB [a]		OV3	OV3	1	1
0.16	0.39	1						
0.15	0.32	2						
0.14	0.26	3	1	2	1	1	3	3
0.13	0.21	4	3	4	3	3	4	4
0.12	0.16	5	4	5	4	4	5	5
0.11	0.13	6	5	6	5	5	6	6
0.10	0.10	7	6	6	6			
0.095	0.082	7½	7	7½	7	7	7	7
0.09	0.069	8	8		8	8	8	8
0.08	0.049	9	9		9	9	9	9
0.07	0.033	10						
0.06	0.021	11						
0.05	0.012	12						
0.04	0.006	Dust						

[a] Air rifle.

A.12.5 Designations for normal shotgun pellets: metric system

Pellet Ø [mm]	Mass [g]	D	B	GB	F Paris	Lyon	NL	I	A/PL
6.0	1.26	Posten IV							
5.75	1.10		4/S/G						
5.5	0.97	6/0				3/0			00
5.42	0.93			5/S/G or 4A [a]					
5.25	0.84	5/0			5/0	00			
5.15	0.79			3A [a]			6/0		
5.0	0.73	4/0			4/0	0			1
4.92/3	0.69		AA						
4.75	0.62	3/0			3/0	1			2
4.57/8	0.55		5/0	A			3/0		
4.50	0.53	00		4B [a]	00	2			4
4.30/2	0.47		4/0	3B [a]			00	2/0	
4.25	0.45	0			0	3			5
4.06/9	0.40		3/0	BB			0		

Pellet Ø [mm]	Mass [g]	D	B	GB	F Paris	Lyon	NL	I	A/PL
4.0	0.37	1			1	4			6
3.90/1	0.34		00	B			1	1	
3.75	0.31	2			2	5	2		7
3.62	0.28		0	1			3	2	
3.50	0.25	3			3	5P			
3.41/3	0.23		2	2			4	3	
3.24/5	0.20	4	3	3	4	6			9
3.04/5	0.16		4	4			G6	4	
3.00	0.16	5			5	7			10
2.87/8	0.14			4½					
2.79	0.13		5	5			K6	5	
2.75	0.12	6			6	7P			11
2.70/2	0.12			5½					
2.59/60	0.10		6	6				6	
2.50/1	0.091	7	6½	6½	7	8			12
2.41/2	0.081		7	7			7	7	
2.25/8	0.067	8			8	8P			14
2.20/1	0.062		8	8			8	8	
2.02/3	0.048		9	9		9	9	9	
2.00	0.046	9			9				
1.78	0.033		10	10			10	10	
1.75	0.031	10			10	10			
1.68	0.028		11	11				11	
1.57	0.022		12	12			11	11	
1.50/2	0.020	11			11	11	12		
1.25/6	0.011	12			12				
1.00/02	0.006	Dust			Dust				

3A, 3B and 4A are the same as AAA, BBB and AAAA respectively.

B   Glossary

B.1   English ⇒ German ⇒ French

English	German	French
Absolute value	Betrag (eines Vektors) (m)	valeur absolue (f)
acceleration	Beschleunigung (f)	accélération (f)
acceleration due to gravity - of free fall	Erdbeschleunigung (f)	accélération terrestre (f)
acceptance test	Abnahmeprüfung (f)	épreuve de recette (f)
accuracy	Präzision (f)	précision (f)
action	Schloss (n)	batterie (f)
action	Verschluss (m)	sertissage (m), culasse (f) fermeture (f)
action face	Stoßboden (m)	bascule (f), tonnère (m)
action lever	Verschlusshebel (m)	levier de culasse (m)
active area	Wirkfläche (f)	surface active (f)
addend	Summand (m)	terme d'une addition (m)
adjustable trigger	verstellbarer Abzug (m)	détente réglable (f)
adjustment	Justierung (f), Einstellung (f)	réglage (m)
adjustment screw	Einstellschraube (f)	molette de réglage, vis de - (f)
aim	visieren	viser, pointer
aim	Ziel (n)	cible (f)
aim	zielen	viser
aiming	Zielen (n)	pointage (m), visée (f)
aiming point	Zielmarke (f)	chiffre de visée
air column	Luftsäule (f)	colonne d'air (f)
air drag	Luftwiderstand (m)	résistance à l'air (f)
air drag coefficient	Luftwiderstandsbeiwert (m)	coefficient de résistance à l'air (m)
air pistol	Luftpistole (f)	pistolet à air (comprimé) (m)
air resistance	Luftwiderstand (m)	résistance à l'air (f)
air rifle	Luftgewehr (n)	carabine à air comprimé (f)
air-rifle shooting	Luftgewehrschießen (n)	tir à air comprimé (m)
air-pellet	Luftgewehrgeschoss (n)	plomb Diabolo (m)
ajustable	verstellbar	réglable
alarm pistol	Alarmpistole (f)	pistolet d'alarme (m)
ammunition	Munition (f)	munition (f)
ammunition code, - marking	Munitionskennzeichnung (f)	marquage des munitions (m)
ammunition plant	Munitionsfabrik (f)	cartoucherie (f)
amorphous	amorph	amorphe
angle	Winkel (m)	angle (m)
angle of attack	Auftreffwinkel (m)	angle d'arrivée (m)
angle of departure	Abschusswinkel (m)	angle de projection (m)

English	German	French
angle of descent	Fallwinkel (m)	angle de chute (m)
angle of incidence	Anstellwinkel (m)	angle d'incidence (m)
angle of jump	Abgangsfehlerwinkel (m)	angle de relèvement (m)
angle of projection	Wurfwinkel (m)	angle de projection (m)
angle of transition	Umschlagwinkel (m)	angle de transition (m)
angle of twist	Drallwinkel (m)	angle des rayures (m)
angular acceleration	Winkelbeschleunigung (f)	accélération angulaire (f)
angular displacement	Drehwinkel (m)	déplacement (m) angulaire
angular displacement	Drehwinkel (m)	rotation (f)
angular frequency	Kreisfrequenz (f)	fréquence angulaire (f)
angular momentum	Drehimpuls (m)	moment cinétique (m)
angular velocity	Winkelgeschwindigkeit (f)	vitesse angulaire (f)
antimony lead	Hartblei (n)	plomb durci (m)
anvil	Amboss (m)	enclume (f)
aperture	Lochblende (f)	oeilleton (m)
aperture foresight	Ringkorn (n)	guidon annulaire (m)
aperture sight	Diopter (n)	dioptre (m)
approximation	Näherung (f)	approximation (f)
arc of a circle	Kreisbogenstück (n)	arc de cercle (m)
archery	Bogensport (m)	tir à l'arc (m)
area of extravasation	Zone der Extravasation (f)	surface d'extravasion
arm	Waffe (f)	arme (f)
armed lever	Kipphebel (m)	culbuteur (m)
ascending branch	aufsteigender Ast (m)	branche ascendante (f)
assault rifle	Sturmgewehr (n)	fusil d'assaut, fusil mitrailleur (m)
autoloading shotgun	Selbstladeflinte (f)	fusil de chasse automatique (m)
autoloading rifle	Selbstladegewehr (n)	fusil automatique (m)
automatic fire	Seriefeuer (n)	rafale (f), feu de série (m)
semi-automatic pistol	Selbstladepistole (f)	pistolet automatique (m)
automatic rifle	Selbstladegewehr (n)	fusil automatique (m)
automatic shotgun	Selbstladeflinte (f)	fusil de chasse automatique
axis of sight	Visierlinie (f)	ligne de mire (f)
axis of the bore	Seelenachse (f)	âme (f)
axle of cylinder	Trommelachse (f)	pivot de barillet, axe de - (m)
Backlash	Abzugsspiel (n), Triggerstop (m)	immobilisation de détente (f)
ball powder	Kugelpulver (n)	poudre sphérique (f)
ballistics	Ballistik (f)	balistique (f)
bar	Stange (f)	tige (f)
barrel	Lauf (m)	canon (m)
barrel axis	Laufachse (f)	âme (f)
barrel length	Lauflänge (f)	longueur du canon (f)
barrel rib	Laufschiene (f)	bande (de visée) (f)
barrel time	Laufdurchgangszeit (f)	temps de parcours de l'âme (m)
belted case	Gürtelhülse (f)	douille à ceinture (f)
bending stress	Biegespannung (f)	contrainte de flexion (f)
berdan primer	Berdanzündhütchen (n)	amorce à berdan (f)
bevel	abkanten	rogner

English	German	French
big-game hunting	Hochwildjagd (f)	grande chasse (f)
black powder	Schwarzpulver (n)	poudre noir (f)
blank cartridge	Blindpatrone (f)	cartouche à blanc (f)
blank cartridge	Platzpatrone (f)	cartouche à blanc (f)
blasting cap	Sprengkapsel (f)	détonateur (m)
blind shell	Blindgänger (m)	raté, non éclaté
blowback	Rückstoß (m)	recul (m)
blowback operated gun	Rückstoßlader (m)	automatique par recul
blowback system	Masseverschluss (m)	culasse à inertie (f)
blown out primer	Zündhütchenausfaller (m)	décapsulage d'amorce (m)
blue	brünieren	bronzer, brunir
blueing	Brünierung (f)	bronzage, brunissage (m)
boattail	Heckkonus (m)	bi-ogivale
boiling point	Siedepunkt (m)	point d'ébullition (m)
bolt	Schloss (n)	batterie (f)
bolt	Riegel (m)	verrou (m), verrouillage (m)
bolt	Verschluss (m)	sertissage (m), culasse (f) fermeture (f)
bolt handle	Verschlusshebel (m)	levier de culasse (m)
bolt head	Verschlusskopf (m)	tête de culasse (f)
bore	Laufbohrung (f)	alésage du canon (m)
	Bohrung (f), Kaliber (n)	trou, perçage (m) calibre (m)
bottleneck case	Schulterhülse (f)	douille à épaulement (f)
boundary layer	Grenzschicht (f)	couche limite (f)
boundary condition	Randbedingung (f)	condition aux limites (f)
boundary effect	Randeffekt (m)	effet aux limites (m)
bow	Bogen (m)	arc (m)
brand in	einbrennen	cuire (au four)
brass	Messing (n)	laiton (m)
brass-coated	vermessingt	laitonné
brass-plated	vermessingt	laitonné
break-action gun	Kipplauf (m)	canon basculant (m)
breech	Schloss (n), Verschluss (m)	batterie (f) culasse (f), fermeture (f)
breech block	Verschlussblock (m)	bloc culasse (m)
breech loader	Hinterlader (m)	arme à chargement par la culasse
breech lock	Blockverschluss (m)	
brightness	Lichtstärke (f)	luminosité (f)
brown	brünieren	bronzer, brunir
browning	Brünierung (f)	bronzage, brunissage (m)
buckshot	Postenschrot (n)	chevrotine (f)
bulk modulus	Kompressionsmodul (m)	module de compressibilité (m)
bullet, projectile	Geschoss (n)	projectile (m)
bullet jacket	Geschossmantel (m)	chemise (de projectile) (f)
bullet weight	Geschossgewicht (n)	poids du projectile (m)
burn in	einbrennen	cuire (au four)
burning chamber	Brennkammer (f)	chambre de combustion (f)
burning rate	Abbrandgeschwindigkeit (f)	vitesse de combustion (f)
burning rate	Brenngeschwindigkeit (f)	vitesse de combustion (f)
burst	platzen	éclater
barrel burst	Laufsprengung (f)	éclat de canon

butt (of weapon) .... Kolben (Gewehr) (m) .... crosse (f)
butt (behind target) .... Kugelfang (m) .... pare-balles (m), butte (f)
buttplate .... Schaftkappe (f) .... plaque de couche (f)

Cage .... Treibkäfig (m) .... sabot (m)
calibre .... Kaliber (n) .... calibre (m)
calibrate .... kalibrieren, eichen .... calibrer, étalonner
calibrated .... geeicht .... calibré, étalonné
calibrating press .... Kalibrierpresse (f) .... presse à calibrer (f)
calibration .... Eichung (f) .... étalonnage (m)
cam .... Mitnehmer (m) .... accrochage (m)
                                             Nocken (m)   ergot, came (f)
cannelure .... Würgerille (f) .... cannelure (f)
carbine .... Karabiner (m) .... carabine (f), mousqueton (m)
carbon dioxide .... Kohlendioxid (n) .... dioxyde de carbone (f)
carrier .... Ladelöffel (m) .... auget (m)
cartridge .... Patrone (f) .... cartouche (f)
cartridge belt .... Patronengürtel (m) .... ceinture cartouchière (f)
cartridge case .... Patronenhülse (f) .... douille (f)
cartridge feed .... Patronenzufuhr (f) .... alimentation (de cartouche) (f)
case .... Hülse (f) .... douille (f)
case head .... Hülsenboden (m) .... culot (m)
case neck .... Hülsenhals (m) .... collet (m)
case rim .... Hülsenrand (m) .... bordure de la douille (f)
case-rupture .... Hülsenreißer (m) .... rupture de la douille, (f)
catch .... Verriegelung (f) .... verrou (m), verrouillage (m)
cattle killer cartridge .... Schlachtpatrone (f) .... cartouche pour appareils (f)
                                                                                                       d'abattage (m)
cavity .... Kavität (f) .... cavité (f)
centre of gravity .... Schwerpunkt (m) .... centre de masse (m)
   centre of mass ....   centre de gravité (m)
centre of rotation .... Drehpunkt (m) .... centre de rotation (m)
centre of pressure .... Angriffspunkt der Druckkräfte (m) . centre de pression (m)
centrefire .... Zentralzündung (f) .... percussion centrale (f)
centrefire ignition .... Zentralfeuerung (f) .... percussion centrale (f)
degree celsius .... Grad Celsius (m) .... degré Celsius (m)
chamber .... Patronenlager (n) .... chambre à cartouches (f)
chamber burst .... Patronenlagersprengung (f) .... bascule brisée (f)
chamfer .... abkanten .... rogner
chance of hit .... Treffererwartung (f) .... probabilité de toucher (f)
change of state .... Zustandsänderung (th.dyn.) (f) ..... changement d'état (m)
charge .... Ladung (f) .... charge (f)
cheekpiece .... Schaftbacke (f) .... busc (m)
chilled shot .... Hartschrot (m) .... plomb durci (m)
choke .... Choke (m) .... choke (m)
choke-bore .... Chokebohrung (f) .... alésage du choke (m)
choked barrel .... Lauf mit Chokebohrung (m) .... canon avec choke (m)
choked fire .... Schmauchfeuer (n)
chrome .... Chrom (m) .... chrome (m)
chromium-plated .... verchromt .... chromé
circular motion .... Kreisbewegung (f), Dreh- .... mouvement circulaire (m)
circumferential velocity .... Umfangsgeschwindigkeit (f) .... vitesse tangentielle (f)
clay bird, clay-pigeon .... Tontaube (f), Wurftaube (f) .... pigeon (d'argile) (m)

cleaning kit	Reinigungsgerät (n)	nécessaire de nettoyage (m)
closed physical system	abgeschlossenes System (n)	système autonome, -fermé (m)
closed season arm	Schonzeitwaffe (f)	arme pour la chasse fermée (f)
cock	spannen	armer
cocking lever	Spannhebel (m)	levier d'armement (m)
cocking piece	Spannstück (n)	dispositif d'armement (m)
coefficient	Beiwert (m)	coefficient (m)
coercive	zwingend	obligatoirement
cohesive force	Kohäsivkraft (f)	force de cohésion (f)
coil spring	Spiralfeder (f)	ressort à boudin (m)
collision	Stoß (physikalisch) (m)	choc (m)
combustion	Verbrennung (f)	combustion (f)
competition	Wettkampf (m)	compétition (f), match (m)
compressibility	Kompressibilität (f)	compressibilité (f)
compressive force	Druckkraft (f)	force de compression (f)
compressive strain	Druckspannung (f)	contrainte de compression (f)
continuity equation	Kontinuitätsgleichung (f)	équation de continuité (f)
verification of trigger pull weight	Abzugsgewichtkontrolle (f)	peson (m)
conversion rate	Messfolge (f)	
cook-off	Selbstzündung (f)	auto-allumage (m)
coordinate system	Koordinatensystem (n)	système de coordonnées (m)
copper	Kupfer (n)	cuivre (m)
copper-plated	verkupfert	coupré
counterweight	Laufgewicht (n)	contrepoids (m)
covering	Abdeckung (f)	recouvrement (m)
crack	platzen	éclater
crest	Helm (m)	casque (m)
critical defect	kritischer Fehler (m)	défaut critique (m)
cross product	Vektorprodukt (n)	produit vectoriel (m)
cross-section ratio	Querschnittsverhältnis (n)	rapport d'expansion (m)
cross-sectional area	Querschnittsfläche (f)	section (f)
crosswind	Querwind (m), Seiten-	vent latéral (m)
crossover	Übergang (Zustand) (m)	transition (f)
crossways	quer	transversal, latéral
crystalline	kristallin (fest)	cristallin
crystalline structure	Gitterstruktur (f)	structure cristalline (f)
cup wad	Becherpfropfen (m)	bourre à cuvette, - à jupe (f)
curvature	Krümmung (f)	courbure (f)
curved trajectory	gekrümmte Bahn (f)	trajectoire courbe (f)
cushion wad	Pfropfen mit Dämpfung (m)	bourre à coussin (f)
cut case	Hülsenabreißer (m)	douille rompue (f)
cylinder	Trommel (f), Walze (f)	barillet (m)
cylinder bore (barrel)	Zylinderlauf (m)	canon cylindrique (m)
cylinder lock	Kammerverschluss (m)	
Damascus barrel	Damastlauf (m)	canon damas (m)
damp	dämpfen	amortir
damped	gedämpft	amorti
damping	Dämpfung (f)	amortissement (m)
finishing shot	Fangschuss (m)	coup de grâce (m)
decapper	Zündhütchenausstoßer (m)	desamorceur (m)
decapping press	Zündhütchenausstoßer (m)	desamorceur (m)

deceleration	Verzögerung (f)	décélération (f)
deceleration coefficient	Verzögerungskoeffizient (m)	coefficient de décélération (m)
decreased velocity	verminderte Geschwindigkeit (f)	vitesse réduite (f)
deflagration	Deflagration (f)	déflagration (f)
deform	deformieren	déformer
deformation projectile	Deformationsgeschoss (n)	projectile expansif (m)
degree of freedom	Freiheitsgrad (m)	degré de liberté (m)
delayed blowback system	verzögerter Masseverschluss (m)	culasse à ouverture retardée (f)
dense	dicht	étanche, dense
density	Dichte (f)	masse volumique (f)
depression	Unterdruck (m)	dépression (f)
derivative	Differenzialquotient (m)	dérivée (f)
descending branch	absteigender Ast (m)	branche descendante (f)
detonating fuze, detonating cord	Sprengschnur (f)	cordeau détonant (m)
detonation	Detonation (f)	détonation (f)
detonation rate	Detonationsgeschwindigkeit (f)	vitesse de détonation (f)
detonator	Zünder (m)	détonateur, allumeur (m), fusée (f)
deviation	Abweichung (f)	déviation, écart (m)
deviation of point of impact	Treffpunktabweichung (f)	écart du point d'impact (m)
diagonally	diagonal	diagonale
difference quotient	Differenzenquotient (m)	différence (f)
differential	Differenzial (n)	différentiel (m)
differential coefficient	Differenzialquotient (m)	dérivée (f)
dilution	Verdünnung (f)	dilution (f)
dimension	Ausdehnung (f)	dimension (f), extension
dimensional analysis	Dimensionanalyse (f)	analyse dimensionelle (f)
disc	Scheibe (f)	bague (f), cible (f), disque (m)
disc powder	Scheibenpulver (n)	poudre à disque (f)
dispersion	Streuung (f)	dispersion (f), écart (m) groupement (m)
variation in length	Längenstreuung (f)	dispersion en portée (f)
displacement	Verdrängung (f), Wegdifferenz (f)	déplacement (m)
distance	Entfernung (f)	distance (f)
disturbance	Störung (f)	dérangement (m)
dog	Mitnehmer (m)	accrochage (m)
dot product	Skalarprodukt (Vektoren) (n)	produit scalaire (m)
double base powder	Nitroglycerinpulver (n)	poudre à double base (f)
double pull trigger	Druckpunktabzug (m)	détente à bossette (f)
double rifle	Doppelbüchse (f)	fusil à deux canon (m)
double set trigger	Stecher (m)	double détente (f)
double trigger	Doppelabzug (m)	double détente (f)
dove tail	Schwalbenschwanz (m)	queue d'arronde (f)
drag coefficient	Widerstandsbeiwert (m)	coefficient de résistance (m)
drilling	Drilling (m)	drilling, fusil à trois canons (m)
driving force	erregende Kraft (Schwingung) (f)	force d'exitation (f)
drop	Fallhöhe (f)	hauteur de chute (f)
dry	hart (Abzug)	sec
dry firing, dry practice	Trockentraining (n)	entrainement à sec (m)
dry trigger	trockener Abzug (m)	détente sèche (f)
dud	Blindgänger (m)	raté, non éclaté
dummy cartridge, drill round	Manipulierpatrone (f)	cartouche de manipulation (f)

English	German	French
dustproof	staubdicht	étanche aux poussières
Ear defenders	Gehörschutz (m)	protection anti-bruit (f)
edged weapon	Blankwaffe (f)	arme blanche (f)
effect	Wirkung (f)	effet (m)
efficiency	Wirkungsgrad (m)	rendement (m)
ejection port	Auswurföffnung (f)	ouverture d'éjection (f)
ejector	Auswerfer (m)	éjecteur (m)
elevation angle	Erhebungswinkel (m)	élévation (f)
end of burning	Brennschluss (m)	fin de combustion (f)
end-burning velocity	Brennschlussgeschwindigkeit (f)	vitesse en fin de combustion (f)
energy	Energie (f)	énergie (f)
energy assumption	Energieübernahme (f)	gain d'énergie (m)
energy balance	Energiebilanz (f)	bilan énergétique (m)
energy drop	Energieverlust (m)	perte d'énergie (f)
energy of rotation	Drehenergie (f)	énergie de rotation (f)
energy release	Energieabgabe (f)	dissipation d'énergie (f)
engraving resistance	Einpresswiderstand (m)	résistance à l'enfoncement (f)
equations of state	Zustandsgleichung (f)	équation d'état (f)
equations of motion	Bewegungsgleichungen (f)	équations de mouvement (f)
equilibrium	Gleichgewicht (n)	équilibre (m)
equilibrium of forces	Kräftegleichgewicht (n)	équilibre des forces (m)
equipped, be - with	ausgerüstet sein mit	équipé de
exit hole	Ausschussloch (n)	trou de sortie du projectile (m)
exit velocity	Austrittsgeschwindigkeit (f)	vitesse de sortie (f)
expansion ratio	Entspannungsverhältnis (n)	rapport de détente (m)
explosion heat	Explosionswärme (f)	chaleur d'explosion (f)
explosion temperature	Explosionstemperatur (f)	température d'explosion (f)
extension	Auslenkung (f)	déviation (f), écart
exterior ballistics	Außenballistik (f)	balistique extérieure (f)
external force	erregende Kraft (Schwingung) (f)	force d'exitation (f)
extractor	Auszieher (m)	extracteur (m)
Face	Stirnfläche (f)	face (f)
failure	Versager (m)	raté (m)
fall	Neigung (f)	pente (f)
falling block bolt action	Fallblockverschluss (m)	culasse à bloc basculant (f)
fastening element	Befestigungselement (n)	élément d'attache (m)
feed lever	Zubringerhebel (m)	auget (m)
feed system	Zuführsystem (n)	système d'alimentation (m)
feedback	Rückkopplung (f)	rétroaction (f)
feeder	Ladelöffel (m)	auget (m)
final position	Endlage (f)	position finale (f)
final velocity	Zielgeschwindigkeit (f), End-	vitesse terminale (f)
fire	Schießen (n)	tir (m)
firing from closed bolt	aufschießend	tirer à culasse fermée
firing from open bolt	zuschießend	tirer à culasse ouverte
firing pin	Schlagbolzen (m)	marteau, percuteur (m)
firing pin	Zündstift (m)	pointeau d'amorçage (m), poinçon (m)
firing pin spring	Schlagbolzenfeder (f)	ressort de percuteur (m)
firing table	Schusstafel (f)	table de tir (f)

flake powder	Blättchenpulver (n)	poudre à paillettes (f)
flare	Leuchtgeschoss (n)	balle éclairante (f)
flash hider	Feuerscheindämpfer (m)	cache-flamme (m)
flash hole	Zündkanal (m)	cheminée (f)
flat-nose bullet	Flachkopfgeschoss (n)	balle nez plat (f)
flat trajectory	gestreckte Flugbahn (f)	trajectoire tendue (f)
flier	Randschrot (m)	
flintlock	Steinschloss (n)	platine à silex (f)
flintlock ignition	Steinschlosszündung (f)	percussion à silex (f)
flow	Strömung (f)	écoulement, flux (m)
flow against	anströmen	écouler
flow cross-section	Strömungsquerschnitt (f)	section d'écoulement (f)
flow resistance	Strömungswiderstand (m)	résistance à l'écoulement (f)
fluid dynamics	Fluiddynamik (f)	dynamique des fluides (f)
fluid motion	Fluss (m)	flux (m)
fluid velocity	Strömungsgeschwindigkeit (f)	vitesse d'écoulement (f)
fold crimp	Sternverschluss (m)	fermeture étoile, - stellaire (f)
follower lever	Zubringerhebel (m)	auget (m)
forced oscillation	erzwungene Schwingung (f)	oscillation forcée (f)
fore-end	Vorderschaft (m)	devant, fût (m)
forearm	Vorderschaft (m)	devant, fût (m)
foresight	Korn (n)	guidon (m)
foresight base	Kornträger (m)	porte-guidon (m)
foresight protector	Kornschutz (m)	protège-guidon (m)
fragment	Splitter (Geschoss) (m)	éclat (projectile) (m)
fragmentating projectile	Zerlegungsgeschoss (n)	projectile à fragmentation (m)
frame	Rahmen (m)	carcasse (f)
frame of reference	Bezugssystem (n)	système de référence (m)
freezing point	Erstarrungspunkt (m)	point de solidification (m)
frequency of rotation	Drehfrequenz (f)	fréquence de rotation (f)
frequency response	Frequenzgang (m)	bande passante (f)
friction coefficient	Reibungszahl (f)	coefficient de frottement (m)
friction sensitivity	Reibempfindlichkeit (Spr.st.) (f)	sensibilité au frottement (f)
frictional force	Reibungskraft (f)	force de frottement (f)
frictionless	reibungsfrei	sans frottement
foresight ramp	Kornsattel (m)	porte-guidon (m)
boundary frequency	Grenzfrequenz (f)	fréquence limite (f)
fulcrum	Drehpunkt (Hebel) (m)	articulation, pivot (m)
full automatic fire	Dauerfeuer (n)	feu en rafales (m)
full choke	Vollchoke (m)	plein choke
full choked barrel	Vollchokelauf (m)	canon à plein choke (m)
full metal jacket bullet	Vollmantelgeschoss (n)	balle chemisée (f)
funnel-shaped	trichterförmig	en forme d'entonnoir
fuze	Zünder (m)	détonateur, allumeur (m), fusée (f)
Game	Wild (n)	gibier (m)
gap	Spalt (m)	fente (f)
gas constant R	Gaskonstante R (f)	constante des gaz parfaits (f)
gas operated weapon	Gasdrucklader (m)	arme fonctionnant par emprunt de gas
gas port	Gasentnahmedüse (f)	évent (m)
gas pressure	Gasdruck (m)	pression de gaz (f)

English	German	French
gaseous	gasförmig	gazeux
gauge (Schrot)	Kaliber (n)	calibre (m)
gauge pressure	Überdruck (bez. Atmosphäre) (m)	surpression (f)
gear (notch)	Schaltzahn (Revolver) (m)	dent de rochet (f)
gilding metal	Tombak (m)	tombac (m)
groove	Nut (f)	encoche (f)
groove depth	Zugtiefe (f)	profondeur de rayure (f)
grooves	Züge (mp)	rayage, rayure (f)
gun	Waffe (f), Gewehr (n)	arme (f), fusil (m)
gun case	Waffenkoffer (m)	valise pour arme (f)
gunpowder	Schießpulver (n)	poudre (f)
barrel wear	Laufabnutzung (f)	usure du canon (f)
gunmaker	Büchsenmacher (m)	armurier (m)
gunshot residue	Schmauchspuren (fp)	residues de tir (mp)
gunsmith	Büchsenmacher (m)	armurier (m)
gyro effect	Kreiseleffekt (m)	effet gyroscopique (m)
gyroscope	Kreisel (m)	gyroscope (m)
Hair trigger	Stecher (m)	double détente (f)
hair trigger lock	Stecherschloss (n)	platine à double détente (f)
hammer	Hahn (m)	chien (m)
hammer	Hammer (m), Schlagstück (n)	marteau (m)
hammer	hämmern	marteler
hammering	Hämmerung (f)	martelage (m)
handgun	Faustfeuerwaffe (f)	arme de poing (f)
hand support	Handstütze (f)	champignon, pommeau (m)
hand-grenade	Handgranate (f)	grenade à main (f)
handicraft	Handarbeit (f)	travail à la main, -manuel (m)
handloading	Wiederladen (n)	rechargement (m)
hard-core bullet	Hartkerngeschoss (n)	balle noyau dure, - perforante (f)
hard shot	Hartschrot (m)	plomb durci (m)
hardness	Härte (f)	dureté (f)
harmonic motion	harmonische Schwingung (f)	oscillation harmonique (f)
head-first	kopfvoran	la tête la première
heat	Wärmeenergie (f)	chaleur (f)
heat of formation	Bildungsenergie (thermodyn.) (f)	chaleur de formation (f)
heating	Erhitzung (f)	échauffement (m)
helical spring	Spiralfeder (f)	ressort à boudin (m)
helmet	Helm (m)	casque (m)
hexogen	Hexogen (n)	hexogène (m)
high-power cartridge	Hochleistungspatrone (f)	cartouche à haut rendement (f)
hit	Treffer (m)	coup portant, touché (m)
hole	Loch (n)	trou
hollow	hohl	creux
hollow charge	Hohlladung (f)	charge creuse (f)
hollow-point bullet	Hohlspitzgeschoss (n)	balle à pointe creuse (f)
hood	Haube (f)	coiffe, calotte
hunting	Jagd (f)	chasse (f)
hunting knife	Jagdmesser (n)	couteau de chasse (m)
hunting shooting	jagdliches Schießen (n)	tir de chasse (m)
hunting weapoin	Jagdwaffe (f)	arme de chasse (f)

Igniter cord	Anzündlitze (f)	cordon d'allumage (m)
ignition(system)	Zündung (f)	amorçage (m), percussion (f) allumage (m)
ignitor	Zünder (m)	détonateur, allumeur (m), fusée (f)
impact	Einschlag (m)	impact (m)
impact fuze	Aufschlagzünder (m)	fusée percutante (f)
impact sensitivity	Schlagempfindlichkeit (f)	sensibilité à l'impact (f)
impact time	Stoßzeit (f)	durée du choc (f)
incendiary bullet	Brandgeschoss (n)	balle incendiaire (f)
incendiary compound	Brandsatz (m)	matière incendiaire (f)
inch	Zoll (Maßeinheit) (m)	pouce (m) (mesure)
incident mass	eintretende Masse (f)	masse incidente (f)
incompressible	dichtebeständig	incompressible
induce	anregen	induire, provoquer
inertia	Trägheit (f)	inertie (f)
initial position	Anfangslage (f)	position initiale (f)
initial resistance	Ausziehwiderstand (m)	résistance d'extraction (f)
initial velocity	Anfangsgeschwindigkeit (f)	vitesse initiale (f)
insert foresight,	Wechselkorn (n)	guidon interchangeable (m)
interchangeable barrel	Wechsellauf (m)	canon interchangable (m)
interchangeable foresight	Wechselkorn (n)	guidon interchangeable (m)
intermediate ballistics	Abgangsballistik (f)	balistique intermédiaire (f)
inviscid	reibungsfrei	sans frottement
Jacket	Mantel (m)	manteau (m), chemise (f)
jacketed	ummantelt	blindé
jacketed bullet	Mantelgeschoss (n)	balle chémisée (f)
Keyhole	Querschläger (m)	ricochet (m)
kinematic viscosity	kinematische Zähigkeit (f)	viscosité cinématique (f)
kinetic energy	Bewegungsenergie (f) kinetische Energie (f)	énergie cinétique (f)
kneeling	kniend	à genoux
knurling	Randrierung (f)	cannelure (f)
Laminated powder	Blättchenpulver (n)	poudre à paillettes (f)
laminated spring	Blattfeder (f)	ressort à lame (m)
land diameter	Felddurchmesser (m)	diamètre du champ des rayures (m)
lands	Felder (np)	champs des rayures (m)
lanyard	Riemen (m)	bretelle (f)
large body	ausgedehnter Körper (m)	volumineux
latch	Riegel (m), Verriegelung (f)	verrou (m), verrouillage (m)
latching	Verriegelung (f)	verrou (m), verrouillage (m)
lateral	seitlich	latéral
lateral deviation	Seitenabweichung (f)	écart lateral (m)
lateral dispersion	Seitenstreuung (f)	dispersion lateral (f)
lateral spread	Seitenstreuung (f)	dispersion lateral (f)
lattice linkage force	Gitterbindungskraft (f)	force de cohésion moléculaire (f)
law of conservation	Erhaltungssatz (m)	loi de conservation (f)

law of conservation of momentum Impulssatz (m) .... loi de la conservation de la
                                                                                                  quantité de
                                                                                                  mouvement (f)
lead .... Blei (n) .... plomb (m)
lead azide .... Bleiazid (n) .... azoture de plomb (f)
lead fouled barrel .... Laufverbleiung (f) .... plombage du canon (m)
lead styphnate .... Bleitrizinat (n) .... trinitrorésorcinate
                                                                                                  de plomb (m)
leaded, lead fouled .... verbleit .... plombé
length .... Länge (f) .... longueur (f)
length deviation .... Längenabweichung (f) .... écart en portée (m)
length of sight, -of sighting line ..... Visierlänge (f) .... longueur de la visée, (f)
                                                                                                  - de la ligne de mire (f)
lever .... Hebel (m) .... levier (m)
lever action gun .... Unterhebelrepetierer (m) .... fusil à levier de sous-garde (m)
lever arm .... Hebelarm (m) .... bras de levier (m)
lift force .... Auftriebskraft (f) .... portance (f)
lift force coefficient .... Auftriebsbeiwert (m) .... coefficient de portance (m)
limiting velocity .... Grenzgeschwindigkeit (f) .... vitesse limite (f)
line of action .... Wirkungslinie,-richtung (f) .... cintrer
line of departure .... Schussrichtung (f) .... direction de tir (f)
line of sight .... Visierlinie (f) .... ligne de mire (f)
linear deviation .... Längenabweichung (f) .... écart en portée (m)
linear momentum .... Impuls (m) .... quantité de mouvement (f)
strain .... Dehnung (f) .... allongement (m), dilatation (f)
liquid .... flüssig .... liquide
liquid propellant rocket .... Flüssigtreibstoff-Rakete (f) .... roquette à propulsion
                                                                                                  liquide (f)
loading density .... Ladedichte (f) .... densité de chargement (f)
lock .... Schloss (n) .... batterie (f)
lock .... Verriegelung (f) .... verrou (m), verrouillage (m)
lock .... Verschluss (m) .... sertissage (m), culasse (f)
                                                                                                  fermeture (f)
lock plate .... Schlossträger (m) .... platine (f)
locked .... verriegelt .... verrouillé
locking groove .... Verriegelungsnut (f) .... encoche de verrouillage (f)
locking lever .... Verschlusshebel (m) .... levier de culasse (m)
locking lug .... Verriegelungswarze (f) .... tenon (m)
low-pass filter .... Tiefpassfilter (n) .... filtre passe-bas (m)
lug .... Nase (f) .... ergot (m)
lug .... Nocken (m), Vorsprung (m) .... came (f)
lug-locking system .... Warzenverriegelung (f) ....

Machine gun .... Maschinengewehr (n) .... mitrailleuse (f)
machine pistol .... Maschinenpistole (f) .... pistolet mitrailleur (m)
magazine .... Magazin (n) .... chargeur, magasin (m)
magazine follower, follower .... Zubringer (m) .... auget, chargeur, (m)
    .... transporteur (m)
magazine housing .... Magazingehäuse (n) .... puits du chargeur (m)
magazine spring .... Magazinfeder (f) .... ressort du chargeur (m)
main defect .... Hauptfehler (m) .... défaut majeur (m)
main spring .... Schlagfeder (f) .... ressort de percussion (m)
mark .... Stempel (m) .... poinçon (m)

mass element	Masseteilchen (n)	éléments de masse (m)
mass flow	austretende Masse (f)	flux de masse (m)
match	Lunte (f)	mèche (f)
match	Wettkampf (m)	compétition (f), match (m)
match	Zündschnur (f)	cordon allumeur (m)
match rifle	Wettkampfgewehr (n)	carabine de compétition (m)
match-lock	Luntenschloss (n)	platine à mèche (f)
match-trigger	Matchabzug (m)	détente de match (f)
material composition	stoffliche Zusammensetzung (f)	composition (chimique) (f)
maximum amplitude	Schwingungsbauch (m)	amplitude maximale (f)
maximum pressure	Maximaldruck (m)	pression maximale (f)
mean	Mittelwert (m)	moyenne (f)
mean deviation	mittlere Abweichung (f)	écart moyen (m)
mean radius (of dispersion)	mittlerer Streukreisradius (m)	rayon moyen de groupement (m)
measure in radians	Bogenmaß (n), im - messen	mesurer en radians
measurement techniques	Messtechnik (f)	technique de mesure (f)
measured quantity	Messgröße (f)	grandeur mesurée (f)
melting point	Schmelzpunkt (m)	point de fusion (m)
mercury fulminate	Knallquecksilber (n)	fulminate de mercure (m)
micrometer sight	Mikrometerdiopter (m)	dioptre micrométrique (m)
misfire	Versager (m)	raté (m)
missile	Rakete (f)	fusée (f), missile (m), roquette (f)
modified choked barrel	Halbchokelauf (m)	canon à demi choke (m)
modulus of elasticity	Elastizitätsmodul (m)	module d'élasticité (f)
moment of a force	Drehmoment (n)	moment (m)
moment of inertia	Trägheitsmoment (n)	moment d'inertie (m)
moment of momentum	Drehimpuls (m)	moment cinétique (m)
rotation	Rotationsbewegung (f)	mouvement de rotation (m)
mushroom	aufpilzen	épanouir en forme de champignon
muzzle	Mündung (f)	bouche (f)
muzzle blast	Mündungsknall (m)	bruit de bouche (m)
muzzle brake	Mündungsbremse (f)	frein de bouche (m)
muzzle energy	Mündungsenergie (f)	énergie de bouche (f)
muzzle flash (secondary -)	Mündungsfeuer (n)	feu de bouche (m)
muzzle loader	Vorderlader (m)	arme (f) à chargement par la bouche
muzzle momentum	Mündungsimpuls (m)	quantité de mouvement initiale (f)
muzzle velocity	Mündungsgeschwindigkeit (f)	vitesse initiale (f)
Narrow	eng	étroit
narrow channel	gerader Einschusskanal (m)	canal étroit incurvé
natural	gesetzmäßig	naturel, légitime
natural frequency	Eigenfrequenz (f)	fréquence propre (f)
negative gauge pressure	Unterdruck (bez. Atmosphäre) (m)	dépression (f)
Newton's laws of motion	Newton'sche Axiome (np)	axiomes de Newton (m)
nickel	Nickel (m)	nickel (m)
nickel(plated) shot	Nickelschrot (m)	plomb nickelé (m)
nickel-plated	vernickelt	nickelé
notch	Kerbe (f), Kimme (f)	encoche (f), hausse (f),

		cran de mire (m)
nozzle	Düse (f)	tuyère (f)
numerical value	Zahlenwert (m)	valeur numérique (f)
nut	Mutter (f), Schraubenmutter (f)	écrou (m)
nutation	Nutation (f)	nutation (f)
Object	Ziel (n)	cible (f)
octogen	Oktogen (n)	octogène (m)
octogonal	achtkantig	octogonal
octogonal barrel	Achtkantlauf (m)	canon octogonal (m)
off-flow angle	Abströmwinkel (m)	angle de séparation (m)
open sight	offene Visierung (f)	visée ouverte (f)
opposite	entgegengesetzt gleich	opposé, de signe contraire
order of magnitude	Größenordnung (f)	ordre de grandeur (m)
overdamped	stark gedämpft	suramorti
overshooting	überschwingen	surosciller
overturn	kippen, umkippen	basculer, renverser
overturning moment	Kippmoment (n)	moment de basculement (m)
overturning moment coefficient	Momentenbeiwert (m)	coefficient du moment de basculement (m)
oxidated, oxidized	verrostet	rouillé
Packing	Verpackung (f)	emballage (m)
palm support	Handballenauflage (f)	appui paume (m), repose-
paper patch	Papierumwicklung (f)	enveloppe de papier (f)
parabolic	parabelförmig	parabolique
parameter	Kennwert (m)	paramètre (m)
partial rotation	Teildrehung (f)	rotation partielle (f)
pass, overtake	überholen	dépasser
pattern	Trefferbild (n)	image des touchés (f)
peep	Loch (n)	trou (m)
penetration	Durchschlag (m)	pénétration (f)
penetration capacity	Eindringvermögen (n)	pouvoir de pénétration (m)
penetration depth	Eindringtiefe (f)	profondeur de pénétration (f)
penetration power	Durchschlagskraft (f)	puissance de perforation (f)
percussion ignition	Perkussionszündung (f)	percussion à capsule (f)
percussion rifle	Perkussionsgewehr (n)	fusil à percussion (m)
permanent cavity	Wundhöhle (f)	cavité permanente (f)
permanent cavity	bleibende Wundhöhle (f)	cavité permanente (f)
perpendicular	vertikal	vertical
petard	Knallpatrone (f)	pétard (m)
phenomenon	Erscheinung (f)	phénomène (m)
physical pendulum	physikalisches Pendel (n)	pendule physique (m)
picric acid	Pikrinsäure (f)	acide picrique (m)
pin	Stift (m)	goupille (f)
pin fire (ignition)	Stiftzündung (f)	percussion à broche (f)
pin fire ignition	Lefaucheuxzündung (f)	percussion à broche (f)
pistol	Pistole (f)	pistolet (f)
pistol cartridge	Pistolenpatrone (f)	cartouche pour pistolets (f)
pistol grip	Pistolengriff (m)	poignée pistolet, crosse - (f)
pistol shooter	Pistolenschütze (m)	pistolier (m)
piston	Kolben, Druckkolben (m)	piston (m)

pitch	kippen	basculer
pitching	Querstehen (n)	se mettre en travers
pivot	Schwenkachse (f)	pivot (m)
pivot friction	Lagerreibung (f)	friction (f) du pivot
plane	ebene Platte (f)	plan (m), plaque plane (f)
plastic case	Plastikhülse (f)	douille (en) plastique (f)
plastic cup wad	Plastikbecherpfropfen (m)	bourre à jupe en plastique (f)
plastic shotshell	Plastikschrotpatrone (f)	cartouche de chasse en plastique (f)
plastic wad	Plastikpfropfen (m)	bourre en plastique (f)
plate	ebene Platte (f)	plan (m), plaque plane (f)
point of aim	Haltepunkt (m)	point de mire (m)
point of impact	Auftreffpunkt, Treffpunkt (m)	point d'impact (m)
point of separation	Ablösepunkt (Grenzschicht) (m)	point de séparation (m)
pointed bullet	Spitzgeschoss (n)	balle pointue (f)
post sight	Balkenkorn (n)	guidon à lame rectangulaire (m)
potassium chlorate	Kaliumchlorat (n)	chlorate de potassium (m)
potassium nitrate	Kaliumnitrat (n)	nitrate de potassium (m)
potassium sulphide	Kaliumsulfid (n)	sulphide de potassium (m)
potential energy	potenzielle Energie (f)	énergie potentielle (f)
powder	Pulver (n)	poudre (f)
powder scale	Pulverwaage (f)	balance à poudre (f)
power	Leistung (physikalisch) (f)	puissance (f)
practice ammunition	Übungsmunition	munition d'exercice (f)
sign	Vorzeichen (n)	signe (m)
precession	Präzession (f)	précession (f)
precision	Präzision (f)	précision (f)
precursor	Vorläufer (m)	précurseur (m)
predetermined breaking point	Sollbruchstelle (f)	amorce de rupture (f)
pressure	Druck (m)	pression (f)
pressure drag	Druckwiderstand (m)	résistance à la compression (f)
pressure energy	Druckenergie (f)	énergie de compression (f)
pressure gauge	Druckaufnehmer (m), Gasdruckmesser (m)	jauge de pression (f)
pressure gradient	Druckgefälle (n)	gradient de pression (m)
pressure transducer	Druckgeber (m)	transducteur de pression (m)
primary (muzzle) flash	Feuer aus Mündung (n)	feu de bouche (m)
primer cap	Zündhütchen (n)	capsule d'amorçage (f)
primer composition	Zündsatz (m)	composition d'amorçage (f)
primer sensitivity	Zündhütchenempfindlichkeit (f)	sensibilité d'amorce (f)
progressive burning powder	progressives Pulver (n)	poudre progressive (f)
progressive twist	progressiver Drall (m)	pas progressif (m)
prone	liegend	couché
proof firing	Beschussprobe (f)	épreuve (f)
proof mark	Beschussstempel (m)	poinçon d'épreuve (m)
propellant	Treibmittel (n)	charge propulsive (f)
proper motion	Eigenbewegung (f)	mouvement propre (m)
ratio of forces	Kräfteverhältnis (n)	rapport des forces (m)
pull (the trigger)	Abziehen (n)	départ (m)
pulse duration	Pulsdauer (f)	durée de pulsation (f)
quantity of heat	Wärmemenge (f)	quantité de chaleur (f)
Range	Schussdistanz, Schussweite (f)	portée (f)

range	Entfernung (f)	distance (f)
range table	Schusstafel (f)	table de tir (f)
ranging	Einschießen (n)	tir de réglage (m)
rapid-fire pistol	Schnellfeuerpistole (f)	pistolet rafaleur (m)
rate of fire	Kadenz (f)	cadence (f)
ratio	Verhältniszahl (f)	rapport (m)
rear sight	Kimme (f), Visier (n)	cran de mire (m), hausse (f)
rear sight slide	Kimmenblatt (n)	feuillet (m)
return spring	Rückholfeder (f)	ressort récupérateur (m)
recaper	Zündhütcheneinsetzgerät (n)	amorceur (m)
recapping press	Zündhütcheneinsetzgerät (n)	amorceur (m)
receiver	Rahmen (m)	carcasse (f)
receiver	Verschlussgehäuse (n)	boîtier de culasse (m)
recoil	Rückstoß (m)	recul (m)
recoil energy	Rückstoßenergie (f)	énergie du recul (f)
recoil operated gun	Rückstoßlader (m)	automatique par recul
recoil pad, recoil reducer	Rückstoßminderer (m)	plaque de couche antirecul (f)
recuperator spring	Vorholfeder (f)	ressort récupérateur (m)
reference plane	Bezugsfläche (f)	surface de référence (f)
reference point	Bezugspunkt (m)	point de référence (m)
reference value	Bezugsgröße (f)	valeur de référence (f)
relation	Beziehung (f)	relation (f)
relaxation time	Zeitkonstante (f)	temps de relaxation (m)
release the safety catch	entsichern	armer
reloading	Wiederladen (n)	rechargement (m)
reloading press	Wiederladepresse (f)	presse à recharger (f)
reloading tool	Wiederladewerkzeug (n)	outillage pour le rechargement (m)
remaining velocity	Restgeschwindigkeit (f)	vitesse restante (f)
repeating rifle	Repetierbüchse (f)	carabine à répétition (f)
repeating shotgun	Repetierflinte (f)	fusil à répétition (m)
report	Knall (m)	détonation (f)
residual energy	Restenergie (f)	énergie résiduelle (f)
resistance to tearing	Reißfestigkeit (f)	résistance au déchirement (f)
resisting force	Widerstandskraft (f)	résistance
resonance	Resonanz (f)	résonance (f)
rest position	Ruhelage (f)	position de repos (f)
resting	ruhend	au repos
retardation	Verzögerung (f)	rétardation (f)
reversible	umkehrbar	réversible
revolver	Revolver (m)	revolver (m)
revolver cartridge	Revolverpatrone (f)	cartouche pour revolvers (f)
rib	Rippe (f)	ailette, nervure (f)
ricochet	Abpraller (m)	ricochet (m)
shoulder weapon	Langwaffe (f), Büchse (f)	arme d'épaule (f)
rifle scope	Zielfernrohr (n)	lunette de visée (f)
rifle-shotgun	Büchsflinte (f)	carabine (m)
rifled	gezogen (Lauf)	rayé
rifling	Züge (mp)	rayage, rayure (f)
rifling twist	Drall (Waffe) (m)	rayure (f)
right-handed system	Rechtssystem (Koordinaten) (n)	système orienté à la main droite (m)
rigid	starr	rigide

English	German	French
rigid object	starrer Körper (m)	corps rigide (m)
rim	Rand (m)	bourrelet (m)
rimfire cartridge	Randfeuerpatrone (f)	cartouche à percussion annulaire (f)
rimfire ignition	Randfeuerzündung (f)	percussion annulaire (f)
rimmed cartridge	Randpatrone (f)	cartouche à bourrelet (f)
rimmed case	Randhülse (f)	douille en bourrelet (f)
ring foresight	Ringkorn (n)	guidon annulaire (m)
rivet	Niet (m)	rivet (m)
rocket	Rakete (f)	fusée (f), roquette (f)
rod	Stange (f)	tige (f)
roll crimp	zugebördelt	
roll crimp	Bördelung (f)	sertissage à rondelle
rotating bolt locking system	Drehverschluss (m)	culasse tournante, - rotative
rotational axis, pivot	Drehachse (f)	axe de rotation, pivot (m)
rotational kinetic energy	Rotationsenergie (f)	énergie cinétique de rotation (f)
rotational symmetry	rotationssymetrisch	symmétrique par rapport à une axe
round	Schuss (m)	coup, tir (m)
round-nose bullet	Rundkopfgeschoss (n)	balle à tête ronde (f)
running boar	laufender Keiler (m)	sanglier courant (m)
rust	Rost (m)	rouille (f)
rust preventing agent	Rostschutzmittel (n)	solution antirouille (f)
anticorrosive	rostschützend	antirouille
rusty	rostig	rouillé
Sabot	Treibspiegel (m)	sabot (m)
safety	Sicherheit (f)	sûreté, sécurité (f)
safety catch	Sicherheitsflügel (m)	levier de sûreté (m)
safety device	Sicherung (f)	mécanisme de sécurité (m)
safety fuze	Schwarzpulverzündschnur (f)	mèche de sureté (f)
safety pin	Sicherheitsstift (m), Vorstecker (m)	goupille de sûreté (f)
safety slide	Sicherheitsschieber (m)	sûreté (f)
sample	Muster (n)	échantillon (m)
sand-blast	sandstrahlen	sabler
scope mount	Zielfernrohrmontage (f)	montage pour lunette (m)
scope sight	Zielfernrohr (n)	lunette de visée (f)
score	Treffer (m)	coup portant, touché (m)
score gauge, scorer	Schusslochprüfer (m)	axe jaugeur (m), jauge (f)
screw	Schraube (f)	vis (f)
sectional density	Querschnittsbelastung (f)	masse par unité de section (f)
semi-jacketed bullet	Teilmantelgeschoss (n)	balle demi blindée (f)
semi-automatic	halbautomatisch	semi-automatique
sensing device	Fühler (m)	jauge, capteur (m)
sensitivity test	Empfindlichkeitsprüfung (f)	épreuve de sensibilité (f)
sensor	Aufnehmer (m)	sonde (f), capteur (m), jauge (f)
separation	Ablösung (f)	séparation (f)
serial number of a weapon	Waffennummer (f)	numéro de série de l'arme (m)
set screw	Einstellschraube (f)	molette de réglage, vis de - (f)
setting	Einstellung (f)	réglage (m)
shadowgraph	Schattenaufnahme (f)	
shape stable	formstabil	indéformable

English	German	French
shear force	Schubkraft (f)	force de poussée (f)
shear rate	Geschwindigkeitsgefälle (n)	gradient de vitesse (m)
shear stress	Scherspannung (f), Schub-	tension de cisaillement (f)
shock wave	Stoßwelle (f)	onde de choc (f)
shoot	Schießen (n)	tir (m)
accuracy	Treffsicherheit (f)	assurance de toucher (f)
wound channel	Schießkanal (m)	tunnel de tir (m)
shooting range	Schießstand (m)	stand de tir (m)
short-range ammunition	Kurzbahnmunition (f)	munition à courte portée (f)
short rifle	Stutzer (m)	carabine (f)
shot	Schrot (m)	grenaille (f)
shot	Schuss (m)	coup, tir (m)
shotgun	Flinte (f), Jagdflinte (f)	fusil de chasse (m)
		- à canon lisse (m)
shot pellet	Schrotkugel (f)	plomb (de chasse) (m)
shot sheaf	Schrotgarbe (f)	gerbe (f)
shot start pressure	Ausziehdruck (m)	pression d'extraction (f)
shotgun cartridge, shotshell	Schrotpatrone (f)	cartouche de chasse (f)
shoulder stabilized	schulterstabilisiert	stabilisé par l'épaule
shoulder weapon	Schulterwaffe (f)	arme d'épaule (f)
shrapnel	Schrapnell (n)	shrapnel (m)
side lock	Seitenschloss (n)	contre-platine (f)
side-by-side shotgun	Doppelflinte (f)	fusil à canon juxtaposés (m)
sideways	seitwärts	de côté
sight	zielen, visieren	viser, pointer
sighting	Zielen (n)	pointage (m), visée (f)
sighting error	Zielfehler (m)	erreur de visée (f)
sights	Visierung (f)	hausse, visée (f)
signal pistol	Alarmpistole (f)	pistolet d'alarme (m)
silencer	Schalldämpfer (m)	silencieux (m)
dynamic -	dynamische -	- dynamique
geometric -	geometrische -	- géométrique
kinematic -	kinematische -	- cinétique
simple pendulum	mathematisches Pendel (n)	pendule mathématique (m)
single action	single action	simple action (f)
single base powder (NC powder)	Nitrocellulosepulver (n)	poudre à simple base (f)
single-shot fire	Einzelfeuer (n)	feu coup par coup (m)
single-shot weapon	Einzellader (m)	fusil à un coup (m)
sinusoidal	sinusförmig	sinusoïdal
size	kalibrieren	calibrer, étalonner
sizing press	Kalibrierpresse (f)	presse à calibrer (f)
slaughter cartridge	Schlachtpatrone (f)	cartouche pour appareils d'abattage(f)
slender	schlank	svelte
slenderness	Schlankheit (f)	sveltesse (f)
slenderness ratio	Schlankheitsgrad (m)	degré de sveltesse (m)
slide	Schieber (m)	poussoir (m), tirette (f)
slope	Neigung (f)	pente (f)
slot	Nut (f)	encoche (f)
small bore rifle	Kleinkalibergewehr (n)	carabine de petit calibre (f)
smokeless powder	rauchloses Pulver (n)	poudre pyroxylée (f)
smoothing	Oberflächenbehandlung (f)	lissage (m)
soft lead	Weichblei (n)	plomb doux (m)

English	German	French
soft-nose bullet	Teilmantelgeschoss (n)	balle demi blindée (f)
solid	fest (Aggregatszustand)	solide
solid (homogeneous) bullet	Vollgeschoss (n)	projectile plein (m)
solid angle	räumlicher Winkel (m)	angle solide (m)
solid body	fester Körper (m)	corps solide (m)
solid propellant rocket	Feststoff-Rakete (f)	roquette à propulsion solide (f)
spatial expansion	räumliche Ausdehnung (f)	volume (m)
speed	Geschwindigkeitsbetrag (m)	vitesse, célérité (f)
spin	Drall (Geschoss) (m)	rotation (f)
spin damping	Drehzahlabnahme (f)	diminution du nombre de tours (f)
spin-stabilized	drallstabil	gyrostabilisé
spitzer bullet	Spitzgeschoss (n)	balle pointue (f)
splinter	Splitter (Holz, Glas) (m)	éclat (bois, verre) (m)
split case	Hülsenreißer (m)	rupture de la douille (f)
spotting scope	Spektiv (n)	longue vue (f)
sporter, sporting arm	Sportwaffe (f)	arme de sport (f)
spread	Streuung (f)	dispersion (f), écart (m)
sprocket	Schaltzahn (Revolver) (m)	dent de rochet (f)
square foresight	Rechteckkorn (n)	guidon rectangulaire (m)
stagnation point	Staupunkt (m)	point de stagnation (m)
stagnation pressure	Staudruck (m)	pression de la retenue (f)
stainless	nicht rostend	antirouille
stainless	rostfrei	inoxydable
standard deviation	Standardabweichung (f)	écart-type (m)
standing	stehend	debout
star crimp	Sternverschluss (m)	fermeture étoile, - stellaire (f)
state of matter	Aggregatszustand (m)	état de matière (m)
state variables	Zustandsgröße (f)	variables d'état (f)
steel	Stahl (m)	acier (m)
steel case	Stahlhülse (f)	douille (en) acier (f)
steel-core bullet	Stahlkerngeschoss (n)	balle à noyau en acier (f)
steel-jacketed bullet	Stahlmantelgeschoss (n)	balle blindée (f)
steepness	Steilheit (Griff) (f)	pente, inclinaison (f)
stick grenade	Stielhandgranate (f)	grenade à manche (f)
stock	Schaft (m)	crosse (f)
stopgap	Notbehelf (m)	solution de fortune (f)
strap	Riemen (m)	bretelle (f)
streamline	Stromfaden (m)	ligne de courant (f)
stress	Spannung (f)	tension (f)
striker	Schlagstück (n)	marteau (m)
stuck bullet	Laufstecker (m)	balle restée (dans le canon) (f)
stud	Nase (f), Vorsprung (m)	ergot (m)
subcalibre	Unterkaliber (n)	sous-calibré
submachine gun	Maschinenpistole (f)	pistolet mitrailleur (m)
subsonic	Unterschall-	subsonique
suction effect	Saugeffekt (m)	effet de succion (m)
superposed rifle-shotgun	Bockbüchsflinte (f)	fusil à canons superposés (m)
superposition principle	Prinzip der Unabhängigkeit von Bewegungen (n)	principe de superposition (m)
supersonic	Überschall-	suprasonique
pivot	Lager (n)	support (m)

surface of a cone	Kegelmantel (m)	surface du cône (f)
sympathetic detonation	Detonationsübertragung (f)	détonation par influence (f)
system of units	Maßsystem (n)	système d'unités (m)
Tangential velocity	Bahngeschwindigkeit (f)	vitesse linéaire (f)
tapered, conical	konisch	conique
tappet	Mitnehmer (m)	accrochage (m)
target	Scheibe (f), Ziel (n)	cible (f), disque (m) bague (f)
target patch	Schusspflaster (n)	palette (de tir) (f)
target practices	Zielübungen (f)	exercices de visée (m)
target transport device	Scheibenzuganlage (f)	rameneur (de cible) (m)
tear, (wear and tear)	Verschleiß (m)	usure (f)
tensile stress	Zugspannung (f)	contrainte de traction (f)
terminal ballistics	Endballistik (f)	balistique terminale (f)
test-firing	Einschießen (n)	tir de réglage (m)
thermal conduction	Wärmeleitung (f)	conduction thermique (f)
thermal energy	Wärmeenergie (f)	énergie thermique (f)
thermal state	Wärmezustand (m)	état thermique (m)
thermic development	Wärmeentwicklung (f)	production de chaleur (f)
thermodynamic state variables	thermodynamische Zustandsgrößen (f)	variables d'état
thickness of the wall	Wandstärke (f)	épaisseur de parois (f)
thin firing tapes	Stoppinen (f)	étoupille (f)
three-shot mechanism	Dreischussautomatik (f)	mécanisme à trois coups (m)
three-barrelled gun	Drilling (m)	drilling, fusil à trois canons (m)
thrust	Schub (Rakete) (m)	poussée (f)
thumbhole	Daumenloch (n)	trou de pouce (m)
thumbhole stock	Schaft mit Daumenloch (m)	crosse à trou de pouce (f)
thumbrest	Daumenauflage (f)	repose-pouce (m)
tie force	Zugkraft (f)	force de traction (f)
tight	dicht	étanche, dense
time of flight	Flugzeit (f)	temps de vol (m)
timing axis	Zeitachse (f)	axe du temps (m)
tip-up lock	Kipplaufverschluss (m)	culasse tombante, - basculante (f)
tip-up rifle	Kipplaufbüchse (f)	carabine expresse (f)
toggle action bolt	Kniegelenkverschluss (m)	culasse à genouillère (f)
tombac	Tombak (m)	tombac (m)
tombac-plated steel	tombakplattierter Stahl (m)	acier plaqué de tombac (m)
top	Kreisel (Spielzeug -) (m)	toupie (f)
torque	Drehmoment (n)	moment (m)
torsional pendulum	Torsionspendel (n)	pendule de torsion (m)
tracer	Leuchtsatz (m)	composition lumineuse (f)
tracer bullet	Leuchtspurgeschoss (n)	balle traçante (f)
tractability	Folgsamkeit (f)	(l'axe du projectile suivant la
tractable	folgsam	tangente de la trajectoire)
trajectory	Flugbahn (f)	trajectoire (f)
transient pressure	Druckschwankung (f)	fluctuation de pression (f) variation -
transonic	transsonisch	transsonique
trigger	Abzug (m)	détente (f)

trigger bar .... Abzugsstange (f) .... tige de détente
trigger verification device .... Abzugsprüfgerät (n) .... contrôleur de détente (m)
trigger device, - mechanism .... Abzugseinrichtung (f) .... système de détente (m)
trigger guard .... Abzugsbügel (m) .... pontet de sous-garde (m)
trigger lever, trigger latch .... Abzugshebel (m) .... gâchette, queue de détente (f)
trigger point .... Druckpunkt (m) .... cran d'arrêt (m)
trigger pull weight .... Abzugsgewicht (n) .... poids de départ, - détente (m)
triggerstop .... Abzugsbegrenzung (f) -stop (m) .... immobilisation de détente
trigger system .... Abzugssystem (n) .... système de détente (m)
trinitrotoluene .... Trinitrotoluol (n) .... trinitrotoluène (m)
tripod .... Dreibein (n) .... trépied (m)
truncate .... abbrechen .... tronquer
tube magazine, tubular magazine ... Röhrenmagazin (n) .... chargeur tubulaire,
                                                                                                                                  magasin tubulaire (m)
tubular powder .... Röhrchenpulver (n) .... poudre tubulaire (f)
tumble .... taumeln .... tomnoyer
twilight performance .... Dämmerungsleistung (f) .... puissance crépusculaire (f)
twist .... Drall (Waffe) (m) .... rayure (f)
twist direction .... Drallrichtung (f) .... direction du pas (f)
twist, twist length .... Dralllänge (f) .... pas des rayures (m)

Underdamped .... schwach gedämpft .... sous-amorti
uniform .... gleichförmig .... uniforme
unit .... Maßeinheit (f) .... unité de mesure (f)

Valve .... Ventil (n) .... soupape (f)
vermin .... Raubzeug (jagbare Nager) (n) ...... rongeur (m)
varmint rifle .... Schonzeitwaffe (f) .... arme pour la chasse fermée
vault .... wölben .... direction de l'action (f)
velocity .... Geschwindigkeit (f) .... vitesse (f)
velocity of descent .... Fallgeschwindigkeit (f) .... vitesse de chute (f)
velocity of sound .... Schallgeschwindigkeit (f) .... vitesse du son (f)
velocity of waves .... Ausbreitungsgeschwindigkeit (f) ... vitesse de propagation (f)
ventilated barrel rib .... ventilierte Laufschiene (f) .... bande ventilée (f)
vertex .... Scheitel (m) .... sommet (m)
vertex distance .... Scheiteldistanz (f) .... distance du sommet (f)
vertex height .... Scheitelhöhe (f) .... hauteur du sommet (f)
vertical deviation .... Höhenabweichung (f) .... écart vertical (m)
vertical dispersion .... Höhenstreuung (f) .... dispersion verticale (f)
viscosity .... Zähigkeit (dynamische) (f) .... viscosité (f)
vivacity .... Lebhaftigkeit (f) .... vivacité (f)

Wad .... Pfropfen (m) .... bourre (f)
wadcutter .... Wadcutter(-geschoss) (n) .... wad-cutter
washer .... Unterlegscheibe .... bague (f)
spirit level .... Wasserwaage (f) .... niveau à bulle (m)
waterproof .... wasserdicht .... étanche
wax .... Wachs (n) .... cire (f)
weapon .... Waffe (f) .... arme (f)
wear .... Abnutzung (f) .... usure (f)
wedged .... verkeilt .... calé
weight .... Gewicht (n) .... poids (m)
wheel shot .... Hohlschuss (Schrot) (m) .... coup creux (m)
wheel-lock .... Radschloss (n) .... platine à rouet (f)

wheel-lock ignition .... Radschlosszündung (f) .... percussion à rouet (f)
with rest .... aufgelegt .... avec appui, - support
with support .... aufgelegt .... avec appui, - support
work .... Arbeit (f) .... travail (m)

Yaw .... taumeln .... tomnoyer
yaw .... Präzessionsbewegung (f) .... mouvement de précession (m)
yaw .... Pendelung (f) .... mouvement pendulaire (m)
Young's modulus .... Elastizitätsmodul (m) .... module d'élasticité (m)

Zeroing .... Einschießen (n) .... tir de réglage (m)

B.2   German – English – French

Abbrandgeschwindigkeit (f) .... burning rate .... vitesse de combustion (f)
abbrechen .... truncate .... tronquer
Abdeckung (f) .... covering .... recouvrement (m)
Abgangsballistik (f) .... intermediate ballistics .... balistique intermédiaire (f)
Abgangsfehlerwinkel (m) .... angle of jump .... angle de relèvement (m)
abgeschlossenes System (n) .... closed physical system .... système autonome, - fermé (m)
abkanten .... chamfer, bevel .... rogner
Ablösepunkt (Grenzschicht) (m) .... point of separation .... point de séparation (m)
Ablösung (f) .... separation .... séparation (f)
Abnahmeprüfung (f) .... acceptance test .... épreuve de recette (f)
Abnutzung (f) .... wear .... usure (f)
Abpraller (m) .... ricochet .... ricochet (m)
Abschusswinkel (m) .... angle of departure .... angle de projection (m)
absteigender Ast (m) .... descending branch .... branche descendante (f)
Abströmwinkel (m) .... off-flow angle .... angle de séparation (m)
Abweichung (f) .... deviation .... déviation, écart (m)
Abziehen (n) .... pull (the trigger) .... départ (m)
Abzug (m) .... trigger .... détente (f)
Abzugsbegrenzung (f) .... triggerstop .... immobilisation de détente
   Abzugsstop (m)
Abzugsbügel (m) .... trigger guard .... pontet de sous-garde (m)
Abzugseinrichtung (f) .... trigger device, - mechanism .... système de détente (m)
Abzugsgewicht (n) .... trigger pull weight .... poids de départ, -détente (m)
Abzugsgewichtkontrolle (f) .... verification of trigger pull weight .. peson (m)
Abzugshebel (m) .... trigger lever, trigger latch .... gâchette, queue de détente (f)
Abzugsprüfgerät (n) .... trigger verification device .... contrôleur de détente (m)
Abzugsspiel (n) .... backlash .... backlash
Abzugsstange (f) .... trigger bar .... tige de détente
Abzugssystem (n) .... trigger system .... système de détente (m)
achtkantig .... octogonal .... octogonal
Achtkantlauf (m) .... octogonal barrel .... canon octogonal (m)
Aggregatszustand (m) .... state of matter .... état de matière (m)
Alarmpistole (f) .... alarm pistol, signal pistol .... pistolet d'alarme (m)
   dynamische - ....   dynamic - ....   - dynamique
   geometrische - ....   geometric - ....   - géométrique
   kinematische - ....   kinematic - ....   - cinétique
Amboss (m) .... anvil .... enclume (f)

German	English	French
amorph	amorphous	amorphe
Anfangslage (f)	initial position	position initiale (f)
Anfangsgeschwindigkeit (f)	initial velocity	vitesse initiale (f)
Angriffspunkt der Druckkräfte (m)	centre of pressure	centre de pression (m)
anregen	induce	induire, provoquer
anströmen	flow against	affluer
Anstellwinkel (m)	angle of incidence	angle d'incidence (m)
Anzündlitze (f)	igniter cord	cordon d'allumage (m)
Arbeit (f)	work	travail (m)
aufgelegt	supported	avec appui, -support
Aufnehmer (m)	sensor	sonde (f), capteur (m)
aufpilzen	mushroom	s'épanouir (en forme de champignon)
aufschießend	firing from closed bolt	tirer à culasse fermée
Aufschlagzünder (m)	impact fuze	fusée percutante (f)
aufsteigender Ast (m)	ascending branch	branche ascendante (f)
Auftreffpunkt (m)	point of impact	point d'impact (m)
Auftreffwinkel (m)	angle of attack	angle d'arrivée (m)
Auftriebskraft (f)	lift force	portance (f)
Auftriebsbeiwert (m)	lift force coefficient	coefficient de portance (m)
Ausbreitungsgeschwindigkeit (f)	velocity of waves	vitesse de propagation (f)
Ausdehnung (f)	dimension	dimension (f), extension
ausgedehnter Körper (m)	large body	volumineux
ausgerüstet sein mit	be equipped with	équipé de
Auslenkung (f)	extension	déviation (f), écart
Ausschussloch (n)	exit hole	trou de sortie (m)
Außenballistik (f)	exterior ballistics	balistique extérieure (f)
austretende Masse (f)	mass flow	flux de masse (m)
Austrittsgeschwindigkeit (f)	exit velocity	vitesse de sortie (f)
Auswerfer (m)	ejector	éjecteur (m)
Auswurföffnung (f)	ejection port	ouverture d'éjection (f)
Ausziehdruck (m)	shot start pressure	pression d'extraction (f)
Auszieher (m)	extractor	extracteur (m)
Ausziehwiderstand (m)	initial resistance	résistance d'extraction (f)
Bahngeschwindigkeit (f)	tangential velocity	vitesse linéaire (f)
Balkenkorn (n)	post sight	guidon à lame rectangulaire (m)
Ballistik (f)	ballistics	balistique (f)
Becherpfropfen (m)	cup wad	bourre à cuvette, -à jupe (f)
Befestigungselement (n)	fastening element	élément d'attache (m)
Beiwert (m)	coefficient	coefficient (m)
Berdanzündhütchen (n)	berdan primer	amorce Berdan (f)
Beschleunigung (f)	acceleration	accélération (f)
Beschussprobe (f)	proof firing	épreuve (f)
Beschussstempel (m)	proof mark	poinçon d'épreuve (m)
Betrag (eines Vektors) (m)	absolute value	valeur absolue (f)
Bewegungsenergie (f)	kinetic energy	énergie cinétique (f)
Bewegungsgleichungen (f)	equations of motion	équations de mouvement (f)
Beziehung (f)	relation	relation (f)
Bezugsfläche (f)	reference plane	surface de référence (f)
Bezugsgröße (f)	reference value	valeur de référence (f)
Bezugspunkt (m)	reference point	point de référence (m)
Bezugssystem (n)	frame of reference	système de référence (m)

Biegespannung (f)	bending stress	contrainte de flexion (f)
Bildungsenergie (thermodyn.) (f)	heat of formation	chaleur de formation (f)
Blankwaffe (f)	edged weapon	arme blanche (f)
Blättchenpulver (n)	flake (laminated) powder	poudre à paillettes (f)
Blattfeder (f)	laminated spring	ressort à lame (m)
Blei (n)	lead	plomb (m)
Bleiazid (n)	lead azide	azoture de plomb (f)
bleibende Wundhöhle (f)	permanent cavity	cavité permanente (f)
Bleitrizinat (n)	lead styphnate	trinitrorésorcinate de plomb (m)
Blindgänger (m)	dud, blind shell	raté, non éclaté
Blindpatrone (f)	blank cartridge	cartouche à blanc (f)
Blockverschluss (m)	breech lock	culasse à bloc (f)
Bockbüchsflinte (f)	superposed rifle-shotgun	fusil à canons superposés (m)
Bogen (m)	bow	arc (m)
Bogenmaß (n), im - messen	measure in radians	mesurer en radians
Bogensport (m)	archery	tir à l'arc (m)
Bohrung (f)	hole, aperture	trou, perçage (m)
Bördelung (f)	roll crimp	sertissage à rondelle
Brandgeschoss (n)	incendiary bullet	balle incendiaire (f)
Brandsatz (m)	incendiary compound	matière incendiaire (f)
Breitenstreuung (f)	lateral spread	dispersion en largeur (f)
Brenngeschwindigkeit (f)	burning rate	vitesse de combustion (f)
Brennkammer (f)	burning chamber	chambre de combustion (f)
Brennschluss (m)	end of burning	fin de combustion (f)
Brennschlussgeschwindigkeit (f)	end-burning velocity	vitesse en fin de combustion (f)
brünieren	brown, blue	bronzer, brunir
Brünierung (f)	browning, blueing	bronzage, brunissage (m)
Büchse (f)	rifle	arme à canon rayé (f)
Büchsenmacher (m)	gunmaker, gunsmith	armurier (m)
Büchsflinte (f)	rifle-shotgun	carabine (m)
Choke (m)	choke	choke (m)
Chokebohrung (f)	choke-bore	alésage du choke (m)
Chrom (m)	chrome	chrome (m)
Damastlauf (m)	damascus barrel	canon damas (m)
Dämmerungsleistung (f)	twilight performance	puissance crépusculaire (f)
dämpfen	damp	amortir
Dämpfung (f)	damping	amortissement (m)
Dauerfeuer (n)	full automatic fire	feu en rafales (m)
Daumenauflage (f)	thumbrest	repose-pouce (m)
Daumenloch (n)	thumbhole	trou de pouce (m)
Deflagration (f)	deflagration	déflagration (f)
Deformationsgeschoss (n)	deformation projectile	projectile expansif (m)
deformieren	deform	déformer
Dehnung (f)	strain	allongement (m), dilatation (f)
Detonation (f)	detonation	détonation (f)
Detonationsgeschwindigkeit (f)	detonation rate	vitesse de détonation (f)
Detonationsübertragung (f)	sympathetic detonation	détonation par influence (f)
dicht	tight, dense, sealed	étanche, dense
Dichte (f)	density	masse volumique (f)

dichtebeständig .... incompressible .... incompressible
Differenzial (n) .... differential .... différentiel (m)
Differenzialquotient (m) .... derivative .... dérivée (f)
                                       differential quotient .... quotient différentiel (m)
Differenzenquotient (m) .... difference quotient .... différence (f)
Dimensionanalyse (f) .... dimensional analysis .... analyse dimensionelle (f)
Diopter (n) .... aperture sight .... dioptre (m)
Doppelabzug (m) .... double trigger .... double détente (f)
Doppelbüchse (f) .... double rifle .... fusil à deux canon (m)
Doppelflinte (f) .... side-by-side shotgun .... fusil à canon juxtaposés (m)
Drall (Geschoss) (m) .... spin .... rotation (f)
Drall (Waffe) (m) .... twist; rifling .... rayure (f)
Drallänge (f) .... twist, twist length .... pas des rayures (m)
Drallrichtung (f) .... twist direction .... direction du pas (f)
drallstabil .... spin-stabilized .... gyrostabilisé
Drallwinkel (m) .... angle of twist .... angle des rayures (m)
Drehachse (f) .... rotational axis, pivot .... axe de rotation, pivot (m)
Drehbewegung (f) .... circular motion .... mouvement circulaire (m)
Drehenergie (f) .... energy of rotation .... énergie de rotation (f)
Drehfrequenz (f) .... frequency of rotation .... fréquence de rotation (f)
Drehimpuls (m) .... angular momentum, ... moment cinétique (m)
                                       moment of momentum
Drehmoment (n) .... torque; moment of a force .... moment (m)
Drehpunkt (m) .... centre of rotation .... centre de rotation (m)
Drehpunkt (Hebel) (m) .... fulcrum .... articulation, pivot (m)
Drehverschluss (m) .... rotating bolt locking system .... culasse tournante, - rotative
Drehwinkel (m) .... angular displacement .... déplacement (m) angulaire
Drehzahlabnahme (f) .... spin damping .... diminution du nombre de tours
Dreibein (n) .... tripod .... trépied (m)
Dreischussautomatik (f) .... three-shot mechanism .... mécanisme à trois
                                                                                                                         coups (m)
Drilling (m) .... drilling, three-barrelled gun .... drilling (m),
                                                                                                                         fusil à trois canons (m)
Druck (m) .... pressure .... pression (f)
Druckaufnehmer (m) .... pressure gauge .... jauge de pression (f)
Druckenergie (f) .... pressure energy .... énergie de compression (f)
Druckgeber (m) .... pressure transducer .... transducteur de pression (m)
Druckgefälle (n) .... pressure gradient .... gradient de pression (m)
Druckkolben (m) .... piston .... piston (m)
Druckkraft (f) .... compressive force .... force de compression (f)
Druckpunkt (m) .... trigger point .... cran d'arrêt (m)
Druckpunktabzug (m) .... double pull trigger .... détente à bossette (f)
Druckschwankung (f) .... transient pressure .... fluctuation de pression (f)
                                                                                               variation de - (f)
Druckspannung (f) .... compressive strain .... contrainte de compression (f)
Druckwiderstand (m) .... pressure drag .... résistance à la compression (f)
Durchschlag (m) .... penetration .... pénétration (f)
Durchschlagskraft (f) .... penetration power .... puissance de perforation (f)
Düse (f) .... nozzle .... tuyère (f)

Ebene Platte (f) .... plane plate .... plan (m), plaque plane (f)
eichen .... calibrate .... étalonner
Eichung (f) .... calibration .... étalonnage (m)

Eigenbewegung (f)......................... proper motion .... mouvement propre (m)
Eigenfrequenz (f) .... natural frequency..................... fréquence propre (f)
einbrennen................................... burn in, brand in .... cuire (au four)
Eindringtiefe (f)............................... penetration depth .... profondeur de pénétration (f)
Eindringvermögen (n).................... pénétration capacity .... pouvoir de pénétration (m)
Einpresswiderstand (m)................. engraving resistance .... résistance à l'enfoncement (f)
Einschießen (n) .... test-firing, ranging, zeroing............ tir de réglage (m)
Einschlag (m) .... impact .... impact (m)
Einstellschraube (f) .... set screw, adjustment screw .... molette de réglage, vis de - (f)
Einstellung (f) .... adjustment, setting......................... réglage (m)
eintretende Masse (f).................... incident mass................................. masse incidente (f)
Einzelfeuer (n)................................ single-shot fire................................ feu coup par coup (m)
Einzellader (m)............................... single-shot weapon........................ arme à un coup (f)
Elastizitätsmodul (m) .... Young's modulus............................ module d'élasticité (m)
                                                modulus of elasticity
Empfindlichkeitsprüfung (f) .... sensitivity test................................. épreuve de sensibilité (f)
Endballistik (f) .... terminal ballistics .... balistique terminale (f)
Endgeschwindigkeit (f).................. final velocity (speed)...................... vitesse terminale (f)
Endlage (f)..................................... final position.................................. position finale (f)
Energie (f)...................................... energy .... énergie (f)
Energieabgabe (f).......................... energy release .... dissipation d'énergie (f)
Energiebilanz (f) .... energy balance .... bilan énergétique (m)
Energieübernahme (f)................... energy assumption......................... gain d'énergie (m)
                                                                                                prise d'énergie (f)
Energieverlust (m)......................... energy drop.................................... perte d'énergie (f)
eng... narrow... étroit
Entfernung (f)................................. distance, range............................... distance (f)
entgegengesetzt gleich .... opposite .... opposé, de signe contraire
entsichern .... release the safety catch.................. armer
Entspannungsverhältnis (n)............ expansion ratio .... rapport de détente (m)
Erdbeschleunigung (f).................... acceleration due to gravity .... accélération terrestre (f)
                                                - of free fall
Erhaltungssatz (m) .... law of conservation .... loi de conservation (f)
Erhebungswinkel (m)...................... elevation angle .... élévation (f)
Erhitzung (f).................................... heating .... échauffement (m)
erregende Kraft (Schwingung) (f).. driving force, external - .... force d'exitation (f)
Erscheinung (f)............................... phenomenon .... phénomène (m)
Erstarrungspunkt (m) .... freezing point .... point de solidification (m)
erzwungene Schwingung (f) .... forced oscillation............................ oscillation forcée (f)
Explosionstemperatur (f)................ explosion temperature .... température d'explosion (f)
Explosionswärme (f)....................... explosion heat................................ quantité de chaleur (f)

Fallblockverschluss (m) .... falling block bolt action .... culasse à bloc basculant (f)
Fallgeschwindigkeit (f)................... velocity of descent......................... vitesse de chute (f)
Fallhöhe (f)..................................... drop.. hauteur de chute (f)
Fallwinkel (m)................................. angle of descent............................. angle de chute (m)
Fangschuss (m) .... finishing shot................................. coup de grâce (m)
Felddurchmesser (m)..................... land diameter................................. diamètre du champ des
                                                                                                rayures (m)
Felder (np)..................................... lands .... champs des rayures (m)
fest (Aggregatzustand).................. solid .... solide
fester Körper (m)........................... solid body .... corps solide (m)
Feststoff-Rakete (f) .... solid propellant rocket................... roquette à propulsion solide (f)

Feuer aus Mündung (n)	primary (muzzle) flash	feu de bouche (m)
Feuerscheindämpfer (m)	flash hider	cache-flamme (m)
Flachkopfgeschoss (n)	flat-nose bullet	balle nez plat (f)
Flinte (f)	shotgun	fusil à canon lisse (m)
Flugbahn (f)	trajectory	trajectoire (f)
Flugzeit (f)	time of flight	temps de vol (m)
Fluiddynamik (f)	fluid dynamics	dynamique des fluides (f)
Fluss (m)	fluid motion	flux (m)
flüssig	liquid	liquide
Flüssigtreibstoff-Rakete (f)	liquid propellant rocket	roquette à propulsion liquide
folgsam	tractable	(l'axe du projectile suivant la tangente de la trajectoire)
Folgsamkeit (f)	tractability	
formstabil	shape stable	indéformable
Freiheitsgrad (m)	degree of freedom	degré de liberté (m)
Frequenzgang (m)	frequency response	bande passante (f)
Fühler (m)	sensing device	jauge, capteur (m)
Gasdruck (m)	gas pressure	pression de gaz (f)
Gasdrucklader (m)	gas operated weapon	arme fonctionnant par emprunt de gaz (f)
Gasdruckmesser (m)	pressure gauge	jauge de pression (f)
Gasentnahmedüse (f)	gas port	évent (m)
gasförmig	gaseous	gazeux
Gaskonstante R (f)	gas constant R	constante des gaz parfaits (f)
gedämpft	damped	amorti
schwach -	underdamped	sous -
stark -	overdamped	sur -
geeicht	calibrated	calibré, étalonné
Gehörschutzmittel (n)	ear defenders	casque anti-bruit (m)
gekrümmte Bahn (f)	curved trajectory	trajectoire courbe (f)
gerader Einschusskanal (m)	narrow channel (NC)	- incurvée
Geschoss (n)	bullet, projectile	projectile (m)
Geschossgewicht (n)	bullet weight	poids du projectile (m)
Geschossmantel (m)	bullet jacket	chemisage du projectile (m)
Geschwindigkeitsgefälle (n)	shear rate	gradient de vitesse (m)
Geschwindigkeit (f)	velocity	vitesse (f)
Geschwindigkeitsbetrag (m)	speed	vitesse, célérité (f)
gesetzmäßig	natural	naturel, légitime
gestreckte Flugbahn (f)	flat trajectory	trajectoire tendue (f)
Gewicht (n)	weight	poids (m)
gezogen (Lauf)	rifled	rayé
Gitterbindungskraft (f)	lattice linkage force	force de cohésion moléculaire (f)
Gitterstruktur (f)	crystalline structure	structure cristalline (f)
gleichförmig	uniform	uniforme
Gleichgewicht (n)	equilibrium	équilibre (m)
Grad Celsius (m)	degree celsius	degré Celsius (m)
Grenzfrequenz (f)	boundary frequency	fréquence limite (f)
Grenzgeschwindigkeit (f)	limiting velocity	vitesse limite (f)
Grenzschicht (f)	boundary layer	couche limite (f)
Größenordnung (f)	order of magnitude	ordre de grandeur (m)
Gürtelhülse (f)	belted case	douille à ceinture (f)

Hahn (m)	hammer	chien (m)
halbautomatisch	semi-automatic	semi-automatique
Halbchokelauf (m)	modified choked barrel	canon à demi choke (m)
Haltepunkt (m)	point of aim	point de mire (m)
Hammer (m)	hammer	marteau (m)
hämmern	hammer	marteler
Hämmerung (f)	hammering	martelage (m)
Handarbeit (f)	handicraft, handiwork	travail à la main, travail manuel
Handballenauflage (f)	palm support	appui paume, repose-paume
Handgranate (f)	hand-grenade	grenade à main (f)
Handstütze (f)	hand support	champignon, pommeau (m)
harmonische Schwingung (f)	harmonic motion	oscillation harmonique (f)
hart (Abzug)	dry	sec
Hartblei (n)	antimony lead	plomb durci (m)
Härte (f)	hardness	dureté (f)
Hartkerngeschoss (n)	hard-core bullet	balle à noyau dure, - perforante (f)
Hartschrot (m)	chilled shot, hard shot	plomb durci (m)
Haube (f)	hood	coiffe, calotte
Hauptfehler (m)	main defect	défaut majeur (m)
Hebel (m)	lever	levier (m)
Hebelarm (m)	lever arm	bras de levier (m)
Heckkonus (m)	boattail	bi-ogivale
Helm (m)	helmet, crest	casque (m)
Hexogen (n)	hexogen	hexogène (m)
Hinterlader (m)	breech loader	arme à chargement par la culasse (f)
Hochleistungspatrone (f)	high-power cartridge	cartouche à haut rendement (f)
Hochwildjagd (f)	big-game hunting	grande chasse (f)
Höhenabweichung (f)	vertical deviation	écart vertical (m)
Höhenstreuung (f)	vertical dispersion	dispersion verticale (f)
hohl	hollow	creux
Hohlladung (f)	hollow charge	charge creuse (f)
Hohlschuss (Schrot) (m)	wheel shot	coup creux (m)
Hohlspitzgeschoss (n)	hollow-point bullet	balle à point creuse (f)
Hülse (f)	case	douille (f)
Hülse mit Schulter (f)	bottlenecked case	douille en forme de bouteille
Hülsenabreißer (m)	cut case	douille rompue (f)
Hülsenboden (m)	case head	culot (m)
Hülsenhals (m)	case neck	collet de la douille (m)
Hülsenrand (m)	case rim	bordure de la douille (f)
Hülsenreißer (m)	split case, case-rupture	rupture de la douille (f)
Impuls (m)	linear momentum	quantité de mouvement (f)
Impulssatz (m)	law of conservation of momentum	loi de la conservation de la quantité de mouvement
Innenballistik (f)	interior ballistics	balistique intérieure (f)
innere Energie (f)	internal energy	énergie interne (f)
innere Reibung (f)	internal friction	frottement interne (m)
Irisblende (f)	iris diaphragm	diaphragme iris (m)
Jagd (f)	hunting	chasse (f)

Jagdflinte (f) .... shotgun .... fusil de chasse (m)
jagdliches Schießen (n) .... hunting shooting .... tir de chasse (m)
Jagdmesser (n) .... hunting knife .... couteau de chasse (m)
Jagdwaffe (f) .... hunting weapoin .... arme de chasse (f)
Justierung (f) .... adjustment .... réglage (m)

Kadenz (f) .... rate of fire .... cadence (f)
Kaliber (n) .... calibre, bore, size .... calibre (m)
Kaliber (Schrot) (n) .... gauge .... calibre (m)
Kaliberquerschnittsfläche (f) .... cross-sectional area .... section (f)
kalibrieren .... calibrate, size .... calibrer, étalonner
Kalibrierpresse (f) .... calibrating press, sizing press .... presse à calibrer (f)
Kaliumchlorat (n) .... potassium chlorate .... chlorate de potassium (m)
Kaliumnitrat (n) .... potassium nitrate .... nitrate de potassium (m)
Kaliumsulfid (n) .... potassium sulphide .... sulphide de potassium (m)
Kammerverschluss (m) .... cylinder lock
Karabiner (m) .... carbine .... carabine (f), mousqueton (m)
Kavität (f) .... cavity .... cavité (f)
Kegelmantel (m) .... surface of a cone .... surface du cône (f)
Kennwert (m) .... parameter .... paramètre (m)
Kerbe (f) .... notch .... encoche (f)
Kimme (f) .... rear sight, notch .... hausse (f), cran de mire (m)
Kimmenblatt (n) .... rear sight leaf, rear sight slide .... feuillet (m)
kinetische Energie (f) .... kinetic energy .... énergie cinétique (f)
kippen .... to pitch, to overturn .... basculer
Kipphebel (m) .... armed lever .... culbuteur (m)
Kipplauf (m) .... break-action gun .... canon basculant (m)
Kipplaufbüchse (f) .... tip-up rifle .... carabine expresse (f)
Kipplaufverschluss (m) .... tip-up lock .... culasse tombante, - basculante
Kippmoment (n) .... overturning moment .... moment de
                                                                                                                    basculement (m)
Kleinkalibergewehr (n) .... small bore rifle .... carabine de petit calibre (f)
Knall (m) .... report .... détonation (f)
Knallpatrone (f) .... petard .... pétard (m)
Knallquecksilber (n) .... mercury fulminate .... fulminate de mercure (m)
Kniegelenkverschluss (m) .... toggle action bolt .... culasse à genouillère (f)
kniend .... kneeling .... à genoux
Kohäsivkraft (f) .... cohesive force .... force de cohésion (f)
Kohlendioxid (n) .... carbon dioxide .... dioxyde de carbone (f)
Kolben (Druck) (m) .... piston .... piston (m)
Kolben (Gewehr) (m) .... butt (of weapon) .... crosse (f)
Kompressibilität (f) .... compressibility .... compressibilité (f)
Kompressionsmodul (m) .... bulk modulus .... module de compressibilité (m)
konisch .... tapered, conical .... conique
Kontinuitätsgleichung (f) .... continuity equation .... équation de continuité (f)
Koordinatensystem (n) .... coordinate system .... système de coordonnées (m)
kopfvoran .... head-first .... la tête la première
Korn (n) .... foresight .... guidon (m)
Kornsattel (m) .... foresight ramp .... porte-guidon (m)
Kornschutz (m) .... foresight protector .... protège-guidon (m)
Kornträger (m) .... foresight base .... porte-guidon (m)
Kräftegleichgewicht (n) .... equilibrium of forces .... équilibre des forces (m)
Kräfteverhältnis (n) .... ratio of forces .... rapport des forces (m)

German	English	French
Kraftstoß (m)	impulse	impulsion (f)
Kreisbewegung (f)	circular motion	mouvement circulaire (m)
Kreisbogenstück (n)	arc of a circle	arc de cercle (m)
Kreisel (m)	gyroscope	gyroscope (m)
Kreisel (Spielzeug -) (m)	top	toupie (f)
Kreiseleffekt (m)	gyro effect	effet gyroscopique (m)
Kreisfrequenz (f)	angular frequency	fréquence angulaire (f)
kristallin (fest)	crystalline	cristallin
kritischer Fehler (m)	critical defect	défaut critique (m)
Krümmung (f)	curvature	courbure (f)
Kugelfang (m)	butt (behind target)	pare-balles (m), butte (f)
Kugelpulver (n)	ball powder	poudre sphérique (f)
Kupfer (n)	copper	cuivre (m)
Kurzbahnmunition (f)	short-range ammunition	munition à courte portée (f)
Kurzwaffe (f)	handgun	arme de poing (f)
Ladedichte (f)	loading density	densité de chargement (f)
Ladelöffel (m)	carrier, feeder	auget (m)
Ladung (f)	charge	charge (f)
Lager (n)	pivot	support (m)
Lagerreibung (f)	pivot friction	friction de support (f)
Länge (f)	length	longueur (f)
Längenabweichung (f)	length deviation, linear deviation	écart en portée (m)
Längenstreuung (f)	variation in length	dispersion en portée (f)
Langwaffe (f)	shoulder weapon	arme d'épaule (f)
Lauf (m)	barrel	canon (m)
Lauf mit Chokebohrung (m)	choked barrel	canon avec choke (m)
Laufabnutzung (f)	barrel wear	usure du canon (f)
Laufachse (f)	barrel axis	âme (f)
Laufbohrung (f)	bore	alésage du canon (m)
Laufdurchgangszeit (f)	barrel time	temps de parcours de l'âme (m)
laufender Keiler (m)	running boar	sanglier courant (m)
Laufgewicht (n)	counterweight	contrepoids (m)
Lauflänge (f)	barrel length	longueur du canon (f)
Laufschiene (f)	barrel rib	bande (de visée) (f)
Laufsprengung (f)	barrel burst, premature - -	éclat du canon (m)
Laufstecker (m)	stuck bullet	balle restée (dans le canon) (f)
Laufverbleiung (f)	lead-fouled barrel	plombage du canon (m)
Lebhaftigkeit (f)	vivacity	vivacité (f)
Lefaucheuxzündung (f)	pin fire ignition	percussion à broche (f)
Leistung (physikalisch) (f)	power	puissance (f)
Leuchtgeschoss (n)	flare	balle éclairante (f)
Leuchtsatz (m)	tracer	composition lumineuse (f)
Leuchtspurgeschoss (n)	tracer bullet	balle traçante (f)
Lichtstärke (f)	brightness	luminosité (f)
liegend	prone	couché
Loch (n)	hole, peep	trou (m)
Lochblende (f)	aperture, peep	oeilleton (m)
Luftgewehr (n)	air rifle	carabine à air comprimé (f)
Luftgewehrgeschoss (n)	airgun pellet	plomb Diabolo (m)
Luftgewehrschießen (n)	air-rifle shooting	tir à air comprimé (m)

German	English	French
Luftpistole (f)	air pistol	pistolet à air (comprimé) (m)
Luftsäule (f)	air column	colonne d'air (f)
Luftwiderstandsbeiwert (m)	air drag coefficient	coefficient de résistance à l'air (m)
Luftwiderstand (m)	air drag, - resistance	résistance à l'air (f)
Lunte (f)	match	mèche (f)
Luntenschloss (n)	match-lock	platine à mèche (f)
Magazin (n)	magazine	chargeur, magasin (m)
Magazinfeder (f)	magazine spring	ressort du chargeur (m)
Magazingehäuse (n)	magazine housing	puits du chargeur (m)
Manipulierpatrone (f)	dummy cartridge, drill round	cartouche de manipulation (f)
Mantel (m)	jacket	manteau (m), chemise (f)
Mantelgeschoss (n)	jacketed bullet	balle chémisée (f)
Maschinengewehr (n)	machine gun	mitrailleuse (f)
Maschinenpistole (f)	submachine gun	pistolet mitrailleur (m)
Maßeinheit (f)	unit	unité de mesure (f)
Masseteilchen (n)	mass element	éléments de masse (m)
Masseverschluss (m)	blowback system	culasse à inertie (f)
Maßsystem (n)	system of units	système d'unités (m)
Matchabzug (m)	match-trigger	détente de match (f)
mathematisches Pendel (n)	simple pendulum	pendule mathématique (m)
Maximaldruck (m)	maximum pressure	pression maximale (f)
Messfolge (f)	conversion rate	
Messgröße (f)	measured quantity	grandeur mesurée (f)
Messing (n)	brass	laiton (m)
Messtechnik (f)	measurement techniques	technique de mesure (f)
Mikrometerdiopter (m)	micrometer sight	dioptre micrométrique (m)
Mitnehmer (m)	tappet, cam, dog	accrochage (m)
Mittelwert (m)	mean	moyenne (f)
mittlere Abweichung (f)	mean deviation	écart moyen (m)
mittlerer Streukreisradius (m)	mean radius (of dispersion)	rayon moyen de groupement (m)
Momentenbeiwert (m)	overturning moment coefficient	coefficient du moment de basculement (m)
Mündung (f)	muzzle	bouche (f)
Mündungsbremse (f)	muzzle brake	frein de bouche (m)
Mündungsenergie (f)	muzzle energy	énergie de bouche (f)
Mündungsfeuer (n)	secondary (muzzle) flash	feu de bouche (m)
Mündungsgeschwindigkeit (f)	muzzle velocity	vitesse initiale (f)
Mündungsimpuls (m)	muzzle momentum	quantité de mouvement initiale (f)
Mündungsknall (m)	muzzle blast	bruit de bouche (m)
Mündungsknalldämpfer (m)	silencer	silencieux (m)
Munition (f)	ammunition	munition (f)
Munitionsfabrik (f)	ammunition plant, - factory	cartoucherie (f)
Munitionskennzeichnung (f)	ammunition marking, - code	marquage des munitions (m)
Muster (n)	sample	échantillon (m)
Mutter (f)	nut	écrou (m)
Näherung (f)	approximation	approximation (f)
Nase (f)	stud, lug	ergot (m)

Neigung (f)	slope, fall, gradient	pente (f)
Newton'sche Axiome (np)	Newton's laws of motion	axiomes de Newton (m)
nicht rostend	stainless	antirouille, inoxydable
Nickel (m)	nickel	nickel (m)
Nickelschrot (m)	nickel(plated) shot	plomb nickelé (m)
Niet (m)	rivet	rivet (m)
Nitrocellulosepulver (n)	single base powder (NC powder)	poudre à simple base (f)
Nitroglycerinpulver (n)	double base powder (NG powder)	poudre à double base (f)
Nocken (m)	cam, lug	ergot, came (f)
Normalspannung (f)	normal stress	force de cisaillement (f)
Notbehelf (m)	stopgap	solution de fortune (f)
Nut (f)	groove, slot	encoche (f)
Nutation (f)	nutation	nutation (f)
Oberflächenbehandlung (f)	smoothing	lissage (m)
offene Visierung (f)	open sight	visée ouverte (f)
Oktogen (n)	octogen	octogène (m)
Papierumwicklung (f)	paper patch	enveloppe de papier (f)
parabelförmig	parabolic	parabolique
Patrone (f)	cartridge	cartouche (f)
Patronengürtel (m)	cartridge belt	ceinture cartouchière (f)
Patronenhülse (f)	cartridge case	douille (f)
Patronenlager (n)	chamber	chambre à cartouches (f)
Patronenlagersprengung (f)	chamber burst	bascule brisée (f)
Patronenzufuhr (f)	cartridge feed	alimentation (de cartouche) (f)
Pendelung (f)	yaw	mouvement pendulaire (m)
Perkussionsgewehr (n)	percussion rifle	fusil à percussion (m)
Perkussionszündung (f)	percussion ignition	percussion à capsule (f)
Pfropfen (m)	wad	bourre (f)
Pfropfen mit Dämpfung (m)	cushion wad	bourre à coussin (f)
physikalisches Pendel (n)	physical pendulum	pendule physique (m)
Pikrinsäure (f)	picric acid	acide picrique (m)
Pistole (f)	pistol	pistolet (m)
Pistolengriff (m)	pistol grip	poignée pistolet, crosse - (f)
Pistolenpatrone (f)	pistol cartridge	cartouche pour pistolets (f)
Pistolenschütze (m)	pistol shooter	pistolier (m)
Plastikbecherpfropfen (m)	plastic cup wad	bourre à jupe en plastique (f)
Plastikhülse (f)	plastic case	douille (en) plastique (f)
Plastikpfropfen (m)	plastic wad	bourre en plastique (f)
Plastikschrotpatrone (f)	plastic shotshell	cartouche de chasse en plastique (f)
platzen	burst, crack	éclater
Platzpatrone (f)	blank cartridge	cartouche à blanc (f)
Postenschrot (n)	buckshot	chevrotine (f)
potenzielle Energie (f)	potential energy	énergie potentielle (f)
Präzession (f)	precession	précession (f)
Präzessionsbewegung (f)	yaw	mouvement de précession (m)
Präzision (f)	accuracy, precision	précision (f)
Prinzip der Unabhängigkeit von Bewegungen (n)	superposition principle	principe de superposition (m)
progressiver Drall (m)	progressive twist, - groove	pas progressif (m)
Pulsdauer (f)	pulse duration	durée de pulsation (f)

Pulver (n) .... powder .... poudre (f)
　　progressives - .... progressive burning powder .... poudre lente (f)
　　degressives (offensives) - .... degressive burning powder .... poudre vive (f)
Pulverwaage (f) .... powder scale .... balance à poudre (f)

Quer .... crossways .... transversal, latéral
Querschläger (m) .... ricochet, keyhole .... ricochet (m)
Querschnittsbelastung (f) .... sectional density .... masse par unité de section (f)
Querschnittsfläche (f) .... cross-sectional area .... section (f)
Querschnittsverhältnis (Düse) (n) .. cross-section ratio .... rapport d'expansion (m)
Querstehen (n) .... pitching .... se mettre en travers
Querwind (m) .... crosswind .... vent latéral (m)

Radschloss (n) .... wheel-lock .... platine à rouet (f)
Radschlosszündung (f) .... wheel-lock ignition .... percussion à rouet (f)
Rahmen (m) .... frame, receiver .... carcasse (f)
Rakete (f) .... rocket, missile .... fusée (f), roquette (f)
Rand (m) .... rim .... bourrelet (m)
Randbedingung (f) .... boundary condition .... condition aux limites (f)
Randeffekt (m) .... boundary effect .... effet aux limites (m)
Randfeuerpatrone (f) .... rimfire cartridge .... cartouche à percussion
　　- annullaire (f)
Randfeuerzündung (f) .... rimfire ignition .... percussion annulaire (f)
Randhülse (f) .... rimmed case .... douille en bourrelet (f)
Randpatrone (f) .... rimmed cartridge .... cartouche à bourrelet (f)
Randrierung (f) .... knurling .... cannelure (f)
Randschrot (m) .... flier ....
Raubzeug (jagbare Nager) (n) .... vermin .... rongeur (m)
rauchloses Pulver (n) .... smokeless powder .... poudre pyroxylée (f)
räumliche Ausdehnung (f) .... spatial expansion .... volume (m)
räumlicher Winkel (m) .... solid angle .... angle solide (m)
Rechteckkorn (n) .... square foresight .... guidon rectangulaire (m)
Rechtssystem (Koordinaten) (n) .... right-handed system .... système orienté à la
　　main droite (m)
Reibempfindlichkeit (Sprengst.) (f).. friction sensitivity .... sensibilité au frottement (f)
reibungsfrei .... frictionless .... sans frottement
　　bezüglich innerer Reibung .... inviscid .... non-visqueux
Reibungskraft (f) .... frictional force .... force de frottement (f)
Reibungszahl (f) .... friction coefficient .... coefficient de frottement (m)
Reinigungsgerät (n) .... cleaning kit .... nécessaire de nettoyage (m)
Reißfestigkeit (f) .... resistance to tearing .... résistance au déchirement (f)
Repetierbüchse (f) .... repeating rifle .... carabine à répétition (f)
Repetierflinte (f) .... repeating shotgun .... fusil à répétition (m)
Resonanz (f) .... resonance .... résonance (f)
Restenergie (f) .... residual energy .... énergie résiduelle (f)
Restgeschwindigkeit (f) .... remaining velocity .... vitesse restante (f)
Revolver (m) .... revolver .... revolver (m)
Revolverpatrone (f) .... revolver cartridge .... cartouche pour revolvers (f)
Riegel (m) .... latch, bolt .... verrou (m), verrouillage (m)
Riemen (m) .... sling, strap, lanyard .... bretelle (f)
Ringkorn (n) .... ring foresight, aperture - .... guidon annulaire (m)
Rippe (f) .... rib .... ailette, nervure (f)
Röhrchenpulver (n) .... tubular powder .... poudre tubulaire (f)

Röhrenmagazin (n)......................... tubular magazin, tube magazine..... chargeur tubulaire (m)
Rost (m).. rust... rouille (f)
rostfrei .... stainless .... inoxydable
rostig... rusty.. rouillé
rostschützend................................. anticorrosive .... antirouille
Rostschutzmittel (n) .... rust preventing agent .... solution antirouille (f)
Rotationsbewegung (f).................. rotation .... mouvement de rotation (m)
Rotationsenergie (f)....................... rotational kinetic energy................. énergie cinétique de rotation
rotationssymetrisch .... rotational symmetry .... symmétrique par rapport
 à une axe
Rückholfeder (f)............................. return spring .... ressort récupérateur (m)
Rückkopplung (f) .... feedback .... rétroaction (f)
Rückstoß (m).................................. recoil, blowback .... recul (m)
Rückstoßenergie (f)....................... recoil energy .... énergie du recul (f)
Rückstoßlader (m).......................... recoil operated gun, blowback - -... automatique par recul
Rückstoßminderer (m) .... recoil reducer, - pad....................... plaque de couche antirecul (f)
Ruhelage (f).................................... rest position .... position de repos (f)
ruhend... resting .... au repos
Rundkopfgeschoss (n).................... round-nose bullet........................... balle à tête ronde (f)

Sandstrahlen................................... sand-blast.. sabler
Saugeffekt (m)................................ suction effect .... effet de succion (m)
Schaft (m)....................................... stock .... crosse (f)
Schaft mit Daumenloch (m)........... thumbhole stock .... crosse à trou de pouce (f)
Schaftbacke (f) .... cheekpiece .... appui-joue (m)
Schaftkappe (f)............................... buttplate.. plaque de couche (f)
Schalldämpfer (m).......................... silencer .... silencieux de son (m)
Schallgeschwindigkeit (f) .... velocity of sound .... vitesse du son (f)
Schaltzahn (Revolver) (m)............. gear(notch), sprocket..................... dent de rochet (f)
Schattenaufnahme (f)..................... shadowgraph .... ombrogramme (m)
Scheibe (f)...................................... target, disc .... bague (f), cible (f), disque (m)
Scheibenpulver (n) .... disc powder .... poudre à disque (f)
Scheibenzuganlage (f).................... target transport device .... rameneur (de cible) (m)
Scheitel (m).................................... vertex.. sommet (m)
Scheiteldistanz (f) .... vertex distance............................... distance du sommet (f)
Scheitelhöhe (f).............................. vertex height.................................. hauteur du sommet (f)
Scherspannung (f).......................... shear stress .... tension de cisaillement (f)
Scherviskosität (f) .... dynamic viscosity.......................... viscosité (f)
Schieber (m)................................... slide .... poussoir (m), tirette (f)
Schießen (n) .... shooting, firing .... tir (m)
Schießpulver (n)............................. gunpowder..................................... poudre (f)
Schießstand (m).............................. shooting range .... stand de tir (m)
Schlachtpatrone (f)......................... cattle killer cartridge, slaughter - ... cartouche (f) pour appareils
 d'abattage
Schlagbolzen (m) .... firing pin.. marteau, percuteur (m)
Schlagbolzenfeder (f)..................... firing pin spring............................. ressort de percuteur (m)
Schlagempfindlichkeit (f) .... impact sensitivity .... sensibilité à l'impact (f)
Schlagfeder (f)................................ main spring.................................... ressort de percussion (m)
Schlagstück (n)............................... hammer, striker .... marteau (m)
schlank .... slender .... svelte
Schlankheit (f)................................ slenderness .... sveltesse (f)
Schlankheitsgrad (m) .... slenderness ratio .... degré de sveltesse (m)
Schloss (n)...................................... bolt, lock, breech, action .... batterie (f)

Schlossträger (m)	lock plate	platine (f)
Schmauchspuren (fp)	gunshot residue	residues de tir (mp)
Schmelzpunkt (m)	melting point	point de fusion (m)
Schnellfeuergewehr (n)	assault rifle	fusil mitrailleur (m)
Schnellfeuerpistole (f)	rapid-fire pistol	pistolet rafaleur (m)
Schonzeitwaffe (f)	closed season arm, varmint rifle	arme pour la chasse fermée (f)
Schrapnell (n)	shrapnel	shrapnel (m)
Schraube (f)	screw	vis (f)
Schraubenmutter (f)	nut	écrou (m)
Schrot (m)	shot	grenaille (f)
Schrotflinte (f)	shotgun	fusil de chasse (m)
Schrotgarbe (f)	shot sheaf	gerbe (f)
Schrotkugel (f)	shot pellet	plomb (de chasse) (m)
Schrotpatrone (f)	shotgun cartridge, shotshell	cartouche de chasse (f)
Schub (Rakete) (m)	thrust	poussée (f)
Schubkraft (f)	shear force	force de poussée (f)
Schubspannung (f)	shear stress	tension de cisaillement (f)
Schulterhülse (f)	bottleneck case	douille à épaulement (f)
schulterstabilisiert	shoulder stabilized	stabilisé par l'épaule
Schulterwaffe (f)	shoulder weapon	arme d'épaule (f)
Schuss (m)	round, shot	coup, tir (m)
Schussdistanz (f)	range	portée (f)
Schusskanal (m)	wound channel	trajet du projectile dans les chairs (m)
Schusslochprüfer (m)	score gauge, scorer	axe jaugeur (m), jauge (f)
Schusspflaster (n)	target patch	palette (de tir) (f)
Schussrichtung (f)	line of departure	direction de tir (f)
Schusstafel (f)	firing table, range table	table de tir (f)
Schussweite (f)	range	portée (f)
Schusswinkel (m)	angle of shot	angle de tir (m)
Schwalbenschwanz (m)	dove tail	queue d'arronde (f)
Schwarzpulver (n)	black powder	poudre noir (f)
Schwarzpulverzündschnur (f)	safety fuze	mèche de sureté (f)
Schwenkachse (f)	pivot	pivot (m)
Schwerpunkt (m)	centre of mass, centre of gravity	centre de masse (m), - de gravité (m)
Schwingungsbauch (m)	maximum amplitude	amplitude maximale (f)
Seelenachse (f)	axis of the bore, - of gun	âme (f)
Seitenabweichung (f)	lateral deviation	écart lateral (m)
Seitenschloss (n)	side lock	contre-platine (f)
Seitenstreuung (f)	lateral dispersion, - spread	dispersion lateral (f)
seitlich	lateral	latéral
seitlicher Wind (m)	crosswind	vent latérale (m)
seitwärts	sideways	penché, incliné
Selbstladeflinte (f)	automatic shotgun, autoloading -	fusil de chasse automatique (m)
Selbstladegewehr (n)	automatic rifle, self-loading rifle	fusil automatique (m)
Selbstladepistole (f)	semi-automatic pistol	pistolet automatique (m)
Selbstzündung (f)	cook-off	auto-allumage (m)
Seriefeuer (n)	automatic fire	rafale (f), feu de série (m)
Sicherheit (f)	safety	sûreté, sécurité (f)
Sicherheitsflügel (m)	safety catch	levier de sûreté (m)
Sicherheitsschieber (m)	safety slide	sûreté (f)

German	English	French
Sicherheitsstift (m)	safety pin	goupille de sûreté (f)
Sicherung (f)	safety device	mécanisme de sécurité (m)
Siedepunkt (m)	boiling point	point d'ébullition (m)
single action	single action	simple action
sinusförmig	sinusoidal	sinusoïdal
Skalarprodukt (Vektoren) (n)	dot product	produit scalaire (m)
Sollbruchstelle (f)	predetermined breaking point	amorce de rupture (f)
Spalt (m)	gap	fente (f)
spannen	cock	armer
Spannhebel (m)	cocking lever	levier d'armement (m)
Spannstück (n)	cocking piece	dispositif d'armement (m)
Spannung (f)	stress	tension (f)
Spektiv (n)	spotting scope	longue vue (f)
Spiralfeder (f)	coil spring, helical -	ressort à boudin (m)
Spitzgeschoss (n)	pointed bullet, spitzer -	balle pointue (f)
Splitter (Geschoss) (m)	fragment	éclat (projectile) (m)
Splitter (Holz, Glas) (m)	splinter	éclat (bois, verre) (m)
Sportwaffe (f)	sporter, sporting arm	arme de sport (f)
Sprengkapsel (f)	blasting cap	détonateur (m)
Sprengschnur (f)	detonating fuze	cordeau détonant (m)
Stahl (m)	steel	acier (m)
Stahlhülse (f)	steel case	douille (en) acier (f)
Stahlkerngeschoss (n)	steel-core bullet	balle à noyau en acier (f)
Stahlmantelgeschoss (n)	steel-jacketed bullet	balle blindée (f)
Standardabweichung (f)	standard deviation	écart-type (m)
Stange (f)	rod, bar	tige (f)
starr	rigid	rigide
starrer Körper (m)	rigid object	corps rigide (m)
staubdicht	dustproof	étanche aux poussières
Staudruck (m)	stagnation pressure	pression de la retenue (f)
Staupunkt (m)	stagnation point	point de stagnation (m)
Stecher (m)	hair trigger, double set -	double détente (f)
Stecherschloss (n)	hair trigger lock	platine à double détente (f)
stehend	standing	debout
Steilheit (Griff) (f)	steepness	pente, inclinaison (f)
Steinschloss (n)	flintlock	platine à silex (f)
Steinschlosszündung (f)	flintlock ignition	percussion à silex (f)
Stempel (m)	mark	poinçon (m)
Sternverschluss (m)	star crimp, fold crimp	fermeture étoile, - stellaire (f)
Stielhandgranate (f)	stick grenade	grenade à manche (f)
Stift (m)	pin	goupille (f)
Stiftzündung (f)	pin fire (ignition)	percussion à broche (f)
Stirnfläche (f)	face	face (f)
stoffliche Zusammensetzung (f)	material composition	composition (chimique) (f)
Stoppinen (f)	thin firing tapes	étoupille (f)
Störung (f)	disturbance	dérangement (m)
Stoß (m)	impact	impact, choc (m)
Stoß (physikalisch) (m)	collision	choc (m)
Stoßboden (m)	action face	bascule (f), tônnère (m)
Stoßwelle (f)	shock wave	onde de choc (f)
Stoßzeit (f)	impact time	durée du choc (f)
Streuung (f)	grouping, dispersion, spread	dispersion (f), écart (m) groupement (m)

German	English	French
Stromfaden (m)	streamline	ligne de courant (f)
Strömung (f)	flow	écoulement, flux (m)
Strömungsgeschwindigkeit (f)	fluid velocity	vitesse d'écoulement (f)
Strömungsquerschnitt (f)	flow cross-section	section d'écoulement (f)
Strömungswiderstand (m)	flow resistance	résistance à l'écoulement (f)
Sturmgewehr (n)	assault rifle	fusil d'assault, fusil mitrailleur (m)
Stutzer (m)	short rifle	carabine (f)
Summand (m)	addend	terme d'une addition (m)
Taumeln	yaw, tumble	tomnoyer
Teildrehung (f)	partial rotation	rotation partielle (f)
Teilmantelgeschoss (n)	semi-jacketed bullet, soft-nose -	balle semi-blindée (f)
thermodynamische Zustandsgröße (f)	thermodynamic state variable - parameter of state	variable d'état (m)
Tiefpassfilter (n)	low-pass filter	filtre passe-bas (m)
Tombak (m)	tombac, gilding metal	tombac (m)
tombakplattierter Stahl (m)	tombac-plated steel	acier plaqué de tombac (m)
Tontaube (f)	clay pigeon, clay bird, target	pigeon, - d'argile (m)
Torsionspendel (n)	torsional pendulum	pendule de torsion (m)
Trägheit (f)	inertia	inertie (f)
Trägheitsmoment (n)	moment of inertia	moment d'inertie (m)
transsonisch	transonic	transsonique
Treffer (m)	hit, score	coup portant, touché (m)
Trefferbild (n)	pattern	image des touchés (f)
Treffererwartung (f)	chance of hit	probabilité de toucher (f)
Treffpunkt (m)	point of impact	point d'impact (m)
Treffpunktabweichung (f)	deviation of point of impact	écart du point d'impact (m)
Treffsicherheit (f)	accuracy	assurance de toucher (f)
Treibkäfig (m)	cage	sabot (m)
Treibmittel (n)	propellant	charge propulsive (f)
Treibspiegel (m)	sabot	sabot (m)
trichterförmig	funnel-shaped	en forme d'entonnoir
Triggerstop (m)	triggerstop, backlash	immobilisation de détente
Trinitrotoluol (n)	trinitrotoluene	trinitrotoluène (m)
trockener Abzug (m)	dry trigger	détente sèche (f)
Trockentraining (n)	dry practice, dry firing	entrainement à sec (m)
Trommel (f)	cylinder	barillet (m)
Trommelachse (f)	axle of cylinder	pivot de barillet, axe de - (m)
Ueberdruck (bez. Atmosphäre)	gauge pressure	surpression (f)
Übergang (Zustand) (m)	crossover	transition (f)
überholen	pass, overtake	dépasser
Überschall-	supersonic	suprasonique
überschwingen	overshooting	surosciller
Übungsmunition	practice ammunition	munition d'exercice (f)
Umfangsgeschwindigkeit (f)	circumferential velocity	vitesse tangentielle (f)
umkehrbar	reversible	réversible
umkippen	overturn	renverser, basculer
ummantelt	jacketed	blindé
Umschlagwinkel (m)	angle of transition	angle de transition (m)
umströmter Körper (m)	body surrounded by flow	corps mouillé (m)

Unterdruck (m) .... depression .... dépression (f)
Unterdruck (bez. Atmosphäre) (m) . negative gauge pressure .... dépression (f)
Unterhebelrepetierer (m) .... lever action gun .... fusil à levier de sous-garde (m)
Unterkaliber (n) .... subcalibre .... sous-calibré
Unterlegscheibe .... washer .... bague (f)
Unterschall- .... subsonic .... subsonique

Vektorprodukt (n) .... cross product .... produit vectoriel (m)
Ventil (n) .... valve .... soupape (f)
ventilierte Laufschiene (f) .... ventilated barrel rib .... bande ventilée (f)
verbleit .... leaded, lead fouled .... plombé
Verbrennung (f) .... combustion .... combustion (f)
verchromt .... chromium-plated .... chromé
Verdrängung (f) .... displacement .... déplacement (m)
Verdünnung (f) .... dilution .... dilution (f)
Verhältniszahl (f) .... ratio .... rapport (m)
verkeilt .... wedged .... calé
verkupfert .... copper-plated .... coupré
vermessingt .... brass-coated, brass-plated .... laitonné
verminderte Geschwindigkeit (f) ... decreased velocity .... vitesse réduite (f)
vernickelt .... nickel-plated .... nickelé
Verpackung (f) .... packing .... emballage (m)
verriegelt .... locked .... verrouillé
Verriegelung (f) .... catch, latch, lock, latching .... verrou (m), verrouillage (m)
Verriegelungsnut (f) .... locking groove .... encoche de verrouillage (f)
Verriegelungswarze (f) .... (locking)lug .... tenon (m)
verrostet .... oxidated, oxidized .... rouillé
Versager (m) .... misfire, failure .... raté (m)
Verschleiß (m) .... wear .... usure (f)
Verschluss (m) .... action, bolt, breech, lock .... culasse (f)
                                                                                                                   fermeture (f)
Verschlussblock (m) .... breech block .... bloc culasse (m)
Verschlussgehäuse (n) .... receiver .... boîtier de culasse (m)
Verschlusshebel (m) .... action lever, locking -, bolt handle .. levier de culasse (m)
Verschlusskopf (m) .... bolt head .... tête de culasse (f)
verstellbar .... ajustable .... réglable
verstellbarer Abzug (m) .... adjustable trigger .... détente réglable (f)
verzögerter Masseverschluss (m) ... delayed blowback system .... culasse à ouverture retardée (f)
Verzögerung (f) .... deceleration, retardation .... décélération (f)
Verzögerungskoeffizient (m) .... deceleration coefficient .... coefficient de décélération (m)
Visier (n) .... rear sight .... hausse (f)
visieren .... sight, aim .... viser, pointer
Visierlänge (f) .... length of line of sight .... longueur de la visée, (f)
                                                                                                                   - de la ligne de mire (f)
Visierlinie (f) .... line of sight, axis of sight .... ligne de mire (f)
Visierung (f) .... sights .... hausse, visée (f)
Viskosität (f) .... viscosity .... viscosité (f)
Vollchoke (m) .... full choke .... plein choke
Vollchokelauf (m) .... full choked barrel .... canon à plein choke (m)
Vollgeschoss (n) .... solid (homogeneous) bullet .... projectile plein (m)
Vollmantelgeschoss (n) .... full metal jacket bullet .... balle blindée (f)
Vorderlader (m) .... muzzle loader .... arme à chargement par
                                                                                                                   la bouche

442 Annex B: Glossary

German	English	French
Vorderschaft (m)	forearm, fore-end	devant, fût (m)
Vorholfeder (f)	recuperator spring	ressort récupérateur (m)
Vorläufer (m)	precursor	précurseur (m)
Vorsprung (m)	lug, stud	ergot (m)
Vorstecker (m)	safty pin	goupille, goupille de sûreté (f)
Vorzeichen (n)	sign	signe (m)
Wachs (n)	wax	cire (f)
Wadcutter(-geschoss) (n)	wadcutter	wad-cutter
Waffe (f)	arm, weapon	arme (f)
Waffenkoffer (m)	gun case	valise pour arme (f)
Waffennummer (f)	serial number of a weapon	numéro de série de l'arme (m)
Walze (f)	cylinder	barillet (m)
Wandstärke (f)	thickness of the wall	épaisseur de parois (f)
Wärmeenergie (f)	thermal energy, heat	énergie thermique (f)
Wärmeentwicklung (f)	thermic development	production de chaleur (f)
Wärmeleitung (f)	thermal conduction	conduction thermique (f)
Wärmemenge (f)	quantity of heat	quantité de chaleur (f)
Wärmezustand (m)	thermal state	état thermique (m)
Warzenverriegelung (f)	lug-locking system	
wasserdicht	waterproof	étanche
Wasserwaage (f)	spirit level	niveau à bulle (m)
Wechselkorn (n)	insert foresight, interchangeable -	guidon interchangeable (m)
Wechsellauf (m)	interchangeable barrel	canon interchangable (m)
Wegdifferenz (f)	displacement	déplacement (m)
Weichblei (n)	soft lead	plomb doux (m)
Wettkampf (m)	competition, match	compétition (f), match (m)
Wettkampfgewehr (n)	match rifle	carabine de compétition (m)
Widerstandsbeiwert (m)	drag coefficient	coefficient de résistance (m)
Widerstandskraft (f)	resisting force	résistance (f)
Wiederladen (n)	reloading, handloading	rechargement (m)
Wiederladepresse (f)	reloading press	presse à recharger (f)
Wiederladewerkzeug (n)	reloading tool	outillage pour le rechargement (m)
Wild (n)	game	gibier (m)
Winkel (m)	angle	angle (m)
Winkelbeschleunigung (f)	angular acceleration	accélération angulaire (f)
Winkelgeschwindigkeit (f)	angular velocity	vitesse angulaire (f)
Wirkfläche (f)	active area	surface active (f)
Wirkung (f)	effect	effet (m)
Wirkungsgrad (m)	efficiency	rendement (m)
Wirkungslinie, -richtung (f)	line of action	direction de l'action (f)
wölben	vault	cintrer
Wundhöhle (f)	permanent cavity	cavité permanente (f)
Wurftaube (f)	clay bird, clay-pigeon, target	pigeon (d'argile) (m)
Wurfwinkel (m)	angle of projection	angle de projection (m)
Würgerille (f)	cannelure	cannelure (f)
Zähigkeit (dynamische) (f)	viscosity	viscosité (f)
kinematische -	kinematic viscosity	viscosité cinématique (f)
Zahlenwert (m)	numerical value	valeur numérique (f)
Zeitachse (f)	timing axis	axe du temps (m)

Zeitkonstante (f).............................. relaxation time............................... temps de relaxation (m)
Zentralfeuerung (f)........................ centrefire ignition .... percussion centrale (f)
Zentralzündung (f) .... centrefire... percussion centrale (f)
Zerlegungsgeschoss (n)................. fragmentating projectile .... projectile à fragmentation (m)
Ziel (n).. target, aim, object........................... cible (f)
zielen .... aim, sight .... viser
Zielen (n).. aiming, sighting.............................. pointage (m), visée (f)
Zielfehler (m) .... sighting error .... erreur de visée (f)
Zielfernrohr (n) .... rifle scope, (tele)scope sight........... lunette de visée (f)
Zielfernrohrmontage (f) .... scope mount .... montage pour lunette (m)
                                                                                                            embase de - (f)
Zielgeschwindigkeit (f)................. final velocity.................................. vitesse terminale (f)
Zielmarke (f) .... aiming point .... chiffre de visée
Zielübungen (f) .... target practices .... exercices de visée (m)
Zoll (Maßeinheit) (m) .... inch .... pouce (m) (mesure)
Zone der Extravasation (f) .... area of extravasation .... surface d'extravasion
Zubringer (m)................................ magazine follower, follower .... auget (m), chargeur (m),
                                                                                                            transporteur (m)
Zubringerhebel (m) .... follower lever, feed - .... auget (m)
Zuführsystem (n)............................ feed system..................................... système d'alimentation (m)
Züge (mp)...................................... grooves, rifling .... rayures (fp)
zugebördelt..................................... a roll crimp .... bordé
Zugkraft (f).................................... tie force... force de traction (f)
Zugspannung (f)............................ tensile stress .... contrainte de traction (f)
Zugtiefe (f).................................... groove depth................................... profondeur de rayure (f)
Zünder (m) .... fuze, detonator, ignitor .... détonateur, allumeur (m),
                                                                                                            fusée (f)
Zündhütchen (n)............................ primer cap...................................... capsule d'amorçage (f)
Zündhütchenausfaller (m).............. blown out primer .... décapsulage d'amorce (m)
Zündhütchenausstoßer (m)............. decapper, decapping press............. desamorceur (m)
Zündhütcheneinsetzgerät (n).......... recaper, recapping press .... amorceur (m)
Zündhütchenempfindlichkeit (f) .... primer sensitivity............................ sensibilité d'amorce (f)
Zündkanal (m)................................ flash hole .... cheminée (f)
Zündsatz (m) .... primer composition .... composition d'amorçage (f)
Zündschnur (f)................................ match .... cordon allumeur (m)
Zündstift (m) .... firing pin... poinçon, pointeau d'amorçage
Zündung (f) .... ignition (system), primer................ amorçage (m), percussion (f)
                                                                                                            allumage (m)
zuschießend.................................... firing from open bolt .... tirer à culasse ouverte
Zustandsänderung (thermodyn.) (f)change of state................................ changement d'état (m)
Zustandsgleichung (f) .... equations of state............................ équation d'état (f)
Zustandsgröße (f) .... state variables .... variables d'état (f)
zwingend .... coercive .... obligatoirement
Zylinderlauf (m)............................. cylinder bore (barrel)..................... canon cylindrique (m)

B.3 French ⇒ German ⇒ English

Accélération (f)............................. Beschleunigung (f) .... acceleration
accélération terrestre (f) .... Erdbeschleunigung (f).................... acceleration due to gravity
                                                                                                            - of free fall
à genoux .... kniend .... kneeling

French	German	English
accélération angulaire (f)	Winkelbeschleunigung (f)	angular acceleration
accrochage (m)	Mitnehmer (m)	tappet, cam, dog
acide picrique (m)	Pikrinsäure (f)	picric acid
acier (m)	Stahl (m)	steel
acier plaqué de tombac (m)	tombakplattierter Stahl (m)	tombac-plated steel
affluer	anströmen	flow against
ailette, nervure (f)	Rippe (f)	rib
alésage du canon (m)	Laufbohrung (f)	bore
alésage du choke (m)	Chokebohrung (f)	choke-bore
alimentation (de cartouche) (f)	Patronenzufuhr (f)	cartridge feed
allongement (m), dilatation (f)	Dehnung (f)	strain
allumage (m)	Zündung (f)	ignition(system), primer
allumeur (m)	Zünder (m)	fuze, detonator, ignitor
âme (f)	Laufachse (f), Seelenachse (f)	barrel axis, axis of the bore
amorçage (m)	Zündung (f)	ignition(system), primer
amorce (f)	Zündhütchen (n)	primer
amorce à berdan (f)	Berdanzündhütchen (n)	berdan primer
amorce de rupture (f)	Sollbruchstelle (f)	predetermined breaking point
amorceur (m)	Zündhütcheneinsetzgerät (n)	recaper, recapping press
amorphe	amorph	amorphous
amorti	gedämpft	damped
amortir	dämpfen	damp
amortissement (m)	Dämpfung (f)	damping
amortisseur de son	Schalldämpfer (m)	silencer
amplitude maximale (f)	Schwingungsbauch (m)	maximum amplitude
analyse dimensionelle (f)	Dimensionanalyse (f)	dimensional analysis
angle (m)	Winkel (m)	angle
angle d'arrivée (m)	Auftreffwinkel (m)	angle of attack
angle d'incidence (m)	Anstellwinkel (m)	angle of incidence
angle de chute (m)	Fallwinkel (m)	angle of descent
angle de projection (m)	Abschusswinkel (m)	angle of departure
angle de projection (m)	Wurfwinkel (m)	angle of projection
angle de relèvement (m)	Abgangsfehlerwinkel (m)	angle of jump
angle de transition (m)	Umschlagwinkel (m)	angle of transition
angle des rayures (m)	Drallwinkel (m)	angle of twist
angle de séparation (m)	Abströmwinkel (m)	off-flow angle
angle de tir (m)	Schusswinkel (m)	angle of shot
angle solide (m)	räumlicher Winkel (m)	solid angle
antirouille	rostschützend	anticorrosive
antirouille, inoxydable	nicht rostend	stainless
approximation (f)	Näherung (f)	approximation
appui paume, repose-paume (m)	Handballenauflage (f)	palm support
arc (m)	Bogen (m)	bow
arc de cercle (m)	Kreisbogenstück (n)	arc of a circle
tir à l'arc (m)	Bogensport (m)	archery
arme (f)	Waffe (f)	arm, weapon
arme à canon rayé (f)	Büchse (f)	rifle
arme à chargement par la bouche	Vorderlader (m)	muzzle loader
arme à chargement par la culasse (f)	Hinterlader (m)	breech loader
arme blanche (f)	Blankwaffe (f)	edged weapon
arme d'épaule (f)	Schulterwaffe (f)	shoulder weapon
arme de chasse (f)	Jagdwaffe (f)	hunting weapoin

arme de poing (f).............................. Faustfeuerwaffe (f)......................... handgun
arme de sport (f).............................. Sportwaffe (f)................................. sporter, sporting arm
arme fonctionn. par emprunt de gaz... Gasdrucklader (m)........................... gas operated weapon
arme pour la chasse fermée............ Schonzeitwaffe (f).......................... closed season arm
                                                                                                                                      varmint rifle
armer .... entsichern, spannen .... release the safety catch, cock
armurier (m) .... Büchsenmacher (m) .... gunmaker, gunsmith
articulation, pivot (m) .... Drehpunkt (Hebel) (m)................... fulcrum
assurance de toucher (f) .... Treffsicherheit (f) .... accuracy
au repos .... ruhend... resting
auget (m) ....„... Ladelöffel (m), Zubringer (m)........ carrier, feeder, follower
auget (m) .... Zubringerhebel (m) .... follower lever, feed -
auto-allumage (m).......................... Selbstzündung (f) .... cook-off
automatique par recul..................... Rückstoßlader (m).......................... recoil operated gun,
                                                                                                                                      blowback operated gun
avec appui, - support .... aufgelegt... supported
axe de barillet (m) .... Trommelachse (f) .... axle of cylinder
axe de rotation, pivot (m)............... Drehachse (f).................................. rotational axis, pivot
axe du temps (m)............................ Zeitachse (f).................................... timing axis
axe jaugeur (m), jauge (f)............... Schusslochprüfer (m) .... score gauge, scorer
axiomes de Newton (m) .... Newton'sche Axiome (np)............. Newton's laws of motion
azoture de plomb (f)....................... Bleiazid (n).................................... lead azide

Backlash .... Abzugsspiel (n) .... backlash
bague (f).. Scheibe (f) .... target, disc
bague (f).. Unterlegscheibe.............................. washer
balance à poudre (f)....................... Pulverwaage (f) .... powder scale
balistique (f)................................... Ballistik (f) .... ballistics
balistique extérieure (f) .... Außenballistik (f) .... exterior ballistics
balistique intérieure (f)................... Innenballistik (f).............................. interior ballistics
balistique intermédiaire (f)............. Abgangsballistik (f)........................ intermediate ballistics
balistique terminale (f) .... Endballistik (f) .... terminal ballistics
balle à noyau en acier (f)................ Stahlkerngeschoss (n)..................... steel-core bullet
balle à point creuse, - tête - (f) .... Hohlspitzgeschoss (n) .... hollow-point bullet
balle à tête ronde (f) .... Rundkopfgeschoss (n)..................... round-nose bullet
balle blindée (f) .... Stahlmantelgeschoss (n)................. steel-jacketed bullet
balle chémisée (f)........................... Vollmantelgeschoss (n).................. full metal jacket bullet
balle demi chemisée (f).................. Teilmantelgeschoss (n).................. semi-jacketed bullet
balle éclairante (f) .... Leuchtgeschoss (n)......................... flare
balle incendiaire (f) .... Brandgeschoss (n) .... incendiary bullet
balle nez plat (f) .... Flachkopfgeschoss (n).................... flat-nose bullet
balle à noyau dure, - perforante (f) Hartkerngeschoss (n)...................... hard-core bullet
balle pointue (f).............................. Spitzgeschoss (n)............................ pointed bullet, spitzer -
balle restée (dans le canon) (f) .... Laufstecker (m) .... stuck bullet
balle traçante (f) .... Leuchtspurgeschoss (n).................. tracer bullet
bande (de visée) (f) .... Laufschiene (f) .... barrel rib
bande passante (f)........................... Frequenzgang (m) .... frequency response
bande ventilée (f) .... ventilierte Laufschiene (f).............. ventilated barrel rib
barillet (m)...................................... Trommel (f).................................... cylinder
barillet (m)...................................... Walze (f)... cylinder
bascule (f), tonnère (m).................. Stoßboden (m)................................ action face
bascule brisée (f) .... Patronenlagersprengung (f)............ chamber burst

basculer	kippen	to pitch, to overturn
basculer	umkippen	overturn
batterie (f)	Schloss (n)	bolt, lock, breech, action
bi-ogivale	Heckkonus (m)	boattail
bilan énergétique (m)	Energiebilanz (f)	energy balance
blindé	ummantelt	jacketed
bloc culasse (m)	Verschlussblock (m)	breech block
boîtier de culasse (m)	Verschlussgehäuse (n)	receiver
bordé	zugebördelt	a roll crimp
bordure de la douille (f)	Hülsenrand (m)	case rim
bouche (f)	Mündung (f)	muzzle
bourre (f)	Pfropfen (m)	wad
bourre à coussin (f)	Pfropfen mit Dämpfung (m)	cushion wad
bourre à cuvette, - à jupe (f)	Becherpfropfen (m)	cup wad
bourre à jupe en plastique (f)	Plastikbecherpfropfen (m)	plastic cup wad
bourre en plastique (f)	Plastikpfropfen (m)	plastic wad
bourrelet (m)	Rand (m)	rim
branche ascendante (f)	aufsteigender Ast (m)	ascending branch
branche descendante (f)	absteigender Ast (m)	descending branch
bras de levier (m)	Hebelarm (m)	lever arm
bretelle (f)	Riemen (m)	sling, strap, lanyard
bronzage, brunissage (m)	Brünierung (f)	browning, blueing
bronzer, brunir	brünieren	brown, blue
bruit de bouche (m)	Mündungsknall (m)	muzzle blast
busc (m)	Schaftbacke (f)	cheekpiece
Cache-flamme (m)	Feuerscheindämpfer (m)	flash hider
cadence (f)	Kadenz (f)	rate of fire
calé	verkeilt	wedged
calibre (m)	Kaliber (n)	calibre, bore, gauge (Schrot)
calibré, étalonné	geeicht	calibrated
calibrer, étalonner	kalibrieren	calibrate, size
cannelure (f)	Würgerille (f), Randrierung (f)	cannelure, knurling
canon (m)	Lauf (m)	barrel
canon à demi choke (m)	Halbchokelauf (m)	modified choked barrel
canon à plein choke (m)	Vollchokelauf (m)	full choked barrel
canon avec choke (m)	Lauf mit Chokebohrung (m)	choked barrel
canon basculant (m)	Kipplauf (m)	break-action gun
canon cylindrique (m)	Zylinderlauf (m)	cylinder bore (barrel)
canon damas (m)	Damastlauf (m)	damascus barrel
canon interchangable (m)	Wechsellauf (m)	interchangeable barrel
canon octogonal (m)	Achtkantlauf (m)	octogonal barrel
capteur (m)	Fühler (m)	sensing device
capsule d'amorçage (f)	Zündhütchen (n)	primer cap
carabine (f)	Stutzer (m)	short rifle
carabine (f), mousqueton (m)	Karabiner (m)	carbine
carabine (m)	Büchsflinte (f)	rifle-shotgun
carabine à air comprimé (f)	Luftgewehr (n)	air rifle
carabine à répétition (f)	Repetierbüchse (f)	repeating rifle
carabine de compétition (m)	Wettkampfgewehr (n)	match rifle
carabine de petit calibre (f)	Kleinkalibergewehr (n)	small bore rifle
carabine expresse (f)	Kipplaufbüchse (f)	tip-up rifle
carcasse (f)	Rahmen (m)	frame, receiver

cartouche (f)	Patrone (f)	cartridge
cartouche à blanc (f)	Blindpatrone (f)	blank cartridge
cartouche à blanc (f)	Platzpatrone (f)	blank cartridge
cartouche à bourrelet (f)	Randpatrone (f)	rimmed cartridge
cartouche à haut rendement (f)	Hochleistungspatrone (f)	high-power cartridge
cartouche à percussion annul. (f)	Randfeuerpatrone (f)	rimfire cartridge
cartouche de chasse (f)	Schrotpatrone (f)	shotgun cartridge, shotshell
- en plastique (f)	Plastikschrotpatrone (f)	plastic shotshell
cartouche de manipulation (f)	Manipulierpatrone (f)	dummy cartridge, drill round
cartouche pour appareils d'abattage (f)	Schlachtpatrone (f)	slaughter cartridge
cartouche pour pistolets (f)	Pistolenpatrone (f)	pistol cartridge
cartouche pour revolvers (f)	Revolverpatrone (f)	revolver cartridge
cartoucherie (f)	Munitionsfabrik (f)	ammunition plant, - factory
casque (m)	Helm (m)	helmet, crest
casque anti-bruit (m)	Gehörschutzmittel (n)	ear defenders
cavité (f)	Kavität (f)	cavity
cavité permanente (f)	bleibende Wundhöhle (f)	permanent cavity
cavité permanente (f)	Wundhöhle (f)	permanent cavity
ceinture cartouchière (f)	Patronengürtel (m)	cartridge belt
centre de masse (m)	Schwerpunkt (m)	centre of mass
centre de pression (m)	Angriffspunkt der Druckkräfte (m)	centre of pressure
centre de rotation (m)	Drehpunkt (m)	centre of rotation
centre de gravité (m)	Schwerpunkt (m)	centre of gravity
chaleur d'explosion (f)	Explosionswärme (f)	explosion heat
chaleur de formation (f)	Bildungsenergie (thermodyn.) (f)	heat of formation
chambre à cartouches (f)	Patronenlager (n)	chamber
chambre de combustion (f)	Brennkammer (f)	burning chamber
champignon, pommeau (m)	Handstütze (f)	hand support
champs des rayures (m)	Felder (np)	lands
changement d'état (m)	Zustandsänderung (thermodyn.) (f)	change of state
charge creuse (f)	Hohlladung (f)	hollow charge
charge propulsive (f)	Treibmittel (n)	propellant
charge (f)	Ladung (f)	charge
chargeur (m)	Magazin (n)	magazine
chargeur (m)	Zubringer (m)	magazine follower, follower
chargeur tubulaire, magasin - (m)	Röhrenmagazin (n)	tubular magazine, tube magazine
chasse (f)	Jagd (f)	hunting
cheminée (f)	Zündkanal (m)	flash hole
chemise (de projectile) (f)	Geschossmantel (m)	bullet jacket
chemise (f)	Mantel (m)	jacket
chevrotine (f)	Postenschrot (n)	buckshot
chien (m)	Hahn (m)	hammer
chiffre de visée	Zielmarke (f)	aiming point
chlorate de potassium (m)	Kaliumchlorat (n)	potassium chlorate
choc (m)	Stoß (physikalisch) (m)	collision
choke (m)	Choke (m)	choke
chromé	verchromt	chromium-plated
chrome (m)	Chrom (m)	chrome
cible (f)	Scheibe (f)	target, disc
cible (f)	Ziel (n)	target, aim, object

cintrer	Wirkungslinie,-richtung (f)	line of action
cire (f)	Wachs (n)	wax
coefficient (m)	Beiwert (m)	coefficient
coefficient de décélération (m)	Verzögerungskoeffizient (m)	deceleration coefficient
coefficient de frottement (m)	Reibungszahl (f)	friction coefficient
coefficient de portance (m)	Auftriebsbeiwert (m)	lift force coefficient
coefficient de résistance (m)	Widerstandsbeiwert (m)	drag coefficient
coefficient de résistance à l'air (m)	Luftwiderstandsbeiwert (m)	air drag coefficient
coiffe, calotte	Haube (f)	hood
collet (m)	Hülsenhals (m)	case neck
colonne d'air (f)	Luftsäule (f)	air column
combustion (f)	Verbrennung (f)	combustion
compétition (f), match (m)	Wettkampf (m)	competition, match
composition (chimique) (f)	stoffliche Zusammensetzung (f)	material composition
composition d'amorçage (f)	Zündsatz (m)	primer composition
composition lumineuse (f)	Leuchtsatz (m)	tracer
compressibilité (f)	Kompressibilität (f)	compressibility
condition aux limites (f)	Randbedingung (f)	boundary condition
conduction thermique (f)	Wärmeleitung (f)	thermal conduction
conique	konisch	tapered, conical
constante des gaz parfaits (f)	Gaskonstante R (f)	gas constant R
contrainte de compression (f)	Druckspannung (f)	compressive strain
contrainte de flexion (f)	Biegespannung (f)	bending stress
contrainte de traction (f)	Zugspannung (f)	tensile stress
contre-platine (f)	Seitenschloss (n)	side lock
contrepoids (m)	Laufgewicht (n)	counterweight
contrôleur de détente (m)	Abzugsprüfgerät (n)	trigger verification device
cordeau détonant (m)	Sprengschnur (f)	detonating fuze, detonating cord
cordon allumeur (m)	Zündschnur (f)	match
cordon d'allumage (m)	Anzündlitze (f)	igniter cord
corps mouillé (m)	umströmter Körper (m)	body surrounded by flow
corps rigide (m)	starrer Körper (m)	rigid object
corps solide (m)	fester Körper (m)	solid body
couché	liegend	prone
couche limite (f)	Grenzschicht (f)	boundary layer
coup creux (m)	Hohlschuss (Schrot) (m)	wheel shot
coup de grâce (m)	Fangschuss (m)	finishing shot
coup portant, touché (m)	Treffer (m)	hit, score
coup, tir (m)	Schuss (m)	round, shot
coupré	verkupfert	copper-plated
courbure (f)	Krümmung (f)	curvature
couteau de chasse (m)	Jagdmesser (n)	hunting knife
cran d'arrêt (m)	Druckpunkt (m)	trigger point
cran de mire (m)	Kimme (f)	rear sight, notch
creux	hohl	hollow
cristallin	kristallin (fest)	crystalline
crosse (f)	Kolben (Gewehr) (m)	butt (of weapon)
crosse (f)	Schaft (m)	stock
crosse à trou de pouce (f)	Schaft mit Daumenloch (m)	thumbhole stock
cuire (au four)	einbrennen	burn in, brand in
cuivre (m)	Kupfer (n)	copper
culasse (f)	Verschluss (m)	action, bolt, breech, lock

culasse à bloc basculant (f) .... Fallblockverschluss (m) .... falling block bolt action
culasse à genouillère (f) .... Kniegelenkverschluss (m).............. toggle action bolt
culasse à inertie (f)......................... Masseverschluss (m) .... blowback system
culasse à ouverture retardée (f)...... verzögerter Masseverschluss (m)... delayed blowback system
culasse tombante, - basculante (f).. Kipplaufverschluss (m) .... tip-up lock
culasse tournante, - rotative .... Drehverschluss (m) .... rotating bolt locking system
culbuteur (m)................................... Kipphebel (m) .... armed lever
culot (m).. Hülsenboden (m)............................. case head

Debout .... stehend... standing
décapsulage d'amorce (m).............. Zündhütchenausfaller (m) .... blown out primer
décélération (f) .... Verzögerung (t).............................. deceleration, retardation
défaut critique (m).......................... kritischer Fehler (m)...................... critical defect
défaut majeur (m)........................... Hauptfehler (m).............................. main defect
déflagration (f) .... Deflagration (f).............................. deflagration
déformer.. deformieren .... deform
degré Celsius (m) .... Grad Celsius (m) .... degree celsius
degré de liberté (m)........................ Freiheitsgrad (m)........................... degree of freedom
degré de sveltesse (m).................... Schlankheitsgrad (m) .... slenderness ratio
dense.. dicht.. tight, dense, sealed
densité de chargement (f)............... Ladedichte (f) .... loading density
dent de rochet (f)............................ Schaltzahn (Revolver) (m) .... gear(notch), sprocket
départ (m).. Abziehen (n)................................... pull (the trigger)
dépasser .... überholen .... pass, overtake
déplacement (m)............................. Verdrängung (f).............................. displacement
déplacement (m)............................. Wegdifferenz (f)............................. displacement
déplacement (m) angulaire............. Drehwinkel (m) .... angular displacement
dépression (f).................................. Unterdruck (bez. Atmosphäre) (m) . negative gauge pressure
dépression (f).................................. Unterdruck (m)............................... depression
dérangement (m) .... Störung (f) .... disturbance
dérivée (f).. Differenzialquotient (m)................ derivative,
desamorceur (m) .... Zündhütchenausstoßer (m)............ decapper, decapping press
détente (f).. Abzug (m) .... trigger
détente à bossette (f) .... Druckpunktabzug (m) .... double pull trigger
détente de match (f)........................ Matchabzug (m) .... match-trigger
détente réglable (f) .... verstellbarer Abzug (m) .... adjustable trigger
détente sèche (f) .... trockener Abzug (m) .... dry trigger
détonateur (m)................................. Sprengkapsel (f) .... blasting cap
détonateur (m)................................. Zünder (m)...................................... fuze, detonator, ignitor
détonation (f).................................. Detonation (f) .... detonation
détonation (f).................................. Knall (m)... report
détonation par influence (f)............ Detonationsübertragung (f) .... sympathetic detonation
devant, fût (m)................................ Vorderschaft (m) .... forearm, fore-end
déviation (f), écart .... Auslenkung (f) .... extension
déviation, écart (m) .... Abweichung (f).............................. deviation
diamètre du champ des rayures (m).. Felddurchmesser (m)..................... land diameter
diaphragme iris (m)........................ Irisblende (f).................................. iris diaphragm
différence (f) .... Differenzenquotient (m)................ difference quotient
différentiel (m) .... Differenzial (n).............................. differential
dilution (f) .... Verdünnung (f)............................... dilution
dimension (f), extension................. Ausdehnung (f).............................. dimension
diminution du nombre de tours (f) . Drehzahlabnahme (f)..................... spin damping
dioptre (m)...................................... Diopter (n)...................................... aperture sight

dioptre micrométrique (m)	Mikrometerdiopter (m)	micrometer sight
dioxyde de carbone (f)	Kohlendioxid (n)	carbon dioxide
direction de l'action (f)	wölben	vault
direction de tir (f)	Schussrichtung (f)	line of departure
direction du pas (f)	Drallrichtung (f)	twist direction
dispersion (f)	Streuung (f)	grouping, dispersion, spread
dispersion en largeur (f)	Breitenstreuung (f)	lateral spread
dispersion en portée (f)	Längenstreuung (f)	variation in length
dispersion lateral (f)	Seitenstreuung (f)	lateral dispersion, - spread
dispersion verticale (f)	Höhenstreuung (f)	vertical dispersion
dispositif d'armement (m)	Spannstück (n)	cocking piece
disque (m)	Scheibe (f)	target, disc
dissipation d'énergie (f)	Energieabgabe (f)	energy release
distance (f)	Entfernung (f)	distance, range
distance du sommet (f)	Scheiteldistanz (f)	vertex distance
double détente (f)	Doppelabzug (m)	double trigger
double détente (f)	Stecher (m)	hair trigger, double set -
douille (en) acier (f)	Stahlhülse (f)	steel case
douille (en) plastique (f)	Plastikhülse (f)	plastic case
douille (f)	Patronenhülse (f)	cartridge case
douille rompue (f)	Hülsenabreißer (m)	cut case
douille à ceinture (f)	Gürtelhülse (f)	belted case
douille à épaulement (f)	Schulterhülse (f)	bottleneck case
douille en bourrelet (f)	Randhülse (f)	rimmed case
douille en forme de bouteille (f)	Hülse mit Schulter (f)	bottlenecked case
drilling, fusil à trois canons (m)	Drilling (m)	drilling, three-barrelled gun
durée de pulsation (f)	Pulsdauer (f)	pulse duration
durée du choc (f)	Stoßzeit (f)	impact time
dureté (f)	Härte (f)	hardness
dynamique des fluides (f)	Fluiddynamik (f)	fluid dynamics
Écart (m)	Streuung (f)	grouping, dispersion, spread
écart du point d'impact (m)	Treffpunktabweichung (f)	deviation of point of impact
écart en portée (m)	Längenabweichung (f)	length deviation, linear -
écart lateral (m)	Seitenabweichung (f)	lateral deviation
écart moyen (m)	mittlere Abweichung (f)	mean deviation
écart-type (m)	Standardabweichung (f)	standard deviation
écart vertical (m)	Höhenabweichung (f)	vertical deviation
échantillon (m)	Muster (n)	sample
échauffement (m)	Erhitzung (f)	heating
éclat (bois, verre) (m)	Splitter (Holz, Glas) (m)	splinter
éclat (projectile) (m)	Splitter (Geschoss) (m)	fragment
éclat du canon (m)	Laufsprengung (f)	barrel burst, premature - -
éclater	platzen	burst, crack
écoulement, flux (m)	Strömung (f)	flow
écrou (m)	Mutter (f)	nut
écrou (m)	Schraubenmutter (f)	nut
effet (m)	Wirkung (f)	effect
effet aux limites (m)	Randeffekt (m)	boundary effect
effet de succion (m)	Saugeffekt (m)	suction effect
effet gyroscopique (m)	Kreiseleffekt (m)	gyro effect
éjecteur (m)	Auswerfer (m)	ejector
élément d'attache (m)	Befestigungselement (n)	fastening element

éléments de masse (m)	Masseteilchen (n)	mass element
élévation (f)	Erhebungswinkel (m)	elevation angle
emballage (m)	Verpackung (f)	packing
embase de lunette (f)	Zielfernrohrmontage (f)	scope mount
enclume (f)	Amboss (m)	anvil
encoche (f)	Kerbe (f)	notch
encoche (f)	Nut (f)	groove, slot
encoche de verrouillage (f)	Verriegelungsnut (f)	locking groove
énergie (f)	Energie (f)	energy
énergie cinétique (f)	Bewegungsenergie (f)	kinetic energy
énergie cinétique (f)	kinetische Energie (f)	kinetic energy
énergie cinétique de rotation (f)	Rotationsenergie (f)	rotational kinetic energy
énergie de bouche (f)	Mündungsenergie (f)	muzzle energy
énergie de compression (f)	Druckenergie (f)	pressure energy
énergie de rotation (f)	Drehenergie (f)	energy of rotation
énergie du recul (f)	Rückstoßenergie (f)	recoil energy
énergie interne (f)	innere Energie (f)	internal energy
énergie potentielle (f)	potenzielle Energie (f)	potential energy
énergie résiduelle (f)	Restenergie (f)	residual energy
énergie thermique (f)	Wärmeenergie (f)	thermal energy, heat
entrainement à sec (m)	Trockentraining (n)	dry practice, dry firing
enveloppe de papier (f)	Papierumwicklung (f)	paper patch
épaisseur de parois (f)	Wandstärke (f)	thickness of the wall
épanouir en forme de champignon	aufpilzen	mushroom
épreuve (f)	Beschussprobe (f)	proof firing
épreuve de recette (f)	Abnahmeprüfung (f)	acceptance test
épreuve de sensibilité (f)	Empfindlichkeitsprüfung (f)	sensitivity test
équation d'état (f)	Zustandsgleichung (f)	equations of state
équation de continuité (f)	Kontinuitätsgleichung (f)	continuity equation
équations de mouvement (f)	Bewegungsgleichungen (f)	equations of motion
équilibre (m)	Gleichgewicht (n)	equilibrium
équilibre des forces (m)	Kräftegleichgewicht (n)	equilibrium of forces
équipé de	ausgerüstet sein mit	equipped, be - with
ergot (m)	Nase (f)	stud, lug
ergot (m)	Vorsprung (m)	lug, stud
ergot, came (f)	Nocken (m)	cam, lug
erreur de visée (f)	Zielfehler (m)	sighting error
étalonnage (m)	Eichung (f)	calibration
étalonner	eichen	calibrate
étanche	dicht	tight, dense, sealed
étanche	wasserdicht	waterproof
étanche aux poussières	staubdicht	dustproof
état de matière (m)	Aggregatszustand (m)	state of matter
état thermique (m)	Wärmezustand (m)	thermal state
étoupille (f)	Stoppinen (f)	thin firing tapes
étroit	eng	narrow
douille (f)	Hülse (f)	case
évent (m)	Gasentnahmedüse (f)	gas port
exercices de visée (m)	Zielübungen (f)	target practices
extracteur (m)	Auszieher (m)	extractor
Face (f)	Stirnfläche (f)	face
fente (f)	Spalt (m)	gap

French	German	English
fermeture étoile, - stellaire (f)	Sternverschluss (m)	star crimp, fold crimp
feu coup par coup (m)	Einzelfeuer (n)	single-shot fire
feu de bouche (m)	Feuer aus Mündung (n)	primary (muzzle) flash
feu de bouche (m)	Mündungsfeuer (n)	secondary (muzzle) flash
feu de série (m)	Seriefeuer (n)	automatic fire
feu en rafales (m)	Dauerfeuer (n)	full automatic fire
feuillet (m)	Kimmenblatt (n)	rear sight leaf, - - slide
filtre passe-bas (m)	Tiefpassfilter (n)	low-pass filter
fin de combustion (f)	Brennschluss (m)	end of burning
fluctuation de pression (f)	Druckschwankung (f)	transient pressure
flux (m)	Fluss (m)	fluid motion
flux de masse (m)	austretende Masse (f)	mass flow
force d'exitation (f)	erregende Kraft (Schwingung) (f)	driving force, external -
force de cisaillement (f)	Normalspannung (f)	normal stress
force de cohésion (f)	Kohäsivkraft (f)	cohesive force
force de compression (f)	Druckkraft (f)	compressive force
force de frottement (f)	Reibungskraft (f)	frictional force
force de poussée (f)	Schubkraft (f)	shear force
force de traction (f)	Zugkraft (f)	tie force
force de cohésion moléculaire (f)	Gitterbindungskraft (f)	lattice linkage force
frein de bouche (m)	Mündungsbremse (f)	muzzle brake
fréquence angulaire (f)	Kreisfrequenz (f)	angular frequency
fréquence de rotation (f)	Drehfrequenz (f)	frequency of rotation
fréquence limite (f)	Grenzfrequenz (f)	boundary frequency
fréquence propre (f)	Eigenfrequenz (f)	natural frequency
frottement interne (m)	innere Reibung (f)	internal friction
fulminate de mercure (m)	Knallquecksilber (n)	mercury fulminate
fusée (f)	Zünder (m)	fuze, detonator, ignitor
fusée (f), missile (m), roquette (f)	Rakete (f)	rocket, missile
fusée percutante (f)	Aufschlagzünder (m)	impact fuze
fusil à canon juxtaposés (m)	Doppelflinte (f)	side-by-side shotgun
fusil à canon lisse (m)	Flinte (f)	shotgun
fusil à canons superposés (m)	Bockbüchsflinte (f)	superposed rifle-shotgun
fusil à deux canon (m)	Doppelbüchse (f)	double rifle
fusil à levier de sous-garde (m)	Unterhebelrepetierer (m)	lever action gun
fusil à percussion (m)	Perkussionsgewehr (n)	percussion rifle
fusil à répétition (m)	Repetierflinte (f)	repeating shotgun
fusil à un coup (m)	Einzellader (m)	single-shot weapon
fusil automatique (m)	Selbstladegewehr (n)	automatic rifle, self-loading rifle
fusil d'assault, fusil mitrailleur (m)	Sturmgewehr (n)	assault rifle
fusil de chasse (m)	Jagdflinte (f), Schrotflinte (f)	shotgun
fusil de chasse automatique (m)	Selbstladeflinte (f)	automatic shotgun
fusil mitrailleur (m)	Schnellfeuergewehr (n)	assault rifle
Gâchette (f)	Abzugshebel (m)	trigger lever, trigger latch
gazeux	gasförmig	gaseous
gerbe (f)	Schrotgarbe (f)	shot sheaf
gibier (m)	Wild (n)	game
goupille (f)	Stift (m)	pin
goupille de sûreté (f)	Sicherheitsstift (m)	safety pin
goupille, goupille de sûreté (f)	Vorstecker (m)	safty pin
gradient de pression (m)	Druckgefälle (n)	pressure gradient

gradient de vitesse (m) .... Geschwindigkeitsgefälle (n) .... shear rate
grande chasse (f) .... Hochwildjagd (f) .... big-game hunting
grandeur mesurée (f) .... Messgröße (f) .... measured quantity
grenade à main (f) .... Handgranate (f) .... hand-grenade
grenade à manche (f) .... Stielhandgranate (f) .... stick grenade
grenaille (f) .... Schrot (m) .... shot
groupement (m) .... Trefferbild (n) .... pattern
guidon (m) .... Korn (n) .... foresight
guidon à lame rectangulaire (m) .... Balkenkorn (n) .... post sight
guidon annulaire (m) .... Ringkorn (n) .... ring foresight, aperture -
guidon interchangeable (m) .... Wechselkorn (n) .... insert foresight
guidon rectangulaire (m) .... Rechteckkorn (n) .... square foresight
gyroscope (m) .... Kreisel (m) .... gyroscope
gyrostabilisé .... drallstabil .... spin-stabilized

Hausse (f) .... Kimme (f) .... rear sight, notch
hausse (f) .... Visier (n) .... rear sight
hausse, visée (f) .... Visierung (f) .... sights
hauteur de chute (f) .... Fallhöhe (f) .... drop
hauteur du sommet (f) .... Scheitelhöhe (f) .... vertex height
hexogène (m) .... Hexogen (n) .... hexogen

Image des touchés (f) .... Trefferbild (n) .... pattern
immobilisation de détente .... Abzugsbegrenzung (f) .... triggerstop
immobilisation de détente .... Triggerstop (m) .... triggerstop, backlash
impact (m) .... Einschlag (m) .... impact
impact (m) .... Stoß (m) .... impact
impulsion (f) .... Kraftstoß (m) .... impulse
incompressible .... dichtebeständig .... incompressible
indéformable .... formstabil .... shape stable
induire, provoquer .... anregen .... induce
inertie (f) .... Trägheit (f) .... inertia
inoxydable .... rostfrei .... stainless

Jauge de pression (f) .... Druckaufnehmer (m) .... pressure gauge
jauge de pression (f) .... Gasdruckmesser (m) .... pressure gauge
jauge (m) .... Fühler (m) .... sensing device

Laiton (m) .... Messing (n) .... brass
laitonné .... vermessingt .... brass-coated, brass-plated
latéral .... seitlich .... lateral
légitime .... gesetzmäßig .... natural
levier (m) .... Hebel (m) .... lever
levier d'armement (m) .... Spannhebel (m) .... cocking lever
levier de culasse (m) .... Verschlusshebel (m) .... action lever, bolt handle
levier de sûreté (m) .... Sicherheitsflügel (m) .... safety catch
ligne de courant (f) .... Stromfaden (m) .... streamline
ligne de mire (f) .... Visierlinie (f) .... line of sight, axis of sight
liquide .... flüssig .... liquid
lissage (m) .... Oberflächenbehandlung (f) .... smoothing
loi de conservation (f) .... Erhaltungssatz (m) .... law of conservation
longue vue (f) .... Spektiv (n) .... spotting scope

longueur (f)	Länge (f)	length
longueur de la ligne de mire (f) - de la visée (f)	Visierlänge (f)	length of line of sight
longueur du canon (f)	Lauflänge (f)	barrel length
luminosité (f)	Lichtstärke (f)	brightness
lunette de visée (f)	Zielfernrohr (n)	rifle scope, (tele)scope sight
Magasin (m)	Magazin (n)	magazine
manteau (m)	Mantel (m)	jacket
marquage des munitions (m)	Munitionskennzeichnung (f)	ammunition marking, - code
marteau (m)	Hammer (m)	hammer
marteau (m)	Schlagstück (n)	hammer, striker
marteau, percuteur (m)	Schlagbolzen (m)	firing pin
martelage (m)	Hämmerung (f)	hammering
marteler	hämmern	hammer
masse incidente (f)	eintretende Masse (f)	incident mass
masse par unité de section (f)	Querschnittsbelastung (f)	sectional density
masse volumique (f)	Dichte (f)	density
matière incendiaire (f)	Brandsatz (m)	incendiary compound
mécanisme à trois coups (m)	Dreischussautomatik (f)	three-shot mechanism
mécanisme de sécurité (m)	Sicherung (f)	safety device
mèche (f)	Lunte (f)	match
mèche de sureté (f)	Schwarzpulverzündschnur (f)	safety fuze
mesurer en radians	Bogenmaß (n), im - messen	measure in radians
mitrailleuse (f)	Maschinengewehr (n)	machine gun
module de compressibilité (m)	Kompressionsmodul (m)	bulk modulus
module d'élasticité (m)	Elastizitätsmodul (m)	Young's modulus modulus of elasticity
molette de réglage, vis de - (f)	Einstellschraube (f)	set screw, adjustment screw
moment (m)	Drehmoment (n)	torque moment of a force
moment cinétique (m)	Drehimpuls (m)	angular momentum moment of momentum
moment d'inertie (m)	Trägheitsmoment (n)	moment of inertia
moment de basculement (m)	Kippmoment (n)	overturning moment
montage pour lunette (m)	Zielfernrohrmontage (f)	scope mount
mouvement circulaire (m)	Drehbewegung (f)	circular motion
mouvement circulaire (m)	Kreisbewegung (f)	circular motion
mouvement de précession (m)	Präzessionsbewegung (f)	yaw
mouvement de rotation (m)	Rotationsbewegung (f)	rotation
mouvement pendulaire (m)	Pendelung (f)	yaw
mouvement propre (m)	Eigenbewegung (f)	proper motion
moyenne (f)	Mittelwert (m)	mean
munition (f)	Munition (f)	ammunition
munition à courte portée (f)	Kurzbahnmunition (f)	short-range ammunition
munition d'exercice (f)	Übungsmunition	practice ammunition
Naturel	gesetzmäßig	natural
nécessaire de nettoyage (m)	Reinigungsgerät (n)	cleaning kit
nickel (m)	Nickel (n)	nickel
nickelé	vernickelt	nickel-plated
nitrate de potassium (m)	Kaliumnitrat (n)	potassium nitrate
niveau à bulle (m)	Wasserwaage (f)	spirit level

non-visqueux	reibungsfrei bez. innerer Reibung	inviscid
numéro de série de l'arme (m)	Waffennummer (f)	serial number of a weapon
nutation (f)	Nutation (f)	nutation

Obligatoirement	zwingend	coercive
octogène (m)	Oktogen (n)	octogen
octogonal	achtkantig	octogonal
oeilleton (m)	Lochblende (f)	aperture, peep
ombrogramme (m)	Schattenaufnahme (f)	shadowgraph
onde de choc (f)	Stoßwelle (f)	shock wave
opposé, de signe contraire	entgegengesetzt gleich	opposite
ordre de grandeur (m)	Großenordnung (f)	order of magnitude
oscillation forcée (f)	erzwungene Schwingung (f)	forced oscillation
oscillation harmonique (f)	harmonische Schwingung (f)	harmonic motion
outillage pour le rechargement (m)	Wiederladewerkzeug (n)	reloading tool
ouverture d'éjection (f)	Auswurföffnung (f)	ejection port

Palette (de tir) (f)	Schusspflaster (n)	target patch
parabolique	parabelförmig	parabolic
paramètre (m)	Kennwert (m)	parameter
pare-balles (m), butte (f)	Kugelfang (m)	butt (behind target)
pas des rayures (m)	Dralllänge (f)	twist, twist length
pas progressif (m)	progressiver Drall (m)	progressive twist, - groove
pendule de torsion (m)	Torsionspendel (m)	torsional pendulum
pendule mathématique (m)	mathematisches Pendel (n)	simple pendulum
pendule physique (m)	physikalisches Pendel (n)	physical pendulum
pénétration (f)	Durchschlag (m)	penetration
pente (f)	Neigung (f)	slope, fall, gradient
pente, inclinaison (f)	Steilheit (Griff) (f)	steepness
percussion (f)	Zündung (f)	ignition(system), primer
percussion à broche (f)	Lefaucheuxzündung (f)	pin fire ignition
percussion à broche (f)	Stiftzündung (f)	pin fire (ignition)
percussion à capsule (f)	Perkussionszündung (f)	percussion ignition
percussion à rouet (f)	Radschlosszündung (f)	wheel-lock ignition
percussion à silex (f)	Steinschlosszündung (f)	flintlock ignition
percussion annulaire (f)	Randfeuerzündung (f)	rimfire ignition
percussion centrale (f)	Zentralfeuerung (f)	centrefire ignition
percussion centrale (f)	Zentralzündung (f)	centrefire
perte d'énergie (f)	Energieverlust (m)	energy drop
peson (m)	Abzugsgewichtkontrolle (f)	verification of trigger pull weight
pétard (m)	Knallpatrone (f)	petard
phénomène (m)	Erscheinung (f)	phenomenon
pigeon (d'argile) (m)	Wurftaube (f), Tontaube (f)	clay bird, clay-pigeon, target
pistolet (m)	Pistole (f)	pistol
pistolet à air (comprimé) (m)	Luftpistole (f)	air pistol
pistolet automatique (m)	Selbstladepistole (f)	semi-automatic pistol
pistolet d'alarme (m)	Alarmpistole (f)	alarm pistol, signal pistol
pistolet mitrailleur (m)	Maschinenpistole (f)	submachine gun
pistolet rafaleur (m)	Schnellfeuerpistole (f)	rapid-fire pistol
pistolier (m)	Pistolenschütze (f)	pistol shooter
piston (m)	Druckkolben (m)	piston
pivot (m)	Schwenkachse (f)	pivot

pivot de barillet (m)	Trommelachse (f)	axle of cylinder
plan (m), plaque plane (f)	ebene Platte (f)	plane, plate
plaque de couche (f)	Schaftkappe (f)	buttplate
plaque de couche antirecul (f)	Rückstoßminderer (m)	recoil reducer, - pad
platine (f)	Schlossträger (m)	lock plate
platine à double détente (f)	Stecherschloss (n)	hair trigger lock
platine à mèche (f)	Luntenschloss (n)	match-lock
platine à rouet (f)	Radschloss (n)	wheel-lock
platine à silex (f)	Steinschloss (n)	flintlock
plein choke	Vollchoke (m)	full choke
plomb (de chasse) (m)	Schrotkugel (f)	shot pellet
plomb (m)	Blei (n)	lead
plomb Diabolo (m)	Luftgewehrgeschoss (n)	airgun pellet
plomb doux (m)	Weichblei (n)	soft lead
plomb durci (m)	Hartblei (n)	antimony lead
plomb durci (m)	Hartschrot (m)	chilled shot, hard shot
plomb nickelé (m)	Nickelschrot (m)	nickel(plated) shot
plombage du canon (m)	Laufverbleiung (f)	lead-fouled barrel
plombé	verbleit	leaded, lead fouled
poinçon d'épreuve (m)	Beschussstempel (m)	proof mark
poids (m)	Gewicht (n)	weight
poids de départ, - détente (m)	Abzugsgewicht (n)	trigger pull weight
poids du projectile (m)	Geschossgewicht (n)	bullet weight
poignée pistolet, crosse - (f)	Pistolengriff (m)	pistol grip
poinçon (m)	Stempel (m)	mark
poinçon, pointeau d'amorçage (m)	Zündstift (m)	firing pin
point d'ébullition (m)	Siedepunkt (m)	boiling point
point d'impact (m)	Treffpunkt (m)	point of impact
point d'impact (m)	Auftreffpunkt (m)	point of impact
point de fusion (m)	Schmelzpunkt (m)	melting point
point de mire (m)	Haltepunkt (m)	point of aim
point de référence (m)	Bezugspunkt (m)	reference point
point de séparation (m)	Ablösepunkt (Grenzschicht) (m)	point of separation
point de solidification (m)	Erstarrungspunkt (m)	freezing point
point de stagnation (m)	Staupunkt (m)	stagnation point
pointage (m), visée (f)	Zielen (n)	aiming, sighting
pontet de sous-garde (m)	Abzugsbügel (m)	trigger guard
portance (f)	Auftriebskraft (f)	lift force
porte-guidon (m)	Kornsattel (m)	foresight ramp
porte-guidon (m)	Kornträger (m)	foresight base
portée (f)	Schussdistanz (f), -weite (f)	range
position de repos (f)	Ruhelage (f)	rest position
position finale (f)	Endlage (f)	final position
position initiale (f)	Anfangslage (f)	initial position
pouce (m) (mesure)	Zoll (Maßeinheit) (m)	inch
poudre (f)	(Schieß-)Pulver (n)	gunpowder
poudre à disque (f)	Scheibenpulver (n)	disc powder
poudre à double base (f)	Nitroglycerinpulver (n)	double base powder
poudre à paillettes (f)	Blättchenpulver (n)	flake (laminated) powder
poudre à simple base (f)	Nitrocellulosepulver (n)	single base powder
poudre noir (f)	Schwarzpulver (n)	black powder
poudre lente (f)	progressives Pulver (n)	progressive burning powder
poudre pyroxylée (f)	rauchloses Pulver (n)	smokeless powder

poudre sphérique (f)	Kugelpulver (n)	ball powder
poudre tubulaire (f)	Röhrchenpulver (n)	tubular powder
poudre vive (f)	degressives (offensives) Pulver (n)	degressive burning powder
poussée (f)	Schub (Rakete) (m)	thrust
poussoir (m), tirette (f)	Schieber (m)	slide
pouvoir de pénétration (m)	Eindringvermögen (n)	pénétration capacity
précession (f)	Präzession (f)	precession
précision (f)	Präzision (f)	accuracy, precision
précurseur (m)	Vorläufer (m)	precursor
presse à calibrer (f)	Kalibrierpresse (f)	calibrating press
presse à recharger (f)	Wiederladepresse (f)	reloading press
pression (f)	Druck (m)	pressure
pression d'extraction (f)	Ausziehdruck (m)	shot start pressure
pression de gaz (f)	Gasdruck (m)	gas pressure
pression de la retenue (f)	Staudruck (m)	stagnation pressure
pression maximale (f)	Maximaldruck (m)	maximum pressure
principe de superposition (m)	Überlagerungsprinzip (f)	superposition principle
prise d'énergie (f)	Energieübernahme (f)	energy assumption
probabilité de toucher (f)	Treffererwartung (f)	chance of hit
production de chaleur (f)	Wärmeentwicklung (f)	thermic development
produit scalaire (m)	Skalarprodukt (Vektoren) (n)	dot product
produit vectoriel (m)	Vektorprodukt (n)	cross product
profondeur de pénétration (f)	Eindringtiefe (f)	penetration depth
profondeur de rayure (f)	Zugtiefe (f)	groove depth
projectile (m)	Geschoss (n)	bullet, projectile
projectile à fragmentation (m)	Zerlegungsgeschoss (n)	fragmentating projectile
projectile expansif (m)	Deformationsgeschoss (n)	deformation projectile
projectile plein (m)	Vollgeschoss (n)	solid (homogeneous) bullet
protège-guidon (m)	Kornschutz (m)	foresight protector
puissance (f)	Leistung (physikalisch) (f)	power
puissance crépusculaire (f)	Dämmerungsleistung (f)	twilight performance
puissance de perforation (f)	Durchschlagskraft (f)	penetration power
puits du chargeur (m)	Magazingehäuse (n)	magazine housing
Quantité de chaleur (f)	Wärmemenge (f)	quantity of heat
quantité de mouvement (f)	Impuls (m)	linear momentum
quantité de mouvement initiale (f)	Mündungsimpuls (m)	muzzle momentum
queue d'arronde (f)	Schwalbenschwanz (m)	dove tail
queue de détente (f)	Abzugshebel (m)	trigger lever, trigger latch
loi de la conservation du quantité momentum de mouvement (f)	impulssatz (m)	law of conservation of
Rafale (f)	Seriefeuer (n)	automatic fire
rameneur (de cible) (m)	Scheibenzuganlage (f)	target transport device
rapport (m)	Verhältniszahl (f)	ratio
rapport d'expansion (m)	Querschnittsverhältnis (Düse) (n)	cross-section ratio
rapport de détente (m)	Entspannungsverhältnis (n)	expansion ratio
rapport des forces (m)	Kräfteverhältnis (n)	ratio of forces
raté (m)	Versager (m), Blindgänger (m)	misfire, dud, blind shell
rayage, rayure (f)	Züge (mp)	grooves, rifling
rayé	gezogen (Lauf)	rifled
rayon moyen de groupement (m)	mittlerer Streukreisradius (m)	mean radius (of dispersion)

French	German	English
rayure (f)	Drall (Waffe) (m)	twist, rifling twist
rechargement (m)	Wiederladen (n)	reloading, handloading
recouvrement (m)	Abdeckung (f)	covering
recul (m)	Rückstoß (m)	recoil, blowback
réglable	verstellbar	ajustable
réglage (m)	Einstellung (f)	adjustment, setting
réglage (m)	Justierung (f)	adjustment
relation (f)	Beziehung (f)	relation
rendement (m)	Wirkungsgrad (m)	efficiency
renverser	umkippen	overturn
repose-pouce (m)	Daumenauflage (f)	thumbrest
résidus de tir (mp)	Schmauchspuren (fp)	gunshot residue
résistance (f)	Widerstandskraft (f)	resisting force
résistance à l'air (f)	Luftwiderstand (m)	air drag, - resistance
résistance à l'écoulement (f)	Strömungswiderstand (m)	flow resistance
résistance à l'enfoncement (f)	Einpresswiderstand (m)	engraving resistance
résistance à la compression (f)	Druckwiderstand (m)	pressure drag
résistance au déchirement (f)	Reißfestigkeit (f)	resistance to tearing
résistance d'extraction (f)	Ausziehwiderstand (m)	initial resistance
résonance (f)	Resonanz (f)	resonance
ressort à boudin (m)	Spiralfeder (f)	coil spring, helical -
ressort à lame (m)	Blattfeder (f)	laminated spring
ressort de percussion (m)	Schlagfeder (f)	main spring
ressort de percuteur (m)	Schlagbolzenfeder (f)	firing pin spring
ressort du chargeur (m)	Magazinfeder (f)	magazine spring
ressort récuparateur (m)	Vorholfeder (f)	recuperator spring
ressort récupérateur (m)	Rückholfeder (f)	return spring
rétroaction (f)	Rückkopplung (f)	feedback
réversible	umkehrbar	reversible
revolver (m)	Revolver (m)	revolver
ricochet (m)	Querschläger (m)	ricochet, keyhole
ricochet (m)	Abpraller (m)	ricochet
rigide	starr	rigid
rivet (m)	Niet (m)	rivet
rogner	abkanten	chamfer, bevel
rongeur (m)	Raubzeug (jagbare Nager) (n)	vermin
roquette à propulsion liquide (f)	Flüssigtreibstoff-Rakete (f)	liquid propellant rocket
roquette à propulsion solide (f)	Feststoff-Rakete (f)	solid propellant rocket
rotation (f)	Drall (Geschoss), Drehwinkel (m)	spin, angular displacement
rotation partielle (f)	Teildrehung (f)	partial rotation
rouillé	rostig	rusty
rouillé	verrostet	oxidated, oxidized
rouille (f)	Rost (m)	rust
rupture de la douille (f)	Hülsenreißer (m)	split case, case-rupture
Sabler	sandstrahlen	sand-blast
sabot (m)	Treibkäfig (m), Treibspiegel (m)	cage, sabot
sanglier courant (m)	laufender Keiler (m)	running boar
sans frottement	reibungsfrei	frictionless
se mettre en travers	Querstehen (n)	pitching
sec	hart (Abzug)	dry
section (f)	Querschnittsfläche (f)	cross-sectional area
section d'écoulement (f)	Strömungsquerschnitt (f)	flow cross-section

semi-automatique	halbautomatisch	semi-automatic
sensibilité à l'impact (f)	Schlagempfindlichkeit (f)	impact sensitivity
sensibilité au frottement (f)	Reibempfindlichkeit (Sprengst.) (f)	friction sensitivity
sensibilité d'amorce (f)	Zündhütchenempfindlichkeit (f)	primer sensitivity
séparation (f)	Ablösung (f)	separation
sertissage (m)	Verschluss (m)	action, bolt, breech, lock
sertissage à rondelle	Bördelung (f)	roll crimp
shrapnel (m)	Schrapnell (n)	shrapnel
signe (m)	Vorzeichen (n)	sign
silencieux (m)	Mündungsknalldämpfer (m)	silencer
- dynamique	dynamische -	dynamic -
- géométrique	geometrische -	geometric -
- cinétique	kinematische -	kinematic -
simple action	single action	single action
sinusoïdal	sinusförmig	sinusoidal
solide	fest (Aggregatszustand)	solid
solution antirouille (f)	Rostschutzmittel (n)	rust preventing agent
solution de fortune (f)	Notbehelf (m)	stopgap
sommet (m)	Scheitel (m)	vertex
sonde (f), capteur (m), jauge (f)	Aufnehmer (m)	sensor
soupape (f)	Ventil (n)	valve
sous amorti	schwach gedämpft	underdamped
sous-calibré	Unterkaliber (n)	subcalibre
stabilisé par l'épaule	schulterstabilisiert	shoulder stabilized
stand de tir (m)	Schießstand (m)	shooting range
structure cristalline (f)	Gitterstruktur (f)	crystalline structure
subsonique	Unterschall-	subsonic
sulphide de potassium (m)	Kaliumsulfid (n)	potassium sulphide
support (m)	Lager (n)	pivot
suprasonique	Überschall-	supersonic
sur amorti	stark gedämpft	overdamped
sûreté (f)	Sicherheitsschieber (m)	safety slide
sûreté, sécurité (f)	Sicherheit (f)	safety
surface active (f)	Wirkfläche (f)	active area
surface d'extravasion	Zone der Extravasation (f)	area of extravasation
surface de référence (f)	Bezugsfläche (f)	reference plane
surface du cône (f)	Kegelmantel (m)	surface of a cone
surosciller	überschwingen	overshooting
surpression (f)	Überdruck (bez. Atmosphäre) (m)	gauge pressure
svelte	schlank	slender
sveltesse (f)	Schlankheit (f)	slenderness
symmétrique par rapport à une axe	rotationssymetrisch	rotational symmetry
système autonome, - fermé (m)	abgeschlossenes System (n)	closed physical system
système d'alimentation (m)	Zuführsystem (n)	feed system
système d'unités (m)	Maßsystem (n)	system of units
système de coordonnées (m)	Koordinatensystem (n)	coordinate system
système de détente (m)	Abzugseinrichtung (f)	trigger device, - mechanism
système de détente (m)	Abzugssystem (n)	trigger system
système de référence (m)	Bezugssystem (n)	frame of reference
système orienté à la main droite (m)	Rechtssystem (Koordinaten) (n)	right-handed system
Table de tir (f)	Schusstafel (f)	firing table, range table
technique de mesure (f)	Messtechnik (f)	measurement techniques

température d'explosion (f) .... Explosionstemperatur (f) .... explosion temperature
temps de parcours de l'âme (m) ..... Laufdurchgangszeit (f) .... barrel time
temps de relaxation (m) .... Zeitkonstante (f) .... relaxation time
temps de vol (m) .... Flugzeit (f) .... time of flight
tenon (m) .... Verriegelungswarze (f) .... (locking)lug
tension (f) .... Spannung (f) .... stress
tension de cisaillement (f) .... Scher-, Schubspannung (f) .... shear stress
terme d'une addition (m) .... Summand (m) .... addend
tête la première (f) .... kopfvoran .... head-first
tête de culasse (f) .... Verschlusskopf (m) .... bolt head
tige (f) .... Stange (f) .... rod, bar
tige de détente .... Abzugsstange (f) .... trigger bar
tir (m) .... Schießen (n) .... shoot, fire
tir à air comprimé (m) .... Luftgewehrschießen (n) .... air-rifle shooting
tir de chasse (m) .... jagdliches Schießen (n) .... hunting shooting
tir de réglage (m) .... Einschießen (n) .... test-firing, ranging, zeroing
tirer à culasse fermée .... aufschießend .... firing from closed bolt
tirer a culasse ouverte .... zuschießend .... firing from open bolt
tombac (m) .... Tombak (m) .... tombac, gilding metal
tomnoyer .... taumeln .... yaw, tumble
toupie (f) .... Kreisel (Spielzeug -) (m) .... top
trajectoire (f) .... Flugbahn (f) .... trajectory
trajectoire courbe (f) .... gekrümmte Bahn (f) .... curved trajectory
trajectoire incurvée .... gerader Einschusskanal (m) .... Narrow Channel
trajectoire tendue (f) .... gestreckte Flugbahn (f) .... flat trajectory
trajet du proj. dans les chairs (m)... Schusskanal (m) .... wound channel
transducteur de pression (m) .... Druckgeber (m) .... pressure transducer
transition (f) .... Übergang (Zustand) (m) .... crossover
transporteur (m) .... Zubringer (m) .... magazine follower, follower
transsonique .... transsonisch .... transonic
transversal, latéral .... quer, seitwärts .... crossways, sideways
travail (m) .... Arbeit (f) .... work
travail à la main, travail manuel (m). Handarbeit (f) .... handicraft, handiwork
trépied (m) .... Dreibein (n) .... tripod
trinitrorésorcinate de plomb (m) .... Bleitrizinat (n) .... lead styphnate
trinitrotoluène (m) .... Trinitrotoluol (n) .... trinitrotoluene
tronquer .... abbrechen .... truncate
trou (m) .... Loch (n) .... hole, peep
trou de pouce (m) .... Daumenloch (n) .... thumbhole
trou de sortie du projectile (m) .... Ausschussloch (n) .... exit hole
trou, perçage (m) .... Bohrung (f) .... hole, aperture
tuyère (f) .... Düse (f) .... nozzle

Uniforme .... gleichförmig .... uniform
unité de mesure (f) .... Maßeinheit (f) .... unit
usure (f) .... Abnutzung (f), Verschleiß (m) ...... wear
usure du canon (f) .... Laufabnutzung (f) .... barrel wear

Valeur absolue (f) .... Betrag (eines Vektors) (m) .... absolute value
valeur de référence (f) .... Bezugsgröße (f) .... reference value
valeur numérique (f) .... Zahlenwert (m) .... numerical value
valise pour arme (f) .... Waffenkoffer (m) .... gun case
variables d'état (f) .... Zustandsgröße (f) .... state variables

vent latéral (m)	Querwind (m), Seitenwind (m)	crosswind
verrou (m), verrouillage (m)	Riegel (m), Verriegelung (f)	latch, bolt, catch, lock
verrouillé	verriegelt	locked
vertical	vertikal	perpendicular
vis (f)	Schraube (f)	screw
viscosité (f)	Scherviskosität (f)	dynamic viscosity
	Viskosität (f)	viscosity
	Zähigkeit (dynamische) (f)	viscosity
viscosité cinématique (f)	kinematische Zähigkeit	kinematic viscosity
visée ouverte (f)	offene Visierung (f)	open sight
viser	zielen, visieren	aim, sight
vitesse (f)	Geschwindigkeit (f)	velocity
vitesse angulaire (f)	Winkelgeschwindigkeit (f)	angular velocity
vitesse d'écoulement (f)	Strömungsgeschwindigkeit (f)	fluid velocity
vitesse de chute (f)	Fallgeschwindigkeit (f)	velocity of descent
vitesse de combustion (f)	Abbrandgeschwindigkeit (f)	burning rate
	Brenngeschwindigkeit (f)	burning rate
vitesse de détonation (f)	Detonationsgeschwindigkeit (f)	detonation rate
vitesse de propagation (f)	Ausbreitungsgeschwindigkeit (f)	velocity of waves
vitesse de sortie (f)	Austrittsgeschwindigkeit (f)	exit velocity
vitesse du son (f)	Schallgeschwindigkeit (f)	velocity of sound
vitesse en fin de combustion (f)	Brennschlussgeschwindigkeit (f)	end-burning velocity
vitesse initiale (f)	Anfangsgeschwindigkeit (f)	initial velocity
	Mündungsgeschwindigkeit (f)	muzzle velocity
vitesse limite (f)	Grenzgeschwindigkeit (f)	limiting velocity
vitesse linéaire (f)	Bahngeschwindigkeit (f)	tangential velocity
vitesse réduite (f)	verminderte Geschwindigkeit (f)	decreased velocity
vitesse restante (f)	Restgeschwindigkeit (f)	remaining velocity
vitesse tangentielle (f)	Umfangsgeschwindigkeit (f)	circumferential velocity
vitesse terminale (f)	Endgeschwindigkeit (f)	final velocity (speed)
vitesse, célérité (f)	Geschwindigkeitsbetrag (m)	speed
vivacité (f)	Lebhaftigkeit (f)	vivacity
volume (m)	räumliche Ausdehnung (f)	spatial expansion
volumineux	ausgedehnter Körper (m)	large body
Wad-cutter	Wadcutter(-geschoss) (n)	wadcutter

Bibliography

General

Physics

BECKER E (1974) Technische Strömungslehre. Teubner, Stuttgart
FALK G, RUPPEL W (1973) Mechanik, Relativität, Gravitation. Springer, Berlin Heidelberg New York
FERCHER AF (1992) Medizinische Physik, Physik für Mediziner, Pharmazeuten und Biologen. Springer, Vienna New York
FUNG YC (1993) Biomechanics, Mechanical Properties of Living Tissues. Springer, New York Berlin Heidelberg
HALLIDAY D, RESNICK R, WALKER J (2003) Physik. Wiley-VCH, Weinheim
HERING E, MARTIN R, STOHRER M (1997) Physik für Ingenieure. Springer, Berlin Heidelberg New York
HOERNER S (1958) Fluid-Dynamic Drag. Published by the author
KALIDE W (1980) Technische Strömungslehre. Hanser, Munich
LÜSCHER E (1967) Experimentalphysik Teil 1. Bibliographisches Institut, Mannheim
MAGNUS K (1971) Kreisel. Springer, Berlin Heidelberg New York
SOMMERFELD A (1977) Vorlesungen über theoretische Physik, Band 1, Mechanik. Harri Deutsch, Frankfurt
SZABÒ I (1972) Höhere Technische Mechanik. Springer, Berlin Heidelberg New York
ZIEREP J (1976) Theoretische Gasdynamik. Braun, Karlsruhe

Arms and ammunition

BARNES FC (1993) Cartridges of the World. DBI Books, Northfield IL
EZELL EC (1988) Small Arms Today. Arms and Armour Press, London
HOGG IV (1985) Jane's Directory of Small Arms Ammunition. Jane's Publishing Comp, London
HUON J (1988) Military Rifle & Machine Gun Cartridges. Ironside International Publishers, Alexandria VA
LAMPEL G, MAHRHOLDT R (1981) Waffen-Lexikon. BLV, Munich
LONG D (1986) Combat Ammunition. Citadel Press, Secaucus NJ
JONES TL (2000) Specialty Police Munitions. Paladin Press, Boulder CO
ZUKAS JA, WALTERS WP (1998) Explosive Effects and Applications, Springer-Verlag, New York NY

Ballistics

FARRAR CL, LEEMING DW (1983) Military Ballistics – A Basic Manual. Brassey's Defence Publishers, Headington Hill Hall
HAUCK G (1972) Äußere Ballistik. Militärverlag, Berlin

KNEUBUEHL BP (1998) Geschosse (Bd. 1): Ballistik, Treffsicherheit, Wirkungsweise. Verlag Stocker-Schmid, Dietikon (Motorbuch-Verlag, Stuttgart)
KNEUBUEHL BP (2004) Geschosse (Bd. 2); Ballistik, Wirksamkeit, Messtechnik. Verlag Stocker-Schmid, Dietikon (Motorbuch-Verlag, Stuttgart)
KUTTERER RE (1958) Ballistik. Vieweg & Sohn, Braunschweig
LAMPEL, W, SEITZ G (1983) Jagdballistik. Verlag Neumann-Neudamm, Melsungen
MCCOY RL (1999) Modern Exterior Ballistics, The Launch and Flight of Symmetric Projectiles. Schiffer Military History, Atglen PA
MCSHANE EJ, KELLEY JL, RENO FV (1953) Exterior Ballistics. University of Denver Press, Denver CO
MOLITZ H, STROBEL R (1963) Äußere Ballistik. Springer, Berlin

Wound ballistics, general

BEYER JC (1962) Wound ballistics. Off. of the Surg General, Dep. Army, Washington DC
BIRCHER H (1896) Neue Untersuchungen über die Wirkung der Handfeuerwaffen. Sauerländer, Aarau
BIRCHER H (1897) Atlas zu: Neue Untersuchungen über die Wirkung der Handfeuerwaffen. Sauerländer, Aarau
COOPER GJ, DUDLEY HAF, GANN DS, RODERICK AL, MAYNARD RL (1997) Scientific Foundation of Trauma (Part 1: Mechanisms of Injury). Butterworth-Heinemann, Oxford
HATCHER JS (1927) Pistols and Revolvers and their Use. Plantersville TX
HATCHER JS (1935) Textbook of Pistols and Revolvers, their Ammunition, Ballistics and Use. Plantersville TX
JANZON B (1983) High Energy Missile Trauma, Doctoril Thesis. Göteborg University
SEEMAN T (pub.) (1979) Wound Ballistics, Proceed. 3rd Int. Symposium 1978. Acta Chir Scand Suppl 489
SEEMAN T (pub.) (1982) Wound Ballistics, Proceed. 4th Int. Symposium 1981. Acta Chir Scand Suppl 508
SEEMAN T (pub.) (1988) Wound Ballistics, Proceed. 5th Int. Symposium 1985. J Trauma 28 Suppl
SELLIER K, KNEUBUEHL B (2001) Wundballistik und ihre ballistischen Grundlagen. Springer, Berlin Heidelberg New York
WANG ZG, LIU YQ (pub.) (1990) Proceed. 6th Int Symposium on Woundballistics 1988. J Trauma (China) 6(2) Suppl

Wound ballistics and forensics

BRODGON BG (1998) Forensic radiology. CRC Press, Boca Raton FL
DI MAIO VJM (1999) Gunshot Wounds, Practical Aspects of Firearms, Ballistics and Forensic Techniques. CRC Press, Boca Raton FL
POLLAK S (2003) Schussverletzungen. in: MADEA B (pub.), Praxis Rechtsmedizin, Springer, Berlin Heidelberg New York
ROTHSCHILD MA (1999) Freiverkäufliche Schreckschusswaffen; Medizinische, rechtliche und kriminaltechnische Bewertung. Schmidt-Römhild, Lübeck
SELLIER K (1982) Schußwaffen und Schußwirkungen I. Schmidt-Römhild, Lübeck
SELLIER K (1977) Schußwaffen und Schußwirkungen II. Schmidt-Römhild, Lübeck
THALI MJ, DIRNHOFER R, VOCK P (2009) The Virtopsy Approach: 3D Optical and Radiological Scanning and Reconstruction in Forensic Medicine. Taylor & Francis

Wound ballistics and surgery

COUPLAND RM (1993) War Wounds of Limbs. Butterworth-Heinemann, Oxford
DUFOUR D, JENSEN SK, OWEN-SMITH M, SALMELA J, STENING GF, ZETTERSTRÖM B (1988) Surgery for Victims of War. International Committee of the Red Cross, Geneva
GANZONI N (1975) Die Schussverletzung im Krieg. Verlag Hans Huber, Bern Stuttgart Vienna
LA GARDE LA (1991) Gunshot Injuries. Lancer Militaria, Mt. Ida, Arkansas, (Reprint. Originally published by Wm. Wood and Co, New York NY, 1916)
MAHONEY PF, RYAN JM, BROOKS AJ, SCHWAB CW (2005) Ballistic Trauma, A Practical Guide. Springer, London

History of handguns and handgun ammunition

BOCK G, WEIGEL W, SEITZ G (1978) Handbuch der Faustfeuerwaffen. Neumann-Neudamm, Melsungen
GÖTZ H-D (1972) Mit Pulver und Blei. Goldmann, Munich
SCHMIDT R (1968) Die Handfeuerwaffen, Textband und Tafelband. Akademische Druck- und Verlagsanstalt, Graz
SCHWEIZERISCHER SCHÜTZENVEREIN (pub.) (1976) Hand- und Faustfeuerwaffen schweizerischer Ordonnanz 1817–1975. Huber, Frauenfeld

International Conventions

BEST G (1980) Humanity in Warfare, The Modern History of the International Law of Armed Conflicts. Weidenfeld and Nicolson, London
DEUTSCHES ROTES KREUZ (1988) Die Genfer Rotkreuz-Abkommen vom 12. August 1949 und die beiden Zusatzprotokolle vom 8. Juni 1977. Schriften des Deutschen Roten Kreuzes Bonn
DUELFFER J (1978) Regeln gegen den Krieg? Die Haager Friedenskonferenzen von 1899 und 1907 in der internationalen Politik. Ullstein
LAUN R (1947) Die Haager Landkriegsordnung. Schroedel Verlag, Hannover
SANDOZ Y (1975) Des armes interdites en droit de la guerre. Dissertation Université de Neuchâtel

Articles and papers

ABELE G (1956) Die durch die Geschoßwucht gegebene Gefährdungsgrenze bei Kopfschüssen in Bezug auf das neue Waffengesetz. Kongr. f. Gerichtl u. Soziale Med, Marburg
ADAMS DB (1982) Wound ballistics: A review. Mil Med 147:831–834
AEBI F, BRÖNNIMANN E, MAYOR J (1977) Some Observations on the Behaviour of Small Calibre Projectiles in Soap Targets. Acta Chir Scand Suppl 477
ALLEN IV, SCOTT R, TANNER JA, (1982) Experimental high-velocity missile head injury. Injury 14:183–193
ALLEN IV, KIRK J, MAYNARD RL, COOPER GK, SCOTT R, CROCKARD A (1983) Experimental penetrating head injury: Some aspects of light microscopical and ultrastructural abnormalities. Acta Neurochir (Vienna) 32:99–104
AMATO JJ, BILLY LJ, GRUBER RP, LAWSON NS (1970) Vascular injuries. An experimental study of high and low velocity missile wounds. Arch Surg 101:167–174
AMATO JJ, RICH NM, BILLY LJ, GRUBER RP, LAWSON NS (1971) High-velocity arterial injury: A study of the mechanism of injury. J Trauma 11:412–416

AMATO JJ, RICH NM (1972) Temporary cavity effects in blood vessel injury by high velocity missiles. J Cardiovasc Surg (Torino) 13:147–155
BARACH E, TOMLANOVICH M, NOWAK R (1986) Ballistics: A Pathophysiologic Examination of the Wounding Mechanisms of Firearms: Part I. J Trauma 26:225–235
BARACH E, TOMLANOVICH M, NOWAK R (1990) Ballistics: A Pathophysiologic Examination of the Wounding Mechanisms of Firearms: Part II. J Trauma 30:374–383
BARRY TP, LINTON PC (1977) Biophysics of rotary mower and snowblower injuries of the hand: high vs. low velocity "missile". J Trauma 17:211–214
BEEBE GW, DE BAKEY MB (1952) Battle casualties. CC Thomas, Springfield IL
BELLAMY RF, ZAJTCHUK R (1991a) Assessing the effectiveness of conventional weapons, in: BELLAMY RF, ZAJTCHUK R, pub., Textbook of military medicine: conventional warfare: ballistic, blast and burn injuries, Part 1, Vol 3. Washington DC, Office of the Surgeon General, Department of the Army, pp 53–82
BELLAMY RF, ZAJTCHUK R (1991b) The physics and biophysics of wound ballistics, in: BELLAMY RF, ZAJTCHUK R pub., Textbook of military medicine: conventional warfare: ballistic, blast and burn injuries, Part 1, Vol 3. Washington DC, Office of the Surgeon General, Department of the Army, pp 107–162
BERG StO (1974) Supersonic gunshot wounds. Criminologist 9:51–57
BERG St (1955) Zur Frage der Bestimmung des Geschoßkalibers aus den Maßen der Knochenschußlücke bei Schädelschüssen. Deut Z gerichtl Med 43:575–579
BERG St (1959) Veränderungen der Textiloberfläche bei Nahschüssen. Arch Kriminol 124:5–8, 17–22
BERG St (1964) Die Durchschlagskraft von Pistolengeschossen im menschlichen Körper. Arch Kriminol 134:17–23
BERLIN R, GELIN LE, JANZON B, LEWIS D, RYBECK B, SANDEGARD J, SEEMAN T (1976) Local effects of assault rifle bullets in live tissues. Acta Chir Scand Suppl 459:1–84
BERLIN R, JANZON B, RYBECK B, SANDEGARD J, SEEMAN T (1977a) Local effects of Assault Rifle Bullets in Live Tissues, Part II: Further Studies in Live Tissues and Relations to Some Simulant Media. Acta Chir Scand Suppl 477:1–48
BERLIN R, JANZON B, RYBECK B, SEEMAN T (1977b) Local effects of assault rifle bullets in live tissues and relations to simulant media, Proceed. 3rd Int Symposium on Ballistics, Karlsruhe
BERLIN R, JANZON B, RYBECK B, SEEMAN T (1979) Retardation of spherical missiles in live tissue. Acta Chir Scand Suppl 489:91–100
BERLIN R, JANZON B, RYBECK B, SCHANTZ B, SEEMAN T (1982) A proposed standard methodology for estimating the wounding capacity of small calibre projectiles or other missiles. Acta Chir Scand Suppl 508:11–29
BERLIN R, JANZON B, LIDÉN E, NORDSTRÖM G, SCHANTZ B, SEEMAN T, WESTLING F (1988a) Terminal Behaviour of Deforming Bullets. J Trauma 28 Suppl:58–62
BERLIN R, JANZON B, LIDÉN E, NORDSTRÖM G, SCHANTZ B, SEEMAN T, WESTLING F (1988b) Wound Ballistics of Swedish 5.56 mm Assault Rifle AK 5. J Trauma 28 Suppl:75–83
BERNDT (1897) Die Zahl im Kriege. Statistische Zahlen aus der neueren Kriegsgeschichte. Vienna
BETZ P, PANKRATZ H, PENNING R, EISENMENGER W (1993) Homicide with a captive bolt pistol. Am J Forensic Med Path 14:54–57
BEVEL T, GARDNER RM (2002) Bloodstain pattern analysis, CRC, Boca Raton FL, 2nd ed, pp 211–219
BHOOPAT T (1995) A case of internal bevelling with an exit gunshot wound to the skull. Forensic Sci Int 17:97–101
BIR CA, STEWART SJ (2005) Skin Penetration Assessment of Kinetic Energy Munitions. Symposium on Non Lethal Weapons, London
BIR CA, STEWART SJ, WILHELM M (2005) Skin Penetration Assessment of Less Lethal Kinetic Energy Munitions. J Forensic Sci 50, No. 6:1–4
BLACK AN, BURNS BD, ZUCKERMAN S (1941) An experimental study of the wounding mechanism of high velocity missiles. Brit Med J 20 (December):872-874

BLIN F, ROUGE D, BRAS P, ARBUS L (1984) Aspects médico-légaux des blessures par projectiles de petit calibre. J Méd Lég Droit Méd 27:77–84

BOCK H, NEU M, BETZ P, SEIDL S (2002) Unusual craniocerebral injury caused by a pneumatic nail gun. Int J Legal Med 116:279–281

BOLLIGER SA, THALI MJ, ROSS S, BUCK U, NAETHER S, VOCK P (2008) Virtual autopsy using imaging: bridging radiologic and forensic sciences. A review of the Virtopsy and similar projects. Eur Radiol 18(2):273–282

BOND SJ, SCHNIER GC, MILLER FB (1996) Air-powered guns: too much firepower to be a toy. J Trauma 41:674–678

BONTE W, KIJEWSKI H, (1976) Textilveränderungen durch mündungsfernen Austritt von Pulvergasen. Z Rechtsmed 77:223–231

BOWYER GW (1995) Afghan war wounded: Application of the Red Cross wound classification. J Trauma 38:64–67

BOWYER GW (1997) Management of small fragment wounds in modern warfare: a return to Hunterian principles? Ann Roy Coll Surg 79:175–182

BOWYER GW, STEWART MPM, RYAN JM (1993) Gulf war wounds: Application of the Red Cross wound classification. Injury 24:597–600

BRATZKE H, PÖLL W, KADEN B (1985) Ungewöhnliche Handlungsfähigkeit nach Kopfsteckschuss. Arch Kriminol 175:31–39

BREITENECKER R, SENIOR W (1967) Shotgun patterns. I. An experimental study on the influence of intermediate targets. J Forensic Sci 12:193–204

BREITENECKER R (1969) Shotgun wound patterns. Am J Clin Pathol 52:258–269

BREITENECKER R (1973) Über den Schrotschuss. Beitr Gerichtl Med 30;38–57

BRUCHEY WJ, FRANK DE (1983) Police Handgun Ammunition: Incapacitation Effects Volume I: Evaluation. NIJ Report 100-83, US Department of Justice

BRUESCHWEILER W, BRAUN M, DIRNHOFER R, THALI MJ (2003) Analysis of patterned injuries and injury-causing instruments with forensic 3D/CAD supported photogrammetry (FPHG): an instruction manual for the documentation process. Forensic Sci Int 132(2):130–138

BRÜNING W (1934) Die Untersuchung und Beurteilung von Selbstmörderschusswaffen. Deut Z Gerichtl Med 23:71–82

V. BRUNS N (1898a) Inhumane Kriegsgeschosse. Arch Klin Chir 57:602–607

V. BRUNS N (1898b) Über die Wirkung der Bleispitzgeschosse: 'Dum Dum' Geschosse. Tübingen

V. BRUNS N (1899) Über die Wirkung der neuesten Armeegeschosse: Hohlspitzengeschosse. Tübingen

BUCK U, NAETHER S, BRAUN M, BOLLIGER SA, FRIEDERICH H, JACKOWSKI C, AGHAYEV E, CHRISTE A, VOCK P, DIRNHOFER R, THALI MJ (2007) Application of 3D documentation and geometric reconstruction methods in traffic accident analysis: with high resolution surface scanning, radiological MSCT/MRI scanning and real data based animation. Forensic Sci Int 170(1):20–28

BUSCH W (1874) Fortsetzung der Mittheilungen über Schussversuche. Langenbecks Arch Surg 17:155–189

BUSCH W (1873) Verhandlungen des 2. Chirurgen-Congresses in Berlin

BUTLER EG, PUCKETT WO, HARVEY EN, MCMILLEN JH (1945) Experiments on Head Wounding by High Velocity Missiles. J Neurosurg 2:358–363

BYERLY WG, PENDSE D (1971) War surgery in a forward surgical hospital in Vietnam: A continuing report. Mil Med 136:221–226

BYERS CM, MCCRAE D (1971) Determining wound ballistics. Am Rifleman

CALLENDER GR, FRENCH R (1935) Wound ballistics. Studies in the mechanism of wound production by rifle bullets. Mil Surgeon 77:177–201

CALLENDER GR (1943) Woundballistics. War Med 3:337–350

CARANTA R, LEGRAIN D (1993) L'efficacité des munitions d'armes de poing. 3ème édition, Crépin-Leblond, Paris

CARELESS CM (1982) The resistance of human skin to compressive cutting. Med Sci Law 22:99–106
CEDERBERG A, ROKKANEN P (1982) Remote effects of pressure waves in missile trauma. The intra-abdominal pressure changes in anesthetized pigs wounded in one thigh. Acta Chir Scand Suppl 508:167–173
CHARTERS ACr III, CHARTERS AC (1976) Wounding Mechanism of Very High Velocity Projectiles. J Trauma 16:464–470
CHENG XY, FENG TS, LIU YQ, MA YY, WU BJ, FU RX, XIE GP, LI M, CHEN ZC, WANG DT, XU GW (1988) Wounding Properties of Steel Pellets With Different Velocities and Quality on Soft Tissue of Dogs. J Trauma 28 Suppl:33–36
CHOWANIEC C, KOBEK M, JABLONSKI C, KABIESZ-NENICZKA S, KARCZEWSKA W (2008) Case study of fatal gunshot wounds from non-lethal projectiles. Forensic Sci Int
CLARK SL, WARD JW (1943) The Effects of Rapid Compression Waves on Animals submerged in Water. Surg Gynecol Obstet 74:403–412
CLEMEDSON CJ (1953) Om projektilverkan ur medicinsk synpunkt, Tävlingsskr. Lab. v. Försvar. Fors'anst, Linköping
CLEMEDSON CJ (1953) On the effect of projectiles from a medical viewpoint (Swedish), K Krigs vet Akad Handl 6:311–360
CLEMEDSON CJ, FALCONER L, FRANKENBERG L, JÖNSSON A, WENNERSTRAND J (1973) Head injuries caused by small calibre, high velocity bullets; an experimental study. Z Rechtsmed 73:103–114
CLYNE AJ (1954) Missile Wounds in Malaya. Brit Med J 10:48–78
COE JI (1982) External bevelling of entrance wounds by handguns. Am J Forensic Med Path 3:215–219
COOPER GJ, RYAN JM (1990) Interaction of penetrating missiles with tissues: some common misapprehensions and implications for wound management. Brit J Surg 77:606-610
CORTHÉSY RM (1985) Schallgeschwindigkeit in Seife, GRD, Fachsektion 273, internal report
CORTHÉSY RM (1987) Viskosität von Seife, Pilotversuch, GRD, Fachsektion 273, internal report
COUPLAND RM (1991) The Red Cross wound classification, ICRC, Geneva
COUPLAND RM (1993a) Hand grenade injuries among civilians. J Am Med Ass 270:624-626
COUPLAND RM (1993b) Management principles. In: COUPLAND RM, War wounds of limbs: surgical management, Oxford, Butterworth Heinemann, pp 12-35
COUPLAND RM (1999a) Clinical and legal significance of fragmentation of bullets in relation to size of wounds: retrospective analysis. Brit Med J 319:403-406
COUPLAND RM (1999b) The effects of weapons and the Solferino cycle. Brit Med J 319:864-865
COUPLAND RM, KORVER A (1991) Injuries from antipersonnel mines: the experience of the International Committee of the Red Cross. Brit Med J 303:1509-1512
COUPLAND RM, KNEUBUEHL BP, ROWLEY DI, BOWYER GW (2000) Wound ballistics, surgery and the law of war, Trauma 2000 2:1–10
COWGILL JP (1975) The Newest Look At Handgun Ballistics. Am Rifleman, Oct:38–41
CRAGG J (1967) Nail gun fatality. Brit Med J 4:784
CROCKARD HA, BROWN FD, JOHNS LM, MULLAN S (1977) An experimental cerebral missile injury model in primates. J Neurosurg 46:776–783
CZURSIEDEL H (1937) Ein Selbstmord mittels eines Bolzenschussapparates. Deut Z ges Gerichtl Med 28:132–133
DAHLGREN B, BERLIN R, JANZON B, NORDSTRÖM G, NYLÖF U, RYBECK B, SCHANTZ B, SEEMAN T (1979) The extent of muscle tissue damage following missile trauma one, six and twelve hours after the infliction of trauma, studied by the current method of debridement. Acta Chir Scand Suppl 489:137–144
DAHLGREN B, ALMSKOG B, BERLIN R, NORDSTRÖM G, RYBECK B, SCHANTZ B, SEEMAN T (1982) Local effects of antibacterial therapy (benzyl-penicillin) on missile wound infection rate and tissue devitalisation when debridement is delayed for twelve hours. Acta Chir Scand Suppl 508:271–279

DEMUTH WE jr (1966) Bullet velocity and design as determinants of wounding capability – an experimental study. J Trauma 6:222–232
DEMUTH WE jr (1971) The mechanism of shotgun wounds. J Trauma 11:219–229
DEMUTH WE jr, SMITH JM (1966) High velocity bullet wounds of muscle and bone: The basis of rational early treatment. J Trauma 6:744–755
DI MAIO VJM, SPITZ WU (1970) Injury by birdshot. J Forensic Sci 15:396–402
DI MAIO VJM, SPITZ WU (1972) Variations in wounding due to unusual firearms and recently available ammunition. J Forensic Sci 17:377–386
DI MAIO VJM, ZUMWALT RE (1977) Rifle wounds from high velocity, center-fire hunting ammunition. J Forensic Sci 22:132–140
DI MAIO VJM, PETTY CS (1978) The Effectiveness of the Sawed-off Shotgun. Forensic Sci Gazette 9(2):5–6
DI MAIO VJM, JONES JA, CARUTH III WW, ANDERSON LL, PETTY CS (1974) A comparison of the wounding effects of commercially available handgun ammunition suitable for police use. FBI Law Enforcement Bulletin, pp 3–8
DI MAIO VJM, JONES JA, CARUTH III, WW, ANDERSON LL, PETTY CS (1975) The Effectiveness of Snub-Nose Revolvers and Small Automatic Pistols. FBI Law Enforcement Bulletin, June
DI MAIO VJM, COPELAND AR, BESANT–MATTHEWS PE, FLETCHER LA, JONES A (1982) Minimal velocities necessary for perforation of skin by air gun pellets and bullets. J Forensic Sci 27:894–898
DIMOND FC jr, RICH NM (1967) M-16 rifle wounds in Vietnam. J Trauma 7:619–625
DIRNHOFER R, JACKOWSKI C, VOCK P, POTTER K, THALI MJ (2006) VIRTOPSY: Minimally invasive, imaging-guided virtual autopsy. Radiographics Sep-Oct 26(5):1305-1333
DITTMANN W (1989) Wundballistische Untersuchungen zur Klinik der Schädel-Hirn-Schussverletzungen. Wehrmed Monatsschr 33:3–14
DIXON DS (1980) Determination of direction of fire from graze gunshot wounds. J Forensic Sci 25:272–279
DIXON DS (1982) Keyhole lesions in gunshot wounds of the skull and direction of fire. J Forensic Sci 27:555–566
DIXON DS (1984a) Determination of direction of fire from graze gunshot wounds of internal organs. J Forensic Sci 29:331–335
DIXON DS (1984b) Exit keyhole lesion and direction of fire in a gunshot wound of the skull. J Forensic Sci 29:336–339
DOBBYN RC, BRUCHEY WJ, SHUBIN LD (1975) An Evaluation of Police Handgun Ammunition: Summary Report. US Department of Justice
DOTZAUER G (1979) Verletzungen durch Hochgeschwindigkeitsgeschosse. Hefte Unfallheilk 138:85–90
DUBIN HC (1974) A cavitation model for kinetic energy projectiles penetrating gelatin. Bal. Res. Lab. Memo Rept. 2423
DZIEMIAN AJ (1952) Wound ballistics of the .30 caliber M2 rifle ball. Chem. Corps Med Lab. Research Rep. No. 106, March
DZIEMIAN AJ (1958) A provisional Casualty Criteria for Fragments and Projectiles. Army Chemical Center, MD, CWLR 2226, Aug
DZIEMIAN AJ (1958) The penetration of steel spheres into tissue models. US Army Chem. Warfare Lab. Techn. Rep. CWLR 2226, Aug
DZIEMIAN AJ, HERGET CM (1950) Physical aspects of primary contamination of bullet wounds. Mil Surgeon 106:294–299
DZIEMIAN AJ, MENDELSON JA, LINDSEY D (1961) Comparison of the Wounding Characteristics of some Commonly Encountered Bullets. J Trauma 1:341–353
EISENMENGER W, WILSKE J, STIEFEL D (1989) Short-stop Munition. Z Rechtsmed 103:137–145
ELBEL H (1958) Studien zur Entstehung der Stanzverletzungen bei absoluten Nahschüssen. Med Welt 20:343–345
V. ESMARCH F (1899) Open letter. Deutsche Revue, Jan

FACKLER ML (1986) Discussion of "A Study of .22 Caliber Rimfire Exploding Bullets: Effects in Ordnance Gelatin". J Forensic Sci 31:801–802
FACKLER ML (1987a) Kritisch betrachtet: Spezialmunition für Faustfeuerwaffen. IWR 3
FACKLER ML (1987b) Personal communication
FACKLER ML (1989) Annotations to: "The effect of the THV bullet in animal tissue" by PJT Knudsen (Z Rechtsmed 101:219–227, 1989). Z Rechtsmed 103:69–71
FACKLER ML (1991) Theodor Kocher and the scientific Foundation of Wound Ballistics. Surg Gynecol Obstet 172:153–160
FACKLER ML, MALINOWSKI JA (1985) The Wound Profile: A Visual Method for Quantifying Gunshot Wound Components. J Trauma 25:522–529
FACKLER ML, MALINOWSKI JA (1987) Discussion of "Winchester Silvertip Ammunition – A Study in Ordnance Gelatin" (by RAGSDALE et al.). J Forensic Sci 32:837–838
FACKLER ML, MALINOWSKI JA (1988a) Internal Deformation of the AK-74; A Possible Cause for its Erratic Path in Tissue. J Trauma 28 Suppl:72–75
FACKLER ML, MALINOWSKI JA (1988b) Ordnance Gelatin for Ballistic Studies. Am J Forensic Med Path 9:218–219
FACKLER ML, KNEUBUEHL BP, (1990) Applied Wound Ballistics: What's new and what's true. J Trauma (China) 6(2) Suppl:32–37
FACKLER ML, SURINCHAK JS, MALINOWSKI JA, BOWEN RE (1984a) Bullet Fragmentation: A Major Cause of Tissue Disruption. J Trauma 24:35–39
FACKLER ML, SURINCHAK JS, MALINOWSKI JA, BOWEN RE (1984b) Wounding Potential of the Russian AK-74 Assault Rifle. J Trauma 24:263–266
FACKLER ML, BELLAMY RF, MALINOWSKI JA (1986) Wounding Mechanism of Projectiles Striking at More than 1.5 km/sec. J Trauma 26:250–254
FACKLER ML, BELLAMY RF, MALINOWSKI JA (1988a) A Reconsideration of the Wounding Mechanism of Very High Velocity Projectiles – Importance of Projectile Shape. J Trauma 28 Suppl:63–67
FACKLER ML, BELLAMY RF, MALINOWSKI JA (1988b) The Wound Profile: Illustration of the Missile-tissue Interaction. J Trauma 28 Suppl:21–29
FACKLER ML, MALINOWSKI JA HOXYE StW, JASON A (1990) Wounding Effects of the AK-47 Rifle Used by Patrick Purdy in the Stockton, California, Schoolyard Shooting of January 17, 1989. Am J Forensic Med Path 11:185–189
FALLER-MARQUARDT M (1991) Zur Analyse von Bolzenschuß-Stanzverletzungen. Beitr Gerichtl Med 49:193–200
FALLER-MARQUARDT M, BOHNERT M, POLLAK S (2004) Detachment of the periosteum and soot staining of its underside in contact shots to the cerebral cranium. Int J Legal Med 118:343–347
FENG TS, MA Y, FU RX, LI M (1988) The Wounding Characteristics of Spherical Steel Fragments in Live Tissues. J Trauma 28 Suppl:37–40
FEUCHTWANGER MM (1982) High velocity missile injuries: A Review. J Roy Soc Med 75:966–969
FINCK PA (1965) Ballistic and forensic pathologic aspects of missile wounds. Conversion between Anglo-American and metric system. Mil Med 130:545–569
FINNIE JW (1993) Pathology of experimental traumatic craniocerebral missile injury. J Comp Pathol 108:93–101
FREYTAG E (1963) Autopsy findings in head injuries from firearms, statistical evaluation of 254 cases. Arch Pathol 76:215–225
FRITZ E (1933) Randabsprengungen an der Einschußseite des Schädelknochens bei Nahschüssen aus mehrschüssigen Faustfeuerwaffen. Deut Z Gerichtl Med 43:575–579
FRYC O, KROMPECHER T (1979) Überlebenszeit und Handlungsfähigkeit bei tödlichen Verletzungen. Beitr Gerichtl Med 37:389–392
FU RX MA YY, FENG TS, LI M (1988) An Estimation of the Physical Characteristics of Wounds Inflicted by Spherical Fragments. J Trauma 28 Suppl:85–88

GAWLICK H, KNAPPWORST J (1972) Zielballistische Untersuchungsmethoden an Jagdbüchsengeschossen. Ballistisches Laboratorium der Dynamit Nobel AG Werk Stadeln (company publication)

GERBER AM, MOODY RA (1972) Craniocerebral missile injuries in the monkey: An experimental physiological model. J Neurosurg 36:43–49

V. GIERKE HE (1968) Response of the body to mechanical forces. An overview. Ann NY Acad of Sci 152:172–186

GÖRANSSON AM, INGVAR DH, KUTYNA F (1988) Remote Cerebral Effects on EEG in High-energy Missile Trauma. J Trauma 28 Suppl:204–205

GRÄMIGER M (2010) Beurteilung der Morphologie von Schussverletzungen im Körperinnern mit dem Computermodell VeMo-S einschliesslich dessen Validierung. Doktorthesis, Med Fakultät Universität Bern

GREENBERG SW, GONZALES D, GURDJIAN ES, THOMAS LM (1968) Changes in physical properties of bone between the in vivo, freshly dead and embalmed conditions. Proc 12th Stapp Car Crash Conf, pp 271–279

GREENWOOD C (1980) The Political Factors. Gun Digest, 34th Ed

GROßE PERDEKAMP M, VENNEMANN B, MATTERN D, SERR A, POLLAK S (2005) Tissue defect at the gunshot entrance wound, what happens to the skin? Int J Legal Med 119:217–222

GRUNDFEST H (1945) Penetration of Steel Spheres into Bone. Nat. Research Counc, Missiles Cas. Rep. No.10

GRUNDFEST H, KORR HI, MCMILLEN JH, BUTLER EG (1945) Ballistics of the Penetration of Human Skin by Small Spheres. Off. Sci. Research a. Develop

GUARESCHI G (1933) Experimenteller Beitrag zu der vermeintlichen Schraubwirkung von Geschossen in Geweben. Deut Z Gerichtl Med 22:322–327

GURNEY RW (1943) The initial velocities of fragments from bombs, shells and grenades. Army Ballistic Research Laboratory BRL Report 405

GURNEY RW (1944) A new casualty criterion. BRL Report No. 498, Aberdeen

HABERDA A (1919) Eduard R v. Hofmanns Lehrbuch der gerichtlichen Medizin. 10. Aufl, Bd 1, Urban & Schwarzenberg, Berlin Vienna, pp 326–357

HADERSDORFER H (1958) Können herabfallende Geschosse Verletzungen verursachen? Arch Kriminol 122:191–194

HAGELIN KW, RÖCKERT HO, SEEMAN T (1982) Optical spectroscopy on damaged muscle tissue in vivo by incident light and in slides of the same area by transmitted light. Acta Chir Scand Suppl 508:245–249

HAGELIN KW, RÖCKERT HO, SEEMAN T, SUNDSTRÖM E, ÖRTENDAL T (1982) Optical properties of muscle tissue in high-energy missile wounds. Acta Chir Scand Suppl 508:235–243

HAIN JR Fatal arrow wounds. J Forensic Sci 34:691–693

HAMPTON OP jr (1961) The indications for debridement of gunshot (bullet) wounds of the extremities in civilian practice. J Trauma 1:368–372

HARGER JH, HUELKE DF (1970) Femoral fractures produced by projectiles; the effects of mass and diameter on target damage. J Biomech 3:487–493

HARRIS W, LUTERMAN A, CURRERI PW (1983) BB and pellet guns – Toys or deadly weapons? J Trauma 23:566–569

HARVEY EN, MCMILLEN JH (1947) An Experimental Study of Shock Waves Resulting from the Impact of HV Missiles on Animal Tissues. J Exp Med 85:321–328

HARVEY EN, BUTLER EG, MCMILLEN JH, PUCKETT W (1945) Mechanism of Wounding. War Med 8:91–104

HARVEY EN, WHITEELEY AH, GRUNDFEST H, MCMILLEN JH (1946) Piezoelectric Crystal Measurements of Pressure Changes in the Abdomen of deeply anaesthetized Animals during the Passage of HV Missiles. Mil Surgeon 98:509–528

HARVEY EN, KORR JM, OSTER G, MCMILLEN JH (1947) Secondary Damage in Wounding Due to Pressure Changes Accompanying the Passage of High Velocity Missiles. Surgery 21:218–239

HAUSBRANDT F (1943) Experimentelle Studien zur Entstehungsmechanik und Morphologie einiger Nahschusszeichen. Deut Z Gerichtl Med 38:45–76

HEATON LD (1962) Wound Ballistics in WW II Supplemented by Experiences in the Korean War, Off. Surg General, Dep. of the Army, Washington

HENN R, LIEBHARDT E (1969) Zur Topik außerhalb des Schußkanals gelegener Hirnrindenblutungen. Arch Kriminol 143:188–191

HENRY IG (1967) The GURNEY formula and related approximations for high-explosive deployment of fragments, AD-813398, Hughes Aircraft Co. Report PUB-189

HINRICSSON H (1940) On the effects of bullets (Swedish). Sv. Läkartidn:321–327

HINRICSSON H (1952) New insights into the effects of bullets (Swedish). Nord Med 47:185–186

V. HOFMANN E (1898) Atlas der gerichtlichen Medizin. Kontur- oder Ringelschuss, Lehmann, Munich, Tafel 20

HOLMSTRÖM A, LARSSON J, LEWIS DH (1982) Microcirculatory and biochemical studies of skeletal muscle tissue after high energy missile trauma. Acta Chir Scand Suppl 508:257–259

HOPKINSON DAW, WATTS JC (1963) Studies in experimental missile injuries of skeletal muscle. Proc Roy Soc Med 56:461–468

HUBBS K, KLINGER D (2004) Impact Munitions Database of Use and Effects. Research Report, submitted to the US Department of Justice, Document No. 204433, February

HUDSON P (1981) Multishot firearm suicide. Examination of 58 cases. Am J Forensic Med Path 2:239–242

HUELKE DF, DARLING JH (1964) Bone fractures produced by bullets. J Forensic Sci 9, 461–469

HUELKE DF, BUEGE LJ, HARGER JH (1967) Bone Fractures Produced by High Velocity Impacts. Am J Anat 120:123–131

HUELKE DF, HARGER JH, BUEGE LJ, DINGMANN HG, HARGER DR (1968a) An experimental study in Bio-Ballistics. J Biomech 1:97–105

HUELKE DF, HARGER JH, BUEGE LJ, DINGMANN HG (1968b) An experimental Study in Bio-Ballistics: Femoral Fractures produced by Projectiles, II. Shaft Impact. J Biomech 1:313–321

ISFORT A (1961) Bolzenschußverletzungen. Deut Z Gerichtl Med 52:60–69

JACOB B, HUCKENBECK W, DALDRUP T, HAARHOFF K, BONTE W (1990) Suicides by starter's pistols and air guns. Am J Forensic Med Path 11:285–290

JAMES SH, SUTTON TP (1998) Bloodstain patterns produced by high-velocity impact. In: James SH, Eckert WG (pub.) Interpretation of bloodstain evidence at crime scenes. CRC, Boca Raton FL, 2nd ed, pp 67–83

JANSSEN W, STIEGER W (1964) Verletzungen durch Bolzenschuß-Apparate unter besonderer Berücksichtigung der Spurenmerkmale. Arch Kriminol 134:26–37;96–102

JANSSON I, ERIKSSON R, JORFELDT L, LILJEDAHL, LOVÉN L, RAMMER L, LENNQUIST S (1982) Post-traumatic respiratory failure after high velocity missile injury of the limb in anaesthetized pigs. Acta Chir Scand Suppl 508:345–349

JANZON B (1975) Calculations of the Behaviour of Small Calibre, Spin Stabilized Projectiles Penetrating a Dense Medium. FOA rapport C 20041-M2, Mar

JANZON B (1977) Approximate theory on the cavity formation on projectile impact in dense media. FOA Rapport C 20196-D4, Aug

JANZON B (1982a) Soft soap as a tissue simulant medium for woundballistic studies investigated by comparative firing with assault rifles AK and M16A1 into live, anaesthetized animals. Acta Chir Scand Suppl 508:79–88

JANZON B (1982b) Edge, size and temperature effects in soft soapblock simulant targets used for wound ballistic studies. Acta Chir Scand Suppl 508:105–122

JANZON B, SEEMAN T (1985) Muscle devitalisation in high-energy missile wounds, and its dependence on energy transfer. J Trauma 25:138–144

JANZON B, BERLIN R, NORDSTRAND I, RYBECK B, SCHANTZ B (1979) Drag and Tumbling Behaviour of Small Calibre Projectiles in Tissue Simulant. Acta Chir Scand Suppl 498:57–70

JANZON B, SCHANTZ B, SEEMAN T (1988) Scale Effects in Ballistic Wounding. J Trauma 28 Suppl:29–32

JAUHARI M, BANDYOPADHYAY A (1976) Ballistique des blessures: Modèle mathématique relatif à la rupture d'une membrane de peau sous l'impact d'un projectile sphérique. Sci. et Techn. de l'Armement 50:1–18

JAUHARI M, BANDYOPADHYAY A (1975) Measurement of Volume of Permanent Cavity in Wound Ballistics Studies. J Indian Acad Forensic Sci 14(2):57–58

JAUHARI M, BANDYOPADHYAY A (1976a) Wound Ballistics: An Analysis of a Bullet in Gel. J Forensic Sci 21:616–624

JAUHARI M, BANDYOPADHYAY A (1976b) Wound Ballistics: Analysis of Motion of .303 SPA Bullet in 20% Gelatin Gel. J Indian Acad Forensic Sci 15(1):6–14

JAUHARI M, BANDYOPADHYAY A (1976c) Effect of Bullet Instability on Wound Ballistics Parameter. J Indian Acad Forensic Sci 15(1):89–90

JAUHARI M, BANDYOPADHYAY A (1976d) Wound Ballistics. An Evaluation of the Wounding Power of .380 ball MK2, K.F Revolver Cartridge on the Basis of Gelatin Gel Data. J Indian Acad Forensic Sci 15(2):1–6

JAUHARI M, MAHANTA P (1978a) Regression analysis of penetration data in gel at low velocities for non-deformable spherical projectiles of steel. Police Research and Develop, Jan–Mar:21–23

JAUHARI M, MAHANTA P (1978b) Wound Ballistics: Study of the Rupture of Human Skin Membrane under the Impact of a Projectile. 8th Int Meet of Int Ass Forensic Sci, Wichita KS, USA

JAUHARI M, SINHA JK (1962) Wounding effect of a spherical shot falling under gravity. J Forensic Sci 7:346–350

JAUHARI M, CHATTERJI SM, SEN A (1983) Shotgun wound patterns. Forensic Sci Int 22:123–130

JEANQUARTIER R (1996a) Wirkung kleiner Splitter gegen weiche Ziele. Bericht Nr. 1337, FA26

JEANQUARTIER R (1996b) Wirkung kleiner Splitter gegen Faserwerkstoffe. Bericht Nr. 1274, FA 26

JEANQUARTIER R, KNEUBUEHL BP (1983) Beschuß von weichen Zielen mit Modellsplittern. Eidg. Munitionsfabrik Thun, internal report

JOHANSSON L, HOLMSTRÖM A, LENNQUIST S, NORRBY K, NYSTRÖM PO (1982a) Intramural haemorrhage of the intestine as an indirect effect of missile trauma. Acta Chir Scand 148:15–19

JOHANSSON L, HOLMSTRÖM A, NORRBY K, NYSTRÖM PO, LENNQUIST S (1982b) Intramural haemorrhage of the intestine as an indirect effect of abdominal missile trauma. Acta Chir Scand Suppl 508:175–177

JONES TL (2000) Specialty Police Munitions. Paladin Press, Boulder CO

JOSSELSON AR, JOHNSON WD, WAGNER DO, GARNER DD, JOHNSON FB, LUNDY DR (1985) A Study of .22 Caliber Rimfire Exploding Bullets: Effects in Ordnance Gelatin. J Forensic Sci 30:760–772

JOURNEE C (1907) Rapport entre la force vive des balles et la gravité des blessures, qu'elles peuvent causer. Revue d'Artillerie 70:81–120

JOURNEE C, PIEDELIEVRE R (1928) Pénétration des plombs de chasse et des chevrotines dans le corps humain. Ann méd lég 8:225–231

JOURNEE C, GUY R, PIEDELIEVRE R (1930) Les projectiles, vecteurs de microbes. Ann méd lég 10:667–672

JUDE R, PIEDELIEVRE R (1925) La pression du liquide céphalo-rachidien dans les blessures par coup de feu de crane. Ann méd lég 5:411–417

JUVIN P, BRION F, TEISSIÈRE F, DURIGON M (1999) Prolonged activity after an ultimately fatal gunshot wound to the heart. Am J Forensic Med Path 20:10–12

KARGER B (1995a) Penetrating gunshots to the head and lack of immediate incapacitation. I. Wound ballistics and mechanisms of incapacitation. Int J Legal Med 108:53–61

KARGER B (1995b) Penetrating gunshots to the head and lack of immediate incapacitation. II. Review of case reports. Int J Legal Med 108:117–126

KARGER B (2003) Schussverletzungen; Pfeilschussverletzungen. In: BRINKMANN B, MADEA B (pub.), Handbuch gerichtliche Medizin, Bd 1. Springer, Berlin Heidelberg New York:683–688

KARGER B, BRINKMANN B (1997) Multiple gunshot suicides: Potential for physical activity and medico-legal aspects. Int J Legal Med 110:188–192
KARGER B, KNEUBUEHL BP (1996) On the physics of momentum in ballistics: Can the human body be displaced or knocked down by a small arms projectile? Int J Legal Med 109:147–149
KARGER B, TEIGE K (1995) Suizid mit einem Bolzensetzwerkzeug: Wundballistik und Einschußmorphologie. Arch Kriminol 195:153–158
KARGER B, NÜSSE R, SCHROEDER G, WÜSTENBECKER S, BRINKMANN B (1996) Backspatter from experimental close-range shots to the head. I. Macrobackspatter. Int J Legal Med 109:66–74
KARGER B, NÜSSE R, TRÖGER HD, BRINKMANN B (1997a) Backspatter from experimental close-range shots to the head. II. Microbackspatter and the morphology of bloodstains. Int J Legal Med 110:27–30
KARGER B, TEIGE K, BRINKMANN B (1997b) Laceration of the thoracic aorta from a .22 lr bullet. Int J Legal Med 110:92–94
KARGER B, PUSKAS Z, RUWALD B, TEIGE K, SCHUIRER G (1998) Morphological findings in the brain after experimental gunshots using radiology, pathology and histology. Int J Legal Med 111:314–319
KARGER B, SUDHUES H, KNEUBUEHL BP, BRINKMANN B (1998) Experimental Arrow Wounds: Ballistics and Traumatology. J Trauma, 45(3):495–501
KARGER B, BRATZKE H, GRAß H, LASCZKOWSKI G, LESSIG R, MONTICELLI F, WIESE J, ZWEIHOFF RF (2004) Crossbow homicides. Int J Legal Med 118:332–336
KIJEWSKI H (1979) Möglichkeiten zur Bestimmung von Kaliber, Geschossart und -geschwindigkeit aus der Morphologie des Schusskanals im Schädelknochen. Arch Kriminol 164:107–121
KIRKPATRICK JB, DIMAIO VJM (1978) Civilian gunshot wounds of the brain. J Neurosurg 48:185–198
KLAGES U, WEITHOENER D (1973) Untersuchungen zur Perforationsgröße bei Schussverletzungen der Schädelkalotte durch Kleinkaliberwaffen. Z Rechtsmed 73:35–44
KLAUE R (1949) Die indirekten Frakturen der vorderen Schädelgrube beim Schädeldachschuss. Deut Z Nervenheilk 161:167–193
KNAPPWORST J (1976) Untersuchung der Geschosswirkung durch quantitative Energieabgabemessungen im Zielmedium Gelatine. In: "Munition", Hsg. Bundeskriminalamt, Wiesbaden
KNEUBUEHL BP (1981) Kaliberverkleinerung bei Handfeuerwaffen. ASMZ 5:301–306
KNEUBUEHL BP (1981) Zum Massenpunktsmodell der äußeren Ballistik. Forschung in den Fachstellen der GRD, internal report
KNEUBUEHL BP (1982) Physical and Mathematical Backgrounds of the Stability Theory of a Spinning Shell. J of Ballistics, 6(2):1385–1398
KNEUBUEHL BP (1983) Calculation of the Stability and the Tractability Factors of a Spinning Shell. Proceedings of the 7th Int Symposium on Ballistics
KNEUBUEHL BP (1985) c_w-Bestimmung in Freifluganlagen. Lecture notes, "Bestimmung aussenballistischer Kenngrössen", Weil/Rhein
KNEUBUEHL BP (1986) Die Vorgänge in der Waffe (Innenballistik). Schweizer Waffen-Magazin 2/86–12/86
KNEUBUEHL BP (1990a) Die Gefährlichkeit von Schrot nach Durchschuss von Glas. Arch Kriminol 185 (1/2):27–34
KNEUBUEHL BP (1990b) Die Wirkung einer Schusswaffe. Int Waffen-Magazin 1-2/90–9/90
KNEUBUEHL BP (1993) Ballistische Beurteilung der 35 mm MR-35 Punch. GRD, FA 26, internal report
KNEUBUEHL BP (1994a) Faustfeuerwaffen-Patronen und ihre Probleme. Int Waffen-Magazin 4/94–10/94
KNEUBUEHL BP (1994b) Some Progress in Small Arms Wound Ballistics Research. Proceedings European Small Arms Symposium, Shrivenham
KNEUBUEHL BP (1995a) Etappen der Wundballistik. Schweiz Z Mil Med, 72(2):41–43

KNEUBUEHL BP (1995b) Die Munitionswirkung und die Grundsätze der Verhältnismässigkeit. Sonderbeilage Allg. Schweiz. Militärzeitschrift, ASMZ 10:24–25
KNEUBUEHL BP (1997) Bogen, Armbrust und andere Federwaffen. Int Waffen-Magazin 12/97–5/98
KNEUBUEHL BP (1998a) Physikalisch-ballistische Grundlagen zur Wirksamkeit der Gas- und Schreckschusswaffen. Bericht Nr. 1436, Gruppe Rüstung, FA 26
KNEUBUEHL BP (1998b) Geschosse Bd. 1 - Ballistik, Treffsicherheit, Wirkungsweise. Verlag Stocker-Schmid, Dietikon (Motorbuch-Verlag, Stuttgart), pp 90ff
KNEUBUEHL BP (1999a) Untersuchungen zur Wirksamkeit von Kurzwaffengeschossen. Bericht Nr. 1485, Gruppe Rüstung, FA 26
KNEUBUEHL BP (1999b) Splitterballistik und Splitterwundballistik. Bericht Nr. 1486, Gruppe Rüstung, FA 26
KNEUBUEHL BP (1999c) Das Abprallen von Geschossen aus forensischer Sicht. Dissertation ESC-IPS, Université de Lausanne, (148 S)
KNEUBUEHL BP (2000) Studie über die Wirksamkeit und Wirkung von Splittern geringer Masse (Grundlagenbericht VeMo-S, 2. Teil). Gruppe Rüstung, FA 26
KNEUBUEHL BP (2001) The Ballistics of "Hornussen". Proceedings 19th International Symposium on Ballistics, Vol. 1, Interlaken
KNEUBUEHL BP (2002a) Wundballistik – Grundlagen und Anwendungen Teil 1. Schweiz Z Mil Katastrophenmed, 79(2):40–44
KNEUBUEHL BP (2002b) Wundballistik – Grundlagen und Anwendungen Teil 2. Schweiz Z Mil Katastrophenmed, 79(4):108–111
KNEUBUEHL BP (2004) Geschosse Bd. 2 - Ballistik, Wirksamkeit, Messtechnik. Verlag Stocker-Schmid, Dietikon (Motorbuch-Verlag, Stuttgart), pp 27–28
KNEUBUEHL BP, FACKLER ML (1990) New Methods of Measurements in Wound Ballistics Research. J Trauma (China) 6(2) Suppl:92–95
KNEUBUEHL BP, GLARDON MJ (2010) Polizeigeschosse und ihre Deformation. Rechtsmedizin 20:80–84
KNEUBUEHL BP, MAISSEN E (1978) Endballistik von Handfeuerwaffen. GRD, Fachabteilung 27, internal report
KNEUBUEHL BP, SELLIER K (1992) Wound Ballistics: A New Understanding of the Behaviour of a Bullet in a Dense Medium. Proceedings 13th Int Symposium on Ballistics, Vol. 3, Stockholm
KNEUBUEHL BP, THALI MJ (2003) The evaluation of a synthetic long bone structure as a substitute for human tissue in gun shot experiments. Forensic Sci Int 138:44–49
KNEUBUEHL BP, THALI MJ (2005) Wundballistik und Virtopsy. Kriminalistik 59(12):44–49
KNUDSEN PJT (1988) The effect of the THV bullet in animal tissue. Z Rechtsmed 101:219–227, (1989, 103:69–71)
KNUDSEN PJT, SOERENSEN OH (1994) The initial yaw of some commonly encountered military rifle bullets. Int J Legal Med 107:141–146
KNUDSEN PJT, SVENDER J (1994) Doppler Radar velocity measurements for wound ballistics experiments. Int J Legal Med 107:1–6
KNUDSEN PJT, THEILADE P (1993) Terminal Ballistics of the 7.62 mm NATO Bullet: Autopsy findings. Int J Legal Med 106:61–67
KOCHER T (1875a) Über die Sprengwirkung der modernen Kleingewehrgeschosse. Corresp Blt Schweizer Ärzte 5:3–7:29–33:69–74
KOCHER T (1875b) Neue Beiträge zur Kenntnis der Wirkungsweise der modernen Kleingewehrgeschosse. Corresp Blt Schweizer Ärzte 9:65–71:104–109:133–137
KOCHER T (1880) Über Schusswunden. Experimentelle Untersuchungen über die Wirkungsweise der modernen Kleingewehrgeschosse. FCW Vogel, Leipzig
KOCHER T (1895a) Die Verbesserungen der Geschosse vom Standpunkt der Humanitaet. Atti Dell'XI. Congresso Medico Internazionale, Roma, 29. Marzo–5. Aprile, 1894, Vol. I, Parte Generale; Ripamonte & Colombo, Roma:320–325

KOCHER T (1895b) Zur Lehre von den Schusswunden durch Kleinkalibergeschosse. G Fischer & Co, Cassel

KOKINAKIS W, NEADES D, PIDDINGTON M, ROECKER E (1979) A gelatin methodology for estimating vulnerability of personnel to military rifle systems. Acta Chir Scand Suppl 489:35–55

KÖNIG HG, SCHMIDT V (1989) Beobachtungen zur Ausbreitungsgeschwindigkeit und Entstehungsursache von Berstungsfrakturen beim Schuß. Beitr Gerichtl Med 47:247–255

KOOPS E, PÜSCHEL K, KLEIBER M, JANSSEN W, MÖLLER G (1987) Todesfälle durch sogenannte Bolzenschußgeräte. Beitr Gerichtl Med 45:103–107

KORAC Z, KELENC D, HANCEVIC J, BASKOT A, MIKULIC D (2002) The application of computed tomography in the analysis of permanent cavity: A new method in terminal ballistics. Acta Clin Croat 41:205–209

KRAULAND W (1952) Zur Handlungsfähigkeit Kopfschussverletzter. Acta Neurochir 2:233–239

KRÖNLEIN RU (1899) Beitrag zur Lehre der Schädel-Hirnschüsse aus unmittelbarer Nähe mittels des schweizerischen Repetiergewehrs Modell 1889. Arch Klein Chir 59:67–76

KUHL J, JANSSEN W (1977) Vergleichende Untersuchungen zur Perforationsgröße durch großkalibrige Handfeuerwaffen am menschlichen Schädel. Arch Kriminol 160:91–104

KUUSELA T, KURRI J, TIKKA S, CEDERBERG A, ROKKANEN P (1982) Estimation of the extent of high velocity missile wounds in soft tissue with ultrasonography – An experimental study with special reference to the detection of X-ray negative bodies. Acta Chir Scand Suppl 508:251–255

LAGARDE LA (1892) NY Med J 56:458

LAGARDE LA (1893) Report of a series of experiments conducted at Frankfort Arsenal, in connection with the Ordnance Department of the US Army. Report to the Surgeon General for the Secretary of War, Washington: Govt. Printing Office:73–95

LAGARDE LA (1895) NY Med Rec17:25

LAGARDE LA (1914) Gunshot Injuries. John Bale Sons and Danielson Ltd, London

LARSSON J, LENNQUIST S, LOVÉN L, LEWIS DH, LILJEDAHL SO (1982) Systemic metabolic effects observed in muscle tissue after high energy missile trauma. Acta Chir Scand Suppl 508:323–326

LI M, MA Y-Y, FU R-X, FENG T-S (1988) The Characteristics of the Pressure Waves Generated in the Soft Target by Impact and its Contribution to Indirect Bone Fractures. J Trauma 28 Suppl:104–109

LIEBEGOTT G (1948/49) Seltener kombinierter Selbstmord und seine versicherungsrechtliche Auswirkung. Deut Z ges Gerichtl Med 39:351–355

LIGNITZ E, KOOPS E, PÜSCHEL K (1988) Tod durch Bolzenschussgeräte – eine retrospektive Analyse von 34 Fällen aus Berlin und Hamburg. Arch Kriminol 182:83–93

LIU YQ, WU BJ, XIE GP, CHEN Z C, TANG Z, WANG C (1982) Wounding effects of two types of bullets on soft tissue of dogs. Acta Chir Scand Suppl 508:211–221

LIU,Y, GUO R, WU B, LI S, WANG D (1988a) Pressure variation in Temporary Cavities Trailing three Different Projectiles Penetrating Water and Gelatin. J Trauma 28 Suppl:9–13

LIU Y, LI S, WU B, WANG D, JIANG S, CHENG X, MA Q, GUO R, CHEN Z (1988b) Characteristics of Cavities Trailing Different Projectiles Penetrating Water. J Trauma 28 Suppl:13–16

LIU Y, CHEN X, LI S, CHEN SX, GUO R, WANG D,FU X, JIANG S, XU G (1988c) Wounding Effects of Small Fragments of Different Shapes at Different Velocities on Soft Tissues of Dogs. J Trauma 28 Suppl:95–98

LIVER JD, WHITTY GF (1988) Wound-Dynamic Studies in Australia. J Trauma 28 Suppl:54–56

LORENZ R (1948) Der Schusskanal im Röntgenbilde. Deut. Z Gerichtl Med 39:435–448

LOVEN L, LARSSON L, LUND N, LENNQUIST S (1982) Changes in serum phosphate and erythrocyte 2,3-diphosphoglycerate after missile injury of the limb. Acta Chir Scand Suppl 508:261–263

LUFF K (1956) Beobachtungen über die Druck- und Sogwirkung von Geschossen nach Knochen- und Weichteildurchschüssen. Deut Z Gerichtl Med 45:414–419

LUFF K (1968) Untersuchungen zur Frage des Druckdifferenzausgleichs im Schußkanal. Beitr Gerichtl Med 24:108–113
LUFF K, RONNET A (1958) Über den Nachweis und die Fixierung der Geschoßwirkung von Handfeuerwaffen mittels Alginaten. Deut Z Gerichtl Med 47:603–608
MA Y-Y, FENG T-S, FU R-X, LI M (1988) An Analysis of the Wounding Factors of Four Different Shapes of Fragments. J Trauma 28 Suppl:230–235
MACPHERSON D (1994) Bullet Penetration. Ballistic Publications, Box 772, El Segundo CA
MADEA B, HENßGE C, STAAK M (1987) Möglichkeiten der Prioritätsdiagnostik bei Schädelschüssen. Arch Kriminol 180:41–46
MARCINKOWSKI T, PRZYBYLSKI Z, STOCHAJ M (1971) Experimentelle Untersuchungen über die Durchschlagskraft von Geschossen aus der Pistole P64. Krim Forens Wiss 5:261–263
MARKERT K, RÖMER G (1973) Penetrierende Schädelverletzungen durch Luftdruckwaffen. Krim Forens Wiss 12:107–114
MARSH TO, BROWN ER, BURKHARDT RP, DAVIS JH (1989) Two six-shot suicides in close geographic and temporal proximity. J Forensic Sci 34:491–494
MARSHALL EP, SANOW EJ (1992) Handgun Stopping Power, the definitive study. Paladin Press Boulder Colorado
MARTY W, SIGRIST T, VONLANTHEN B, WYLER D (1994a) Histolgischer Nachweis von Zündsatzelementen in Schmauchspuren am Hauteinschuss. Rechtsmedizin 4:110–112
MARTY W, SIGRIST TH, WYLER D (1994b) Measurement of Skin Temperature at the Entry Wound by Means of Infrared Thermography. Am J Forensic Med Path 15(1):1–4
MATHESON JM (1968) Missile wounds since the Second World War. J R Army Med Corps 114:11
MATTOO BN (1984) Discussion of "Minimal Velocities Necessary for Perforation of Skin by Air Gun Pellets and Bullets" (by DI MAIO). J Forensic Sci 29:700–703
MATTOO BN, WANI AK (1969) Casualty criteria for wounds from firearms with special reference to shot penetration. J Forensic Sci 14:120–128
MATTOO BN, WANI AK, ASGEKAR MD (1974) Casualty criteria for wounds from firearms with special reference to shot penetration II. J Forensic Sci 19:585–589
MATUNAS EA (1984) "Power Index Rating". Gun Digest 84, DBI Books
MAURER H (1961) Verletzungen durch Schussapparate. Beitr Gerichtl Med 21:48–66
MAYER RM (1932) Über typische Schädelschrägeinschüsse und die Bestimmung des Einschusswinkels. Deut Gerichtl Med 18:419–425
MEDDINGS DR (1997) Weapons injuries during and after periods of conflict: Retrospective analysis. Brit Med J 315:417-1420
MEDDINGS DR, O'CONNOR SM (1999) Circumstances around weapon injury in Cambodia after departure of a peacekeeping force: Prospective cohort study. Brit Med J 319:412-415
MEIXNER K (1923) Schussverletzungen durch Handfeuerwaffen. Arch Kriminol 75:87–108
MEIXNER K, WERKGARTNER A (1928) Schussverletzungen im Straßenkampf. Beitr Gerichtl Med 7:32–48
MELLOR SG, EASMON CSF, SANFORD JP (1997) Wound contamination and antibiotics. In: RYAN JM, et al. (pub.), Ballistic Trauma. London, Arnold, pp 61-71
MENDELSON JA, GLOVER JL (1966) Experimental study of sphere and shell-fragment wounds of soft tissues. Edgewood Arsenal Techn. Rep EATR 4003, Aug
MENDELSON JA, GLOVER JL (1967) Sphere and shell fragment wounds of soft tissues: Experimental study. J Trauma 7:889–914
MENZIES RC, ANDERSON LE (1980) The Glaser Safety and the Velex/Velet Exploding Bullet. J Forensic Sci 25:44–52
MERKEL J, MAILÄNDER R (1993) Über ein neues Verfahren zur Sicherung von Schmauchspuren an Schusshänden. Arch Kriminol 191:139–150
METTER D, SCHULZ E (1983) Morphologische Merkmale der Schusswunden in Leber und Milz. Z Rechtsmed 90:167–172
MILLAR R, RUTHERFORD WH, JOHNSTON S, MALHOTRA VJ (1975) Injuries caused by rubber bullets: A report on 90 patients. Brit J Surg 62:480–486

MISSLIWETZ J (1987) Zur Grenzgeschwindigkeit bei der Haut. (Eine experimentelle ballistische Untersuchung mit Geschossen vom Kaliber 4 mm und 4.5 mm). Beitr Gerichtl Med 65:411–432
MISSLIWETZ J (1990) Ungewöhnliche Handlungsfähigkeit bei Herzdurchschuss durch Schrotgarbe. Arch Kriminol 185:129–135
MISSLIWETZ J, WIESER I (1985a) Medizinische und technische Aspekte der Waffenwirkung, I. Bogen und Armbrust. Beitr Gerichtl Med 43:437–444
MISSLIWETZ J, WIESER I (1985b) Medizinische und technische Aspekte der Waffenwirkung, II. Fernöstliche Waffen. Beitr Gerichtl Med 43:445–455
MISSLIWETZ J, WIESER I (1990) Medizinische und technische Aspekte der Waffenwirkung, III. Schwarzpulvervorderladerwaffen. Beitr Gerichtl Med 48:685–696
MOOSBERG L (2003) Does the Swedish Use of the 12.7 mm Multipurpose Projectile Undermine the St. Petersburg Declaration? Master Thesis, Uppsala University
DE MUTH WE, NICHOLAS GG, MUNGER BL (1978) Buckshot wounds. J Trauma 18:53–57
NN (1970) Vorschlag über Verwundungs-Kriterium bei Luftdruckwaffen, Bericht Hembrug (Niederlande), Feb.
NN (1975–80) An evaluation of police handgun ammunition: Summary report. (Law Enforc. Standards Lab, LESL), LESP-Rpt-0101.01 DOJ
NADJEM H, POLLAK S (1993) Kombinierte Suizide unter Verwendung von Viehbetäubungsapparaten. Med Sachverst 89:29–33
NADJEM H, POLLAK S (1999) Einschussbefunde bei Verwendung von Viehbetäubungsapparaten ohne Schmauchabzugkanälchen. Arch Kriminol 203:91–102
NAUDE GP, BONGARD FS (1996) From deadly weapon to toy and back again: The danger of air rifles. J Trauma 41:1039–1043
NENNSTIEL R (1990) Once again – spheres in gelatin. J Trauma (China) 6(2) Suppl:119–123
NORDSTRAND I, JANZON B, RYBECK B (1979) Break up behaviour of some small calibre projectiles when penetrating a dense medium. Acta Chir Scand Suppl 489:81–90
NYSTRÖM PO, JOHANSSON L, LENNQUIST S (1982) Effects of saline irrigation on bacterial populations in the abdomen subjected to missile trauma. Acta Chir Scand Suppl 508:289–293
OEHMICHEN M (1992) Neuropathologie des Kopfschusses. In: SATERNUS KS (pub.) Kopfschuss. Spannungsfeld zwischen Medizin und Recht. Rechtsmedizinische Forschungsergebnisse, Bd 2, Schmidt-Römhild, Lübeck, pp 37–50
OEHMICHEN M, KÖNIG HG, STAAK M (1985) Morphologie des Hirnschusses. Beitr Gerichtl Med 43:55–62
OGILVIE WH (1944) Forward Surgery in Modern War. Butterwoths London, 2.Aufl
OGSTON A (1899) Continental criticism of English rifle bullets. Brit Med J:752–757
ORDOG GJ, WASSERBERGER J, BALASUBRAMANIAM S (1988) Shotgun Wound Ballistics. J Trauma 28:624–631
ORLOWSKI T, DOMANIECKI J, BADOWSKI A (1982a) Effect of missile velocity on the pathophysiology in injuries. Acta Chir Scand Suppl 508:315–321
ORLOWSKI T, PIECUCH T, DOMANIECKI J, BADOWSKI A (1982b) Mechanisms of development of shot wounds caused by missiles of different initial velocity. Acta Chir Scand Suppl 508:123–127
OTTOSON R (1964) Cavitation produced by different projectiles. Mil Med 129:1017–1024
OTTOSON R (1960) Studies regarding projectile effects, II. Projectile retardation in soft media. Comparison between water, liver and gelatin as targets for shots with spherical and cylindrical bullets (Swedish). Nat Def Research Inst, Dep 1, Report 10243-8506
OTTOSON R (1961) Studies regarding projectile effects, III. Comparison between the effects of various projectiles. Nat Def Research Inst, Dep 1, Report 1933-8506
OTTOSON R, PETTERSON H, SKARIB T (1960) Studies regarding projectile effects I, Shots against gelatin blocks with ammununition m/41 and spherical balls. Nat Def Research Inst, Dept 1, Report. 1174-8506
OWEN-SMITH MS (1981) High velocity missile injuries. London, Edward Arnold

PADDLE BM (1988) A Scanning Fluorometer for Imaging Ischaemic Areas in Traumatized Muscle. J Trauma 28 Suppl:189–193
PALTAUF A (1890) Über die Einwirkung von Pulvergasen auf das Blut und einen neuen Befund beim Nahschuss. Wien Klin Wschr 3, 984–991:1015–1017
PANGHER J (1909) Mitth üb Geg d Art u Gen-Wesen, S 615
PANKRATZ H, FISCHER H (1985) Zur Wundballistik des Krönlein-Schusses. Z Rechtsmed 95:213–215
PANKRATZ H, STEINBACH T, STIEFEL D (1986) Suizid mit Bolzensetzwerkzeug. In: EISENMENGER W, LIEBHARDT E, SCHUCK M (pub.) Medizin und Recht. Festschrift für Wolfgang Spann, Springer, Berlin Heidelberg New York, pp 246–250
PETERS CE (1990a) Recommended Experiments on the Physical Mechanism in Wound Ballistics. J Trauma (China) 6(2) Suppl:34–41
PETERS CE (1990b) A Mathematical-Physical Model of Wound Ballistics. J Trauma (China) 6(2) Suppl:303–318
PETERS CE (1990c) Common Misconceptions about the Physical Mechanism in Wound Ballistics. J Trauma (China) 6(2) Suppl:319–326
PETERSEN BL (1991) External beveling of cranial gunshot entrance wounds. J Forensic Sci 36:1592–1595
PETERSOHN F (1967) Über die Aktions- und Handlungsfähigkeit bei schweren Schädeltraumen. Deut Z Gerichtl Med 59:259–270
PFAEHLER J (1977) Endballistik der Infanteriemunition. Proc. 3rd Int Symp on Ballistics, Karlsruhe
PIEDELIEVRE R (1926) Le transport des débris de vêtements par les projectiles et leur pénétration dans la peau. Ann Méd lég 6:87–95
POLLAK S (1977) Zur Morphologie der Bolzenschussverletzung. Z Rechtsmed 80:153–165
POLLAK S (1980) Zur Morphologie der Einschusswunden im Palmar- und Plantarbereich. Z Rechtsmed 86:41–47
POLLAK S (1982) Zur Makro- und Mikromorphologie der durch Faustfeuerwaffen erzeugten Einschusswunden. Beitr Gerichtl Med 40:493–520
POLLAK S (1987) Morphologische Besonderheiten bei Schussverletzungen der Aorta. Wien Klin Wschr 99:732–736
POLLAK S, LINDERMANN A (1990) Verletzungsbilder und Röntgenbefunde nach Schüssen mit selten verwendeter Flintenlaufmunition. Beitr Gerichtl Med 48:507–518
POLLAK S, RITT F (1992) Vergleichende Untersuchungen an Einschusslücken in Rumpf- und Extremitätenknochen mit vorwiegend spongiöser Struktur. Beitr Gerichtl Med 50:363–372
POLLAK S, ROPOHL D (1991) Morphologische und morphometrische Aspekte des Kontusionsringes «Schürfsaumes» an Einschusswunden. Beitr Gerichtl Med 49:183–191
POLLAK S, STELLWAG-CARION C (1988) Morphologische Besonderheiten bei absoluten Nahschüssen auf bekleidete oder bedeckte Körperregionen. Beitr Gerichtl Med 46:401–407
PROKOSCH E (1995a) The science of wound ballistics. In: PROKOSCH E, The technology of killing: A military and political history of antipersonnel weapons. London, Zed Books, pp 10–29
PROKOSCH E (1995b) Bringing the 1899 Dum-Dum ban up to date. In: PROKOSCH E, The technology of killing: A military and political history of antipersonnel weapons. London, Zed Books, pp 190–193
PSCHYREMBEL (2007) Klinisches Wörterbuch. 261. Aufl, de Gruyter, Berlin, pp 1726–1727
PUCKETT WO (1946) The Wounding Effect of Small High-Velocity Fragments as Revealed by High Speed Radiography. J Elisha Mitchell Sc Soc 62:59–64
PUCKETT WO, GRUNDFEST H, MCELROY WP, MCMILLEN JH (1946) Damage to peripheral Nerves by High Velocity Missiles without a Direct Hit. J Neurosurg 3:294–305
PUCKETT WO, MCELROY WP, HARVEY EN (1946) Studies on Wounds of the Abdomen and Thorax Produced by High Velocity Missiles. Mil Surgeon 98:427–439
RAGSDALE BD (1987) Discussion of "Winchester Silvertip Ammunition – A Study in Ordnance Gelatin" (with FACKLER). J Forensic Sci 32:838–840

RAGSDALE BD, JOSSELSON AR (1986) Winchester Silvertip Ammunition – A Study in Ordnance Gelatin. J Forensic Sci 31:855–868
RAGSDALE BD, JOSSELSON A (1988a) Experimental Gunshot Fractures. J Trauma 28 Suppl:109–115
RAGSDALE BD, JOSSELSON A (1988b) Predicting Temporary Cavity Size from Radial Fissure Measurements in Ordnance Gelatin. J Trauma 28 Suppl:5–9
RAGSDALE BD, SOHN SS (1988) Comparison of the Terminal Ballistics of Full Metal Jacket 7.62 mm M80 (NATO) and 5.56 mm M193 military bullets: A study in ordnance gelatin. J Forensic Sci 33:676–696
RESCHELEIT T, ROTHSCHILD MA, SCHNEIDER V (2001) Zur Frage der Differenzierung von Ein- und Ausschuss bei auf fester Unterlage anliegenden bekleideten Opfern. Rechtsmedizin 11:212–216
RICH NM (1968a) Vietnam Missile Wounds Evaluated in 750 Patients. Mil Med 133:9–22
RICH NM (1968b) Wounding Power of Various Ammunition. Resident Physician:72–74
RICH NM, JOHNSON EV, DIMOND FC (1967) Wounding Power of Missiles Used in Vietnam. JAMA 199:157–168
RISSE M, WEILER G (1988) Beitrag zur hydrodynamischen Geschoßwirkung bei Schädel-Sprengschüssen. Arch Kriminol 182:75–82
ROBENS W, KÜSSWETTER K (1982) Fracture typing to human bone by assault missile trauma. Acta Chir Scand Suppl 508:223–227
RONGXIANG F, YUYUAN M, TIANSHUN F, MING L (1988) An estimation of the physical characteristics of wounds inflicted by spherical fragments. J Trauma 28 Suppl:85–88
ROSE SC, FUJISAKI CK, MOORE EE (1988) Incomplete fractures associated with penetrating trauma: Etiology appearance, and natural history. J Trauma 28:106–109
ROSS AH (1996) Calibre estimation from cranial entrance defect measurements. J Forensic Sci 41:629–633
ROTHENHÄUSLER H, SENF H (1975) Untersuchung des endballistischen Verhaltens der Infanteriemunition AP M61 beim Durchschlagen von Dural-Platten. Beitr ball Forsch V 4/75, Nov
ROTHSCHILD MA, KNEUBUEHL BP (1996) Physikalische Grundlagen zur Messung von Gasdruck und Energiestrom bei Schreckschusswaffen. Arch Kriminol 198:151–159
ROTHSCHILD MA, SCHNEIDER V (2000) Gunshot wound to the head with full recovery. Int J Legal Med 113:349–351
ROTHSCHILD MA, JUNGK J, SCHNEIDER V (1998) Thermische Verletzungen durch den Feuerstrahl von Schreckschusswaffen. Rechtsmedizin 9:9–13
RUNGE H (1906) Schießlehre für Infantererie. Berlin
RYAN JM, RICH NM, BURRIS DG, OSCHNER MG (1997) Biophysics and pathophysiology of penetrating injury. In: RYAN JM et al (pub.) Ballistic Trauma. London, Arnold, pp 31-46
RYBECK B (1974) Missile wounding and hemodynamic effects of energy absorption. Acta Chir Scand Suppl 450
RYBECK B, JANZON B (1976) Absorption of missile energy in soft tissue. Acta Chir Scand 142, 201–207
SAHLI E (1990) Viskositätsmessungen an Gelatine und Seife. GRD, Fachsektion 936, internal report
SCEPANOVIC D (1979) Steel ball effect – Investigation of shooting at blocks of soap. Acta Chir Scand Suppl 489:71–80
SCEPANOVIC D, ALBREHT M (1982) Effects of small calibre arms projectiles in soap. Acta Chir Scand Suppl 508:49–60
SCEPANOVIC D, ALBREHT M, ERDELJAN D (1982) A method for predicting effects of military rifles. Acta Chir Scand Suppl 508:29–37
SCEPANOVIC D, ALBREHT M, ERDELJAN D, MILIVOJEVIC V, PETROVIC M, CUK V, DJUKNIC M (1988) Evaluation of the New Type of Military Bullet and Rifling. J Trauma 28 Suppl:68–72

SCHANTZ B (1979) Aspects on the choice of experimental animals when reproducing missile trauma. Acta Chir Scan Suppl 489:121–130
SCHANTZ B (1982) Is the missile wound a model suitable for general trauma studies? Acta Chir Scand Suppl 508:159–166
SCHMIDT V, GÖB J (1981) Selbsttötung mit ungewöhnlichen Schussapparaten. Arch Kriminol 167:11–20
SCHRADER G (1942) Selbstmord durch 5 Herzschüsse. Beitr Gerichtl Med 16:117–120
SCHULTE-HOLTHAUSEN J (1961) Geschoßwirkung. Wehrmed Mitt 5:72–73
SCHULTE-HOLTHAUSEN J (1965) Wundballistik bei Kriegsschußverletzungen. Wehrmed Mschr 9:99–104
SCHUMACHER M, OEHMICHEN M, KÖNIG HG, EINIGHAMMER H, BIEN S (1985) Computer tomographic studies on wound ballistics of cranial gunshot injuries. Beitr Gerichtl Med 43:95–101
SCHYMA C, SCHYMA P (1997a) Der praktische Schusshandnachweis. Die PVAL-Methode im Vergleich zu Abzügen mit Folie. Rechtsmedizin 7:152–156
SCHYMA C, SCHYMA P (1997b) Über Verletzungsmöglichkeiten durch Gummigeschosse aus der Selbstverteidigungswaffe MR 35 Punch. Arch Kriminol 200:87–94
SELLIER K (1969) Einschussstudien an der Haut. Beitr Gerichtl Med 25:265–270
SELLIER K (1975) Schädigungen und Tod infolge Schussverletzungen. In: MUELLER B (pub.) Gerichtliche Medizin. 2. Aufl, Bd 1, Springer, Berlin Heidelberg New York, pp 563–608
SELLIER K (1976) Die Eindringtiefe von Bleikugeln in weiches Gewebe. Arch Kriminol 158:75–185
SELLIER K (1979) Bemerk. zur Arbeit v. TAUSCH et al.: «Experiments on the penetration power of various bullets...». Z Rechtsmed 83:163–168
SELLIER K (1983) Geschoßwirkung im lebenden Körper. Die Pirsch 35:1830–1833
SELLIER K, KNÜPLING H (1969) Über die Eindringtiefe von Geschossen in Knochen. Arch Kriminol 144:155–160
SELLIER K, KNEUBUEHL B (2001) Wound ballistics and the scientific background. Springer, Berlin Heidelberg New York, 2nd ed., pp 355–358
SIEGMUND B (2006) Untersuchung der Geschosswirkung in der sehr frühen Phase unter besonderer Berücksichtigung der Hochgeschwindigkeitsmunition. Inaugural dissertation, Göttingen
SIGHTS WP jr (1969) Ballistics Analysis of Shotgun Injuries to the Central Nervous System. J Neurosurg 31:25–33
SIGRIST T, KNÜSEL HP, MARKWALDER C, RABL W (1992) Der «innere Abstreifring» – ein Einschussmerkmal auf Leistenhaut. Arch Kriminol 189:91–99
SILLIPHANT WH, BEYER JC (1955) Wound Ballistics. Mil Med 117:238
SLESINGER EG (1943) War Wounds and Injuries. Arnold, London
SMIALEK JE, SPITZ WU (1976) Short range ammunition a possible anti hijacking device. J Forensic Sci 21:856–861
SMITH HW, WHEATLEY KK (1984) Biomechanics of femur fractures secondary to gunshot wounds. J Trauma 24:970–977
SMITH OC, BERRYMAN HE, LAHREN CH (1987) Cranial fracture patterns and estimate of direction from low velocity gunshot wounds. J Forensic Sci 32:1416–1421
SMITH OC, HARRUFF RC (1988) Evidentiary Value of the Contents of Hollow-Point-Bullets. J Forensic Sci 33:1052–1057
SMITH S (1943) Voluntary acts after a gunshot wound of the brain. Police J 16:108–110
SPENCER C, C (1908) Gunshot Wounds. Hodder and Stoughton, London
SPERRAZZA J (1962) Casualty criteria for wounding soldiers. BRL, Techn. Mitt. Nr. 1486, Jun
SPERRAZZA J, KOKINAKIS W (1965) Criteria for incapacitation soldiers with fragments and flechettes. BLR Report Nr. 1269, Jan
SPERRAZZA J, KOKINAKIS W (1968) Ballistic limits of tissue and clothing. Ann NY Acad of Sci 152:163–167

SPIERS EM (1975) The Use of the Dum Dum Bullet in Colonial Warfare. J Imp Commonw Hist Vol.4
STAUFFER E, KNEUBUEHL BP (1992) Rekonstruktion eines Militärunfalls, das fehlende Projektil. Zentralbl Rechtsmed 38(1)
STEINDLER RA (1980) Air gun pellet penetration. Med Sci Law 20:93–98
STEWART GM (1961) Eye Protection against Small High Speed Missiles. Am J Ophthalmol 51:51–80
STIEFEL, D (1992) SD 88 vom SAPL – die neue Polizeipistole? Polizei Verkehr Technik 3:76–80
STONE CI (1987) Observations and statistics relating to suicide weapons. J Forensic Sci 32:711–716
STONE CI (1992) Characteristics of firearms and gunshot wounds as markers of suicide. Am J Foren Med Path 13:275–280
STONER HB (1982) Pathophysiogical responses to multiple injuries. Acta Chir Scand Suppl 508:303–308
STRASSMANN F (1885) Lehrbuch der gerichtlichen Medicin. Enke, Stuttgart, pp 376–385
STRASSMANN G (1919) Versuche zur Unterscheidung von Ein- und Ausschuss. Arch Kriminol 17:308–319
SUNESON A (1989) Distant pressure wave effects on nervous tissues by high-energy missile impact. Doctor Thesis, Göteborg
SUNESON A, HANSSON H-A, SEEMAN T (1987) Peripheral high-energy missile hits cause pressure changes and damage to the nervous system. J Trauma 27:782–789
SUNESON A, HANSSON H-A, SEEMAN T (1988) Central and Peripheral Nervous Damage following High-energy Missile Wounds in the Thigh. J Trauma 28 Suppl:197–203
SUNESON A, HANSSON H-A, LYCKE H-A, SEEMAN T (1989) Pressure Wave Injuries to Rat Dorsal Root Ganglion Cells in Culture Caused by High-Energy Missiles. J Trauma 29:10–18
SUNESON A, HANSSON, H-A, SEEMAN T (1990a) Pressure Wave Injuries to the Nervous System Caused by High-Energy Missile Extremity Impact: Part I. Local and Distant Effects on the Peripheral Nervous System. A light and electron microscopic study on pigs. J Trauma 30:281–294
SUNESON A, HANSSON H-A, KJELLSTRÖM BT, LYCKE H-A, SEEMAN T (1990b) Pressure Waves Caused by High-Energy Missile Impair Respiration of Cultured Dorsal Root Ganglion Cells. J Trauma 30:484–488
SUNESON A, HANSSON H-A, SEEMAN T (1990c) Pressure Wave Injuries to the Nervous System Caused by High-Energy Missile Extremity Impact: Part II. Distant Effects on the Central Nervous System. A light and electron microscopic study on pigs. J Trauma 30:295–306
SWAN KG, SWAN RC, LEVINE MG, ROCKO JM (1983) The US M-16 rifle versus the Russian AK-47 rifle. Am Surgeon 49:472–476
SYKES LN, CHAMPION HR, FOUTY WJ (1988) Dum Dums, hollow-points, and devastators: Techniques designed to increase wounding potential of bullets. J Trauma 28:618–623
TABACK N, COUPLAND RM (2007) The science of human security. Medicine, Conflict and Survival 23(1):3-9
TAUSCH D, SATTLER W, WEHRFRITZ K, WEHRFRITZ G, WAGNER H-J (1978) Experiments on the penetration power of various bullets into skin and muscle tissue. Z Rechtsmed 81:309–328
TAYLOR J (1948) African Rifles and Calibres. Gun Room Press, Highland Park NJ
TEIGE K, JAHNKE R, GERLACH D, KEMPERS B, FISCHER (1986) Die Verteilung textiler Fasern im Schußkanal. Z Rechtsmed 96:183–197
THALI MJ, DIRNHOFER R (2004) Forensic radiology in German-speaking area. Forensic Sci Int, 144(2-3):233–242
THALI MJ, KNEUBUEHL BP, ZOLLINGER U (1998) Zur Dynamik des Kopfschusses: Hochgeschwindigkeitsstudien am «Haut-Schädel-Gehirn-Modell». 77. Jahrestagung der DGRM

THALI MJ, KNEUBUEHL BP, ZOLLINGER U, DIRNHOFER R (1999a) Gunshot wounds on the head: Studies of the dynamic effects with a "Skin-Skull-Brain-Model". 51st Meeting Am Acad Forensic Sci, Orlando FL

THALI MJ, KNEUBUEHL BP, DIRNHOFER R, ZOLLINGER U (1999b) Skin entrance wound on the head: High-speed-photography studies of the dynamic effects with an artificial "Skin-Skull-Brain-Model". 15th Triennial Meeting Int Ass Forensic Sci, Los Angeles

THALI MJ, KNEUBUEHL BP, KOENIGSDORFER U, ZOLLINGER U, DIRNHOFER R (2000) The evaluation of a synthetic tubing for human vascular tissue in gunshot experiments. 52nd Meeting Am Acad. Forensic Sci, Reno

THALI MJ, KNEUBUEHL BP, VOCK P, ALLMEN G, DIRNHOFER R (2002a) High-speed documented experimental gunshot to a skull-brain model and radiologic virtual autopsy. Am J Forensic Med Path 23(3):223-228

THALI MJ, KNEUBUEHL BP, ZOLLINGER U, DIRNHOFER R (2002b) The "Skin-Skull-Brain-Model": A new instrument for the study of gunshot effects. Forensic Sci Int 125:178–189

THALI MJ, KNEUBUEHL BP, ZOLLINGER U, DIRNHOFER R (2002c) A study of the morphology of gunshot entrance wounds, in connection with their dynamic creation, utilizing the "Skin-Skull-Brain-Model". Forensic Sci Int 125:190–194

THALI MJ, BRAUN M, BRUESCHWEILER W, DIRNHOFER R (2003a) "Morphological imprint": Determination of the injury-causing weapon from the wound morphology using forensic 3D/CAD-supported photogrammetry. Forensic Sci Int 132(3):177–181

THALI MJ, BRAUN M, DIRNHOFER R (2003b) Optical 3D surface digitizing in forensic medicine: 3D documentation of skin and bone injuries. Forensic Sci Int 137(2-3):203–208

THALI MJ, BRAUN M, WIRTH J, VOCK P, DIRNHOFER R (2003c) 3D surface and body documentation in forensic medicine: 3D/CAD Photogrammetry merged with 3D radiological scanning. J Forensic Sci 48(6):1356–1365

THALI MJ, KNEUBUEHL BP, ZOLLINGER U, DIRNHOFER R (2003d) A high speed study of the dynamic bullet – body interactions produced by grazing gunshots with full metal jacketed and lead projectiles. Forensic Sci Int 132:93–98

THALI MJ, YEN K, SCHWEITZER W, VOCK P, BOESCH C, OZDOBA C, SCHROTH G, ITH M, SONNENSCHEIN M, DOERNHOEFER T, SCHEURER E, PLATTNER T, DIRNHOFER R (2003e) Virtopsy, a new imaging horizon in forensic pathology: Virtual autopsy by postmortem multislice computed tomography (MSCT) and magnetic resonance imaging (MRI) – a feasibility study. J Forensic Sci 48(2):386–403

THALI MJ, YEN K, VOCK P, OZDOBA C, KNEUBUEHL BP, SONNENSCHEIN M, DIRNHOFER R (2003f) Image-guided virtual autopsy findings of gunshot victims performed with multi-slice computed tomography and magnetic resonance imaging and subsequent correlation between radiology and autopsy findings. Forensic Sci Int 138(1-3):8-16

THALI MJ, BRAUN M, BUCK U, AGHAYEV E, JACKOWSKI C, VOCK P, SONNENSCHEIN M, DIRNHOFER R (2005) Virtopsy – scientific documentation, reconstruction and animation in forensic: Individual and real 3D data based geometric approach including optical body/object surface and radiological CT/MRI scanning. J Forensic Sci 50(2):428–442

THALI MJ, JACKOWSKI C, OESTERHELWEG L, ROSS SG, DIRNHOFER R (2007) Virtopsy – the Swiss virtual autopsy approach. Leg Med (Tokyo) 9(2):100–104

THORESBY FP (1966) "Cavitation", the wounding process of the high velocity missile, a review. J Royal Army MedCorps 112: 89–99

THORESBY FP, DARLOW HM (1967) The mechanism of primary infection of bullet wounds. Brit J Surg 54:359–361

TIAN HM, HUANG MJ LIU YQ, WANG Z G (1982) Primary bacterial contamination of wound track. Acta Chir Scand Suppl 508:265–269

TIKKA S (1982) The contamination of missile wounds with special reference to early antimicrobial therapy. Acta Chir Scand Suppl 508:281–287

TIKKA S, CEDERBERG A, LEVÄNEN J, LÖTJÖNEN V, ROKKANEN P (1982) Local effects of three standard assault rifle projectiles in live tissue. Acta Chir Scand Suppl 508:61–77

TIKKA S, CEDERBERG A, ROKKANEN P (1982a) Remote effects of pressure waves in missile trauma. The intra-abdominal pressure changes in anaesthetized pigs wounded in one thigh. Acta Chir Scand. Suppl 508:167–173

TIKKA S, LÖTJÖNEN V, CEDERBERG A, ROKKANEN P (1982b) The behaviour of three standard small calibre projectiles in soap blocks. Acta Chir Scand Suppl 508:89–104

TILLETT CW, ROSE HW, HERGET C (1962) High Speed Photographic Study of Perforating Ocular Injury by the BB. Am J Ophthalmol 54:675–688

TSCHIRHART DL, NOGUCHI TT, KLATT EC (1991) A simple histochemical technique for the identification of gunshot residue. J Forensic Sci 36, 543–547

US CONGRESS, OFFICE OF TECHNOLOGY ASSESSMENT (1992) Police Body Armor Standards and Testing, Vol. II. Appendices OTA-ISC-535, Washington DC, US Government Printing Office, Appendix A, 10, Sept

UNITED NATIONS, GENERAL ASSEMBLY (1978) Working Paper on Certain Small Calibre Weapons and Projectiles. Submitted by Mexico, Sweden and Zaire, Sept

UNITED NATIONS, GENERAL ASSEMBLY (1979) Report of the Informal Working Group on Small Calibre Weapon Systems. April

UNITED NATIONS, GENERAL ASSEMBLY (1979) Draft Resolution on Small Calibre Weapon Systems, Submitted by Egypt, Ireland, Jamaica, Mexico, Sweden, Switzerland, Uruguay. Sept

UNITED NATIONS, GENERAL ASSEMBLY (1980) Small Calibre Projectiles Working Paper, submitted by Sweden. Sept

UNITED NATIONS, SWEDISH DELEGATION (1979) Draft Proposal on the Regulation of the Use of Small Calibre Weapon Systems. March

VITALE V, BERGH AK (1984) Letters to the Editor: Further Discussion of "Minimal Velocity necessary for Penetration of Skin". J Forensic Sci 29:378

WALCHER K (1929) Über Bewusstlosigkeit und Handlungsunfähigkeit. Deut Z Gerichtl Med 13:313–322

WANG Z, TANG C, CHEN X, SHI T (1988) Early Pathomorphologic Characteristics of the Wound Track Caused by Fragments. J Trauma 28 Suppl:89–95

WANG Z, SHI K, WANG C, SUN D, HUA J, LI T, XU Z (1990) Measurement and Analysis of the pressure waves at the moment of wounding by steel spheres and fragments. J Trauma (China) 6(2) Suppl:96–100

WANG ZG, FENG JX, LIU YQ (1982a) Pathomorphological observations of gunshot wounds. Acta Chir Scand Suppl 508:185–195

WANG ZG, QIAN CW, ZHAN DC, SHI TZ, TANG CG (1982b) Pathological changes of gun shot wounds at various intervals after wounding. Acta Chir Scand Suppl 508:197–210

WARKEN D (1982) Untersuchungen zum Pendelverhalten ausgewählter kleinkalibriger Munition. Forschungsbericht Frauenhofer-Gesellschaft, EMI-AFB 8/20

WARNIMENT DC (1983) Letter to the Editor: Discussion of "Minimal Velocities Necessary for Perforation of Skin by Air Gun Pellets and Bullets". J Forensic Sci 28:551

WARNIMENT DC (1984) More Discussion of "Minimal Velocities Necessary for Perforation of Skin by Air Gun Pellets and Bullets". J Forensic Sci 29:963–964

WASCHER RA, GWINN BC (1995) Air rifle pellet injury to the heart with retrograde caval migration. J Trauma 38:379–381

WATKINS FP, PEARCE BP, STAINER MC (1982) Assessment of terminal effects of high velocity projectiles using tissue simulants. Acta Chir Scand Suppl 508:39–47

WATKINS FP, PEARCE BP, STAINER MC (1988) Physical effects of the penetration of head simulants by steel spheres. J Trauma 28 Suppl:40–54

WEEDN VW, MITTLEMAN RE (1984) Stud guns revisited. Report of a suicide and literature review. J Forensic Sci 29:670–680

WEHNER HD, SELLIER K (1981) Shockwave-induced compound action potentials in the peripheral nerve. Z Rechtsmed 86:239–243

WEHNER HD, SELLIER K (1982) Compound action potentials in the peripherial nerve induced by shock waves. Acta Chir Scand Suppl 508:179–184

WEIGEL W (1975) Die Wirkung von Revolver- und Pistolengeschossen auf den menschlichen Körper. Personal communication
WEIMANN W (1931) Über atypische Einschussöffnungen am Schädel. Deut Z Gerichtl Med 16:345–351
WERKGARTNER A (1924) Eigenartige Hautverletzungen durch Schüsse aus angesetzten Selbstladepistolen. Beitr Gerichtl Med 6:148–166
WERKGARTNER A (1928) Schürfungs- und Stanzverletzungen der Haut am Einschuss durch die Mündung der Waffe. Deut Z Gerichtl Med 11:154–168
WILLIAMS RL, STEWART GM (1964) Ballistic Studies in Eye Protection. Am J Ophthalmol 58:453–464
WILSON LB (1921) Dispersion of bullet energy in relation to wound effects. Mil Surgeon 49:241–251
WIND G, FINLEY RW, RICH NM (1988) Three-dimensional Computer Graphic Modeling of Ballistic Injuries. J Trauma 28 Suppl:16–20
WIRTH J, MARKERT K, STRAUCH H (1983) Ungewöhnliche Suizide mit Viehbetäubungsgeräten. Z Rechtsmed 90:53–59
WOJAHN H (1968) CO–Hb-Konzentration im Schusskanal als Zeichen des Nahschusses. Beitr Gerichtl Med 24:190–193
WOLF AW, BENSON DR, SHOJI H, HOEPRICH P, GILMORE A (1978) Autosterilisation in Low velocity Bullets. J Trauma 18:63
WOODRUFF CE (1898) The causes of the explosive effect of modern small-calibre bullets. NY Med J 67:593–601
WOODWARD AA (1971) The use of models in the study of wound ballistics. AMRL-Tr-71-29, Paper No. 32, Dec
WORLD MEDICAL ASSOCIATION (1996) 48th General Assembly, Statement: *Weapons and their Relation to Life and Health*
WULLENWEBER R, SCHNEIDER V, GRUMME T (1977) A computer-tomographic examination of cranial bullet wounds. Z Rechtsmed 80:227–246
YEN K, THALI MJ, PESCHEL O, KNEUBUEHL BP, ZOLLINGER U, DIRNHOFER R (2003) Blood-Spatter Patterns, Hands Hold Clues for the Forensic Reconstruction of the Sequence of Events. Am J Forensic Med Path 24(2):1–9
ZHANG D, QIAN C, LIU Y, SHI T, LI D, HUANG M (1988) Morphopathologic Observations on High-velocity Steel Bullet Wounds at Various Intervals after Wounding. J Trauma 28 Suppl:98–104
ZIPERMAN HH (1961) The management of soft tissue missile wounds in war and peace. J Trauma 1:361–367

Photo credits

Chap.	Figures	Sources
2	1–6, 8, 9, 10–32, 34, 46, 48–58	B. Kneubuehl, Thun, Switzerland
	7, 33, 47	armasuisse, W+T, Thun, Switzerland
	35, 43–45	SIG, Schweizerische Industriegesellschaft, Neuhausen, Switzerland
	36–42	M. Rindlisbacher, photographer, Brenzikofen, Switzerland

Bibliography

Chap.	Figures	Sources
3	1	N. Ganzoni, Schaffhausen, Switzerland
	2	From: BIRCHER H, Atlas, 1897
	3, 7, 12–17, 23, 37–39, 48, 49, 52, 53	armasuisse, W+T, Thun, Switzerland
	4–6, 8–10, 19, 24–27, 29, 33–36, 45–47, 50, 55	B. Kneubuehl, Thun, Switzerland
	11	M. Rindlisbacher, photographer, Brenzikofen, Switzerland
	18, 31, 32, 56	K. Sellier, Bonn, Germany
	28, 43, 44, 51	M. Fackler, Hawthorne FL, USA
	30	L. Haag, Carefree AZ, USA
	40–42	J. Knappworst, RUAG, Fürth, Germany
	54	M. Thali, IRM Universität Zürich, Switzerland
	57, 58	From: BRUCHEY WJ, FRANK DE, 1983
	59, 60	Arbeitsgruppe VeMo-S (B. Kneubuehl, Thun, Switzerland)
4	1–3	From: BRUCHEY WJ, FRANK DE, 1983
	4, 5, 10–12, 17, 18, 21–24, 29–33, 38, 40–43, 45–47, 49–56	B. Kneubuehl, Thun, Switzerland
	6	From work by STURDIVAN
	7–8	From: HUELKE DF, HARGER J et al, 1968a,b
	9, 13–15, 19–20, 34–36, 39	armasuisse, W+T, Thun, Switzerland
	11, 16	M. Rindlisbacher, photographer, Brenzikofen, Switzerland
	25–28, 37, 48	K. Sellier, Bonn, Germany
	44	Internationales Waffen-Magazin, Zürich/Switzerland
5	1, 3, 4, 7, 8, 10, 11	M. Rothschild, Köln, Germany
	2, 6, 9, 24, 25, 27–29	B. Kneubuehl, Thun, Switzerland
	5	armasuisse, W+T, Thun, Switzerland
	12–21, 30–44	M. Thali, IRM Universität Zürich, Switzerland
	22, 23, 26	Documentation of criminal cases, by kind permission of the court concerned
6	1a, 2a–b, 3a–b, 4a–c, 5a–b, 6a, 7a, 8	R. Coupland, International Committee of the Red Cross, Geneva, Switzerland
	1b, 2c, 3c, 4d, 5c, 6b, 7b, 9	armasuisse, W+T, Thun, Switzerland
7	1, 2, 3	B. Kneubuehl, Thun, Switzerland.

Index

A

Abdominal cavity 116
Abrasions 248
Acceleration 4, 6
Action 4 (RUAG) 130
Ageing of soap 144
Air
   density 72
   resistance 72
   temperature 72
Air embolism 278
AK-47 61
AK-74 61
Alarm pistol 52, 210, 282
   open wounds from - 212
Ammunition
   for pistols 41
Angle of attack 71
Angle of incidence **71**, 76, 80, 98, 106, 118
Angular acceleration 6
   velocity 6, 10
Animal experiments 155
Anti-tank rocket 209
Aorta 223, 271
AP-8 46
Armour piercing bullet 46
Arrow 66
   stabilization 78
   wounds 283
Arteriae carotis 220
Artillery guns 53
Assault rifle 61
Auto-ignition temperature 38
Automatic fire 44
Automatic weapon 58, 61
Autopsy 254

B

Backspatter 255
Bacteria 135
Bacterium prodigiosum 135
Ball powder 66
Ballistic paradox (third) 201
Ballistic trauma 306
   documenting - 317
Ballistics tables 73
Barnes Triple-Shock 48
Barrel
   rifled 55
   smooth-bore 55
Bean Bag 244
Belted ammunition 63
Belted case 40
Berdan primer 39
Bernoulli's equation 22, 27
Billiard-ball effect (shot) 277
Black powder 37, 44, 321, 324
Blank cartridge 52
Blood
   droplets, very fine 255
   poisoning 278
   pressure 220, 224
   trickles 254
   vessels simulant 153
Bloodstain pattern
   exit wounds 256
Bloom 138
Blowback operated weapon 57
Body cavity 256
Bolt 56
Bolt action 57
Bone 131, 216
   damage 198
Bone fracture
   remote from the wound channel 225
   funnel-shaped defect 274
Bone fragments 132, 266, 272, 315
   as secondary projectiles 133
   simulant 151
   splinters 199
Booster 53
Boundary layer 21
Boxer primer 39
Boyle's law 17

Brain 266
Brain injuries 267
Brain tissue 267
Break action 57
Breastbone, fracture of - 248
Breech block 56
Breech-loading principle 330
Brenneke slug 49
   Super Sabot 49
   TAG 48
   TUG 47
Bulk viscosity 20
Bullet 34, 35
   compressed 128
   contaminated by bacteria 135
   heating 134
   hole in bone 272
   hunting - 228
   sterilized 133
   temperature 133, 134
   tracer - 228
   wipe 257
Bullet base 134
Bullet wounds
   to bones 272
   "high velocity" 213
   "low-velocity" 213
Bullet, elongated 322
Bullet, specials
   50 Brown. multipurpose 226
   Action 1, 4, 5 (RUAG) 205
   Action 4 (RUAG) 42
   Alpha 202
   Arcane (SFM) 42, 203
   dum-dum 326
   EMB (Hirtenberger) 205
   Gold Dot (Federal) 205
   Golden Saber (Rem.) 42, 205
   Hydra Shok (Federal) 42, 205
   KTW 42
   Mark II 326
   Mark IV 328
   PT plastic training 43, 204
   QD P.E.P. (MEN) 42, 205
   Quik-Shok (Triton) 207
   Ranger (Win.) 205
   Razor Ammo (MagSafe) 207
   RWS cone-point (RUAG) 229
   Safety Slug (Glaser) 43
   Silvertip (Win.) 205
   Skarp 202
   THV (SFM) 42, 203
   Triplex 43

Bullets for silhouette pistols 103
Bullets with secondary projectiles 206
Burn
   due to bullets 136
Burning rate 66
Bursting fractures 269
Butterfly fracture 152, 275

C

C.I.P. 41, 52
Cadavers 93, 156
Calcium content in bone 198
Calibre 34, 55
Calibre reduction 323
Canister rounds 325
Cannelure 45, 128
Captive bolt pistol 284
Cartesian system 3
Cartridge parts 34
Cartridges
   identifying 34
Case 40
Casualty criterion 180
Cattle 155
Cavitation 95, 133
Cavity pulsation 110
Cell cultures 157, 222
Cell damage 220
Celsius 15
Central regulatory failure 277
Central skin defect 257
Centre of rotation 10
Chamber 40, 56
Characteristic velocities 193
Chassepot 88
Close range shot 255
Clothing 254
Collapse of the temporary cavity 273
Combining CT and MRI 289
Combustion 38, 66
Comparison of effectiveness criteria 174
Compressibility 19
Computer Man 159
Computer tomography 287
Concussion 248
Conditions
   for experiments 96
Conservation of momentum 166
Contact shot 255, 262, 276
Contaminated wound 314
Continuity equation 21, 27
Contusion ring 258

Cook off 57
Coordinate system
  ballistic 3
  Cartesian 3
Copper-alloy bullet 104
Cotton linters 37
Coup contrecoup 220
Crack length method 140
Cranium 267
Crime scene 253
  documenting 289
Crush 314
Curved path 5
Cylindrical bullets 104
Cylindrical wave 215

D

D'Alembert's principle 13
Danger area 251
Dangerosity 250
De Laval
  nozzles 27
  pressure 27
  velocity 27
Death by shock 219, 278
Deceleration 4
  characteristic - 186
  measuring techniques 30
  of the bullet 93
Deflagration 66
Deformation 8
  of the bullet tail 129
  time 101
Deforming bullet 101, 126, 204
  for handguns 129
Deforming full metal-jacketed bullet 207
Density 19
Desiccation 258
Destruction 8
  of tissue 93
Detonation 66
Diabolo pellet 281
Differential quotients 4
Differentiation
  between entry and exit wound 260
Direction of shot 266
Displacement model 84
Displacement of tissue 111
Distant shot 264
Document nature of wounds 305
Documentation techniques 286
Documenting wounds 286

Dogs 156
Doppler principle 31
Double rifle 63
Double set trigger 58
Double-action (DA) 59, 60
Double-action only (DAO) 60
Drag coefficient 22, 120
  for 7 mm steel spheres 194
  for soap, gelatine, biological tissue 194
Drag factor 94
Drag stabilization 78
Dreyse needle gun 88
Drill hole fracture 200
Dum-dum bullet 103, 326
Duration of a pulse 110
Dynamic shooting 127
Dynamic viscosity 19

E

ECG-changes 223
Effect 163
  factors of 164
Effect of height 72
Effectiveness 163, 201, 202, 212, 339
  definition 177
  measures of - 165
  unit of measure 177
Effectiveness criteria (military) 185
Energy 9
  balance 69
  density 188
  flow density 282
  flux 26
  kinetic - 9
  of a jet 28
  of rotation 11
  potential - 9
  thresholds 180
  transfer 184, 315
  transferred 93, 122
  flux density 26, 209, 210
Entry wound 255, 257, 312
Equation of state for gases 17
  of Sturdivan 248
  of motion 12
Examination, forensic 253
Excitation of nerves 218
Exit wound 255, 256, 260
Expected kinetic energy 184
Experiments using animal 137
Explosion
  heat 38

temperature 38
Explosive bullet 205
Exponential curve 109
Exponential energy decrease 236
Exponential function 118
Extension to the model 109
Exterior ballistics 65
   calculations 72
External ricochet 270
Extravasation zone 112, 113, 258, 266

F

Fahrenheit 15
FAL rifle 61
Falling block 57
Filler 276
Fin stabilization 78
Fin-stabilized projectile 50, 103
Firing pin 56, 57
Flake powder 66
Flash
   primary 71, 259
   secondary 56, 71, 259
Flechettes 50, 78, 103
Flobert 39
Flobert cartridge 323
Flow 21
   stationary 21
Flow forces 22
Flowability 97
Fluid
   frictionless 21
Fluid jets 26
FN 303 ("non-lethal") 244
Foot-pound-force 8
Force 7
Fracture in bone 131
Fracture mechanism 315
Fragment 53, 232
   prefabricated 54
Fragment wound 232, 312
   aspects 312
   cranio-cerebral 310
   entry wound diameter 235
   equation of energy 234
   experimental verification 236
   form of the wound channel 234
   multiple 310
   multiple small fragments 313
   penetration depth 238
   wound pattern 232
fragmentation bullet 101, 205

fragmentation munitions 53
fragmenting bullet 126
Fragments 81, 87
full metal-jacketed bullet 89, 97, 126
fully automatic fire 45
functions
   of the case 40
fuse 54

G

G 3 rifle 61
Gas cavity 255
Gas gangrene 314
Gas jet 209
Gas operated 61
Gas operated weapon 57, 333
Gas-powered weapons 281
Gel strength 138
Gelatine 15, 89, 96, **138**, 179, 184, 187
   producing - 139
Globe fracture 269
Glycerine soap 15, 89, 96, **143**, 179
Goats 156
Grazing shot 261
   direction of shot 262
Grooves 55
Gunshot residues 255, 259
Gyroscope 74, 98
Gyroscopic
   effect 118
   stability 98
   stability number 77
   stabilization 106
Gyrostatic moment 76

H

Hague Peace Conference 1899 328
Hand grenade 53
Handgun bullet 87, 103
   tumbling 233
   unstable 258
Handguns 58, 186
Hazard probability 252
Head model
   WATKINS 114
   THALI ET al. 153
Heat 16
   damage to the tissue 136
   energy 17
   engines 17
High explosive 38
High-speed cameras 30

Hirtenberger ABC (bullet) 47
History of firearms 321
Hollow bone 132, 273
  simulant for - 152
Hollow charge 50
Hollow-point bullet 101, 126
Horse cadavers 155
Hunting bullet 228
Hunting rifle 63
Hydraulic pressure 89
  in the bone 131

I

Impact energy 188
Impact velocity 127
Impala (bullet) 48, 51
Imperial/US units 4
Incapacitation 182, 279
  absolute 280
Incendiary bullet 325
Inelastic collision 166
Infection 135, 278
Inhalation of blood 278
Initial shock wave 111
Interaction bullet – body 291
Interaction projectile – tissue 305
Interior ballistics 65, 323
  calculations 68
Interior rebounding shot 270
Intermediate ballistics 65
Intermediate range shot 264
Internal energy 17
International Committee of the Red Cross
  (ICRC) 2, 232
International Conventions
  additional protocols (1977) 338
  Geneva Conventions (1864) 334
  Geneva Conventions (1949) 337
  Hague Convention (1899) 227, 305, 336
  St. Petersburg Declaration (1868) 227,
    325, 328, 335
  Regulations of The Hague (1907) 337
International law 226
International treaties 45
Irreversible deformation 109
Isentropic gas 20

J

Jacketed bullet 326
Jet 26, 208
  velocity 28

K

Kidneys 266, 271
  tearing 248
Kinematics 4
Kinetic energy 9, 69, 93, 165
Krönlein shot 116, 269

L

Laceration 248, 314
Lands 55
Lateral deflection 74
Lateral forces 70
Law of conservation 11
  of energy 12
Laws
  of collisions 166
  of deceleration 108
Lead styphnate 38
Length of the wound track 312
Less-than-lethal projectiles 240
Lift force coefficient 23, 120
Limbs 152
Linters 37
Liquid jet 209
Liver 187, 266, 271
  tearing 248
Living force 165
Local energy transfer 93
Locked system 60
Long bullet track 307
Long weapons 58
Low impact velocity 129, 192
Low sectional density 247
Luger Parabellum 333
Lung 271

M

M 16 assault rifle 122
M14 rifle 61
Mach number 20, 23, 108
Machine gun 63, 333
Magazine 61
Magnetic resonance imaging 287
Margin of distension 258
Mark II bullet 326
Marrow 198, 273
MARTEL's law **84**, 117, 145, 179
Mass
  gravitational 7
  inertial 7

Measures of effectiveness
   80 J criterion 180
   comparison 174
   knockout value (Taylor) 173
   measure of effectiveness (Sellier) 169
   military purposes 179
   power index rating (Matunas) 171
   relative incapacitation index 170
   relative stopping power 168
   shots in clay (Caranta and Legrain) 173
   stopping power 167
   street results (Marshall and Sanow) 174
   summary 175
   volume of the shot channel in wood (Weigel) 168
   wound trauma incapacitation (MacPherson) 175
Medical conference in Rome 90
Medical/biological aspect 96
MEN bevel bullet 48
Mental state of the victim 164
Mercury fulminate 38, 323
Military small arms 44
Military surgeons 88
Mind, influence of - 164
Model
   physical-mathematical 157
Moment of inertia 10
   calculating 33
   measuring techniques 32
Momentum 7
Mortars 53
Motion of the bullet 97, 137
Movement
   rotational 5
   of the bullet 138
Multiple hits 259
Multipurpose bullet 226
Muscle 137, 187
Musket, smoothbored 322
Muzzle 26
   brake 56
   pressure 67
Muzzleloaders 322
Muzzle-target distance 262

N

Nammo Forex (bullet) 47
Narrow channel **97**, 118, 119, 132, 254
   length of the - 120
Natural fragment 105
Navier-Stokes equations 25, 192

Near-contact shot 264
Nerve 218
Neutral 75
Newtonian fluid 24
Nitrocellulose 37
   powder 324
Non-lethal projectiles 240
Normal force coefficient 120
Nozzle 26
Number of pulses 111
Nutation 74, 75, 77, 118

O

Osteoporotic bones 198
Overkill 181
Overstabilization 79
Overturning moment 98
   coefficient 23
Oxygen consumption 222

P

Paraplegia 314
Parasympathetic nervous system 220
Pathophysiological effects 306
Pathophysiology 316
Pendulum, physical 32
Penetration depth 94, 186, 193
Penetration process 191
Percussion cap 323
Percussion ignition 323
Periosteum 282
   simulant 152
Permanent wound channel **112**, 113, 142, 265
Photoelectric barriers 30
Physical ballistic aspect 96
Physical model 107
Physical state of the victim 164
Pigs 156
Pipe bomb 54
Pistol 58, 331
Plastic cartridge cases 50
Plastic shot 244
Plugging model 83
Point mass 7
Point
   of application 8
   of impact 164
Posture of the victim 254
Potential effect 163
Potential energy 9, 17
Pounds per square inch 8

Powder
  combustion 67
  geometric form 66
  nitrocellulose 38, 44
  smokeless 37
Precession 74, 75, 98, 118
  motion 77
Prefabricated fragments 82
Pressure
  amplitude 111
  changes 213
    in blood vessels 223
  coefficient 22
  curve in skull 114
  energy 17
  fluctuations 116
  waves 213
Primary biological effect 213
Primer 34, 38
Principle of proportionality 88, 201
Probability of hit 182
Probability of incapacitation 182
Projectile
  constant sectional density 118
Projectile, non lethal
  12-bore Bean Bag 247
  12-bore rubber ball 241
  12-bore rubber shot 244
  12-bore rubber sphere 241
  12-bore sabot 241
  35 mm MR-35 Punch 241
  38 Spl Schirneker 243
  38 Spl Short Stop 243, 247
  44 mm Flash Ball 241
  Bean Bag 244
  CQT 246
  known thresholds 248
  non-hazardous 250
  projectiles used in sport 248
Propagation velocity 210, 213
Propellant 34, 36
Proper motion of a bullet 32, 73
Proportionality
  in the effect 334
  of the means 334
Protective equipment 252
Psychological element 165
PUPPE's rule 274

R

Radar 30
Radiological documentation 286

Rain 74
Rate of fire 61
Reactions 279
Real wounds 307
Reconstruction 253, 291
  of crimes 154
  of shooting incidents 292
Red Cross wound classification 317
Relative incapacitation index 160
Remote effects 213
Repeater 58
Repeating
  pistol 331
  rifle 330
Residual energy 236
Residual velocity 193
Resistance, effective - 22
Retardation coefficient 94, **108**, 117, 194
Revolver 58, 331
Reynolds number 25, 108
Rib, fracture 248
Ribcage 270
Ricochet 253, 312
Rifle 58, 322
Rifle bullet 87, 212, 259
  tumbling 233
Rifled barrel 56
Rifleman wound model 160
Rifle-shotgun 63
Rimfire cartridge 39
Rimmed case 40
Rotation 6
  in context with "stopping power" 166
Rotational
  energy 136
  movement 5
  velocity 166
RUAG Ammotec partition bullet 47
Rubber shot 244

S

SAAMI 41
Safety devices 58
Salvo Squeeze Bore 206
Scanner, 3D 286
Secondary biological effect 213
Sectional density 65, 94, 103, 191, 202, 233, 323
  reduced - 241
Self-loading weapon 58
Semi-automatic fire 44
Semi-automatic weapon 58

Semi-jacketed bullet 101, 126, 205
  non-deforming - 208
  hollow-point bullets 205
Seven-hole powder 67
Shape-stable bullet 126
Sheep 156
Shock
  anaphylactic 278
  cardiogenic 278
  hypovolaemic 278
  neurogenic 278
  septic 278
Shock wave 213, 214, 222, 314
  caused by projectile 217
  changes to the EEG 223
Shot 230
Shot wound
  abdomen (ricochet) 309
  in thigh 307
  of the head 308
  right shoulder 309
  shoulder (ricochet) 310
Shotgun 48, 50, 58, 63
  autoloading 63
  double-barrelled 63
Shotgun cartridge 48
Shotgun wound, internal morphology 276
Shotgun wound, morphology 276
Shoulder
  case 40
Shoulder stabilization 78, 102, 104, 205
Shrapnel shell 88
SI unit 3
  for temperature 15
Sidewind 74
Sierra Matchking 127
Silhouette shooting 127
Simulants 15, 113, 136
  advantages and disadvantages 150
  comparing 147
Simulated
  blood vessels 225
  wounds 307
Single-action (SA) 59
Single-shot weapon 58
SINOXID 38
SINTOX 39
Skin-skull-brain-model 153, 297
Skull
  fractures 248
  injuries 268
  radial cracks 269
  secondary fractures 269

Slug 49, 231, 275
  balle blondeau 231
Smokeless powder 324
Soap see glycerine soap
Soap to conduct measurements 146
Softair gun 250
Solid bullet 97, 126
  copper 104, 130
  deforming 205
Sound wave 214
Special gas constant of air 17
Specific
  gas volume 38
  heat capacity 16
Speed of sound 20, 214
  in glycerine soap 144
  measuring techniques 33
Sphere 104
Spherical wave 215
Spin-stabilized bullet 6, 76
Spleen 266, 271
Springfield 45-70 88
Squeezing lead 128
Stability 75
Stable flight 76
Stagnation pressure 22
Staphylococcus aureus 135
State of movement 71
States of matter 16
Steel cubes 201
Steel sphere 123, 200
  firing into bone 195
Sten submachine gun 61
Sterility 133
Stopping power 92, 167
Straight wound channel 102
Stress 8
Styx 46
Sub-calibre ammunition 50
Submachine gun 58, 61
Surface documentation 286
Surface temperature
  of a bullet 135
Sympathetic nervous system 220
Synthetic models 298
Système International d'Unités 3

T

Temperature of a bullet 133
Temporary cavity 92, 95, **110**, 112, 198, 213, 254, 255, 265
  second - 101

volume of the - 140
Tensile 8
Terminal ballistics 65
Thermal effects
   at the entry wound 259
Thermal energy 69
Thin sheet 84
Thompson submachine gun 61
Three-dimensional model 159
Threshold energy density
   for eyes 200
   of skin 188
Threshold for skin 250
Threshold velocity 187, 189
   for bones 195
   for eyes 200
Throwback velocity 166
Time fuse 54
Time of impact 71
Tissue damage 307
   significant 311
   volume 311
Tissue destroyed 10
Torque 10
Tracer bullets 46, 227
Tractability 79
Trajectory in a vacuum 13
Trigger 57
   double set 58
   bar 57
   spring 58
Tumble 75
Tungsten sphere 194
Types of wound channel 107

U

Unstable 75
   bullet 106
Uzi submachine gun 61

V

Vacuum 101
Vacuum in the wound channel 256
Vagus nerve 220
Validation 306
Velocity 4
   as a function of distance 14
   measuring techniques 30
   profile 158
VeMo-S 160
Vetterli rifle 88
Vietnam War 122

Virtopsy 286
Viscosity 19, 25
Viscous fluid 23
Vivacity of the powder 67
Volume as a function of distance 145
Vulnerability index 169
   Function 159

W

Wadcutter bullet 104
Wave front 111, 216
Weight 8
   of the bullet 72
Winchester Fail-Safe 47
Winchester Partition Gold 47
Work 8, 17
Wound 165
Wound ballistics 1, 65, 306
   and international law 226
   studies 305
Wound channel **97**, 112, 199, 254
   first section 97
   in the body 267
   second section 99
   third section 100
Wound excision 316
Wound infection 133
Wound profile 141, 313
Wound tracks, real 311
Wounding criterion (NATO) 184
Wounding potential 282

Y

Yaw angle 110
Young's modulus 110, 224

Calibres and bullets

10.3 mm steel spheres 197
12/70 63, 64
19th Century 88
22 extra L.R. 44
22 L.R. 43, 60, 64, 66, 192
22 Long 43
22 Mag. 60
22 rimfire 115
22 Winchester Magnum 44
223 Rem. 46, 64, 339
308 Win. 46, 64, 69
315 52
32 S & W Long WC 44

320 short 52
338 Lapua 46
338 Lapua Magnum 63, 226
35 R 52
357 Magnum 206, 243
38 Spl 192, 206, 243, 246
38 Spl Short Stop 243
38 Spl Wadcutter 44
380 R 52
4.6 × 30 52
44 Rem. Mag. 203
45 Auto 173, 191
45 R 52
454 Casull 203
5.45 × 39
5.45 × 39 (Kalashnikov) 61, 46, 332
5.56 × 45 46, 63, 332, 339
5.56 × 45 M193 129
5.7 × 28 51
50 A.E. 203
50 Browning 47, 63, 226
6 mm Flobert 52
6 mm Norma BR 64
6.35 Browning 115, 191
6.35 mm steel spheres 196
6.8 mm Rem. SPC 51
7.62 × 39 129
7.62 × 39 (Kalashnikov) 45, 61, 332
7.62 × 51 129, 332
7.62 × 54 R 40, 46, 63
7.62 mm NATO 46, 121
7.65 Browning 115, 135, 191
8 mm 52
8 mm Lebel 40, 332
9 mm Brown. short 191
9 mm Browning 66
9 mm Luger 115, 173, 191, 206, 246, 255
9 mm PA 52
9.3 × 74 R 40